Symmetry of Many-Electron Systems

This is Volume 34 of
PHYSICAL CHEMISTRY
A series of monographs
Edited by ERNEST M. LOEBL, *Polytechnic Institute of New York*

A complete list of the books in this series appears at the end of the volume.

Symmetry of Many-Electron Systems

I. G. KAPLAN
L. Ya. Karpov Physico-Chemical Institute
Moscow, USSR

TRANSLATED BY

J. GERRATT
Department of Theoretical Chemistry
University of Bristol, England

ACADEMIC PRESS New York San Francisco London 1975

A Subsidiary of Harcourt Brace Jovanovich, Publishers

ACADEMIC PRESS, INC.
111 Fifth Avenue, New York, New York 10003

United Kingdom Edition published by
ACADEMIC PRESS, INC. (LONDON) LTD.
24/28 Oval Road, London NW1

Library of Congress Cataloging in Publication Data

Kaplan, Il'iâ Grigor'evich.
Symmetry of many-electron systems.

(Physical chemistry, a series of monographs, no. 34)
Translation of Simmetriiâ mnogoėlektronnykh sistem.
Bibliography: p.
1. Mathematical physics. 2. Nuclear physics. 3. Groups, Theory of. 4. Quantum theory. I. Title. II. Series.
QC20.K3713 539.7'2112 73-803
ISBN 0-12-397150-0

PRINTED IN THE UNITED STATES OF AMERICA

Symmetry of Many-Electron Systems. Translated from the original Russian edition entitled Simmetriiâ mnogoėlektronnykh sistem, published by Izdatel'stvo "Nauka" Moscow, 1969.

Contents

Chapter II The Permutation Group

Part 1. General Considerations

Part 2. The Standard Young–Yamanouchi Orthogonal Representation

Part 3. The Nonstandard Representation

Chapter III Groups of Linear Transformations

Part 1. Continuous Groups

Part 2. The Three-Dimensional Rotation Group

Part 3. Point Groups

Chapter IV Tensor Representations and Tensor Operators

SYMMETRY AND QUANTAL CALCULATIONS

Chapter V Principles of the Application of Group Theory to Quantum Mechanics

Chapter VI Classification of States

Translator's Note

The original edition of this book first appeared in 1969. Professor Kaplan has now added to it a considerable amount of extra material for the English edition. This consists largely of Section 8.10, which contains a description of the Hartree–Fock self-consistent field method, a survey of methods of treating electron correlation, and a description of the theory of configurations of nonorthogonal orbitals. Tables 4-8 in Appendix 5 have also been expanded and in Chapter I, Section 1.8, a description of semidirect products has been inserted. I have also taken the opportunity of correcting a number of small misprints and errors.

Preface to Russian Edition

Group-theoretical methods are nowadays widely used in a large variety of physical and chemical problems. Group theory enables one to examine qualitatively a number of properties of a particular system of interest without having to resort to a quantitative calculation, and at the same time its use simplifies such a calculation substantially. The fundamental basis for applying the abstract apparatus of group theory to particular problems in the structure of matter is the symmetry properties of the objects under investigation. In the case of atomic and molecular systems one may divide this symmetry into two types: (a) symmetry with respect to spatial transformations and (b) symmetry with respect to permutations of identical particles. Both types of symmetry stem from the properties of the Schrödinger equation for many-electron systems.

The influx of group-theoretical methods into quantum mechanics occurred in roughly three stages. The first stage (1927–1930) was characterized by the rapid adoption of group-theoretical methods into quantum mechanics as the subject was being formulated during those years. The basic papers by Wigner must be mentioned here primarily, as well as those by Bethe, Weyl, and van der Waerden.

However, the unfamiliar mathematics of group theory, and also the possibility of obtaining many results by direct calculation, caused a number of physicists to dislike group-theoretical methods; in the literature of that period one may even find the expression "the group pestilence"! When Slater published his determinantal method for calculations on many-electron systems in 1929, interest in group theory waned for a time (except in the field of molecular vibrations and in certain problems in the theory of solids). The simplicity and visuality of the determinantal method led to the elucidation of a number of problems in the theory of atoms, molecules, and solids. However, attempts to apply this method to complicated electronic configurations ran into virtually insuperable calculational difficulties.

The third stage was characterized by the wide application of group-theoretical methods to problems in quantum mechanics following the appearance of a series of classic papers by Racah (1942–1949). Racah developed new techniques based upon the fractional parentage method for constructing wavefunctions. It turns out that certain groups of continuous transformations from among the so-called Lie groups, which had not been previously employed in physics, are useful in classifying states and in constructing wavefunctions. Racah's papers on atomic spectroscopy, followed by those of Jahn, Edmonds, and others on nuclear shell theory, laid the ground for the application of continuous groups and permutation groups to other fields of physics and also to chemistry. The recent successes in classifying elementary particles serve as an excellent example of the effectiveness of group theory in dealing with intricate physical problems.

This book is devoted to the use of group-theoretical methods in quantal calculations on many-electron systems. Applications of group theory to solids are completely neglected, since a separate monograph would be needed to give an account of this. The main emphasis is directed toward the application of group-theoretical methods to the classification and calculation of molecular states. The fractional parentage method is explained in detail, because this technique is now widespread in atomic and nuclear spectroscopy. The method is here suitably adapted for application to molecular systems. The technique of using coordinate wavefunctions when dealing with spin-independent interactions is fully developed in Chapters VII and VIII, and its utility in problems in quantum chemistry is demonstrated. A large number of tables, useful in actual calculations, is given in Appendixes 1–7.

No previous knowledge of group theory is necessary in order to read this book, but some acquaintance with basic quantum mechanics is assumed. In Chapters I-IV, which are devoted to the mathematical apparatus of group theory, I address myself particularly to those readers who are interested in practical applications of the theory in quantal calculations. The permutation group is described in considerably greater detail than is the custom in the majority of textbooks, since it is a basic tool in calculations with coordinate wavefunctions.

It gives me great pleasure to thank A. S. Kompaneyets for his moral support and helpful advice during the writing of this book. I thank E. D. Trifonov for reading it in manuscript form and for making a number of useful comments, and also my wife, Lara, for her invaluable help in the shaping and editing of the manuscript.

MATHEMATICAL APPARATUS

No one can consider himself
an expert in arithmetic
without a knowledge of fractions.

Cicero

CHAPTER I

Basic Concepts and Theorems of Group Theory

PART 1. Properties of Group Operations

1.1. Group Postulates

A set of elements $A, B, \ldots,$ is said to form a group $\mathbf{G}$ if it satisfies the following four conditions:

(1) A "multiplication law" for the elements is specified, i.e., a rule is prescribed according to which each pair of elements P, Q is placed into correspondence with some element R which is also contained in the same set. Element R is termed the product of the elements P and Q, and is written in the form

$$R = PQ. \tag{1.1}$$

(2) The product of the factors is associative:

$$P(QR) = (PQ)R, \tag{1.2}$$

i.e., in order to specify a product uniquely, it is sufficient to specify the order of the factors.

(3) Among the elements of the group there is a unit element E possessing the property

$$EQ = QE = Q \tag{1.3}$$

for any Q belonging to the group (the condition that the element Q belongs to the group $\mathbf{G}$ is denoted symbolically as follows: $Q \in \mathbf{G}$).

(4) For every element $Q \in \mathbf{G}$ there exists an inverse element $Q^{-1} \in \mathbf{G}$ that satisfies the equation

$$Q^{-1}Q = QQ^{-1} = E. \tag{1.4}$$

The element which is the inverse of a product of elements is given by

$$(PQ)^{-1} = Q^{-1}P^{-1}, \tag{1.5}$$

as can easily be seen on multiplying PQ by $Q^{-1}P^{-1}$ and using the associative rule for products.

The product of elements in a group is generally noncommutative, i.e.,

$$PQ \neq QP.$$

If all elements of a group satisfy the equation $PQ = QP$, the group is said to bc *Abelian*. The *cyclic* groups are a particular case of Abelian groups in which all the elements are obtained by successively raising the power of one of the elements, i.e., the n elements of a cyclic group can be represented as follows:

$$A, A^2, A^3, \ldots, A^n \equiv E. \tag{1.6}$$

The following theorem is valid for the elements of a group.

If G_a runs through all the elements of a group $\mathbf{G}$ and G_0 is some fixed element of $\mathbf{G}$, then the product G_0G_a (or G_aG_0) also runs through all the elements of the group and, moreover, does so once only.

Any element G_b of the group can, in fact, be obtained by multiplying G_0 from the right by $G_a = G_0^{-1}G_b$. Furthermore, the product G_0G_a cannot occur more than once since if $G_0G_a = G_0G_b$, then by multiplying this equation from the left by G_0^{-1} we would obtain $G_a = G_b$. Hence for different G_a all G_0G_a are different.

From this theorem it follows that an arbitrary function of the group elements when summed over all the elements of the group is an invariant:

$$\sum_{G_a} f(G_a) = \sum_{G_a} f(G_0G_a) = \sum_{G_b} f(G_b). \tag{1.7}$$

1.2. Examples of Groups

(a) The simplest example of a group is formed by the set of all rational numbers (excluding zero) with respect to the operation of multiplication. Multiplication of these numbers is associative by definition, the role of unit element is played by unity, and every number possesses an inverse.

(b) The set of all vectors in three-dimensional space constitutes a group with respect to the operation of addition. In this case the operation of multiplying the elements of the group is vector addition. The operation of vector addition possesses the associative property, and the unit element is represented by a vector of zero length. Mutually inverse elements of the group are represented by vectors equal in magnitude and opposite to one another in direction.

(c) As a more complicated example we consider the permutations of N objects. We number the objects by the integers 1 to N. As is well known, we can form from N numbers $N!$ different permutations, which we represent by the symbol

$$P = \begin{pmatrix} 1 & 2 & 3 & \cdots & N \\ i_1 & i_2 & i_3 & \cdots & i_N \end{pmatrix}, \tag{1.8}$$

where i_k stands for the number which, as a result of the permutation, takes the place of the number k. The product of two permutations P_2P_1 is also a permutation, the effect of which is equivalent to the action of P_1 followed by that of P_2.

One can associate with every permutation (1.8) an inverse permutation

$$P^{-1} = \begin{pmatrix} i_1 & i_2 & i_3 & \cdots & i_N \\ 1 & 2 & 3 & \cdots & N \end{pmatrix}. \tag{1.9}$$

The effect of applying successively a permutation and its inverse is to leave the objects in their original positions, i.e., this forms the identical permutation, which we denote by I.

The $N!$ permutations of N objects thus form a group which is called the *permutation group* or the *symmetric group*, and which we shall denote by π_N.‡ It is convenient to decompose the elements of a permutation group into products of *cycles*. A permutation is called a cycle if it can be written in the form§

$$P_{i_1 i_2 \ldots i_k} \equiv \begin{pmatrix} i_1 & i_2 & i_3 & \cdots & i_k & i_{k+1} & \cdots & i_N \\ i_2 & i_3 & i_4 & \cdots & i_1 & i_{k+1} & \cdots & i_N \end{pmatrix} \tag{1.10}$$

Thus the six elements of the group π_3 can be written as the cyclic permutations

$$I, \quad P_{12}, \quad P_{13}, \quad P_{23}, \quad P_{123}, \quad P_{132}. \tag{1.11}$$

(d) Another example of a group is formed by the set of spatial transformations of an equilateral triangle under which the triangle is sent into itself. Such transformations are called symmetry operations (asymmetric bodies cannot be made to coincide with themselves under any such transformations apart from the identity). There are six distinct nonequivalent symmetry operations for an equilateral triangle and for these we choose the following (see Fig. 1.1):

E. The identity operation, which leaves the triangle unchanged.

$\mathbf{C}_3, \mathbf{C}_3^2$. Clockwise rotations by 120° and 240°, respectively, about an axis perpendicular to the plane of the triangle and passing through its center of gravity.

$\sigma_v^{(1)}, \sigma_v^{(2)}, \sigma_v^{(3)}$. Reflections in planes perpendicular to the plane of the triangle and passing through its medians.

All other symmetry operations are equivalent to one of those just enumerated. Thus, for example, a rotation by 180° about an axis passing through a median is equivalent to a reflection σ_v.

‡Translator's note: The symmetric group is often denoted in the literature by S_N.

§In Chapter II the notation $(i_1 i_2 \ldots i_k)$ is used for a cycle.

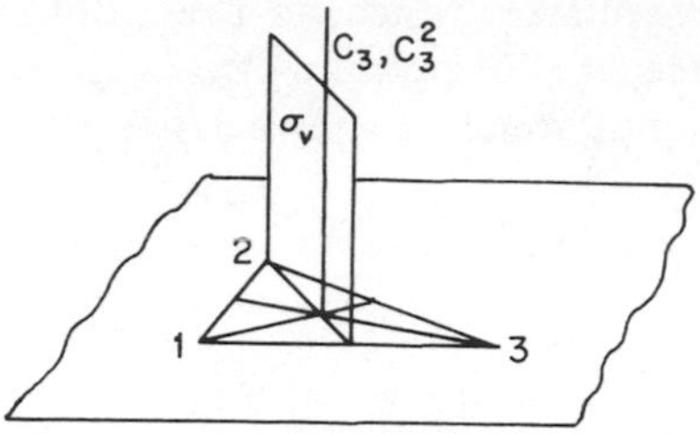

Figure 1.1

Let us now determine the products of all possible pairs of the six symmetry operations of the equilateral triangle given in Fig. 1.1, and let us put the results in the form of a table (Table 1.1). The operations which act on the triangle first are placed on the top line of the table. From this can be seen that the successive application of any two operations is equivalent to a single operation from the same set. For every operation Q there is an inverse operation Q^{-1}

Table 1.1

Multiplication Table for the Group $\mathbf{C}_{3v}$

	E	C_3	C_3^2	$\sigma_v^{(1)}$	$\sigma_v^{(2)}$	$\sigma_v^{(3)}$
E	E	C_3	C_3^2	$\sigma_v^{(1)}$	$\sigma_v^{(2)}$	$\sigma_v^{(3)}$
C_3	C_3	C_3^2	E	$\sigma_v^{(2)}$	$\sigma_v^{(3)}$	$\sigma_v^{(1)}$
C_3^2	C_3^2	E	C_3	$\sigma_v^{(3)}$	$\sigma_v^{(1)}$	$\sigma_v^{(2)}$
$\sigma_v^{(1)}$	$\sigma_v^{(1)}$	$\sigma_v^{(3)}$	$\sigma_v^{(2)}$	E	C_3^2	C_3
$\sigma_v^{(2)}$	$\sigma_v^{(2)}$	$\sigma_v^{(1)}$	$\sigma_v^{(3)}$	C_3	E	C_3^2
$\sigma_v^{(3)}$	$\sigma_v^{(3)}$	$\sigma_v^{(2)}$	$\sigma_v^{(1)}$	C_3^2	C_3	E

which leads to the identity transformation E. By using the table one easily verifies that the associative rule is obeyed. The symmetry operations of a triangle therefore form a group which is usually denoted by $\mathbf{C}_{3v}$. The group $\mathbf{C}_{3v}$ is one of the so-called *point groups* (see Chapter III, Part 3).

1.3. Isormorphism and Homomorphism

Groups are classified either as *finite* or *infinite* depending upon whether they contain a finite or infinite number of elements. The number of elements in a group is termed its *order*. Two groups $\mathbf{G}$ and $\mathbf{G}'$ of the same order are termed *isomorphic* if a one-to-one correspondence between their elements can be defined such that if the element $A' \in \mathbf{G}'$ corresponds to $A \in \mathbf{G}$ and $B' \in \mathbf{G}'$ corresponds to $B \in \mathbf{G}$, then the element $C' = A'B'$ corresponds to $C = AB$. Isomorphic groups are identical as far as their abstract group properties are concerned, although the actual meaning of their respective elements may be quite different.

The groups $\mathbf{C}_{3v}$ and $\boldsymbol{\pi}_3$ introduced in the previous section form an example of isomorphous groups. Thus we can regard each symmetry operation of the triangle as the corresponding permutation of its vertices, and so derive a one-to-one correspondence between the elements of the two groups:

$$\begin{array}{lcccccc} \mathbf{C}_{3v}: & E & C_3 & C_3{}^2 & \sigma_v^{(1)} & \sigma_v^{(2)} & \sigma_v^{(3)} \\ \boldsymbol{\pi}_3: & I & P_{132} & P_{123} & P_{23} & P_{13} & P_{12} \end{array}$$

This is a particular case of a more general theorem (*Cayley's theorem*) which states that any finite group of order n is isomorphous with a subgroup (see next section) of the permutation group $\boldsymbol{\pi}_n$.

If for each element A of a group $\mathbf{G}$ there is a corresponding element A' of a group $\mathbf{G}'$ so that if $AB = C$, then $A'B' = C'$, the two groups are said to be *homomorphic*. In contrast to isomorphism, homomorphism does not require a one-to-one correspondence between the elements of the groups, and one element of the group $\mathbf{G}'$ may correspond to several elements of $\mathbf{G}$.

1.4. Subgroups and Cosets

If it is possible to select from a group $\mathbf{G}$ some set of elements which, with the same multiplication law, form a group $\mathbf{H}$ among themselves, then $\mathbf{H}$ is called a subgroup of $\mathbf{G}$. Every group possesses a trivial subgroup that consists of the one unit element of the group. Henceforth when referring to a subgroup a nontrivial subgroup will always be implied.

Let the element G_1 belong to a finite group $\mathbf{G}$, but not to a subgroup of $\mathbf{G}$, $\mathbf{H}$, which consists of h elements. We multiply all the elements of $\mathbf{H}$ (e.g., from the left) by G_1 and obtain some set of h elements which we denote by $G_1\mathbf{H}$. Now, no element of $G_1\mathbf{H}$ belongs to $\mathbf{H}$. Otherwise we would have for some two elements $H_a, H_b \in \mathbf{H}$ the equation $G_1H_a = H_b$, or $G_1 = H_bH_a^{-1}$, i.e., $G_1 \in \mathbf{H}$, in contradiction with our initial assumption. We now take some element $G_2 \in \mathbf{G}$ not belonging to either $\mathbf{H}$ or $G_1\mathbf{H}$ and form the set $G_2\mathbf{H}$. In an analogous way one can show that $G_1\mathbf{H}$ and $G_2\mathbf{H}$ do not possess any elements in common. If $\mathbf{H}$, $G_1\mathbf{H}$, and $G_2\mathbf{H}$ do not exhaust all the elements of the group, we form $G_3\mathbf{H}, \ldots$, etc., until all the elements of $\mathbf{G}$ are divided into m sets:

$$\mathbf{H},\quad G_1\mathbf{H},\quad G_2\mathbf{H},\quad \ldots,\quad G_{m-1}\mathbf{H}. \tag{1.12}$$

The order of the group is therefore given by $g = mh$. As a result, we arrive at a theorem known as *Lagrange's theorem*. This states that the order of a subgroup of a finite group is a divisor of the order of the group.

From this theorem it obviously follows that a group whose order is a prime number does not possess any subgroups.

The division of the group elements into the sets (1.12) is uniquely determined on specifying the subgroup $\mathbf{H}$ since any element of the set $G_k\mathbf{H}$ may play the role of G_k. Thus let $G_k' = G_k H_a$, where H_a is an arbitrary element of the subgroup $\mathbf{H}$. The set of elements $G_k'\mathbf{H} \equiv G_k H_a \mathbf{H} = G_k\mathbf{H}$, the last equality following from the theorem of Section 1.1.

The sets (1.12) are called the *left cosets* of the subgroup $\mathbf{H}$. *Right cosets* are defined in an analogous manner. The number of cosets of a subgroup is called the *index* of the subgroup. With the exception of $\mathbf{H}$, none of the cosets $G_k\mathbf{H}$ forms a subgroup since they do not contain the unit element.

As an example we consider the set of permutations I and P_{12} from the group $\boldsymbol{\pi}_3$ [see (1.11)]. This set forms a subgroup which we denote by $\boldsymbol{\pi}_2$. The index of the subgroup m is equal to $g/h = 6/2 = 3$. There are therefore three cosets. The left cosets of the subgroup $\boldsymbol{\pi}_2$ are given by

$$\begin{aligned} \boldsymbol{\pi}_2 &: \quad I, P_{12}; \\ P_{13}\boldsymbol{\pi}_2 &: \quad P_{13}, P_{123}; \\ P_{23}\boldsymbol{\pi}_2 &: \quad P_{23}, P_{132}. \end{aligned}$$

1.5. Conjugate Elements. Classes

Two elements A and B are said to be *conjugate* if $A = QBQ^{-1}$, where Q is an element of the same group. If two elements A and B are each conjugate to a third element C, then they are also conjugate to each other. Thus it follows from $A = QCQ^{-1}$ and $B = RCR^{-1}$ that $C = Q^{-1}AQ$, whence $B = RQ^{-1}AQR^{-1} = (RQ^{-1})A(RQ^{-1})^{-1}$. The elements of a group therefore fall into sets, in each of which the elements are conjugate to one another. Such sets are called *classes* of the group. A class is determined by specifying one of its elements; thus, knowing A, we can obtain the other elements by forming $G_b A G_b^{-1}$, where G_b runs through all the elements of the group (some elements of the class, however, may be repeated by this). By using the multiplication table for the group $\mathbf{C}_{3v}$ it is not difficult to satisfy oneself that its six elements fall into three classes: E; $C_3, C_3{}^2$; and $\sigma_v^{(1)}$, $\sigma_v^{(2)}$, and $\sigma_v^{(3)}$.

We quote without proof the following further properties of classes.

(a) The unit element forms a class by itself.

(b) Except for the unit class, classes do not form subgroups since they do not contain the unit element.

(c) Every element of an Abelian group forms a class by itself.

(d) The number of elements in a class is a divisor of the order of the group.

1.6. Invariant Subgroups. Factor Groups

Consider the set of elements

$$G_0 H_a G_0^{-1}, \tag{1.13}$$

where G_0 is some fixed element of a group **G** and H_a runs through all the elements of a subgroup **H**. It is easy to verify that the products (1.13) satisfy all four group postulates and consequently form a subgroup of **G**. This subgroup is said to be *conjugate to the subgroup* **H**. The subgroup **H** is called an *invariant subgroup* if for any choice of the element G_0, the conjugate subgroup coincides with **H**, i.e., for any $G_0 \in \mathbf{G}$,

$$G_0 \mathbf{H} G_0^{-1} = \mathbf{H}, \tag{1.14}$$

From this equation it follows that

$$G_0 \mathbf{H} = \mathbf{H} G_0, \tag{1.15}$$

i.e., the left cosets of an invariant subgroup coincide with the right cosets.

Let the index of an invariant subgroup be m, and let us form the m left (say) cosets of **H**:

$$\mathbf{H},\quad G_1\mathbf{H},\quad G_2\mathbf{H},\quad \ldots,\quad G_{m-1}\mathbf{H}. \tag{1.16}$$

We now calculate the product of two of the cosets and obtain

$$G_1\mathbf{H} \cdot G_2\mathbf{H} = G_1(\mathbf{H}G_2)\mathbf{H} = (G_1 G_2)\mathbf{H}, \tag{1.17}$$

where use was made of Eq. (1.15) and the fact that the product of a group with itself gives just the same group. The product of two cosets from the set (1.16) is thus another coset from the same set. Furthermore, one can associate with every coset $G_k\mathbf{H}$ an inverse coset $G_k^{-1}\mathbf{H}$, such that their product is equal to the invariant subgroup **H**. The m cosets (1.16) can consequently be regarded as the elements of a group in which the invariant subgroup itself plays the role of unit element. This group, whose order is equal to the index of the subgroup **H**, is called the *factor group of the invariant subgroup* **H** and is denoted by **G**/**H**. A group which does not possess an invariant subgroup is said to be *simple*.

EXAMPLE

The group $\mathbf{C}_{3v}$ possesses the subgroup $\mathbf{C}_3$ which consists of the three elements E, C_3, and $C_3{}^2$. By using the multiplication table for the group (Table 1.1) it is not difficult to show that this subgroup is invariant. The index of the $\mathbf{C}_3$ subgroup is two and hence there are two cosets:

$$\mathbf{C}_3\colon\ E,\quad C_3,\quad C_3{}^2;$$

$$\sigma_v\mathbf{C}_3\colon\ \sigma_v^{(1)},\quad \sigma_v^{(2)},\quad \sigma_v^{(3)}.$$

These form the factor group $\mathbf{C}_{3v}/\mathbf{C}_3$ of order two in which the role of unit element is played by the subgroup $\mathbf{C}_3$. The multiplication table for the factor group has the following form:

	$\mathbf{C}_3$	$\sigma_v\mathbf{C}_3$
$\mathbf{C}_3$	$\mathbf{C}_3$	$\sigma_v\mathbf{C}_3$
$\sigma_v\mathbf{C}_3$	$\sigma_v\mathbf{C}_3$	$\mathbf{C}_3$

1.7. Direct Products of Groups

Let two finite groups $\mathbf{G}_1$ and $\mathbf{G}_2$ possess no elements in common apart from the unit element. In particular these groups might be subgroups of some larger group. We denote their orders by g_1 and g_2, respectively. If the elements of the group $\mathbf{G}_1$ commute with those of $\mathbf{G}_2$, then by forming all possible products of the elements of these groups we obtain a set of g_1g_2 elements which also form a group. The commutation condition ensures that the product of two elements of the set obtained in this way gives, in fact, an element belonging to the same set:

$$A_1A_2 \cdot B_1B_2 = A_1B_1 \cdot A_2B_2 = C_1C_2.$$

The satisfaction of the remaining group postulates is evident. The group which is thus obtained is called the *direct product* of the groups $\mathbf{G}_1$ and $\mathbf{G}_2$ and is denoted by $\mathbf{G}_1 \times \mathbf{G}_2$.

The simplest example of commuting groups are permutation groups whose elements act upon different sets of objects. Thus, for example, we may form the direct product of two $\boldsymbol{\pi}_2$ permutation groups. Let the permutations of the first group act on the numbers 1 and 2 and the permutations of the second upon the numbers 3 and 4. The direct product of the groups $\boldsymbol{\pi}_2 \times \boldsymbol{\pi}_2$ therefore contains the four elements

$$\boldsymbol{\pi}_2 \times \boldsymbol{\pi}_2 \colon \quad I, \quad P_{12}, \quad P_{34}, \quad P_{12} \cdot P_{34}. \tag{1.18}$$

1.8. The Semidirect Product

As in the previous section, we consider two groups $\mathbf{H}$ and $\mathbf{C}$ that do not possess any elements in common apart from the unit element. However, instead of requiring the elements of one group to commute with those of the other, we impose a less restrictive requirement: This is that the group $\mathbf{H}$ shall be invariant with respect to conjugation with any element of the group $\mathbf{C}$, i.e., that

$$C_a\mathbf{H}C_a^{-1} = \mathbf{H} \tag{1.19}$$

for $C_a \in \mathbf{C}$. The set of elements obtained by multiplying the group $\mathbf{H}$ by the group $\mathbf{C}$ forms a group $\mathbf{G}$ which is called the *semidirect product* of the groups $\mathbf{H}$ and $\mathbf{C}$, and is denoted by

$$\mathbf{G} = \mathbf{H} \wedge \mathbf{C}. \tag{1.20}$$

The condition (1.19) ensures that the product of two elements of the set obtained in this way in fact belongs to the same set, since

$$H_aC_a \cdot H_bC_b = H_a(C_aH_bC_a^{-1})C_aC_b = H_aH_c \cdot C_aC_b = H_dC_c. \qquad (1.21)$$

It is not difficult to verify that the remaining group postulates are also satisfied.

It is evident that **H** forms an invariant subgroup with respect to the semidirect product. It is usual to write this invariant subgroup first on the right-hand side of (1.20); i.e., there is a prescribed order for writing the factors of the semidirect product. We observe in the case of the direct product that since the elements of the different subgroups commute, each subgroup is invariant and the order of the factors is immaterial.

There are many groups which can be written in the form of a semidirect product of their subgroups. Thus Altmann showed that all point groups (Altman, 1963a,b), and all the symmetry groups of nonrigid molecules (Altman, 1967) can be represented as semidirect products. As an example we consider the point group $\mathbf{C}_{3v}$. In Section 1.6 it was shown in the example that the group $\mathbf{C}_3$ forms an invariant subgroup of $\mathbf{C}_{3v}$. It is easy to verify that

$$\mathbf{C}_{3v} = \mathbf{C}_3 \wedge \mathbf{C}_s, \qquad (1.22)$$

where the group $\mathbf{C}_s$ consists of two elements: E and σ_v, the reflection operation in a plane passing through the threefold axis.

PART 2. Representations of Groups

1.9. Definition

A group of square matrices which is homomorphic with a given group is said to form a *representation* of the group. The number of rows or columns of the matrices is called the *dimension* of the representation.

It follows from this definition that one can associate with every element Q of a group a matrix $\mathbf{\Gamma}(Q)$ such that corresponding to the product $QP = R$ of the group elements, one has the matrix product‡

$$\mathbf{\Gamma}(Q)\mathbf{\Gamma}(P) = \mathbf{\Gamma}(R). \qquad (1.23)$$

The multiplication of the matrices on the left-hand side of this equation is carried out according to the usual rules of matrix algebra.§ If the matrices of

‡Representations defined by Eq. (1.23) are called *vector representations*. A more general type of representation is possible, the matrices of which satisfy the relation

$$\mathbf{\Gamma}(Q)\mathbf{\Gamma}(P) = \epsilon\mathbf{\Gamma}(R), \qquad |\epsilon| = 1. \qquad (1.23a)$$

Representations such as these are known as *projective* or *ray* representations.

§An element of the product matrix is given by $\Gamma_{ik}(QP) = \sum_l \Gamma_{il}(Q)\Gamma_{lk}(P)$.

a representation are all distinct, the representation is isomorphic with the group. Such a representation is then said to be *faithful*.

The matrices of a representation can be obtained by elimination or by some other method so that they conform to the multiplication table for the group. In physical applications representations usually arise as the result of applying the elements of a group to functions of some coordinates. The groups which occur in physics are either groups of linear spatial transformations or of permutations of particle coordinates. When elements of these groups are applied to functions of the coordinates their effect is to generate a set of new functions which transform linearly into one another under the action of the group elements. We now examine this process in somewhat greater detail.

Let ψ_0 be a coordinate function defined in the configuration space of the system.‡ The application of an element Q of a group $\mathbf{G}$ to ψ_0 converts this function into some other function which we denote by $\psi_Q \equiv Q\psi_0$. If Q runs through all the elements of the group, we obtain g functions. These may not be linearly independent. Let the number of linearly independent functions ψ_i be equal to $f (f \leq g)$. Under the action of the group elements the functions ψ_i transform only into each other, since by virtue of the properties of a group (see the theorem of Section 1.1) no other functions may appear in this set. We thus have

$$Q\psi_k = \sum_{i=1}^{f} \Gamma_{ik}(Q)\psi_i. \tag{1.24}$$

The coefficients $\Gamma_{ik}(Q)$ form a square matrix of order f. It follows from the definition of the matrix $\Gamma(Q)$ that under the action of the group element Q the function ψ_k transforms according to the kth column of the matrix $\Gamma(Q)$. Each element Q of the group has a matrix $\Gamma(Q)$ corresponding to it. Corresponding to the product of two elements QP is a matrix which is the product of the matrices $\Gamma(Q)$ and $\Gamma(P)$, for

$$QP\psi_k = Q\sum_i \Gamma_{ik}(P)\psi_i = \sum_l \left[\sum_i \Gamma_{li}(Q)\Gamma_{ik}(P)\right]\psi_l = \sum_l [\Gamma(Q)\Gamma(P)]_{lk}\psi_l.$$

The matrices $\Gamma(Q)$ consequently form an f-dimensional representation of the group $\mathbf{G}$. The set of f functions ψ_i which define the $\Gamma(Q)$ matrices are said to form a *basis* for the representation.

1.10. Vector Spaces

A vector space of n dimensions consists of the set of all the vectors $\mathbf{x}$ which can be obtained by forming all possible linear combinations of n linearly independent basis vectors $\mathbf{e}_i$:

$$\mathbf{x} = \sum_{i=1}^{n} x_i\mathbf{e}_i. \tag{1.25}$$

‡For a system of N electrons the dimension of the configuration space is equal to $4N$. Each electron is described by three spatial coordinates and one spin coordinate.

x_i is called the *component* of the vector $\mathbf{x}$ in the direction $\mathbf{e}_i$; x_i may also be a complex quantity, in which case the vector space is complex.

Under a linear transformation from one system of basis vectors to another the components of a vector in the old system transform linearly into the components in the new system. Thus if

$$\mathbf{e}_k' = \sum_i s_{ik}\mathbf{e}_i, \tag{1.26}$$

then from

$$\mathbf{x} = \sum_i x_i\mathbf{e}_i = \sum_k x_k'\mathbf{e}_k' = \sum_{k,i} x_k' s_{ik}\mathbf{e}_i,$$

it follows that

$$x_i = \sum_k s_{ik}x_k'. \tag{1.27}$$

If we represent the vector $\mathbf{x}$ as a column matrix X and denote the transformation matrix in (1.26) by S, then eq. (1.27) can be written as

$$X = SX', \tag{1.28}$$

from which

$$X' = S^{-1}X, \tag{1.29}$$

where S^{-1} is the matrix of the inverse transformation.

Instead of transforming the basis vectors, we can carry out a transformation of the vector space by which we associate with each vector $\mathbf{x}$ a vector $\mathbf{y}$:

$$\mathbf{y} = A\mathbf{x}. \tag{1.30}$$

The quantity A can be regarded as an operator which carries the vector $\mathbf{x}$ into the vector $\mathbf{y}$. The analytical form of A is an n-dimensional matrix which connects the components of the vectors:

$$y_i = \sum_k a_{ik}x_k, \tag{1.31}$$

or in matrix notation

$$Y = AX. \tag{1.32}$$

Let the vectors $\mathbf{y}$ and $\mathbf{x}$ be connected to each other by the relation (1.32). We carry out the transformation (1.26) of the basis vectors. Thus in the new coordinate system

$$Y' = A'X'. \tag{1.33}$$

Let us now find the relation between A' and A. For this purpose we pass to the old coordinate system, carry out in it the transformation (1.32), and return to the new system:

$$Y' = S^{-1}Y, \qquad Y = AX, \qquad X = SX'$$

or

$$Y' = S^{-1}ASX'. \tag{1.34}$$

On comparing (1.33) and (1.34), we obtain

$$A' = S^{-1}AS. \tag{1.35}$$

If the basis vectors form an orthonormal set

$$\mathbf{e}_i \cdot \mathbf{e}_j = \delta_{ij}, \tag{1.36}$$

then the scalar product of two vectors is given by

$$\mathbf{x} \cdot \mathbf{y} = \sum_i x_i^* y_i, \tag{1.37}$$

and the square of the length of a vector is equal to the sum of the squared moduli of its components:

$$\mathbf{x} \cdot \mathbf{x} = |\mathbf{x}|^2 = \sum_i |x_i|^2. \tag{1.38}$$

A linear transformation is said to be unitary if it leaves unchanged the scalar product of two vectors. It follows from this definition of a unitary transformation that

$$\sum_k x_k'^* y_k' = \sum_i x_i^* y_i = \sum_{i,k,l} s_{ik}^* s_{il} x_k'^* y_l'. \tag{1.39}$$

Equation (1.39) holds if the condition

$$\sum_i s_{ik}^* s_{il} = \delta_{kl}. \tag{1.40}$$

is fulfilled. However, according to the definition of the inverse of a matrix,

$$\sum_i s_{ki}^{-1} s_{il} = \delta_{kl}, \tag{1.41}$$

and it therefore follows from a comparison of (1.40) and (1.41) that $s_{ki}^{-1} = s_{ik}^*$. This can be symbolically written in the form‡

$$S^{-1} = \tilde{S}^* \tag{1.42}$$

or

$$S\tilde{S}^* = \tilde{S}^* S = I. \tag{1.42a}$$

Matrices which satisfy the condition (1.42) or (1.42a) are called unitary. From (1.42a) it can be seen that in addition to (1.40) the elements of a unitary matrix also obey the relation

$$\sum_k s_{ik}^* s_{lk} = \delta_{il}. \tag{1.40a}$$

In the case of real matrices we have instead of (1.42a)

$$S\tilde{S} = \tilde{S}S = I. \tag{1.43}$$

Matrices which satisfy (1.43) are said to be *orthogonal*. Their elements satisfy

‡$\tilde{S}$ denotes the transpose of the matrix S. Its elements are related to those of S by $\tilde{s}_{ik} = s_{ki}$.

the following orthogonality conditions:

$$\sum_i s_{ik}s_{il} = \delta_{kl}, \qquad \sum_k s_{ik}s_{lk} = \delta_{il}. \tag{1.44}$$

1.11. Reducibility of Representations

The f basis functions for a representation $\mathbf{\Gamma}$ (see Section 1.9) can be regarded as basis vectors in an f-dimensional vector space. We denote this space by $\mathfrak{R}$ and we say that it transforms according to the representation $\mathbf{\Gamma}$. According to Eq. (1.24), the effect of an element Q of a group $\mathbf{G}$ upon a basis vector ψ_k is to generate a vector $Q\psi_k$ which also lies in $\mathfrak{R}$. Since any vector in a vector space can be expressed in the form of a linear combination of the basis vectors, the effect of an element of the group is to convert every vector of the space into another vector which also belongs to $\mathfrak{R}$. The space $\mathfrak{R}$ is therefore invariant under the transformations of $\mathbf{G}$.

According to Eq. (1.35), a linear transformation of the basis vectors

$$\psi_k' = \sum_{i=1}^{f} s_{ik}\psi_i \tag{1.45}$$

converts the matrices of a representation $\mathbf{\Gamma}$ into the matrices

$$\mathbf{\Gamma}'(Q) = S^{-1}\mathbf{\Gamma}(Q)S. \tag{1.46}$$

Two representations which are connected by a relation such as (1.46) are said to be *equivalent*. It is clear that there is an infinite number of equivalent representations. A transformation which converts a representation into an equivalent one is called a *similarity transformation.*

If a similarity transformation can be found which converts all the matrices of a representation into block-diagonal form

$$\mathbf{\Gamma}' = S^{-1}\mathbf{\Gamma}S = \begin{vmatrix} \mathbf{\Gamma}^{(1)} & & & \\ & \mathbf{\Gamma}^{(2)} & & \\ & & \cdots & \\ & & & \mathbf{\Gamma}^{(m)} \end{vmatrix}, \tag{1.47}$$

then the representation is said to be *reducible*. As a result of such a similarity transformation the representation is decomposed into m representations of smaller dimensions. This is written as follows:

$$\mathbf{\Gamma} \doteq \mathbf{\Gamma}^{(1)} + \mathbf{\Gamma}^{(2)} + \cdots + \mathbf{\Gamma}^{(m)}. \tag{1.48}$$

The reducibility of a representation implies that by means of a linear transformation of the basis vectors the space $\mathfrak{R}$ can be divided up into a number of invariant subspaces $\mathfrak{R}^{(\alpha)}$ each of which transforms according to a representation $\mathbf{\Gamma}^{(\alpha)}$. The operations of $\mathbf{G}$ transform the vectors of each subspace among themselves without mixing vectors from different subspaces.

If there is no transformation which reduces the representation matrices to the block-diagonal form (1.47), then the representation is said to be *irreducible*. We note that an irreducible representation of a group obviously forms a representation of a subgroup of the group. However, as far as the subgroup is concerned, this representation may be reducible and may be decomposed into irreducible representations of the subgroup. This process is called *reduction with respect to a subgroup*.

The following set of matrices offers an example of a reducible representation of the group $\mathbf{C}_{3v}$:

$$\Gamma(E) = \begin{vmatrix} 1 & 0 & 0 \\ 0 & 1 & 0 \\ 0 & 0 & 1 \end{vmatrix}, \quad \Gamma(C_3) = \begin{vmatrix} 0 & 0 & 1 \\ 1 & 0 & 0 \\ 0 & 1 & 0 \end{vmatrix}, \quad \Gamma(C_3{}^2) = \begin{vmatrix} 0 & 1 & 0 \\ 0 & 0 & 1 \\ 1 & 0 & 0 \end{vmatrix}, \tag{1.49}$$

$$\Gamma(\sigma_v^{(1)}) = \begin{vmatrix} 1 & 0 & 0 \\ 0 & 0 & 1 \\ 0 & 1 & 0 \end{vmatrix}, \quad \Gamma(\sigma_v^{(2)}) = \begin{vmatrix} 0 & 0 & 1 \\ 0 & 1 & 0 \\ 1 & 0 & 0 \end{vmatrix}, \quad \Gamma(\sigma_v^{(3)}) = \begin{vmatrix} 0 & 1 & 0 \\ 1 & 0 & 0 \\ 0 & 0 & 1 \end{vmatrix}.$$

It is easily verified by direct multiplication that these matrices conform to the multiplication table for the group (see Table 1.1). By means of a transformation such as (1.47) it can be decomposed into two irreducible representations, one of which is one dimensional and the other two dimensional. In the following sections we shall describe methods for obtaining the irreducible representations which occur in the decomposition (1.48).

1.12. Properties of Irreducible Representations

We quote without proof the following important properties of irreducible representations of finite groups.‡

(a) The number of nonequivalent irreducible representations of a group is equal to the number of classes in the group.

(b) The sum of the squares of the dimensions of the nonequivalent irreducible representations is equal to the order of the group, i.e.,

$$f_1{}^2 + f_2{}^2 + \cdots + f_r{}^2 = g, \tag{1.50}$$

where f_α denotes the dimension of the αth irreducible representation. It follows from this that all the irreducible representations of an Abelian group are one-dimensional, since the number of its irreducible representations is equal to the number of elements in the group.

‡The proofs of these statements can be found in any text on group theory, for example, in the works of Bauer (1933), Smirnov (1964), Murnaghan (1938), and Hamermesh (1962).

(c) The dimension of an irreducible representation of a finite group is a divisor of the order of the group.

(d) The following orthogonality relations hold for the matrix elements of irreducible representations:

$$\sum_R \Gamma_{ik}^{(\alpha)}(R)^* \Gamma_{lm}^{(\beta)}(R) = (g/f_\alpha)\delta_{\alpha\beta}\delta_{il}\delta_{km}, \tag{1.51}$$

$$\sum_{\alpha,i,k} (f_\alpha/g)\Gamma_{ik}^{(\alpha)}(R)^* \Gamma_{ik}^{(\alpha)}(Q) = \delta_{RQ}. \tag{1.52}$$

The summation in (1.51) is taken over all the g elements of the group. In (1.52) all $f_\alpha{}^2$ elements of the matrix $\Gamma^{(\alpha)}(R)$ are summed for each irreducible representation $\Gamma^{(\alpha)}$. The number of terms in the sum (1.52) is equal to the order of the group in accordance with eq. (1.50).

1.13. Characters

Let the matrices $\Gamma^{(\alpha)}(R)$ form a representation of a group **G**. The sum of the diagonal elements of a matrix $\Gamma^{(\alpha)}(R)$ is called the character of the element R in the representation $\Gamma^{(\alpha)}$ and is denoted by $\chi^{(\alpha)}(R)$:

$$\chi^{(\alpha)}(R) = \sum_i \Gamma_{ii}^{(\alpha)}(R). \tag{1.53}$$

Each representation is thus typified by a set of g characters.

The characters of equivalent representations are identical, for in accordance with the rule for matrix multiplication,

$$\begin{aligned} \sum_i (S^{-1}\Gamma(R)S)_{ii} &= \sum_{i,l,m} s_{il}^{-1}\Gamma_{lm}(R)s_{mi} \\ &= \sum_{l,m}\Big(\sum_i s_{mi}s_{il}^{-1}\Big)\Gamma_{lm}(R) = \sum_{l,m}\delta_{ml}\Gamma_{lm}(R) = \sum_m \Gamma_{mm}(R). \end{aligned}$$

Hence the specification of a representation by means of a set of characters does not distinguish between equivalent representations. This is extremely useful, since in physical applications only inequivalent representations are important. By specifying the characters of a representation one can, of course, still distinguish nonequivalent representations.

The elements of a group belonging to a given class are connected among themselves by relations analogous to (1.46). Their characters must therefore all be identical. Consequently, one can denote a class by specifying the character of just one of its members. Furthermore, the number of distinct characters of a representation cannot exceed the number of classes in the group.

The characters of irreducible representations satisfy the following orthogonality relations:

$$\sum_R \chi^{(\alpha)}(R)^* \chi^{(\beta)}(R) = g\delta_{\alpha\beta}. \tag{1.54}$$

These relations are obtained from eq (1.51) by putting i equal to k, l equal to m, and summing over k and m on both sides.

Since the characters of all the elements belonging to a given class are equal, Eq. (1.54) can be written in the form

$$\sum_{\mathbf{C}} g_{\mathbf{C}} \chi^{(\alpha)}(\mathbf{C})^* \chi^{(\beta)}(\mathbf{C}) = g\delta_{\alpha\beta}, \tag{1.55}$$

where the sum is taken over all classes $\mathbf{C}$ of the group, and $g_{\mathbf{C}}$ denotes the number of elements in $\mathbf{C}$ ($\sum_{\mathbf{C}} g_{\mathbf{C}} = g$).

If we write $(g_{\mathbf{C}}/g)^{1/2}\chi^{(\alpha)}$ as $a_{\mathbf{C}\alpha}$, then since the number of classes is equal to the number of irreducible representations, the quantities $a_{\mathbf{C}\alpha}$ form a square matrix. The orthogonality relations for the $a_{\mathbf{C}\alpha}$, Eq. (1-55), are then identical with the unitarity condition (1.40) for the first subscript. The matrix $\|a_{\mathbf{C}\alpha}\|$ is therefore unitary, with the unitarity condition (1.40a) also applying to its elements. From this we obtain a second orthogonality relation for the characters of irreducible representations:

$$\sum_{\alpha} \chi^{(\alpha)}(\mathbf{C}_i)^* \chi^{(\alpha)}(\mathbf{C}_k) = (g/g_{\mathbf{C}_i})\delta_{\mathbf{C}_i\mathbf{C}_k}. \tag{1.56}$$

Among the irreducible representations of a group there is always a one-dimensional representation which is generated by a basis function that is invariant under all the operations of the group. All the characters of this representation are equal to unity. The representation is usually called the *totally symmetric* or *unit* representation, and is denoted by the symbol A_1.

1.14. The Calculation of the Characters of Irreducible Representations

In the next section we show that in order to decompose a reducible representation into its irreducible components, it suffices to know just the characters of the representations. Now, a reducible representation is either given directly by its matrices (in which case the determination of its characters is trivial) or its basis functions are given. In this last case the characters are obtained by eliciting the effect of the operations of the group on the basis functions, and then summing the diagonal elements. For irreducible representations there are a number of methods for calculating the characters. These are based upon the properties of finite groups, and a knowledge of the basis functions is unnecessary [see, for example, the works of Murnaghan (1938) and Hamermesh (1962)]. We give here a method for determining the characters of the irreducible representations of any finite group, and which requires a knowledge only of the group multiplication table.‡

Let the elements $R_j \in \mathbf{C}_j$ and $R_k \in \mathbf{C}_k$, where $\mathbf{C}_j$ and $\mathbf{C}_k$ are two arbitrary classes of a group $\mathbf{G}$. We denote the set of all possible products of the form R_jR_k by $\mathbf{C}_j\mathbf{C}_k$. This set may include elements belonging to various

‡A knowledge of the group multiplication table has been shown to be sufficient not only to calculate the characters, but also for a complete determination of the irreducible representations; see the paper by Flodmark and Blokker (1967).

classes of the group. It can be shown that the product of two classes is commutative, and that the elements of a given class occur in the product an identical number of times. This last circumstance allows one to write the product of two classes in the form of a kind of decomposition into classes of the group:

$$\mathbf{C}_j\mathbf{C}_k = \mathbf{C}_k\mathbf{C}_j = \sum_{l=1}^{r} c_{jk}^l \mathbf{C}_l. \tag{1.57}$$

The c_{jk}^l coefficients can be found from the group multiplication table. Replacement of each class by the corresponding matrices of an irreducible representation leads to the equation [see Hamermesh (1962, p.110)]

$$g_j g_k \chi(\mathbf{C}_j)\chi(\mathbf{C}_k) = \chi(\mathbf{C}_1) \sum_{l=1}^{r} c_{jk}^l g_l \chi(\mathbf{C}_l), \tag{1.58}$$

where g_j and g_k are, respectively, the numbers of elements in the classes $\mathbf{C}_j$ and $\mathbf{C}_k$. Equation (1.58) holds for any irreducible representation of the group. Since the dimension of an irreducible representation is equal to the character of the identity class $\mathbf{C}_1$, we obtain from (1.50) an additional relation:

$$\sum_{\alpha=1}^{r} [\chi^{(\alpha)}(\mathbf{C}_1)]^2 = g. \tag{1.59}$$

In groups of low order there is only one way in which, for given r and g, Eq. (1.59) can be satisfied with integral values for $\chi^{(\alpha)}(\mathbf{C}_1)$. Equations (1.58) and (1.59) are sufficient to determine the characters of all the irreducible representations of a group. The characters that are obtained must, of course, satisfy the orthogonality relations (1.55) and (1.56).

As an example we determine the characters of the irreducible representations of the group $\mathbf{C}_{3v}$, whose multiplication table was given in Table 1.1. This group possesses three classes and hence three irreducible representations, the characters of which are to be found. We enumerate the classes of the group as

$$\mathbf{C}_1:\ E;$$
$$\mathbf{C}_2:\ C_3,\ C_3{}^2;$$
$$\mathbf{C}_3:\ \sigma_v^{(1)},\ \sigma_v^{(2)},\ \sigma_v^{(3)};$$

from which $g_1 = 1$, $g_2 = 2$, $g_3 = 3$. From the multiplication table we obtain

$$\mathbf{C}_2\mathbf{C}_2 = 2\mathbf{C}_1 + \mathbf{C}_2, \qquad \mathbf{C}_3\mathbf{C}_3 = 3\mathbf{C}_1 + 3\mathbf{C}_2, \qquad \mathbf{C}_2\mathbf{C}_3 = 2\mathbf{C}_3.$$

Equations (1.58) therefore assume the form

$$\begin{aligned} 4[\chi(\mathbf{C}_2)]^2 &= \chi(\mathbf{C}_1)[2\chi(\mathbf{C}_1) + 2\chi(\mathbf{C}_2)], \\ 9[\chi(\mathbf{C}_3)]^2 &= \chi(\mathbf{C}_1)[3\chi(\mathbf{C}_1) + 6\chi(\mathbf{C}_2)], \\ 6\chi(\mathbf{C}_2)\chi(\mathbf{C}_3) &= 6\chi(\mathbf{C}_1)\chi(\mathbf{C}_3). \end{aligned} \tag{1.58a}$$

In addition, Eq. (1.59),

$$[\chi^{(1)}(\mathbf{C}_1)]^2 + [\chi^{(2)}(\mathbf{C}_1)]^2 + [\chi^{(3)}(\mathbf{C}_1)]^2 = 6, \tag{1.59a}$$

has the unique solution

$$\chi^{(1)}(\mathbf{C}_1) = \chi^{(2)}(\mathbf{C}_1) = 1, \qquad \chi^{(3)}(\mathbf{C}_1) = 2.$$

We substitute in turn the values $\chi(\mathbf{C}_1) = 1$ and $\chi(\mathbf{C}_1) = 2$ into (1.58a) and solve the resulting systems of equations. Solutions that do not satisfy the orthogonality relations (1.55) and (1.56) are rejected. We thus obtain the characters of all three irreducible representations. These are collected in Table 1.2.‡ Instead of using a symbol, we represent a class by one of its elements

Table 1.2

Characters of the Irreducible Representations of $\mathbf{C}_{3v}$

	E	$2C_3$	$3\sigma_v$
$\chi^{(1)}$	1	1	1
$\chi^{(2)}$	1	1	−1
$\chi^{(3)}$	2	−1	0

since the characters of all the elements in a class are identical. The representative elements appear on the top line of the table, the figure in front of them denoting the number in the class.

1.15. The Decomposition of a Reducible Representation

Consider some reducible representation $\mathbf{\Gamma}$. By definition, one can always find a unitary transformation that will bring this to the block-diagonal form (1.47). We assume that no further reduction is possible, i.e., that the representations $\mathbf{\Gamma}^{(\alpha)}$ which appear in the decomposition (1.48) are irreducible. A given irreducible representation, however, may occur several times in the decomposition. Thus in general we have

$$\mathbf{\Gamma} \doteq \sum_{\beta} a^{(\beta)} \mathbf{\Gamma}^{(\beta)}, \tag{1.60}$$

where $a^{(\beta)}$ denotes the number of times the representation $\mathbf{\Gamma}^{(\beta)}$ occurs in the decomposition of $\mathbf{\Gamma}$. Now since a similarity transformation does not alter the characters of a representation, the characters of the representation $\mathbf{\Gamma}$

‡The irreducible representations of the point groups are usually denoted by special symbols. Character tables of the point groups using the Schönfliess notation [adopted, for example, by Landau and Lifshitz (1965)] are given in Appendix 1.

are given by

$$\chi^{(\Gamma)}(R) = \sum_{\beta} a^{(\beta)} \chi^{(\beta)}(R). \tag{1.61}$$

We multiply (1.61) through by $\chi^{(\alpha)}(R)^*$ and sum over all the elements in the group. By virtue of the orthogonality relations (1.54) we obtain

$$a^{(\alpha)} = (1/g) \sum_{R} \chi^{(\Gamma)}(R) \chi^{(\alpha)}(R)^*. \tag{1.62}$$

For actual applications it is convenient to rewrite this relation in the form

$$a^{(\alpha)} = (1/g) \sum_{\mathbf{C}} g_{\mathbf{C}} \chi^{(\Gamma)}(\mathbf{C}) \chi^{(\alpha)}(\mathbf{C})^*. \tag{1.63}$$

If the characters of the representations are known, Eq. (1.63) enables one easily to determine the irreducible representations which occur in the decomposition of a given reducible representation.

As an example we determine the irreducible representations of $\mathbf{C}_{3v}$ which occur in the decomposition of the representation (1.49). The characters of this representation are

	E	$2C_3$	$3\sigma_v$
$\chi^{(\Gamma)}$	3	0	1

Taking the characters of the irreducible representations of $\mathbf{C}_{3v}$ from Table 1.2, we find from formula (1.63) that

$$a^{(1)} = \tfrac{1}{6}(3 + 3) = 1, \qquad a^{(2)} = \tfrac{1}{6}(3 - 3) = 0, \qquad a^{(3)} = \tfrac{1}{6} 6 = 1.$$

The representation (1.49) consequently breaks down into two irreducible representations the one-dimensional representation $\mathbf{\Gamma}^{(1)}$ and the two-dimensional representation $\mathbf{\Gamma}^{(3)}$.

1.16. The Direct Product of Representations

Let us consider two irreducible representations $\mathbf{\Gamma}^{(\alpha)}$ and $\mathbf{\Gamma}^{(\beta)}$ of a group $\mathbf{G}$, each of which is defined by a set of basis functions $\psi_i^{(\alpha)}$ ($i = 1, 2, \ldots, f_\alpha$) and $\psi_k^{(\beta)}$ ($k = 1, 2, \ldots, f_\beta$) respectively. If we construct all possible products of the form $\psi_i^{(\alpha)}\psi_k^{(\beta)}$, we obtain an $f_\alpha f_\beta$-dimensional basis for a representation of the group. This representation is known as the *direct product* of the representations $\mathbf{\Gamma}^{(\alpha)}$ and $\mathbf{\Gamma}^{(\beta)}$ and is denoted by $\mathbf{\Gamma}^{(\alpha)} \times \mathbf{\Gamma}^{(\beta)}$. The matrix elements of the direct product of the two representations are of the form of products of the matrix elements of $\mathbf{\Gamma}^{(\alpha)}$ and $\mathbf{\Gamma}^{(\beta)}$,

$$R\psi_i^{(\alpha)}\psi_k^{(\beta)} = (R\psi_i^{(\alpha)})(R\psi_k^{(\beta)}) = \sum_{l,m} \Gamma_{li}^{(\alpha)}(R)\Gamma_{mk}^{(\beta)}(R)\psi_l^{(\alpha)}\psi_m^{(\beta)}, \tag{1.64}$$

and the characters of this representation are equal to the product of the constituent characters:

$$\chi^{(\alpha\times\beta)}(R) = \sum_{i,k} \Gamma_{ii}^{(\alpha)}(R)\Gamma_{kk}^{(\beta)}(R) = \chi^{(\alpha)}(R)\chi^{(\beta)}(R). \tag{1.65}$$

The matrix $\Gamma^{(\alpha\times\beta)}(R)$, with matrix elements given by (1.64), is called the *direct product* of the matrices $\Gamma^{(\alpha)}(R)$ and $\Gamma^{(\beta)}(R)$.‡ In contrast with the ordinary product of square matrices, where the order of the resulting matrices is the same as that of the matrices that are multiplied, the order of the matrix resulting from the direct product of two matrices is equal to the product of the orders of the matrices that are multiplied. The direct product of two matrices can be represented in block form; thus for matrices of order two

$$A \times B = \left|\begin{array}{c:c} a_{11}B & a_{12}B \\ \hdashline a_{21}B & a_{22}B \end{array}\right| = \left|\begin{array}{cccc} a_{11}b_{11} & a_{11}b_{12} & a_{12}b_{11} & a_{12}b_{12} \\ a_{11}b_{21} & a_{11}b_{22} & a_{12}b_{21} & a_{12}b_{22} \\ a_{21}b_{11} & a_{21}b_{12} & a_{22}b_{11} & a_{22}b_{12} \\ a_{21}b_{21} & a_{21}b_{22} & a_{22}b_{21} & a_{22}b_{22} \end{array}\right|. \qquad (1.66)$$

It is clear that although $A \times B \neq B \times A$, the two can be converted into each other by an appropriate permutation of the rows and columns. The characters of a direct product are therefore independent of the order in which the representations are multiplied. This also follows from formula (1.65).

If the two representations that are multiplied together are identical but possess different bases, we have

$$\chi^{(\alpha\times\alpha)}(R) = [\chi^{(\alpha)}(R)]^2. \qquad (1.67)$$

Thus the characters of the direct product of the representations $\Gamma^{(3)} \times \Gamma^{(3)}$ of $\mathbf{C}_{3v}$ are (see Table 1.2)

	E	$2C_3$	$3\sigma_v$
$[\chi^{(3)}(R)]^2$	4	1	0

We now reduce this representation. Its decomposition into irreducible components is carried out according to formula (1.63) and contains each irreducible representation of $\mathbf{C}_{3v}$ once, i.e.,

$$\Gamma^{(3)} \times \Gamma^{(3)} \doteq \Gamma^{(1)} + \Gamma^{(2)} + \Gamma^{(3)}. \qquad (1.68)$$

If in the direct product $\Gamma^{(\alpha)} \times \Gamma^{(\alpha)}$ the bases of the two representations $\Gamma^{(\alpha)}$ coincide, then the direct product basis is symmetric with respect to a permutation of its two factors. A direct product of this kind is called the *symmetric product* of a basis with itself, and is denoted by $[\Gamma^{(\alpha)}]^2$. Its dimension is less than $f_\alpha{}^2$ and is equal to $f_\alpha(f_\alpha + 1)/2$; the characters of a symmetric product are not given by a product of characters (1.67), but by an equation which is a special case of the more general formula (4.18):

$$[\chi^{(\alpha)}]^2(R) = \tfrac{1}{2}\chi^{(\alpha)}(R^2) + \tfrac{1}{2}[\chi^{(\alpha)}(R)]^2. \qquad (1.69)$$

‡The equivalent term "Kronecker product of matrices" is also used in the literature.

For example, the characters of the symmetric product $[\Gamma^{(3)}]^2$ of $\mathbf{C}_{3v}$ are

	E	$2C_3$	$3\sigma_v$
$[\chi^{(3)}]^2(R)$	3	0	1

Decomposing this representation into its irreducible components, we obtain

$$[\Gamma^{(3)}]^2 \doteq \Gamma^{(1)} + \Gamma^{(3)}. \tag{1.70}$$

The direct product of two irreducible representations is always reducible except when one of the representations is one dimensional. The decomposition of a direct product into irreducible representations

$$\Gamma^{(\alpha)} \times \Gamma^{(\beta)} \doteq \sum_{\tau} a^{(\tau)}\Gamma^{(\tau)} \tag{1.71}$$

is conventionally called the *Clebsch–Gordan series*. The coefficients in the decomposition are found from the general formula (1.62):

$$a^{(\tau)} = (1/g) \sum_R \chi^{(\alpha\times\beta)}(R)\chi^{(\tau)}(R)^* = (1/g) \sum_R \chi^{(\alpha)}(R)\chi^{(\beta)}(R)\chi^{(\tau)}(R)^*. \tag{1.72}$$

We now determine the condition that the totally symmetric representation A_1 appears in the decomposition (1.71). The characters of this representation $\chi^{(A_1)}(R) = 1$ for all elements in the group, and hence by virtue of the orthogonality relations (1.54) we obtain

$$a^{(A_1)} = (1/g) \sum_R \chi^{(\alpha)}(R)\chi^{(\beta)}(R) = \delta_{\beta\alpha^*}, \tag{1.73}$$

where $\Gamma^{(\alpha)*}$ denotes the representation whose matrix elements are the complex conjugates of the matrix elements of $\Gamma^{(\alpha)}$. Such representations are called *complex conjugate representations*. Consequently, the totally symmetric representation occurs in the decomposition of a direct product of irreducible representations if, and only if, these representations are the complex conjugates of one another. In the case of real matrices the totally symmetric representation occurs only in the direct product of an irreducible representation with itself.

In analogy with the direct product of two irreducible representations of a group, one may define a direct product of an arbitrary number of irreducible representations. The characters of such a direct product are equal to the product of the characters of the representations that are multiplied:

$$\chi^{(\alpha\times\beta\times\cdots\times\omega)}(R) = \chi^{(\alpha)}(R)\chi^{(\beta)}(R)\cdots\chi^{(\omega)}(R). \tag{1.74}$$

The direct products of representations that have been considered so far concern one and the same group. If $\mathbf{G}_1 \times \mathbf{G}_2$ is the direct product of two groups (see Section 1.7), and the irreducible representations $\Gamma^{(\alpha)} \in \mathbf{G}_1$, $\Gamma^{(\alpha)} \in \mathbf{G}_2$, then the direct product of the representations $\Gamma^{(\alpha)} \times \Gamma^{(\beta)}$ is an irreducible representation of the group $\mathbf{G}_1 \times \mathbf{G}_2$. By a derivation similar to that which led to (1.65), it can be shown that the element $R = R_1R_2$ of the group $\mathbf{G}_1 \times \mathbf{G}_2$ possesses the character

$$\chi^{(\alpha\times\beta)}(R) = \chi^{(\alpha)}(R_1)\chi^{(\beta)}(R_2). \tag{1.75}$$

The different irreducible representations of $\mathbf{G}_1 \times \mathbf{G}_2$ are obtained by combining in pairs the irreducible representations of $\mathbf{G}_1$ and $\mathbf{G}_2$.

1.17. Clebsch–Gordan Coefficients

The reduction of a direct product representation into its irreducible components (1.71) is accomplished by means of a linear transformation of the basis functions $\psi_i^{(\alpha)}\psi_k^{(\beta)}$ to a set of functions $\psi_t^{(\tau)}$ which do not mix under the operation of the group elements:

$$\psi_t^{(a\tau)} = \sum_{i,k} \psi_i^{(\alpha)}\psi_k^{(\beta)} \langle \alpha i, \beta k \,|\, a\tau t\rangle. \tag{1.76}$$

The index a distinguishes the representations τ should they occur more than once in the decomposition (1.71). The coefficients $\langle \alpha i, \beta k \,|\, a\tau t\rangle$ appearing in this equation are known as *Clebsch–Gordan coefficients.* Since the basis functions are normally chosen to be orthogonal, the Clebsch–Gordan coefficients constitute a unitary matrix‡ which we denote by $C_{\alpha\beta}$. This matrix reduces the direct product. Thus, in agreement with (1.47),

$$C_{\alpha\beta}^{-1}(\Gamma^{(\alpha)} \times \Gamma^{(\beta)})C_{\alpha\beta} = \begin{bmatrix} \Gamma^{(\tau_1)} & & & \\ & \Gamma^{(\tau_2)} & & \\ & & \cdots & \\ & & & \Gamma^{(\tau_m)} \end{bmatrix}. \tag{1.77}$$

The rows of the $C_{\alpha\beta}$ matrix are numbered by the indices i, k which may take on a total of $f_\alpha f_\beta$ values, and the columns by the indices a, τ, and t. For most of the groups that occur in physics one may choose the matrix of Clebsch–Gordan coefficients to be real, so that the orthogonality relations (1.44) apply to its elements:

$$\begin{aligned} &\sum_{a,\tau,t} \langle \alpha i, \beta k \,|\, a\tau t\rangle\langle \alpha\bar{i}, \beta\bar{k} \,|\, a\tau t\rangle = \delta_{i\bar{i}}\delta_{k\bar{k}}, \\ &\sum_{i,k} \langle \alpha i, \beta k \,|\, a\tau t\rangle\langle \alpha i, \beta k \,|\, \bar{a}\bar{\tau}\bar{t}\rangle = \delta_{a\bar{a}}\delta_{\tau\bar{\tau}}\delta_{t\bar{t}} \end{aligned} \tag{1.78}$$

In an orthogonal transformation the inverse of a matrix is equal to its transpose, and therefore the inverse transformation to (1.76) is

$$\psi_i^{(\alpha)}\psi_k^{(\beta)} = \sum_{a,\tau,t} \psi_t^{(a\tau)}\langle a\tau t \,|\, \alpha i, \beta k\rangle, \tag{1.79}$$

‡Basis functions $\psi_t^{(a\tau)}$ that belong to different irreducible representations are automatically orthogonal to one another (see Section 1.19). Even when an irreducible representation is repeated in the decomposition (1.71), its bases can always be orthogonalized to one another.

or in matrix form

$$\Gamma^{(\alpha)} \times \Gamma^{(\beta)} = C_{\alpha\beta}\Big(\sum_{a,\tau} \Gamma^{(a\tau)}\Big) C_{\alpha\beta}^{-1}. \tag{1.80}$$

Equation (1.80) is equivalent to an equation between matrix elements:

$$\Gamma_{i\bar{i}}^{(\alpha)}(R)\Gamma_{k\bar{k}}^{(\beta)}(R) = \sum_{a,\tau} \sum_{t,\bar{t}} \langle \alpha i, \beta k \,|\, a\tau t\rangle \Gamma_{t\bar{t}}^{(a\tau)}(R) \langle a\tau\bar{t} \,|\, \alpha\bar{i}, \beta\bar{k}\rangle. \tag{1.81}$$

We multiply (1.81) by $\Gamma_{t_0\bar{t}_0}^{(\tau_0)}(R)^*$ and sum over all elements in the group. If the representation $\Gamma^{(\tau_0)}$ occurs only once in the decomposition (1.71) so that the sum over $a\tau$ in (1.81) consists of just a single term with τ_0, use of the orthogonality relations (1.51) leads to the equation

$$\langle \alpha i, \beta k \,|\, \tau_0 t_0\rangle \langle \tau_0 \bar{t}_0 \,|\, \alpha\bar{i}, \beta\bar{k}\rangle = (f_{\tau_0}/g) \sum_R \Gamma_{i\bar{i}}^{(\alpha)}(R)\Gamma_{k\bar{k}}^{(\beta)}(R)\Gamma_{t_0\bar{t}_0}^{(\tau_0)}(R)^*. \tag{1.82}$$

Equation (1.82) can be used to calculate the Clebsch–Gordan coefficients if the matrix elements of the irreducible representations of the group are known. If irreducible representations occur more than once in the decomposition (1.71), the matrix of Clebsch–Gordan coefficients is defined only to within a unitary transformation. This case has been considered by Koster (1958).

From the definition of Clebsch–Gordan coefficients it follows that

$$\langle \alpha i, \beta k \,|\, \tau t\rangle \equiv 0$$

for all $\Gamma^{(\tau)}$ that do not appear in the decomposition of the direct product $\Gamma^{(\alpha)} \times \Gamma^{(\beta)}$.

1.18. The Regular Representation

Let ψ_0 be some coordinate function which does not possess any symmetry properties with respect to the operations of a group **G**. The operation of the g elements of the group upon ψ_0 generates g linearly independent functions

$$\psi_R = R\psi_0. \tag{1.83}$$

The functions ψ_R are converted into one another under the operations of the group,

$$Q\psi_R = QR\psi_0 = P\psi_0 \equiv \psi_P. \tag{1.84}$$

The functions (1.83) therefore constitute a g-dimensional basis for a representation of **G**. A representation such as this is called a *regular representation.* According to (1.84), the matrices of the regular representation, with the exception of the matrix of the identity transformation E, possess zero diagonal elements. The identity matrix is always a diagonal unit matrix. From this it follows that the characters of the regular representation are given by

$$\chi(E) = g, \quad \chi(R) = 0 \qquad \text{for} \qquad R \neq E. \tag{1.85}$$

Substituting (1.85) into (1.62) and remembering that for any representation

$\chi^{(\alpha)}(E) = f_\alpha$, we find that

$$a^{(\alpha)} = (1/g)(f_\alpha g) = f_\alpha,$$

i.e., the number of times each irreducible representation occurs in the decomposition of the regular representation is equal to its dimension.

Let us write formula (1.61), which is an expression for the character of a reducible representation in terms of the irreducible representations occurring in it, for the case $R = E$:

$$\sum_\beta a^{(\beta)} f_\beta = f_\Gamma.$$

If the reducible representation is the regular representation, $f_\Gamma = g$ and $a^{(\beta)} = f_\beta$. Hence we arrive at the result that

$$\sum_\beta f_\beta{}^2 = g. \tag{1.86}$$

We have thus proved a property of which we have already made use: The sum of the squares of the dimensions of all the irreducible representations of a group is equal to the order of the group.

We note that in order to construct the regular representation, one does not have to make use of the basis functions (1.83). The elements of the group themselves can be employed as a basis. A knowledge of the multiplication table of the group is therefore sufficient to write down all the matrices of the regular representation.

1.19. The Construction of Basis Functions for Irreducible Representations

The regular representation considered in the previous section is generated by the g linearly independent functions (1.83). The decomposition of this representation into its irreducible components contains all the nonequivalent irreducible representations of the group. We show that this decomposition can be accomplished by constructing the following linear combinations of the basis functions ψ_R:

$$\psi_{ik}^{(\alpha)} = (f_\alpha/g) \sum_R \Gamma_{ik}^{(\alpha)}(R)^* \psi_R, \tag{1.87}$$

where $\Gamma_{ik}^{(\alpha)}(R)$ is a matrix element of the irreducible representation $\Gamma^{(\alpha)}$ corresponding to the operation R. The summation is taken over all g operations in the group.

We apply an arbitrary operation Q of the group to the function (1.87):

$$Q\psi_{ik}^{(\alpha)} = (f_\alpha/g) \sum_R \Gamma_{ik}^{(\alpha)}(R)^* QR\psi_0 = (f_\alpha/g) \sum_P \Gamma_{ik}^{(\alpha)}(Q^{-1}P)^* P\psi_0. \tag{1.88}$$

In this equation we have denoted the operation QR by P and made use of the invariance property of a sum over the group, Eq. (1.7). Furthermore, we write the matrix element of the product as products of matrix elements and make use of the property (1.42) of unitary matrices:

$$\Gamma_{ik}^{(\alpha)}(Q^{-1}P)^* = \sum_l \Gamma_{il}^{(\alpha)}(Q^{-1})^* \Gamma_{lk}^{(\alpha)}(P)^* = \sum_l \Gamma_{li}^{(\alpha)}(Q) \Gamma_{lk}^{(\alpha)}(P)^*. \tag{1.89}$$

Substituting (1.89) into (1.88), we obtain finally

$$Q\psi_{ik}^{(\alpha)} = \sum_l \Gamma_{li}^{(\alpha)}(Q)\psi_{lk}^{(\alpha)}. \tag{1.90}$$

The function $\psi_{ik}^{(\alpha)}$ therefore transforms as the ith column of the irreducible representation $\Gamma^{(\alpha)}$, and the set of f_α functions $\psi_{ik}^{(\alpha)}$ with fixed second index k forms a basis for the irreducible representation $\Gamma^{(\alpha)}$. One can form altogether f_α independent bases corresponding to the number of different values for k. This is to be expected, since in the decomposition of the regular representation, each irreducible representation occurs as many times as its dimension.

From the form of Eq. (1.87) it follows that in order to obtain a basis function for a representation $\Gamma^{(\alpha)}$, it is sufficient to apply the operator

$$\epsilon_{ik}^{(\alpha)} = (f_\alpha/g) \sum_R \Gamma_{ik}^{(\alpha)}(R)^* R \tag{1.91}$$

to some arbitrary function ψ_0. There are $f_\alpha{}^2$ such operators for every irreducible representation. They form f_α sets which differ from one another in the second index. Each of these sets can be used to obtain basis functions for an irreducible representation. Examples of the use of the operators (1.91) in constructing basis functions will be given in the ensuing chapters.

Should the function ψ_0 possess specific symmetry properties with respect to the operations of the group **G**, then in some cases the application of the operator $\epsilon_{ik}^{(\alpha)}$ to it may give zero. Thus let us determine the result of applying $\epsilon_{ik}^{(\alpha)}$ to a function of the form (1.87):

$$\begin{aligned} \epsilon_{ik}^{(\alpha)}\psi_{mn}^{(\beta)} &= (f_\alpha/g) \sum_R \Gamma_{ik}^{(\alpha)}(R)^* R\psi_{mn}^{(\beta)} \\ &= (f_\alpha/g) \sum_l \sum_R \Gamma_{ik}^{(\alpha)}(R)^* \Gamma_{lm}^{(\beta)}(R)\psi_{ln}^{(\beta)} = \delta_{\alpha\beta}\delta_{km}\psi_{in}^{(\alpha)}. \end{aligned} \tag{1.92}$$

In this equation we made use of (1.90) and the orthogonality relations (1.51). The application of $\epsilon_{ik}^{(\alpha)}$ to a basis function of an irreducible representation therefore either gives zero or another basis function belonging to the same irreducible representation. When $i = k$ the application of $\epsilon_{ii}^{(\alpha)}$ to a basis function $\psi_{in}^{(\alpha)}$ gives just the same function once more. Operators which possess such properties are known as *projection operators*. They satisfy the operator equation

$$\epsilon_{ii}^{(\alpha)}\epsilon_{ii}^{(\alpha)} = \epsilon_{ii}^{(\alpha)}. \tag{1.93}$$

An arbitrary function ψ can be represented in the form of a sum of functions each of which transforms according to an irreducible representation of a group,

$$\psi = \sum_{\alpha,i} \psi_{ii}^{(\alpha)}. \tag{1.94}$$

The summation over α in this equation is taken over all the irreducible representations of the group, and the summation i over all the independent bases for the representation $\Gamma^{(\alpha)}$ $(i = 1, 2, \ldots, f_\alpha)$. The functions $\psi_{ii}^{(\alpha)}$ are given by formula (1.87) with $i = k$.

In order to verify (1.94), we substitute expression (1.87) for $\psi_{ii}^{(\alpha)}$ and use the definition (1.53) for a character. We obtain as a result‡

$$\psi = \sum_{\alpha, R} (f_\alpha/g)\chi^{(\alpha)}(R)^* R\psi. \tag{1.95}$$

Taking into account the orthogonality relations (1.56), the sum over α is equal to

$$\sum_\alpha f_\alpha \chi^{(\alpha)}(R) = \sum_\alpha \chi^{(\alpha)}(E)\chi^{(\alpha)}(R)^* = g\delta_{ER}. \tag{1.96}$$

When this is subtituted into (1.95) we obtain the identity $\psi \equiv \psi$, verifying the original equation.

Let the function $\psi_{ik}^{(\alpha)}$ transform according to the ith column of a representation $\Gamma^{(\alpha)}$, the index k specifying the way in which the functions in this basis are constructed [in a particular case, the $\psi_{ik}^{(\alpha)}$ can be constructed by Eq. (1.87), although this is by no means necessary for the argument that follows]. We now show that functions which transform according to different irreducible representations, or according to different columns of the same representation, are orthogonal to one another, i.e., that

$$\mathcal{K} = \int \psi_{ik}^{(\alpha)*}\psi_{mn}^{(\beta)}\, dV = \delta_{\alpha\beta}\delta_{im}\begin{cases} A(\alpha; k, n) & \text{for} \quad k \neq n, \\ 1 & \text{for} \quad k = n, \end{cases} \tag{1.97}$$

where $A(\alpha;k,n)$ is determined by the choice of the bases k and n, but is independent of i and m.

To prove this, we make use of the fact that an integral over all space is invariant under any transformation of the coordinates. The integral (1.97) is therefore unchanged if the integrand is operated upon by some operation of the group:

$$\mathcal{K} = R\int \psi_{ik}^{(\alpha)*}\psi_{mn}^{(\beta)}\, dV = \sum_{\mu,\nu} \Gamma_{\mu i}^{(\alpha)}(R)^*\Gamma_{\nu m}^{(\beta)}(R) \int \psi_{\mu k}^{(\alpha)*}\psi_{\nu n}^{(\beta)}\, dV. \tag{1.98}$$

We sum this equation over all the operations in the group. The integral on the left-hand side is then simply multiplied by the order of the group. Applying the orthogonality relations (1.51) to the expression on the right-hand side, we obtain as a result

$$g\mathcal{K} = \delta_{\alpha\beta}\delta_{im}(g/f_\alpha)\sum_\mu \int \psi_{\mu k}^{(\alpha)*}\psi_{\mu n}^{(\alpha)}\, dV. \tag{1.99}$$

The sum over μ in (1.99) is independent of i and m. If $k = n$, it is equal to

‡We can use the function ψ in place of ψ_0 since the latter is arbitrary.

f_α, due to the orthonormality of the basis functions. When $k \neq n$ we put

$$(1/f_\alpha) \sum_\mu \int \psi_{\mu k}^{(\alpha)*} \psi_{\mu n}^{(\alpha)} \, dV = A(\alpha; k, n),$$

and so arrive at Eq. (1.97).

An arbitrary function can thus be expanded in a set of orthogonal functions $\psi_{ii}^{(\alpha)}$. The $\psi_{ii}^{(\alpha)}$ can be visualized as the components of a vector ψ in the space of the basis vectors for the irreducible representations of a group. The operator $\epsilon_{ii}^{(\alpha)}$ projects the vector ψ onto the direction (αi), i.e., it picks out the component of ψ in this direction. If there is no such component of ψ, then the result of operating with $\epsilon_{ii}^{(\alpha)}$ upon ψ is of course zero. The geometric interpretation of the effect of $\epsilon_{ik}^{(\alpha)}$ upon ψ is a little more complicated. It may be regarded as the projection of a previously rotated vector ψ onto the direction (αi). This rotation orients the component $\psi_{kk}^{(\alpha)}$ in the direction (αi). The action of $\epsilon_{ik}^{(\alpha)}$ upon ψ therefore gives zero if there is no component of ψ in the direction (αk).

CHAPTER II

The Permutation Group

PART 1. General Considerations

2.1. Operations with Permutations

We consider permutations of the N integers $1, 2, \ldots, N$. There are $N!$ such permutations altogether, each of which is denoted by

$$P = \begin{pmatrix} 1 & 2 & 3 & \cdots & N \\ i_1 & i_2 & i_3 & \cdots & i_N \end{pmatrix}. \tag{2.1}$$

Every permutation can be represented in the form of a product of commuting cycles (1.10). For this purpose the numbers 1 and i_1 in (2.1) are taken to form the first two elements of a cycle. These are followed by the number that replaces i_1, the process being continued until the number that replaces 1 is reached. A similar procedure is followed for the remaining elements. For example,

$$\begin{pmatrix} 1 & 2 & 3 & 4 & 5 & 6 \\ 2 & 4 & 5 & 1 & 3 & 6 \end{pmatrix} = (124)(35)(6). \tag{2.2}$$

A cycle, by definition, is invariant under a cyclic permutation of its elements:

$$(i_1 i_2 i_3 \cdots i_k) = (i_2 i_3 \cdots i_k i_1) = (i_3 \cdots i_k i_1 i_2). \tag{2.3}$$

The number of elements in a cycle is called its *length.* It follows from the definition that if we raise a cycle to a power equal to its length, we obtain the identity permutation:

$$(i_1 i_2 \cdots i_k)^k = I. \tag{2.4}$$

From this we see that

$$(i_1 i_2 \cdots i_k)^{-1} = (i_1 i_2 \cdots i_k)^{k-1}. \tag{2.5}$$

A cycle of two elements is commonly known as a *transposition.* It is obvious that

$$(i_1 i_2) = (i_1 i_2)^{-1}. \tag{2.6}$$

The following rules are useful in manipulating permutations:

(a) The product QPQ^{-1}, where Q and P are arbitrary permutations, is equal to the permutation that is obtained by letting the permutation Q act on P (in the sense of a permutation acting upon the arguments of a function). For example,

$$(13)(35)(13) = (15), \qquad (123)(23)(123)^{-1} = (13).$$

(b) The product of two permutations is independent of their order if the cycles which constitute the permutations do not contain any common elements.

(c) Two cycles which possess a common element can be combined by placing this element at the end of one cycle and at the beginning of the other, i.e.,

$$(ik \cdots lm)(mn \cdots q) = (ik \cdots lmn \cdots q). \tag{2.7}$$

For example,

$$(1245)(346) = (5124)(463) = (512463).$$

(d) A cycle that results from multiplying other cycles may contain a number of identical elements. It is then useful to reduce this to cycles of the form

$$(ik \cdots lmi) \equiv (k \cdots lm). \tag{2.8}$$

Every cycle can always be written in the form of a product of transpositions. Such a representation, however, is not unique. Thus, for example,

$$\begin{aligned}(123 \cdots \mathrm{k}) &= (12)(23) \cdots (\mathrm{k}-1\ \ \mathrm{k}) \\ &= (1\ \mathrm{k})(1\ \ \mathrm{k}-1)(1\ \ \mathrm{k}-2) \cdots (12) \\ &= (21)(2\ \mathrm{k})(2\ \ \mathrm{k}-1) \cdots (23) = \cdots.\end{aligned} \tag{2.9}$$

Nevertheless, the number of transpositions which form a given permutation always possesses a unique parity. A permutation is classed as *even* or *odd* according to whether the number of its transpositions is even or odd.

It is not difficult to show that any permutation can be represented as a product of transpositions of the form $(i-1, i)$, where $i-1$ and i are two consecutive numbers. Thus, for example,

$$(245) = (24)(45) = (23)(34)(23)(45).$$

This plays an important role in the determination of explicit forms for the representation matrices of the permutation group.

2.2. Classes

In Part 1 of Chapter I it was shown that the $N!$ permutations of N objects satisfy the four group postulates and therefore form a group, which we denote by the symbol π_N. The even permutations in π_N form a group by themselves known as the *alternating group*, and constitute a subgroup of π_N. In addition to this, π_N possesses $N - 1$ obvious further subgroups: $\pi_{N-1}, \pi_{N-2}, \ldots$, and π_1.

All permutations that are related to one another by the equation $P_i = QP_jQ^{-1}$, where Q is any permutation of π_N, are, by definition, members of the same class. Since the permutation QP_jQ^{-1} is obtained by the action of Q upon P_j (see the previous section), the cyclic structures of P_i and P_j are identical, i.e., the number and lengths of the cycles in the two permutations coincide. The permutations P_i and P_j differ only in the elements forming the cycles. Each class of π_N is therefore characterized by a particular partition of the N elements into cycles. The number of different classes is determined by the number of different ways of partitioning the number N into positive integral components, i.e., is equal to the number of different integral solutions (with the exception of zero) of the equation

$$1\nu_1 + 2\nu_2 + \cdots + N\nu_N = N. \tag{2.10}$$

A set of numbers $\nu_1, \nu_2, \cdots, \nu_N$ which satisfies Eq. (2.10) uniquely defines a class of π_N. We designate a class by the symbol $\{1^{\nu_1}2^{\nu_2} \cdots m^{\nu_m}\}$, where ν_k is the number of cycles of length k which appear in the permutations in the class. The class $\{1^N\}$ corresponds to the identity permutation. Thus the six permutations of the group π_3 are divided into three classes:

$$\begin{aligned}
\{1^3\}&: \quad (1)(2)(3) \equiv I;\\
\{12\}&: \quad (1)(23), \qquad (2)(13), \qquad (3)(12);\\
(3)&: \quad (123), \qquad (132).
\end{aligned}$$

The group π_4 contains five classes: $\{1^4\}$, $\{1^22\}$, $\{2^2\}$, $\{13\}$, and $\{4\}$.

We now derive a formula which connects the number of elements in a class with its cyclic structure. Let $\{1^{\nu_1}2^{\nu_2} \cdots m^{\nu_m}\}$ be an arbitrary class of the group π_N. We place the N numbers in the cycles in their natural order, and let all the $N!$ permutations of the group operate upon this initial permutation, leaving the parentheses in place. Since the order of the cycles is unimportant, it is clear that $\nu_1!\,\nu_2! \cdots \nu_m!$ of the permutations will coincide. Cyclic permutations of the elements within a cycle also lead to identical permutations. For a single cycle of length k there are k such permutations, giving $2^{\nu_2}3^{\nu_3} \cdots m^{\nu_m}$ permutations in all. As a result, the order of the class is given by

$$g_{\{1^{\nu_1}2^{\nu_2}\cdots m^{\nu_m}\}} = (N!/\nu_1!\nu_2!2^{\nu_2} \cdots \nu_m!m^{\nu_m}). \tag{2.11}$$

For example, the class $\{1^22\}$ of the group π_4 contains $4!/(2!2) = 6$ elements, and the class $\{13\}$ contains $4!/3 = 8$ elements.

2.3. Young Diagrams and Irreducible Representations

Since the number of nonequivalent irreducible representations of a group is equal to the number of its classes, the nonequivalent irreducible representations of π_N are defined, as are the classes, by the different partitions of the number N into positive integral components. Each irreducible representation is typified by one such partition. The partitions are usually written in order of decreasing components $\lambda^{(i)}$:

$$\lambda^{(1)} + \lambda^{(2)} + \cdots + \lambda^{(m)} = N, \qquad \lambda^{(1)} \geq \lambda^{(2)} \geq \cdots \geq \lambda^{(m)}. \qquad (2.12)$$

Some of the $\lambda^{(i)}$ in (2.12) may coincide. It is clear that m cannot exceed N. Equation (2.12) can be regarded as an equation for finding the possible partitions of N, and in this sense is entirely equivalent to Eq. (2.10). The partitions (2.12) can be depicted graphically by means of diagrams known as *Young diagrams*, in which each number $\lambda^{(i)}$ is represented by a row of $\lambda^{(i)}$ cells. Young diagrams will be subsequently denoted by the symbol $[\lambda] \equiv [\lambda^{(1)}\lambda^{(2)} \cdots \lambda^{(m)}]$. The presence of several rows of identical length $\lambda^{(i)}$ will be indicated by a power of $\lambda^{(i)}$. For example,

$[\lambda] = [2^2 1^2] =$ 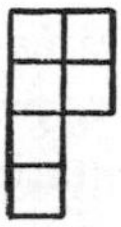

It is obvious that one can form from two cells only two Young diagrams:

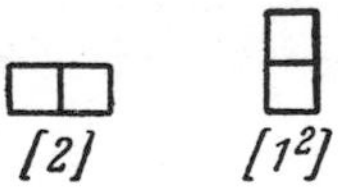

(2.13)

For the group π_3 one can form from three cells three Young diagrams:

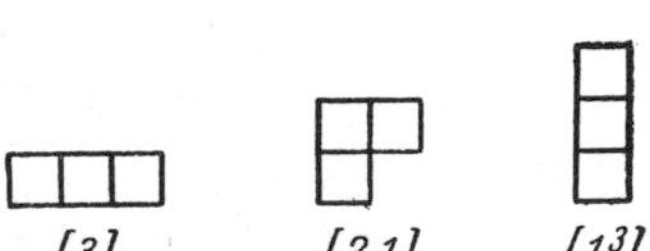

(2.14)

The group π_4 has five Young diagrams:

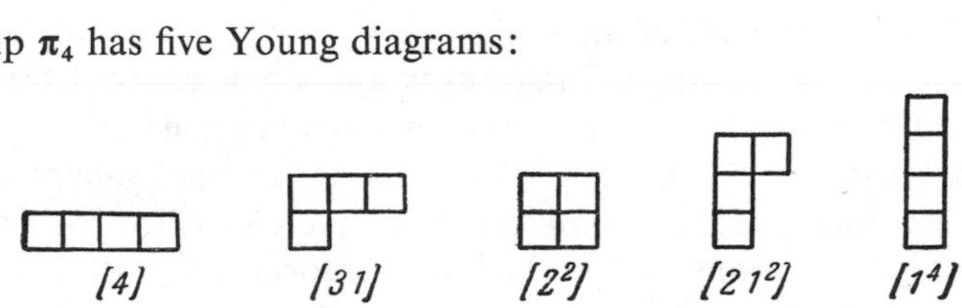

(2.15)

Each Young diagram uniquely corresponds to a specific irreducible representation of the group π_N. The irreducible representations of π_N are therefore usually enumerated by the symbol $[\lambda]$ for the corresponding Young diagrams, and are denoted by $\Gamma^{[\lambda]}$ or simply by $[\lambda]$. As will be shown, the assignment of a Young diagram determines the permutational symmetry of the basis functions for an irreducible representation, i.e., determines the behavior of the basis functions under permutations of their arguments.

PART 2. The Standard Young–Yamanouchi Orthogonal Representation

2.4. Young Tableaux

In general an irreducible representation $\Gamma^{[\lambda]}$ of the group π_N becomes reducible upon passing to the subgroup π_{N-1}, and can be decomposed into irreducible representations $\Gamma^{[\lambda']}$ of this subgroup (reduction with respect to a subgroup). The representations $\Gamma^{[\lambda']}$ into which $\Gamma^{[\lambda]}$ decomposes are determined by all the Young diagrams with $N - 1$ cells that are obtained from the initial diagram upon removing one of its cells (Rutherford, 1947). For example, the irreducible representation of the group π_5, characterized by the Young diagram $\Gamma^{[2^21]}$, splits up into the irreducible representations $\Gamma^{[2^2]}$ and $\Gamma^{[21^2]}$ of π_4:

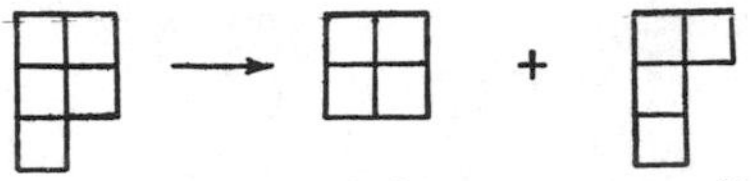

$[2^21]$ $[2^2]$ $[21^2]$

On passing from π_{N-1} to its subgroup π_{N-2}, the representations $\Gamma^{[\lambda']}$ in turn similarly split up into irreducible representations $\Gamma^{[\lambda'']}$ of π_{N-2}, the process continuing until the group π_1 is reached.

Linear combinations can be formed from the f_λ basis functions for an irreducible representation $\Gamma^{[\lambda]}$ such that in this new basis the matrices which correspond to the elements of the subgroups $\pi_{N-1}, \pi_{N-2}, \ldots$, etc. (and which form representations of these subgroups) are all in reduced form. Basis functions of $\Gamma^{[\lambda]}$ chosen in this way simultaneously become basis functions for the representations $\Gamma^{[\lambda']}, \Gamma^{[\lambda'']}, \ldots$, on passing to the subgroups π_{N-1}, $\pi_{N-2}, \ldots$, etc. Each basis function can be uniquely specified by the series of irreducible representations under which it transforms on passing from π_N to the various subgroups. This choice of basis functions can be described graphically as follows. Each basis function is associated with a Young diagram $\Gamma^{[\lambda]}$ the cells of which each contain an integer which may range from one to N. The numbers are distributed among the cells in such a way that when the cell containing the number N is removed, one obtains the Young diagram $\Gamma^{[\lambda']}$ of the irreducible representation of π_{N-1} according to which the basis function transforms on passing to this subgroup. Subsequent removal of the cell containing $N-1$ gives the Young diagram $\Gamma^{[\lambda'']}$, etc., this process continuing until we reach the diagram with just a single cell. It is clear that only those Young diagrams are allowed in which the numbers increase from left to right along the rows and down the columns (otherwise at some stage in the reduction forbidden Young diagrams will occur). We shall refer to a Young diagram of this kind as a *standard Young tableau*. The N numbers can be distributed among the cells of a Young diagram in $N!$ different ways, so that there are $N!$ distinct Young tableaux altogether. Henceforth when speaking of Young tableaux we shall refer only to the standard tableaux, of which there are always less than $N!$

Each basis function of an irreducible representation $\Gamma^{[\lambda]}$ can thus be associated with a Young tableau, and the dimension of the irreducible representation is given by the number of standard Young tableaux, i.e., by the number of different ways of distributing the N numbers in a Young diagram such that they increase from left to right along the rows and down the columns.

For example, we can draw five standard Young tableaux for the representation $\Gamma^{[\lambda]} = \Gamma^{[2^3]}$ of the group π_6 (which we enumerate by the symbol $r^{(i)}$):

$$[\lambda]=[2^3]$$

$$\overset{r^{(1)}}{\begin{array}{|c|c|}\hline 1&2\\ \hline 3&4\\ \hline 5&6\\ \hline\end{array}} \quad \overset{r^{(2)}}{\begin{array}{|c|c|}\hline 1&2\\ \hline 3&5\\ \hline 4&6\\ \hline\end{array}} \quad \overset{r^{(3)}}{\begin{array}{|c|c|}\hline 1&3\\ \hline 2&4\\ \hline 5&6\\ \hline\end{array}} \quad \overset{r^{(4)}}{\begin{array}{|c|c|}\hline 1&3\\ \hline 2&5\\ \hline 4&6\\ \hline\end{array}} \quad \overset{r^{(5)}}{\begin{array}{|c|c|}\hline 1&4\\ \hline 2&5\\ \hline 3&6\\ \hline\end{array}} \tag{2.16}$$

The five basis functions which correspond to these Young tableaux are uniquely specified by the following series of representations of the groups π_5, π_4, π_3, and π_2:

$$\begin{aligned} r^{(1)}&: \quad [2^21] \longrightarrow [2^2] \longrightarrow [21] \longrightarrow [2],\\ r^{(2)}&: \quad [2^21] \longrightarrow [21^2] \longrightarrow [21] \longrightarrow [2],\\ r^{(3)}&: \quad [2^21] \longrightarrow [2^2] \longrightarrow [21] \longrightarrow [1^2],\\ r^{(4)}&: \quad [2^21] \longrightarrow [21^2] \longrightarrow [21] \longrightarrow [1^2],\\ r^{(5)}&: \quad [2^21] \longrightarrow [21^2] \longrightarrow [1^3] \longrightarrow [1^2]. \end{aligned} \tag{2.17}$$

We arrange the Young tableaux in order of increasing deviation of the numbers from their natural order (as read along the rows, and from top to bottom), and we number the basis functions in the same order. We put first the Young tableaux in which the number 2 occurs on the first row, followed by those with 2 on the second row. Of the Young tableaux that have the number 2 on the same row, we place first those in which 3 occurs on the first row, followed by those with 3 on the second row, etc. Succeeding numbers are dealt with similarly.‡ The Young tableau in which the numbers are in their natural order is always first. We refer to this as the *fundamental tableau.* Since each basis function is specified by its Young tableau, one can use the tableaux to number the rows and columns of the irreducible representation according to which the basis functions transform.

If we are only interested in the dimension of an irreducible representation of a permutation group with N large, it is very inconvenient to write out all the possible Young tableaux. Instead, the dimension of an irreducible representation may be calculated by the formula (Murnaghan, 1938; Rutherford, 1947)

$$f_\lambda = \{N! \prod_{i<j} (h_i - h_j)\}/(h_1!\, h_2! \cdots h_m!), \qquad h_i = \lambda^{(i)} + m - i, \tag{2.18}$$

in which $\lambda^{(i)}$ is the length of row i and m is the number of rows in the Young diagram $[\lambda]$. For the representation $\Gamma^{[2^3]}$ formula (2.18) gives $f_\lambda = 5$, in agreement with the number of tableaux. Thus substituting the values of $\lambda^{(i)}$ and m into (2.18), we obtain $h_1 = 4$, $h_2 = 3$, and $h_3 = 2$, and hence $f_\lambda = (6!\,2)/(4!\,3!\,2!) = 5$.

2.5. Explicit Determination of the Matrices of the Standard Representation

The choice of basis functions for the irreducible representations of π_N which was explained in the previous section automatically gives an orthogonal set of functions The is because the basis functions are each characterized by a distinct sequence of irreducible representations according to which they transform on passing from π_N to its subgroups. If the basis functions are also normalized, then in order to determine uniquely the matrices of the resulting orthogonal representation, it is only necessary to fix the phases of the matrix elements. For this purpose it suffices to give just the matrix of a transposition $P_{i-1\,i}$, since all the other permutations in the group can be expressed in the form of a product of transpositions of this kind. Young (see, e.g., Rutherford, 1947) and Yamanouchi (1937) showed that this representation may be chosen to be real, and gave simple rules for constructing the matrices of $P_{i-1\,i}$. According to them, the matrix $\Gamma^{[\lambda]}(P_{i-1\,i})$ has the following nonzero elements:

(a) $\Gamma^{[\lambda]}_{rr}(P_{i-1\,i}) = 1$ if in tableau r the numbers i and $i - 1$ occur on the same row.

‡It is not difficult to satisfy oneself that the order of the Young tableaux in (2.16) obeys this rule.

(b) $\Gamma^{[\lambda]}_{rr}(P_{i-1\,i}) = -1$ if in tableau r the numbers i and $i-1$ occur in the same column.

$$
\text{(c)}\qquad \Gamma^{[\lambda]}(P_{i-1\,i}) =
\begin{array}{c}
\\ \\ r \\ \\ t \\ \\
\end{array}
\left[
\begin{array}{ccccccc}
 & & r & & t & & \\
 & & \vdots & & \vdots & & \\
\cdots & & -1/d & \cdots & [1-(1/d^2)]^{1/2} & \cdots & \\
 & & \vdots & & \vdots & & \\
\cdots & & [1-(1/d^2)]^{1/2} & \cdots & 1/d & \cdots & \\
 & & \vdots & & \vdots & &
\end{array}
\right],
$$

if the tableaux r and t differ only by a permutation of the numbers $i-1$ and i, and if the row in tableau r containing $i-1$ is above that containing i. The letter d denotes the *axial distance* between $i-1$ and i. This is defined as the number of vertical and horizontal steps that one must make in the Young tableau in order to move from $i-1$ to i. Thus in tableau $r^{(4)}$ of (2.16), $d_{23} = 2$ and $d_{34} = 3$.

As an example, we write out the matrix of P_{23} in the irreducible representation $\Gamma^{[2^3]}$, the Young tableaux of which are given in (2.16) (vacant spaces denote zeros):

$$
\Gamma^{[2^3]}(P_{23}) =
\begin{array}{c}
 \\ r^{(1)} \\ r^{(2)} \\ r^{(3)} \\ r^{(4)} \\ r^{(5)}
\end{array}
\begin{array}{ccccc}
r^{(1)} & r^{(2)} & r^{(3)} & r^{(4)} & r^{(4)} \\
-\frac{1}{2} & & \frac{1}{2}\sqrt{3} & & \\
 & -\frac{1}{2} & & \frac{1}{2}\sqrt{3} & \\
\frac{1}{2}\sqrt{3} & & \frac{1}{2} & & \\
 & \frac{1}{2}\sqrt{3} & & \frac{1}{2} & \\
 & & & & -1
\end{array}.
$$

The representation which is defined by these rules is known as *the Young–Yamanouchi standard orthogonal representation*, or simply as the standard representation. Its matrix elements satisfy the following orthogonality relations:

$$
\text{(a)}\qquad \sum_r \Gamma^{[\lambda]}_{rt}(P)\Gamma^{[\lambda]}_{ru}(P) = \delta_{tu}, \qquad \sum_r \Gamma^{[\lambda]}_{tr}(P)\Gamma^{[\lambda]}_{ur}(P) = \delta_{tu}. \tag{2.19}
$$

These follow from the orthogonality of the representation [see Eq. (1.44)].

$$
\text{(b)}\qquad \sum_P \Gamma^{[\lambda]}_{rs}(P)\Gamma^{[\bar{\lambda}]}_{ut}(P) = (N!/f_\lambda)\,\delta_{\lambda\bar{\lambda}}\delta_{ru}\delta_{st}, \tag{2.20}
$$

$$
\sum_{\lambda,r,t} (f_\lambda/N!)\Gamma^{[\lambda]}_{rt}(P)\Gamma^{[\lambda]}_{rt}(Q) = \delta_{PQ}. \tag{2.21}
$$

These equations hold for any finite group; they form a particular case of Eqs. (1.51) and (1.52).

This method of constructing the standard representation of π_N ensures that the matrices are in fully reduced form with respect to the subgroups π_{N-1}, $\pi_{N-2}, \ldots, \pi_2$. This means that if, for example, a permutation P_1 belongs to the group π_{N-2}, then

$$\Gamma^{[\lambda]}_{rt}(P_1) = \delta_{\lambda''_r \lambda''_t} \Gamma^{[\lambda'']}_{r''t''}(P_1), \tag{2.22}$$

where r'' and t'' are the Young tableaux that are obtained from the tableaux r and t by removing the cells containing $N - 1$ and N. If this process leads to different Young diagrams $[\lambda'']$, then the matrix element (2.22) is equal to zero.

The matrices corresponding to permutations of the numbers $n_1 + 1$, $n_1 + 2, \ldots, N$ are diagonal in the Young tableaux for the first n_1 numbers. Denote these tableaux by the symbol r_1. Then the tableau r can be written in the form $r = (r_1 \rho_2)$, where ρ_2 denotes the part of tableau r containing the numbers $n_1 + 1, \ldots, N$. Moreover, the matrix elements are equal for all the r_1 that arise from a particular Young diagram with n_1 cells, i.e.,

$$\Gamma^{[\lambda]}_{r_1\rho_2, \bar{r}_1\bar{\rho}_2}(P_2) = \delta_{r_1\bar{r}_1} \Gamma^{[\lambda]}_{\lambda_1\rho_2, \lambda_1\bar{\rho}_2}(P_2), \tag{2.23}$$

where P_2 denotes a permutation of the numbers $n_1 + 1, \ldots, N$. The matrices for a number of irreducible representations of π_N are given in Appendix 5. It is easily seen that the representations given there satisfy Eqs. (2.19)–(2.23).

As a very simple example, we consider the standard irreducible representations of π_3. This group has three irreducible representations, the Young diagrams of which have already been given [see (2.14)].

The representations $\Gamma^{[3]}$ and $\Gamma^{[1^3]}$ are one dimensional, since there is only one Young tableau for each of them:

$$\begin{array}{|c|c|c|}\hline 1 & 2 & 3 \\ \hline\end{array} \qquad \begin{array}{|c|}\hline 1 \\ \hline 2 \\ \hline 3 \\ \hline\end{array}$$

According to the Young–Yamanouchi rules, the basis function for the $\Gamma^{[3]}$ representation does not change sign under any of the permutations, and generates the *symmetric* representation of π_3. The characters of this representation are obviously all equal to unity. The basis function for the $\Gamma^{[1^3]}$ representation changes sign under odd permutations, and does not change sign under even permutations. A function such as this is said to be *antisymmetric*, and the one-dimensional representation generated by it is also said to be *antisymmetric*. The characters of this representation are equal to $(-1)^p$, where p is the parity of the permutation (p is equal to zero for even permutations, and equal to one for odd permutations). In general, we note that the Young diagram $[N]$ for the group π_N always corresponds to a one-dimensional symmetric representation, and that the diagram $[1^N]$ corresponds to a one-dimensional antisymmetric representation.

The $\mathbf{\Gamma}^{[21]}$ representation possesses two standard Young tableaux:

$$\begin{array}{ccc} r^{(1)} & r^{(2)} & \\ \begin{array}{|c|c|}\hline 1 & 2 \\ \hline 3 & \multicolumn{1}{c}{} \\ \cline{1-1}\end{array} & \begin{array}{|c|c|}\hline 1 & 3 \\ \hline 2 & \multicolumn{1}{c}{} \\ \cline{1-1}\end{array} & d_{23} = 2 \end{array} \tag{2.24}$$

The dimension of the representation is therefore equal to two. The Young tableaux (2.24) enumerate the rows and columns of the matrices of this representation:

$[\lambda] = [21]$

$$\begin{array}{ccc} I & P_{12} & P_{23} \\ \begin{bmatrix} 1 & \\ & 1 \end{bmatrix} & \begin{bmatrix} 1 & \\ & -1 \end{bmatrix} & \begin{bmatrix} -\frac{1}{2} & \frac{1}{2}\sqrt{3} \\ \frac{1}{2}\sqrt{3} & \frac{1}{2} \end{bmatrix} \\ P_{13} = P_{12}P_{23}P_{12} & P_{123} = P_{12}P_{23} & P_{132} = P_{123}^{-1} \\ \begin{bmatrix} -\frac{1}{2} & -\frac{1}{2}\sqrt{3} \\ -\frac{1}{2}\sqrt{3} & \frac{1}{2} \end{bmatrix} & \begin{bmatrix} -\frac{1}{2} & \frac{1}{2}\sqrt{3} \\ -\frac{1}{2}\sqrt{3} & -\frac{1}{2} \end{bmatrix} & \begin{bmatrix} -\frac{1}{2} & -\frac{1}{2}\sqrt{3} \\ \frac{1}{2}\sqrt{3} & -\frac{1}{2} \end{bmatrix} \end{array} \tag{2.25}$$

2.6. The Conjugate Representation‡

We can associate with every standard representation $\mathbf{\Gamma}^{[\lambda]}$ *a conjugate* (or *associated*) representation $\mathbf{\Gamma}^{[\tilde{\lambda}]}$ of the same dimension, the matrices of which $\mathbf{\Gamma}^{[\tilde{\lambda}]}(P)$ differ from those of the standard representation $\mathbf{\Gamma}^{[\lambda]}(P)$ by a factor $(-1)^p$, where p is the parity of the permutation.

The symmetric and antisymmetric representations which were defined in the previous section form the simplest example of conjugate representations. The first representation corresponds to the Young diagram $[N]$ and the second to the diagram $[1^N]$. The Young diagram $[1^N]$ is obtained from $[N]$ by changing the row into a column. In the general case it can be shown that for a representation characterized by the Young diagram $[\lambda]$ the conjugate representation is characterized by a diagram known as the dual Young diagram $[\tilde{\lambda}]$ which is obtained from $[\lambda]$ by interchanging its rows and columns. The rows and columns of the conjugate representation are enumerated by the Young tableaux r of the Young diagram $[\tilde{\lambda}]$. To each Young tableau $r^{(i)}$ there corresponds a tableau $\tilde{r}^{(i)}$ which is obtained from $r^{(i)}$ by interchanging its rows and columns. Since the Young tableaux r are ordered according to the degree with which the numbers in it deviate from their natural order, the ordering of

‡Translator's note: This is sometimes referred to as the *adjoint* representation.

the tableaux $\tilde{r}$ is just the reverse. For example,

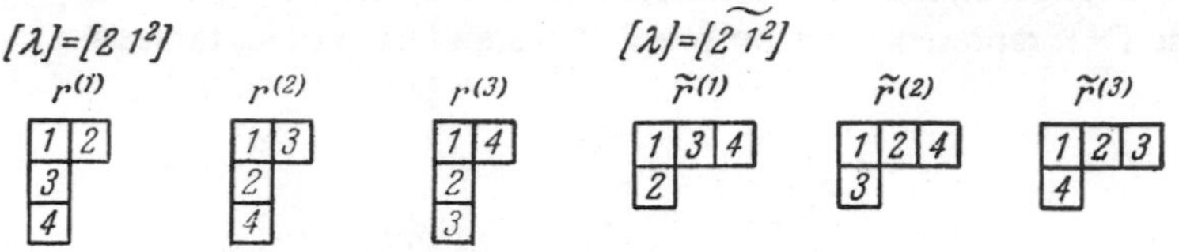

With the Young tableaux $\tilde{r}$ ordered in this way the rules for forming the matrices $\Gamma^{[\tilde{\lambda}]}(P_{i-1\,i})$ are, except for the off-diagonal elements $\Gamma^{[\tilde{\lambda}]}_{\tilde{r}\tilde{t}}(P_{i-1\,i})$, the same as those for the $\Gamma^{[\lambda]}(P_{i-1\,i})$. For two tableaux $\tilde{r}$ and $\tilde{t}$ that differ only by a permutation of $i-1$ and i the off-diagonal elements are given by

$$\Gamma^{[\tilde{\lambda}]}_{\tilde{r}\tilde{t}}(P_{i-1\,i}) = -[1-(1/d^2)]^{1/2}. \tag{2.26}$$

The minus sign in front of the root is introduced in order to satisfy the condition

$$\Gamma^{[\tilde{\lambda}]}(P_{i-1\,i}) = -\Gamma^{[\lambda]}(P_{i-1\,i}).$$

Thus the representation that is conjugate to [21] [see Eq. (2.25)] is characterized by the same Young diagram [21]. The ordering of the Young tableaux $\tilde{r}$, however, is different:

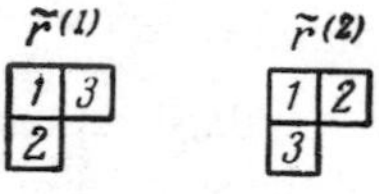

By the rules for constructing the transposition matrices in the standard representation, we obtain, taking (2.26) into account,

$$\Gamma^{[\widetilde{21}]}(P_{12}) = \begin{bmatrix} -1 & \\ & 1 \end{bmatrix}, \qquad \Gamma^{[\widetilde{21}]}(P_{23}) = \begin{bmatrix} \frac{1}{2} & -\frac{1}{2}\sqrt{3} \\ -\frac{1}{2}\sqrt{3} & -\frac{1}{2} \end{bmatrix}. \tag{2.27}$$

We see that these matrices differ by a factor -1 from the corresponding matrices in the standard representation $\Gamma^{[21]}$ [see Eq. (2.25)]. The remaining matrices in the $\Gamma^{[\widetilde{21}]}$ representation are obtained by multiplying the matrices (2.27) together. Consequently, the $\Gamma^{[\widetilde{21}]}$ matrices will differ by a factor $(-1)^p$ from those in the $\Gamma^{[21]}$ representation.

2.7. The Construction of an Antisymmetric Function from the Basis Functions for Two Conjugate Representations

Consider the direct product $\Gamma^{[\lambda_1]} \times \Gamma^{[\lambda_2]}$ of two irreducible representations of π_N. This can be resolved into its irreducible components by a suitable linear transformation of the basis functions. We determine the condition that the antisymmetric representation occurs in this decomposition. The characters of the antisymmetric representation are equal to $(-1)^p$. Substituting this into formula (1.62) and making use of the orthogonality relations (1.54), we find that

$$\begin{aligned} a^{[1^N]} &= (1/N!) \sum_P (-1)^p \chi^{[\lambda_1]}(P) \chi^{[\lambda_2]}(P) \\ &= (1/N!) \sum_P \chi^{[\tilde{\lambda}_1]}(P) \chi^{[\lambda_2]}(P) = \delta_{\tilde{\lambda}_1 \lambda_2}. \end{aligned} \tag{2.28}$$

The antisymmetric representation therefore occurs once only in the reduction of $\Gamma^{[\lambda_1]} \times \Gamma^{[\lambda_2]}$, and only if $[\lambda_2] = [\tilde{\lambda}_1]$. In other words, it is impossible to form an antisymmetric linear combination from products of the basis functions, $\psi_{r_1}^{[\lambda_1]} \psi_{r_2}^{[\lambda_2]}$, unless the representations $\Gamma^{[\lambda_1]}$ and $\Gamma^{[\lambda_2]}$ are conjugate to one another. If $\Gamma^{[\lambda_1]}$ and $\Gamma^{[\lambda_2]}$ are conjugate, then one, and only one, antisymmetric function can be formed in this way. It is easy to verify that this antisymmetric function is given by

$$\psi^{[1^N]} = [1/(f_\lambda)^{1/2}] \sum_r \psi_r^{[\lambda]} \psi_{\tilde{r}}^{[\tilde{\lambda}]}, \tag{2.29}$$

in which the summation is taken over all the f_λ Young tableaux, and the $1/(f_\lambda)^{1/2}$ factor is a normalization constant. Let an arbitrary permutation P operate on the function (2.29). Then

$$\begin{aligned} P\psi^{[1^N]} &= [1/(f_\lambda)^{1/2}] \sum_r \sum_{t,u} \Gamma_{tr}^{[\lambda]}(P) \Gamma_{\tilde{u}\tilde{r}}^{[\tilde{\lambda}]}(P) \psi_t^{[\lambda]} \psi_{\tilde{u}}^{[\tilde{\lambda}]} \\ &= [1/(f_\lambda)^{1/2}] \sum_{t,u} [\sum_r \Gamma_{tr}^{[\lambda]}(P) \Gamma_{ur}^{[\lambda]}(P)] (-1)^p \psi_t^{[\lambda]} \psi_{\tilde{u}}^{[\tilde{\lambda}]}. \end{aligned}$$

The summation over r in this equation gives δ_{tu}, in accordance with Eq. (2.19). As a result, we obtain

$$P\psi^{[1^N]} = (-1)^p \psi^{[1^N]},$$

which demonstrates the correctness of Eq. (2.29).

2.8. Young Operators

In Chapter I it was shown that one can obtain a set of basis functions for an irreducible representation of any finite group by applying the operators (1.91) to some arbitrary function. We shall refer to these operators, when specialized to the standard orthogonal Young–Yamanouchi representation, as *Young operators*, and denote them by $\omega_{rt}^{[\lambda]}$. The *normalized Young operator* is of the form

$$\omega_{rt}^{[\lambda]} = (f_\lambda/N)^{1/2} \sum_P \Gamma_{rt}^{[\lambda]}(P) P. \tag{2.30}$$

The summation over P runs over all the $N!$ permutations in the group π_N. The operator $\omega_{rt}^{[\lambda]}$ differs from the $\epsilon_{ik}^{(\alpha)}$ by the absence of the sign for complex conjugation on the matrix elements (the matrices in a representation $\Gamma^{[\lambda]}$ are all real), and also in the factor in front of the sumation. This factor is chosen so that application of the operator $\omega_{rt}^{[\lambda]}$ to a product of N orthogonal functions $\phi_a(i)$ produces a normalized function (see the following section).

One can form f_λ^2 operators $\omega_{rt}^{[\lambda]}$ for each representation $\Gamma^{[\lambda]}$, and since

$$\sum_\lambda f_\lambda^2 = N!,$$

there are altogether $N!$ such operators for the group π_N. According to Eqs. (2.20) and (2.21), the quantities $(f_\lambda/N!)^{1/2}\Gamma_{rt}^{[\lambda]}(P)$ form an orthogonal matrix of dimension $N!$ Equation (2.30) may therefore be regarded as an orthogonal transformation from $N!$ permutation operators P to $N!$ operators $\omega_{rt}^{[\lambda]}$. The inverse transformation is given by

$$P = \sum_{\lambda, r, t} (f_\lambda/N!)^{1/2}\Gamma_{rt}^{[\lambda]}(P)\omega_{rt}^{[\lambda]}. \tag{2.31}$$

From Eq. (1.92) we can obtain an expression for the product of two Young operators, if we substitute in it $\epsilon_{rt}^{[\lambda]} = (f_\lambda/N!)^{1/2}\omega_{rt}^{[\lambda]}$. As a result, we obtain

$$\omega_{rt}^{[\lambda]}\omega_{us}^{[\tilde{\lambda}]} = (N!/f_\lambda)^{1/2}\delta_{\lambda\tilde{\lambda}}\delta_{tu}\omega_{rs}^{[\lambda]}. \tag{2.32}$$

Upon representing P in the form of a sum (2.31) and using (2.32), we find that

$$P\omega_{rt}^{[\lambda]} = \sum_u \Gamma_{ur}^{[\lambda]}(P)\omega_{ut}^{[\lambda]}, \tag{2.33}$$

i.e., only those operators with the same second index transform into each other under permutations. It follows from (2.33) that a set of f_λ Young operators $\omega_{rt}^{[\lambda]}$ with a fixed second index form a basis for a representation $\Gamma^{[\lambda]}$. The Young operators $\omega_{rt}^{[\lambda]}$ can therefore be regarded as basis vectors for an irreducible representation $\Gamma^{[\lambda]}$.

If the order of the operators on the left-hand side of (2.33) is reversed, we obtain a different result. Using Eqs. (2.31) and (2.32), we find that

$$\omega_{rt}^{[\lambda]}P = \sum_u \Gamma_{tu}^{[\lambda]}(P)\omega_{ru}^{[\lambda]}. \tag{2.34}$$

We consider a few examples of Young operators. The group π_2 possesses two one-dimensional representations, with a single Young operator corresponding to each of them:

$$\omega^{[2]} = (1/\sqrt{2})(I + P_{12}) \tag{2.35}$$

$$\omega^{[1^2]} = (1/\sqrt{2})(I - P_{12}). \tag{2.36}$$

Operator (2.35) is a symmetrizing operator and (2.36) is an antisymmetrizing operator. The Young operators for the one-dimensional representations $\Gamma^{[3]}$ and $\Gamma^{[1^3]}$ of π_3 are similarly constructed:

$$\omega^{[3]} = (1/\sqrt{6})(I + P_{12} + P_{23} + P_{13} + P_{123} + P_{132}), \tag{2.37}$$

$$\omega^{[1^3]} = (1/\sqrt{6})(I - P_{12} - P_{23} - P_{13} + P_{123} + P_{132}). \tag{2.38}$$

The representation $\Gamma^{[21]}$ is two dimensional. Four independent Young operators can therefore be constructed in this case. Taking the matrix elements of the standard representation from Eq. (2.25), we obtain from formula (2.30) the following four operators:

$$\begin{aligned}
\omega_{11}^{[21]} &= (1/\sqrt{12})(2I + 2P_{12} - P_{23} - P_{13} - P_{123} - P_{132}),\\
\omega_{12}^{[21]} &= (1/2)(P_{23} - P_{13} + P_{123} - P_{132}),\\
\omega_{21}^{[21]} &= (1/2)(P_{23} - P_{13} - P_{123} + P_{132}),\\
\omega_{22}^{[21]} &= (1/\sqrt{12})(2I - 2P_{12} + P_{23} + P_{13} - P_{123} - P_{132}).
\end{aligned} \tag{2.39}$$

2.9. The Construction of Basis Functions for a Standard Representation from a Product of *N* Orthogonal Functions

We consider a set of N orthonormal functions $\phi_a(i)$ (the letter i in the argument of the function stands for the set of variables upon which ϕ_a depends):

$$\int \phi_a^* \phi_b \, dV = \delta_{ab}. \tag{2.40}$$

We form the function Φ_0, a product of N functions ϕ_a:

$$\Phi_0 = \phi_1(1)\phi_2(2) \cdots \phi_N(N). \tag{2.41}$$

From Φ_0 we can produce $N!$ different functions in two ways: by permuting the arguments, leaving the functions in place, or by permuting the functions, leaving the order of the arguments unchanged. We distinguish those permutations that permute the functions by placing a bar over the symbols for them. It is not difficult to convince oneself that there is a simple relationship between the two sorts of permutation. Thus

$$\bar{P}\Phi_0 = P^{-1}\Phi_0. \tag{2.42}$$

The permutations P and $\bar{P}$ act upon different objects, and hence commute with one another:

$$\bar{P}Q\Phi_0 = Q\bar{P}\Phi_0. \tag{2.43}$$

The set of permutations P and $\bar{P}$ constitutes two commuting groups, π_N and $\bar{\pi}_N$.

Let us permute the arguments of the function Φ_0. We obtain $N!$ mutually orthogonal functions:

$$\Phi_P = P\Phi_0 = \phi_1(i_1)\phi_2(i_2) \cdots \phi_N(i_N). \tag{2.44}$$

The orthogonality of the functions Φ_P follows from the condition (2.40). The Φ_P form a basis for the regular representation of π_N (see Section 1.18).

Linear combinations of the Φ_P functions which form a basis for a standard orthogonal representation are obtained by applying the Young operator (2.30) to the function Φ_0:

$$\Phi_{rt}^{[\lambda]} = \omega_{rt}^{[\lambda]}\Phi_0 = (f_\lambda/N!)^{1/2} \sum_P \Gamma_{rt}^{[\lambda]}(P)\Phi_P. \tag{2.45}$$

In accordance with (2.33),

$$P\Phi_{rt}^{[\lambda]} = \sum_{u} \Gamma_{ur}^{[\lambda]}(P)\Phi_{ut}^{[\lambda]}. \tag{2.46}$$

Functions $\Phi_{rt}^{[\lambda]}$ with a fixed second index therefore transform into each other under permutations of the arguments [in conformity with the general rule, Eq. (1.90)]. The Young tableau r that corresponds to the first index enumerates the basis functions for the representation $\Gamma^{[\lambda]}$. This tableau thus characterizes the symmetry of the function $\Phi_{rt}^{[\lambda]}$ under permutations of the arguments. The Young tableau t corresponding to the second index enumerates the different bases for $\Gamma^{[\lambda]}$. Tableau t characterizes the symmetry of $\Phi_{rt}^{[\lambda]}$ under permutations of the functions ϕ_a. In order to prove this last statement, we apply a permutation $\bar{P}$ to the function (2.45). By virtue of Eqs. (2.43) and (2.42),

$$\bar{P}\Phi_{rt}^{[\lambda]} = \omega_{rt}^{[\lambda]}\bar{P}\Phi_0 = \omega_{rt}^{[\lambda]}P^{-1}\Phi_0. \tag{2.47}$$

Expressing the operators on the right-hand side of (2.47) in the form (2.34), we finally obtain

$$\bar{P}\Phi_{rt}^{[\lambda]} = \sum_{u} \Gamma_{ut}^{[\lambda]}(P)\Phi_{ru}^{[\lambda]}. \tag{2.48}$$

Functions $\Phi_{rt}^{[\lambda]}$ with a fixed first index therefore transform into each other under permutations of the ϕ_a functions.

In this way one can form from the $N!$ functions (2.44) $N!$ linearly independent functions (2.45), which we denote by $\Phi_{rt}^{[\lambda]}$. If each function $\Phi_{rt}^{[\lambda]}$ is represented by a point on a graph, then the $N!$ points fall into squares, each of which is characterized by a particular Young diagram $[\lambda]$, and contains f_λ^2 points. The rows of the squares are numbered by the Young tableaux r, and the columns by the tableaux t. The functions corresponding to points in a single column of a square transform into each other under permutations of the arguments, and functions that correspond to points on a single row of a square transform into each other under permutations of the ϕ_a.

For example, for $N = 4$, the $4!$ functions $\Phi_{rt}^{[\lambda]}$ fall into five squares, corresponding to the five Young diagrams (2.15), shown in Fig. 2.1.

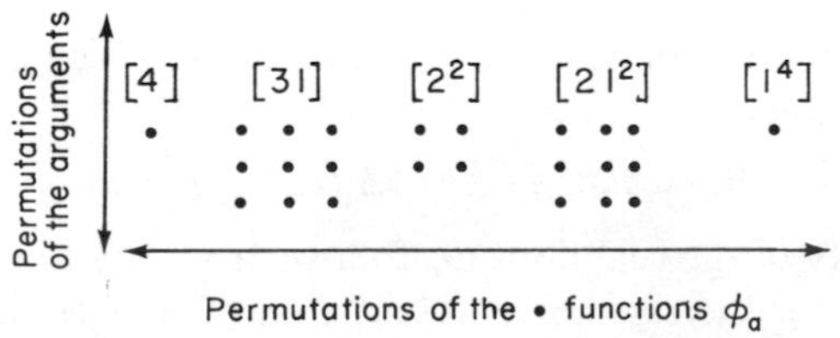

Figure 2.1.

In Section 1.19 it was shown that for any finite group, basis functions that transform according to different columns of an irreducible representation are orthogonal to one another. However, basis functions that transform according to a single column but belong to different bases are no longer orthogonal to each other [see Eq. (1.97)]. The $\Phi_{rt}^{[\lambda]}$ functions, however, satisfy a stronger orthogonality condition than this equation, and are orthogonal with respect to all their indices; i.e.,

$$\int \Phi_{rt}^{[\lambda]*}\Phi_{us}^{[\tilde{\lambda}]}\,dV = \delta_{\lambda\tilde{\lambda}}\delta_{ru}\delta_{ts}. \tag{2.49}$$

The additional orthogonality, as compared to Eq. (1.97), in the second indices follows from the fact that for fixed first indices the functions $\Phi_{rt}^{[\lambda]}$ and $\Phi_{rs}^{[\lambda]}$ belong to a single basis for the group $\tilde{\pi}_N$.

Equation (2.49) can be derived immediately be substituting expression (2.45) for $\Phi_{rt}^{[\lambda]}$ and making use of the orthogonality of the Φ_P and of the orthogonality relations (2.20). Thus,

$$\begin{aligned}\int \Phi_{rt}^{[\lambda]*}\Phi_{us}^{[\tilde{\lambda}]}\,dV &= \frac{(f_\lambda f_{\tilde{\lambda}})^{1/2}}{N!}\sum_{P,Q}\Gamma_{rt}^{[\lambda]}(P)\Gamma_{us}^{[\tilde{\lambda}]}(Q)\int \Phi_P^*\Phi_Q\,dV\\ &= \frac{(f_\lambda f_{\tilde{\lambda}})^{1/2}}{N!}\sum_{P}\Gamma_{rt}^{[\lambda]}(P)\Gamma_{us}^{[\tilde{\lambda}]}(P) = \delta_{\lambda\tilde{\lambda}}\delta_{ru}\delta_{ts}.\end{aligned}$$

Let us assume that m of the functions ϕ_a in Eq. (2.41) are identical; i.e., that

$$\Phi_0 = \phi_1(1)\cdots\phi_1(m)\phi_2(m+1)\cdots\phi_{N-m+1}(N). \tag{2.50}$$

The number of independent functions $\Phi_{rt}^{[\lambda]}$ which may be formed by applying the $f_\lambda{}^2$ Young operators $\omega_{rt}^{[\lambda]}$ to the function (2.50) is less than $f_\lambda{}^2$, since some of the $\Phi_{rt}^{[\lambda]}$ functions are now linearly dependent and some reduce to zero. This is because the function (2.50) is symmetric with respect to permutations of the arguments among the m functions ϕ_1. Only those Young tableaux in which the first m numbers all appear on the same line are allowed. The possible Young diagrams $[\lambda]$ therefore all embody a diagram consisting of a single row with m cells. Since the total number of cells in a Young diagram $[\lambda]$ is equal to N, no column in the diagram may have more than $N - m + 1$ cells (Fig. 2.2).

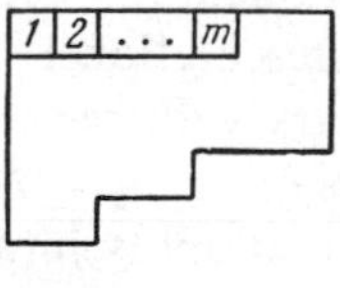

Figure 2.2.

For example, in the case

$$\Phi_0 = \phi_1(1)\phi_1(2)\phi_1(3)\phi_2(4)\phi_3(5)$$

the following Young diagrams are allowed for the group π_5:

$$[\lambda] = [5],\quad [41],\quad [32],\quad [31^2].$$

The Young diagrams $[2^21]$, $[21^3]$, and $[1^5]$ may not be used.

This result does not depend upon the factorization of Φ_0 into a product of functions ϕ_a, and can be put into the following more general form.

The types of permutational symmetry that can be realized for a function of N variables, m of which are identical and the remaining $N - m$ distinct, are those for which the Young diagrams have one row with not less than m cells, and no column with more than $N - m + 1$ cells.

PART 3. The Nonstandard Representation

2.10. Definition

If we restrict the $N!$ permutations of the group π_N to permutations of the numbers $1, 2, \ldots, n_1$, then a standard representation $\Gamma^{[\lambda]}$ splits up into irreducible representations of the group π_{n_1}. However, as far as the $n_2!$ permutations of the numbers $n_1 + 1, n_1 + 2, \ldots, (n_1 + n_2 = N)$ are concerned, a representation $\Gamma^{[\lambda]}$ cannot in general be resolved into irreducible representations $\Gamma^{[\lambda_2]}$ of the group π_{n_2}. This is due to the fact that in the Young tableaux which characterize the symmetry of the basis functions for a representation $\Gamma^{[\lambda]}$ the last n_2 numbers do not themselves form a Young diagram. In Fig. 2.3 the cells containing the last n_2 numbers are shaded, and, as can be seen, these do not form a proper Young diagram.

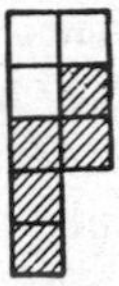

Figure 2.3.

In a number of problems in quantum mechanics it is convenient to work with basis functions for a representation $\Gamma^{[\lambda]}$ which are characterized by a definite permutational symmetry with respect to permutations of both the first n_1 and the last n_2 numbers $(n_1 + n_2 = N)$. Functions such as these form a basis for a representation that splits into irreducible components upon passing from the group π_N to its subgroup $\pi_{n_1} \times \pi_{n_2}$, the direct product of the groups π_{n_1} and π_{n_2}. In contrast to the standard representation, which is reduced with respect to the subgroups $\pi_{N-1}, \pi_{N-2}, \ldots, \pi_2$, a representation which is reduced with respect to a different set of subgroups is said to be *nonstandard*.‡ Every nonstandard representation is defined by the set of subgroups with respect to which it is reduced. A method for obtaining the

‡The concept of nonstandard representation was introduced by Elliott *et al.* (1953), and the general theory of such representations was studied by Kaplan (1961a, 1962a).

matrices of a nonstandard representation in terms of the matrices of the standard representation is described in the following section.

The possible irreducible representations of $\Gamma^{[\lambda_1]} \times \Gamma^{[\lambda_2]}$ into which the representation $\Gamma^{[\lambda]}$ is resolved on reduction with respect to the subgroup $\pi_{n_1} \times \pi_{n_2}$ are given by the decomposition

$$\Gamma^{[\lambda]} \doteq \sum_{\lambda_1, \lambda_2} a(\lambda\lambda_1\lambda_2)\, \Gamma^{[\lambda_1]} \times \Gamma^{[\lambda_2]}. \tag{2.51}$$

The $a(\lambda\lambda_1\lambda_2)$ can be found either by means of a table of characters, remembering that the characters of the representations of $\Gamma^{[\lambda_1]} \times \Gamma^{[\lambda_2]}$ are determined by formula (1.75), or by using Littlewood's theorem (Section 4.4). The basis functions are characterized by Young tableaux for the first n_1 and last n_2 numbers, instead of by tableaux for all N numbers.

For example, on reduction with respect to the subgroup $\pi_3 \times \pi_3$, the representation $\Gamma^{[2^3]}$ is resolved into the following irreducible representations:

$$[2^3] \doteq [21] \times [21] + [1^3] \times [1^3].$$

The basis functions are enumerated by pairs of Young tableaux with three cells:

$$[21]_1[21]_1, \quad [21]_1[21]_2, \quad [21]_2[21]_1, \quad [21]_2[21]_2, \quad [1^3][1^3], \tag{2.52}$$

instead of by the tableaux (2.16). In (2.52) we have denoted a Young tableau $r^{(i)}$ by the symbol for the corresponding diagram together with a suffix i. The Young tableaux are ordered according to the degree with which the numbers in them deviate form their natural order, in accordance with the convention adopted in Section 2.4. In the case of one-dimensional representations the corresponding Young diagram without a suffix is used instead of a tableau.

The standard representation corresponds to the symmetrization of N numbers in succession. In order to characterize completely the symmetry of a basis function in this representation, one has to specify $N - 2$ intermediate Young diagrams. This is equivalent to specifying a Young tableau with N numbers, since the intermediate Young diagrams are determined by successively removing from the tableau the cells containing the numbers $N, N - 1, \ldots$. In general a representation corresponding to the symmetrization of N numbers can be defined in which the numbers are divided up beforehand into k groups, each of which possess its own symmetry scheme. This corresponds to a representation which is reduced with respect to the subgroup

$$\pi_{n_1} \times \pi_{n_2} \times \cdots \times \pi_{n_k}, \qquad \sum_{i=1}^{k} n_i = N. \tag{2.53}$$

In order to characterize completely the symmetry of a basis function in such a representation, in addition to specifying the Young tableaux for each group π_{n_i}, it is further necessary to specify $k - 2$ intermediate Young diagrams, and also the order in which the k groups are symmetrized in the complete Young diagram $[\lambda]$. We enumerate the basis functions in nonstandard representations by the symbol $(r)^A$, where A denotes the type of reduction with respect to subgroups. In the standard representation the Young tableau r plays the role of $(r)^A$.

As an example, we consider the nonstandard representation $\Gamma^{[2^3]}$, which is reduced with respect to the subgroup

$$\pi_2 \times \pi_2 \times \pi_2.$$

Its basis functions are characterized by specifying three Young tableaux with two cells each‡ and one intermediate Young diagram with four cells. Two symmetrization schemes are possible:

$$\text{(a)} \quad (\pi_2 \times \pi_2) \times \pi_2, \qquad \text{and} \qquad \text{(b)} \quad \pi_2 \times (\pi_2 \times \pi_2). \tag{2.54}$$

The basis functions for representations that are symmetrized according to scheme (a) in (2.54) are connected by a linear transformation with those symmetrized according to scheme (b). Let us assume that we have chosen symmetrization scheme (a). Instead of being characterized by a succession of intermediate Young diagrams (2.17), the basis functions for the representation are characterized by the following:

	$[\lambda_1]$	$[\lambda_2]$	$[\lambda_{12}]$	$[\lambda_3]$
$(r^{(1)})^A$	[2]	[2]	$[2^2]$	[2]
$(r^{(2)})^A$	[2]	$[1^2]$	$[21^2]$	$[1^2]$
$(r^{(3)})^A$	$[1^2]$	[2]	$[21^2]$	$[1^2]$
$(r^{(4)})^A$	$[1^2]$	$[1^2]$	$[2^2]$	[2]
$(r^{(5)})^A$	$[1^2]$	$[1^2]$	$[21^2]$	$[1^2]$

(2.55)

2.11. The Transformation Matrix

In order to obtain an explicit form for the matrices of a nonstandard representation, we need to find a linear transformation which connects the basis functions in the standard and nonstandard representations. The coefficients in this transformation form a matrix, which we shall refer to as the *transformation matrix*. The basis functions for a nonstandard representation are enumerated by the symbol $(r)^A$, the explicit form for which may be fairly cumbersome [see, for example, Eq. (2.55)]. It is therefore convenient to use the Dirac notation for the basis functions and matrix elements. Thus

$$|[\lambda](r)^A\rangle \equiv \psi^{[\lambda]}_{(r)^A}, \qquad \langle (r)^A|P|(t)^A\rangle^{[\lambda]} \equiv \Gamma^{[\lambda]}_{(r)^A,\,(t)^A}(P). \tag{2.56}$$

The transformation matrix is defined by the relationship

$$|[\lambda](r)^A\rangle = \sum_r |[\lambda]r\rangle\langle r\,|\,(r)^A\rangle^{[\lambda]} \tag{2.57}$$

in which the summation is taken over all the standard tableaux of the representation $\Gamma^{[\lambda]}$.

We require the transformation matrix to be orthogonal, i.e., that its matrix

‡In this case the Young tableaux correspond uniquely to Young diagrams.

elements satisfy the orthogonality conditions

$$\sum_{(r)^A} \langle r|(r)^A\rangle^{[\lambda]}\langle t|(r)^A\rangle^{[\lambda]} = \delta_{rt},$$
$$\sum_{r} \langle r|(r)^A\rangle^{[\lambda]}\langle r|(t)^A\rangle^{[\lambda]} = \delta_{(r)^A,(t)^A}. \tag{2.58}$$

The orthogonality of this transformation ensures the orthogonality of the nonstandard representation, since the matrices of the standard representation are already orthogonal.

If the transformation matrix is known, the matrices of the nonstandard representation are found from those in the standard representation by using formula (1.35). According to this equation, we obtain

$$\langle (r)^A|P|(t)^A\rangle^{[\lambda]} = \sum_{r,t} \langle (r)^A|r\rangle^{[\lambda]}\langle r|P|t\rangle^{[\lambda]}\langle t|(t)^A\rangle^{[\lambda]}, \tag{2.59}$$

in which we have taken into account the fact that the inverse of an orthogonal matrix is equal to its transpose.

We now derive an expression for the elements of the transformation matrix which connects the basis functions for a standard representation with those for a nonstandard representation that is reduced with respect to the subgroup $\pi_{n_1} \times \pi_{n_2}$ $(n_1 + n_2 = N)$. The first n_1 numbers in the basis functions for the standard representation already possess the required symmetry, and it is therefore necessary to symmetrize only the last n_2 numbers. For this purpose we apply the Young operator $\omega^{[\lambda_2]}_{r_2 t_2}$ from the permutation group of the numbers $n_1 + 1, n_1 + 2, \ldots, n_1 + n_2$ to any one of the basis functions for the standard representation (we refer to this function as a trial function). We write the Young tableaux of the standard representation $\Gamma^{[\lambda]}$ in the form $(r_1\rho_2)$, where r_1 is a standard tableau for the first n_1 numbers and ρ_2 represents the remaining set of cells containing the numbers $n_1 + 1, \ldots, n_1 + n_2$, and which in general does not constitute a Young tableau. The basis function that is obtained, $|[\lambda]r_1r_2\rangle$, possesses the same symmetry with respect to the first n_1 numbers as the trial function, and the symmetry of the last n_2 numbers is characterized by the Young tableau r_2:

$$\begin{aligned}|[\lambda]r_1r_2\rangle &= c'\omega^{[\lambda_2]}_{r_2t_2}|[\lambda](r_1\bar{\rho}_2)\rangle = c\sum_{P_2}\langle r_2|P_2|t_2\rangle^{[\lambda_2]}P_2|[\lambda](r_1\bar{\rho}_2)\rangle \\ &= c\sum_{P_2}\sum_{\rho_2}\langle r_2|P_2|t_2\rangle^{[\lambda_2]}\langle (r_1\rho_2)|P_2|(r_1\bar{\rho}_2)\rangle^{[\lambda]}|[\lambda](r_1\rho_2)\rangle.\end{aligned}$$

The new basis functions are therefore expressed in terms of the old by means of the linear transformation

$$|[\lambda]r_1r_2\rangle = \sum_{\rho_2}|[\lambda](r_1\rho_2)\rangle\langle (r_1\rho_2)|r_1r_2\rangle^{[\lambda]}, \tag{2.60}$$

the matrix of which is expressed in terms of the known matrices of the standard representation‡

$$\langle (r_1\rho_2)|r_1r_2\rangle^{[\lambda]} = c\sum_{P_2}\langle r_2|P_2|t_2\rangle^{[\lambda_2]}\langle (r_1\rho_2)|P_2|(r_1\bar{\rho}_2)\rangle^{[\lambda]}. \tag{2.61}$$

‡A different method for calculating the transformation matrix, based upon a set of recurrence relations, is described by Horie (1964).

The summation in (2.60) is taken over all the Young tableaux $(r_1 p_2)$ in which the first n_1 numbers form the tableau r_1, the permutations P_2 are permutations of the numbers $n_1 + 1, \ldots, n_1 + n_2$ and c is a normalization constant which is determined by the condition that

$$\sum_{p_2} \{\langle (r_1 p_2) | r_1 r_2 \rangle^{[\lambda]}\}^2 = 1. \tag{2.62}$$

The representation matrices according to which the functions (2.60) transform do not depend upon the choice of t_2 and $\bar{p}_2$; these just determine the construction of the basis functions for the nonstandard representation. The transformation matrices do depend upon t_2 and $\bar{p}_2$; however, those obtained from trial functions with different $\bar{p}_2$ differ by a phase factor only. In order to be definite, we shall always choose as $(r_1 \bar{p}_2)$ the Young tableau in which the numbers $n_1 + 1, \ldots, n_1 + n_2$ occur in their natural order as read along successive rows, and as t_2 the fundamental tableau of the representation $[\lambda_2]$. If the application of $\omega_{r_2 t_2}^{[\lambda_2]}$ to the trial function gives zero, we pick as t_2 the tableau immediately following the fundamental.

The rows of the transformation matrix are enumerated by the tableaux $(r_1 p_2)$ and the columns by the tableaux r_1 and r_2. This matrix is diagonal in r_1, and therefore breaks up into blocks. In accordance with Eq. (2.23), all blocks with an r_1 that corresponds to a single Young diagram of n_1 cells are identical. Therefore instead of using r_1 in the designation of the transformation matrix, we may use the symbol for the corresponding Young diagram $[\lambda_1]$:

$$\langle (r_1 p_2) | \bar{r}_1 r_2 \rangle^{[\lambda]} = \delta_{r_1 \bar{r}_1} \langle (\lambda_1 p_2) | (\lambda_1 r_2) \rangle^{[\lambda]}. \tag{2.63}$$

The matrix elements (2.63) are only nonzero for those $[\lambda_1]$ and $[\lambda_2]$ that occur in the decomposition (2.51). If some of the $a(\lambda \lambda_1 \lambda_2)$ in (2.51) are greater than one, further identification of the function (2.60) becomes necessary. This is achieved by specifying the second index t_2 of the particular Young operator $\omega_{r_2 t_2}^{[\lambda_2]}$. Functions $|[\lambda] r_1 r_2(t_2)\rangle$ that differ only in t_2 are not orthogonal to one another, and hence the transformation matrix is no longer orthogonal. The transformation matrix must therefore be orthogonalised in such cases. [See Kaplan (1962a).]‡

‡The symmetry properties of the transformation matrices defined in the present and following sections were studied by Kramer (1967). In analogy with the invariants of the three-dimensional rotation group $\mathbf{R}_3$ (which we refer to as $3n$-j symbols for the moment, see Section 3.7), he introduced invariants of the permutation group. Furthermore, he showed (Kramer, 1968) that these invariants are identical with those of the group of unitary transformations. In the case of electrons the transformation matrices for the permutation group can be expressed in terms of the invariants of the group $\mathbf{SU}_2$, which are just the $3n$-j symbols Thus the matrix (2.61) is equal to a 6-j symbol which depends upon the spins of the subgroups of electrons. The equivalence of the transformation matrices of the permutation group and the invariants of the unitary group was first noted by Kukulin *et al.* (1967).

We now give the matrix for the transformation from the basis (2.16) to the basis (2.52). The necessary matrices for the calculation using formula (2.61) were taken from Tables 1 and 6 of Appendix 5; the matrices $\Gamma^{[2^3]}(P_2)$ which are not given there are obtained by multiplying the transposition matrices. After normalizing according to formula (2.62) we obtain the following orthogonal matrix:

$$\begin{array}{ll} & \begin{array}{ccccc} r_1: & [21]_1 & [21]_1 & [21]_2 & [21]_2 & [1^3] \\ r_2: & [21]_1 & [21]_2 & [21]_1 & [21]_2 & [1^3] \end{array} \\ \langle r\,|\,r_1 r_2\rangle^{[2^3]} = \begin{array}{c} r^{(1)} \\ r^{(2)} \\ r^{(3)} \\ r^{(4)} \\ r^{(5)} \end{array} & \begin{bmatrix} \frac{1}{2} & \frac{1}{2}\sqrt{3} & & & \\ \frac{1}{2}\sqrt{3} & -\frac{1}{2} & & & \\ & & \frac{1}{2} & \frac{1}{2}\sqrt{3} & \\ & & \frac{1}{2}\sqrt{3} & -\frac{1}{2} & \\ & & & & 1 \end{bmatrix}. \end{array} \tag{2.64}$$

In accordance with (2.63), the transformation matrix is divided into blocks, those corresponding to $[\lambda_1] = [2\ 1]$ being identical.

In the case of reduction with respect to the subgroup $\pi_{N-2} \times \pi_2$, the matrix elements (2.61) can be explicitly expressed in terms of the axial distance between the numbers $N-1$ and N. For this purpose the expression given on p. 37 for the transposition matrix $\langle (r_1\rho_2)\,|\,P_{N-1\,N}\,|\,(r_1\bar{\rho}_2)\rangle^{[\lambda]}$ must be substituted into (2.61). The elements of the transformation matrix therefore consist either of unity or of blocks of the form

$$\langle (r_1\rho_2)\,|\,r_1\lambda_2\rangle^{[\lambda]} = \begin{array}{c} \\ \rho_2' \\ \rho_2'' \end{array} \begin{array}{c} \begin{array}{cc} [\lambda_2]:\quad [2] & \qquad [1^2] \end{array} \\ \begin{bmatrix} [(d-1)/2d]^{1/2} & -[(d+1)/2d]^{1/2} \\ [(d+1)/2d]^{1/2} & [(d-1)/2d]^{1/2} \end{bmatrix}, \end{array} \tag{2.65}$$

where d is the axial distance between $N-1$ and N in the tableaux $(r_1\rho_2')$ and $(r_1\rho_2'')$. These tableaux differ only by a permutation of $N-1$ and N, the number $N-1$ occurring above N in tableau $(r_1\rho_2')$.

For example, the matrix of the transformation from the standard representation $\Gamma^{[21^2]}$, the Young tableaux for which were given on p. 34 , to the representation $\Gamma^{[21^2]}$, which is reduced with respect to the subgroup $\pi_2 \times \pi_2$, is of the form

$$\begin{array}{ll} & \begin{array}{ccc} [\lambda_1]: [2] & [1^2] & [1^2] \\ [\lambda_2]: [1^2] & [2] & [1^2] \end{array} \\ \langle r\,|\,\lambda_1\lambda_2\rangle^{[21^2]} = \begin{array}{c} r^{(1)} \\ r^{(2)} \\ r^{(3)} \end{array} & \begin{bmatrix} 1 & & \\ & \frac{1}{3}\sqrt{3} & -\frac{1}{3}\sqrt{6} \\ & \frac{1}{3}\sqrt{6} & \frac{1}{3}\sqrt{3} \end{bmatrix}. \end{array}$$

Let us find the explicit forms for the matrices of the representation $\Gamma^{[21]}$ which is reduced with respect to the subgroup $\pi_1 \times \pi_2$. The Young tableaux for the standard representation were given previously [see Eq. (2.24)] and $d_{23} = 2$. In accordance with (2.65),

$$\langle r|1\lambda_2\rangle^{[21]} = \begin{array}{c} \\ r^{(1)} \\ r^{(2)} \end{array}\!\!\begin{array}{c} [\lambda_2]:\ \ [2] \qquad [1^2] \\ \begin{bmatrix} \frac{1}{2} & -\frac{1}{2}\sqrt{3} \\ \frac{1}{2}\sqrt{3} & \frac{1}{2} \end{bmatrix} \end{array}.$$

The matrices in the nonstandard representation are found from formula (2.59). Instead of (2.25), we now obtain

$[\lambda] = [2\ 1]$

$$\begin{array}{ccc} I & P_{12} & P_{23} \\ \begin{bmatrix} 1 & \\ & 1 \end{bmatrix}, & \begin{bmatrix} -\frac{1}{2} & -\frac{1}{2}\sqrt{3} \\ -\frac{1}{2}\sqrt{3} & \frac{1}{2} \end{bmatrix}, & \begin{bmatrix} 1 & \\ & -1 \end{bmatrix}, \\ P_{13} & P_{123} & P_{132} \\ \begin{bmatrix} -\frac{1}{2} & \frac{1}{2}\sqrt{3} \\ \frac{1}{2}\sqrt{3} & \frac{1}{2} \end{bmatrix}, & \begin{bmatrix} -\frac{1}{2} & \frac{1}{2}\sqrt{3} \\ -\frac{1}{2}\sqrt{3} & -\frac{1}{2} \end{bmatrix}, & \begin{bmatrix} -\frac{1}{2} & -\frac{1}{2}\sqrt{3} \\ \frac{1}{2}\sqrt{3} & -\frac{1}{2} \end{bmatrix}. \end{array} \tag{2.66}$$

As one might expect, the matrix for P_{23} in the nonstandard representation, which is reduced with respect to the subgroup $\pi_1 \times \pi_2$, coincides with the matrix for P_{12} in the standard representation, which is reduced with respect to the subgroup $\pi_2 \times \pi_1$.

2.12. Some Generalizations

In the previous section we considered a method of obtaining the matrix for the transformation from the standard basis to a basis in which the representation is reduced with respect to a subgroup $\pi_{n_1} \times \pi_{n_2}$. The matrix for the transformation to a representation with a more complicated type of reduction can be expressed in terms of products of the transformation matrices (2.61). For example, the basis functions for a representation that is reduced with respect to a subgroup of the form

$$(\pi_{n_1} \times \pi_{n_2}) \times \pi_{n_3} \tag{2.67}$$

are found by applying two Young operators $\omega_{r_2t_2}^{[\lambda_2]}$ and $\omega_{r_3t_3}^{[\lambda_3]}$ in succession to a trial function from the standard representation. The operator $\omega_{r_2t_2}^{[\lambda_2]}$ is defined for the permutation group of the numbers $n_1 + 1, n_1 + 2, \ldots, n_1 + n_2$, and $\omega_{r_3t_3}^{[\lambda_3]}$ for the permutation group of $n_1 + n_2 + 1, n_1 + n_2 + 2, \ldots, n_1 + n_2 + n_3 = N$. The intermediate Young diagrams for the numbers 1 to $n_1 + n_2$ are specified by the symmetry of the trial function. A Young tableau can be divided into three parts, each of which refers to the numbers in one of the groups π_{n_i} in (2.67). We write a tableau in the form $r \equiv (r_1\rho_2\rho_3) \equiv (r_{12}\rho_3) \equiv (r_1\rho_{23})$, where r_{12} corresponds to the Young diagram $[\lambda_{12}]$. As a result, we obtain

$$|[\lambda](r_1r_2)\lambda_{12}r_3\rangle = \sum |[\lambda](r_1\rho_2\rho_3)\rangle\langle(r_1\rho_2\rho_3)|(r_1r_2)\lambda_{12}r_3\rangle^{[\lambda]}, \tag{2.68}$$

where the transformation matrix is equal to a product of two simpler transformation matrices which can be calculated from formula (2.61):

$$\langle(r_1\rho_2\rho_3)|(r_1r_2)\lambda_{12}r_3\rangle^{[\lambda]} = \langle(r_1\rho_2)|r_1r_2\rangle^{[\lambda_{12}]}\langle(r_{12}\rho_3)|r_{12}r_3\rangle^{[\lambda]}. \tag{2.69}$$

The representation that is reduced with respect to

$$\pi_{n_1} \times (\pi_{n_2} \times \pi_{n_3})$$

is derived in two stages. The reduction with respect to $\pi_{n_1} \times \pi_{n_2+n_3}$ is carried out first, followed by reduction with respect to $\pi_{n_2} \times \pi_{n_3}$. We obtain as a result

$$\langle(r_1\rho_2\rho_3)|r_1(r_2r_3)\lambda_{23}\rangle^{[\lambda]} = \sum_{\bar{\rho}_3} \langle r_1\rho_{23}|r_1r_2\bar{\rho}_3\rangle^{[\lambda]}\langle r_2\bar{\rho}_3|r_2r_3\rangle^{[\lambda_{23}]}. \tag{2.70}$$

The transformation matrix which connects two nonstandard representations can be expressed as a product of two matrices, each of which transforms one of the bases to the standard basis:

$$\langle(r)^A|(r)^B\rangle = \sum_r \langle(r)^A|r\rangle\langle r|(r)^B\rangle. \tag{2.71}$$

"Asymmetric" representations are useful in some problems. In these the matrices are defined by two sets of basis functions, each of which is reduced with respect to its own particular series of subgroups (Kaplan, 1961a, 1962a). The matrix elements in such a representation,

$$\langle(r)^A|P|(t)^B\rangle^{[\lambda]}, \tag{2.72}$$

occur, for example, as the coefficients when a permutation P is applied to a function of reduction type B, and the result expressed in terms of functions of type A:

$$P|[\lambda](t)^B\rangle = \sum_{(r)^A} |[\lambda](r)^A\rangle\langle(r)^A|P|(t)^B\rangle^{[\lambda]}. \tag{2.73}$$

The transformation matrix (2.71) is a particular case of the matrix (2.72) with $P = 1$. The following expression forms an obvious generalization of formula (2.59):

$$\langle(r)^A|P|(t)^B\rangle^{[\lambda]} = \sum_{r,t} \langle(r)^A|r\rangle^{[\lambda]}\langle r|P|t\rangle^{[\lambda]}\langle t|(t)^B\rangle^{[\lambda]}. \tag{2.74}$$

2.13. Young Operators in a Nonstandard Representation

By analogy with the definition (2.30) of a Young operator in the standard representation, we have in a nonstandard representation

$$\omega^{[\lambda]}_{(r)^A,(t)^A} = (f_\lambda/N!)^{1/2} \sum_P \langle(r)^A|P|(t)^A\rangle^{[\lambda]}P. \tag{2.75}$$

The matrix elements of the nonstandard representation which appear in this formula are found by means of Eq. (2.59) from the known matrix elements of the standard representation and the transformation matrices. The operator (2.75) can be expressed in terms of standard Young operators for the subgroups, with respect to which the representation $\Gamma^{[\lambda]}$ is reduced, and matrices of the nonstandard representation.

We consider a representation $\Gamma^{[\lambda]}$ with is reduced with respect to the subgroup $\pi_{n_1} \times \pi_{n_2}$ $(n_1 + n_2 = N)$. We derive a relationship between the Young operators of this representation and the Young operators of the standard representations of the groups π_{n_1} and π_{n_2}.

We note first of all that any permutation P of the group π_N can be written in the form of a product

$$P = Q \cdot P_1 \cdot P_2, \tag{2.76}$$

where $P_1 \in \pi_{n_1}$, $P_2 \in \pi_{n_2}$, and Q is a member of one of the cosets of the subgroup $\pi_{n_1} \times \pi_{n_2}$ (see Section 1.4). For example, the subgroup $\pi_2 \times \pi_2$ of the group π_4 consists of four elements, and its index is equal to six. The 24 permutations of π_4 can therefore be divided into six cosets. The first coset is just the subgroup $\pi_2 \times \pi_2$ itself. The permutations in the remaining cosets are found by multiplying the elements of the subgroup by a permutation which does not belong to the subgroup. This division into cosets is, moreover, determined once the subgroup is specified, and does not depend upon the choice of permutations by which the subgroup is multiplied. Denoting $\pi_2 \times \pi_2$ by $\mathbf{H}$, its left cosets are given by

$$\begin{array}{rllll} H: & I, & P_{12}, & P_{34}, & P_{12}\cdot P_{34}; \\ P_{13}\cdot H: & P_{13}, & P_{123}, & P_{134}, & P_{1234}; \\ P_{23}\cdot H: & P_{23}, & P_{132}, & P_{234}, & P_{1342}; \\ P_{14}\cdot H: & P_{14}, & P_{124}, & P_{143}, & P_{1243}: \\ P_{24}\cdot H: & P_{24}, & P_{142}, & P_{243}, & P_{1432}; \\ P_{13}\cdot P_{24}\cdot H: & P_{13}\cdot P_{24}, & P_{1423}, & P_{1324}, & P_{14}\cdot P_{23}. \end{array} \tag{2.77}$$

A Young operator in the representation that is reduced with respect to $\pi_{n_1} \times \pi_{n_2}$ is of the form

$$\omega^{[\lambda]}_{r_1 r_2 t_1 t_2} = (f_\lambda / N!)^{1/2} \sum_P \langle r_1 r_2 | P | t_1 t_2 \rangle^{[\lambda]} P. \tag{2.78}$$

We write P in the form of a product (2.76). The matrix element in (2.78) becomes

$$\langle r_1 r_2 | P | t_1 t_2 \rangle^{[\lambda]} = \sum_{u_1, u_2} \sum_{v_1, v_2} \langle r_1 r_2 | Q | u_1 u_2 \rangle^{[\lambda]} \langle u_1 u_2 | P_1 | v_1 v_2 \rangle^{[\lambda]} \langle v_1 v_2 | P_2 | t_1 t_2 \rangle^{[\lambda]}. \tag{2.79}$$

Since the representation which we have chosen is reduced with respect to permutations of the group $\pi_{n_1} \times \pi_{n_2}$, and hence with respect to permutations of the groups π_{n_1} and π_{n_2}, we have

$$\begin{aligned}\langle u_1u_2 | P_1 | v_1v_2\rangle^{[\lambda]} &= \delta_{v_2u_2}\langle u_1 | P_1 | v_1\rangle^{[\lambda_1]},\\ \langle v_1v_2 | P_2 | t_1t_2\rangle^{[\lambda]} &= \delta_{v_1t_1}\langle v_2 | P_2 | t_2\rangle^{[\lambda_2]}.\end{aligned} \tag{2.80}$$

We now substitute Eqs. (2.79) and (2.80) into (2.78). Taking into account the definition of a Young operator for a standard representation, (2.30), we obtain finally (Jahn, 1954)

$$\omega^{[\lambda]}_{r_1r_2,t_1t_2} = \left(\frac{f_\lambda}{f_{\lambda_1}f_{\lambda_2}}\frac{n_1!\,n_2!}{N!}\right)^{1/2} \sum_{u_1,u_2}\sum_{Q}\langle r_1r_2 | Q | u_1u_2\rangle^{[\lambda]} Q\omega^{[\lambda_1]}_{u_1t_1}\omega^{[\lambda_2]}_{u_2t_2}. \tag{2.81}$$

The set Q in this equation is not unique. One may pick as the Q any set of $N!/(n_1!\,n_2!)$ permutations, one from each coset. The summations over u_1 and u_2 are taken over all the standard tableaux of the representations $\Gamma^{[\lambda_1]}$ and $\Gamma^{[\lambda_2]}$, the allowed pairs $[\lambda_1]$ and $[\lambda_2]$ being determined by expression (2.51).

In Section 2.8 it was shown that the Young operators $\omega^{[\lambda]}_{rt}$ can be regarded as basis vectors for an irreducible representation $\Gamma^{[\lambda]}$. The proof that was given there does not depend upon the particular type of reduction of the representation with respect to subgroups, and therefore holds for any nonstandard representation. Consequently, one may regard Eq. (2.81) as a relation between basis functions, and rewrite it in the form

$$\begin{aligned}|[\lambda]r_1r_2\rangle &= \left(\frac{f_\lambda}{f_{\lambda_1}f_{\lambda_2}}\frac{n_1!\,n_2!}{N!}\right)^{1/2}\\ &\quad\times \sum_{u_1,u_2}\sum_{Q}\langle r_1r_2 | Q | u_1u_2\rangle^{[\lambda]} Q\,|[\lambda_1]u_1\rangle|[\lambda_2]u_2\rangle.\end{aligned} \tag{2.82}$$

The function (2.82) transforms according to the nonstandard representation of π_N which is reduced with respect to the subgroup $\pi_{n_1} \times \pi_{n_2}$. We now transform formula (2.82) so that the function can belong to a representation $\Gamma^{[\lambda]}$ which is reduced with respect to any set of subgroups. For this purpose we multiply the left- and right-hand sides of (2.82) by an element $\langle r_1r_2 | (r)^A\rangle^{[\lambda]}$ of a transformation matrix and sum over r_1 and r_2. The matrix element on the right-hand side of (2.82), however, is first of all put into the form

$$\langle r_1r_2 | Q | u_1u_2\rangle^{[\lambda]} = \sum_{(\bar r)_A}\langle r_1r_2 | (\bar r)^A\rangle^{[\lambda]}\langle(\bar r)^A | Q | u_1u_2\rangle^{[\lambda]}. \tag{2.83}$$

According to the definition of a transformation matrix,

$$\sum_{r_1,r_2}|[\lambda]r_1r_2\rangle\langle r_1r_2 | (r)^A\rangle^{[\lambda]} = |[\lambda](r)^A\rangle. \tag{2.84}$$

In addition, we make use of an orthogonality relation analogous to Eq. (2.58), and finally obtain

$$|[\lambda](r)^A\rangle = \left(\frac{f_\lambda}{f_{\lambda_1}f_{\lambda_2}}\frac{n_1!\,n_2!}{N!}\right)^{1/2}$$
$$\times \sum_{u_1,u_2}\sum_{Q}\langle(r)^A|Q|u_1u_2\rangle^{[\lambda]}Q|[\lambda_1]u_1\rangle|[\lambda_2]u_2\rangle. \qquad (2.85)$$

In the particular case of $[\lambda] = [1^N]$, (2.85) becomes the well-known formula which couples antisymmetric functions of two subsystems [see Condon and Shortley (1964, p. 215)]:

$$\psi^{[1^N]} = \left(\frac{n_1!\,n_2!}{N!}\right)^{1/2}\sum_{Q}(-1)^q Q\psi^{[1^{n_1}]}\psi^{[1^{n_2}]}, \qquad (2.86)$$

where q is the parity of the permutation Q.

A necessary condition for the functions (2.85) and (2.86) to be normalized is that the basis functions which occur in the sum over Q shall be orthogonal to one another for different Q.

CHAPTER III

Groups of Linear Transformations

PART 1. Continuous Groups

3.1. Definition. Distinctive Features of Continuous Groups

In the preceding chapters we considered *discrete groups*, i.e., groups whose elements form a discrete set. A discrete set of elements can always be enumerated by positive integers. However, there is a large class of groups whose elements form a *continuous set*. Each element is characterized by a number of parameters which can vary continuously. Such groups are known as *continuous groups*.

For example, the set of transformations

$$x' = a_1 x + a_2, \tag{3.1}$$

in which the parameters a_1 and a_2 can assume any value on the number scale axis (from $-\infty$ to $+\infty$), forms a two-parameter continuous group. Any variation of the parameters, even if infinitely small, leads to a new element of the group.

We can also formally associate a value of some parameter a with every element G_a of a discrete group. The parameter corresponding to the product of two elements, $G_c = G_a G_b$, is equal to c. The discrete function

$$c = \psi(a, b), \tag{3.2}$$

which places the value of the parameter of the product c into correspondence with the values of the parameters a and b, determines the multiplication table for the group. Thus all the elements of a discrete group can be described by specifying a single parameter which can assume g discrete values. In the case of continuous groups the function (3.2) becomes a continuous function of its arguments, the number of parameters which characterize an element of the group being arbitrary, and even infinite.

In what follows we consider groups of linear transformations in an n-dimensional vector space, the transformation being characterized by a finite number of parameters. These groups form a particular case of *Lie groups*. The group of transformations

$$x_i' = f_i(x_1, \ldots, x_n; a_1, \ldots, a_r), \qquad i = 1, 2, \ldots, n, \tag{3.3}$$

in which the functions are analytic in the parameters a_r, is known as a *r-parameter Lie group*.

The application of two such transformations in succession,

$$x_i' = f_i(x_1 \ldots, x_n; a_1, \ldots, a_r)$$

and

$$x_i'' = f_i(x_1', \ldots, x_n'; b_1, \ldots, b_r),$$

is equivalent to a third transformation,

$$x_i'' = f_i(x_1, \ldots, x_n; c_1, \ldots, c_r),$$

in which the parameters are functions of the parameters of the first two transformations:

$$c_k = \psi_k(a_1, \ldots, a_r; b_1, \ldots, b_r), \qquad k = 1, \ldots, r. \tag{3.4}$$

The transformations (3.3) also have to satisfy the remaining three group postulates.

Thus in the case of the two-parameter group (3.1), Eqs. (3.4) are of the form

$$c_1 = b_1 a_1, \qquad c_2 = b_1 a_2 + b_2. \tag{3.5}$$

The unit element in this group is characterized by the values of the parameters $a_1 = 1$, $a_2 = 0$. According to the definition of an inverse element, the successive application of an element and its inverse must give the identity transformation ($c_1 = 1$, $c_2 = 0$). Hence we find from (3.5) the following relation between the parameters of a transformation and its inverse:

$$\bar{a}_1 = 1/a_1, \qquad \bar{a}_2 = -a_2/a_1.$$

The basic concepts and theorems which were established in Chapter I for discrete groups can immediately be generalized to continuous groups. Only those statements that are based upon the finiteness of the order of the group cease to have any meaning (for example, the statement that the dimensions of the irreducible representations are divisors of the order of the group). The concepts of subgroups, classes of conjugate elements, the reducibility of a representation, etc., are carried over unchanged to continuous groups. Since there is a continuous set of elements in a group, each irreducible representation consists of a continuous set of matrices. The number of nonequivalent irreducible representations is infinite, though these form a discrete series in which the dimensions of all the irreducible representations are finite, i.e., the number of basis functions which transform into each other under the operations of the group is finite.‡

For continuous groups a summation over the elements of a group is replaced by an integration over the domain over which the parameters can vary. The volume element in the integration is chosen so that the integral of an arbitrary continuous function of the group parameters is invariant under any transformation of the group. This is known as *invariant integration*. The mathematical requirement that an integral over the group parameters be invariant is written as

$$\int_R f(R)\, d\tau_R = \int_R f(RQ)\, d\tau_R, \tag{3.6}$$

in which the group operators R and Q are functions of the parameters, and the integration is taken over the whole domain over which the parameters of the group can vary. This equation is the natural extension to a continuous group of relation (1.7) which was proved in Section 1.1.

The replacement of the summations in (1.51) and (1.54) by invariant integrations leads to the following form for the orthogonality relations for continuous groups:

$$\int_R \Gamma_{ik}^{(\alpha)}(R)^* \Gamma_{lm}^{(\beta)}(R)\, d\tau_R = \delta_{\alpha\beta}\delta_{il}\delta_{km}\,(1/f_\alpha) \int_R d\tau_R, \tag{3.7}$$

$$\int_R \chi^{(\alpha)}(R)^* \chi^{(\beta)}(R)\, d\tau_R = \delta_{\alpha\beta} \int_R d\tau_R. \tag{3.7a}$$

Formula (1.62) now becomes

$$a^{(\alpha)} = \left(1 \Big/ \int_R d\tau_R\right) \int_R \chi^{(\Gamma)}(R)\chi^{(\alpha)}(R)^*\, d\tau_R. \tag{3.8}$$

3.2. Examples of Linear Groups

Let us consider a vector space of n dimensions. We subject the space to a linear transformation, the matrix of which is $\{a_{ik}\}$. Every vector in the space $\mathbf{x}$ is transformed into a new vector $\mathbf{x}'$ with components

$$x_i' = \sum_{k=1}^{n} a_{ik} x_k. \tag{3.9}$$

A linear transformation is said to be *nonsingular* if the determinant of the transformation matrix is nonzero. The set of all nonsingular linear transfor-

‡This is true only of groups known as *compact* continuous groups (Pontriagin, 1946), to which practically all the continuous groups that are used in physical applications belong (the Lorentz group is an exception). The condition that the integral (3.6) be invariant cannot be met for noncompact groups, since the integration over the domain of the parameters diverges.

mations in an n-dimensional space forms a group known as the *general linear group*, and which is denoted by $\mathbf{GL}_n$. The product of two linear transformations is, of course, also a linear transformation, the matrix of which is found by multiplying the matrices of the individual transformations. Since the determinant of a transformation is nonzero, there exists for every transformation A an inverse transformation A^{-1}, the successive application of the two transformations leading to the identity transformation. Finally, the associative rule is satisfied.

Since each linear transformation is determined by its matrix, these matrices form an n-dimensional representation of the group $\mathbf{GL}_n$. Generally speaking, the transformation matrices are complex, and therefore in order to characterize a transformation, $2n^2$ real parameters must be specified.

If we restrict the linear transformations to unitary transformations, we obtain the *group of unitary transformations* in an n-dimensional space, which is denoted by the symbol $\mathbf{U}_n$. The elements of the transformation matrices satisfy the unitary conditions

$$\sum_l a_{il}^* a_{kl} = \delta_{ik}. \tag{3.10}$$

This gives n^2 equations among the $2n^2$ parameters,‡ and so the number of independent parameters is equal to n^2. Every element of the unitary group is therefore characterized by specifying n^2 real parameters.

Unitary transformations with determinant equal to unity form the group of unitary unimodular transformations, which is denoted by $\mathbf{SU}_n$ (the symbol is derived from the other name for this group: *the special unitary group*). In addition to satisfying relations (3.10), the elements of the transformation matrices of $\mathbf{SU}_n$ must satisfy the condition that the determinant of a transformation is equal to unity. Every transformation in the group $\mathbf{SU}_n$ is therefore characterized by specifying $n^2 - 1$ real parameters.

The set of all real unitary transformations, i.e., of all orthogonal transformations, forms a subgroup of the unitary group which is called the *orthogonal group* $\mathbf{O}_n$. The orthogonality conditions [see Eq. (1.44)] give $n + \frac{1}{2}n(n-1)$ equations among the n^2 parameters of a matrix. The elements of the orthogonal group are therefore characterized by specifying $n(n-1)/2$ parameters. Since the determinant of a matrix is equal to that of its transpose, it follows from Eq. (1.43) that the square of the determinant of an orthogonal transformation is equal to one, and so the determinant itself can only assume two values, plus or minus one. An orthogonal transformation with determinant equal to plus one corresponds to a *rotation of the space* about the origin of coordinates. An orthogonal transformation with determinant equal to minus

‡For $i = k$ there are n equations for the moduli of the matrix elements; for $i \neq k$ there are $n(n-1)/2$ equations for the real parts of the matrix elements, and as many again for the imaginary parts.

one constitutes a combination of a rotation and an *inversion of the space* with respect to the origin of coordinates.

If the orthogonal transformations are restricted to those whose determinant is equal to plus one, we obtain the *rotation group in n-dimensional space* $\mathbf{R}_n$. For $n = 3$, $\mathbf{R}_n$ becomes the rotation group in three-dimensional space $\mathbf{R}_3$, which is widely used in physics.

Every subgroup of the orthogonal group is called a *point group*. The rotation group $\mathbf{R}_3$ is a continuous point group. Finite point groups, the elements of which consist of combinations of rotations through definite angles and of reflections in planes, are used in the theory of molecules and crystals. The various point groups are classified in Part 3 of this chapter.

If the groups of linear transformations described earlier are placed in the order in which they are contained in each other, we obtain the following scheme:

$$\mathbf{GL}_n \supset \mathbf{U}_n \supset \begin{Bmatrix} \mathbf{O}_n \\ \mathbf{SU}_n \end{Bmatrix} \supset \mathbf{R}_n.$$

3.3. Infinitesimal Operators

Any finite transformation of a continuous group can be represented as a succession of infinitely small transformations. The theory of Lie groups (Pontriagin, 1946; Racah, 1965) shows that a finite transformation is uniquely determined by specifying the infinitely small transformations. All the matrices of an irreducible representation, moreover, can be expressed in terms of the matrices which correspond to infinitely small transformations in the same representation. In short, a continuous group is completely characterized by its infinitely small transformations.

Let us rewrite Eq. (3.3) in a more compact form, denoting the set of r parameters a_ρ by the letter a and the set of n coordinates x_i by the letter x:

$$x_i' = f_i(x; a). \tag{3.3a}$$

It is possible to write another, equivalent expression for the x_i' which represents the identity operation:

$$x_i' = f_i(x'; 0). \tag{3.3b}$$

Instead of differentiating (3.3a), an infinitely small change in the x_i' can be obtained by introducing an infinitely small parameter in (3.3b):

$$x_i' + dx_i' = f_i(x'; \delta a). \tag{3.11}$$

Expanding the right-hand side of (3.11) in a McLaurin series and confining ourselves to terms of the first order of smallness, we obtain

$$dx_i' = \sum_{\rho=1}^{r} [(\partial f_i(x', a)/\partial a_\rho)]_{a=0}\, \delta a_\rho. \tag{3.12}$$

We put

$$[(\partial f_i(x, a)/\partial a_\rho)]_{a=0} = u_{i\rho}(x). \tag{3.13}$$

Then, since

$$x_i + dx_i = x_i + \sum_{\rho=1}^{r} u_{i\rho}(x)\delta a_\rho, \tag{3.14}$$

the specification of the r vectors (3.13) in an n-dimensional space defines an infinitely small change in the position of any point in this space. Let us determine the change in an arbitrary function $F(x)$ for an infinitely small change in x:

$$\begin{aligned} dF(x) &= \sum_{i=1}^{n} \frac{\partial F}{\partial x_i} dx_i = \sum_{i=1}^{n} \frac{\partial F}{\partial x_i} \sum_{\rho=1}^{r} u_{i\rho}(x)\, \delta a_\rho \\ &= \sum_{\rho=1}^{r} \delta a_\rho \left(\sum_{i=1}^{n} u_{i\rho}(x) \frac{\partial}{\partial x_i} \right) F = \sum_{\rho=1}^{r} \delta a_\rho I_\rho F, \end{aligned} \tag{3.15}$$

in which I_ρ denotes the operator

$$I_\rho = \sum_{i=1}^{n} u_{i\rho}(x)\, (\partial/\partial x_i). \tag{3.16}$$

The operators I_ρ are called the *infinitesimal operators* of the group. Their number is equal to the number of group parameters. According to Eq. (3.15), an infinitely small change in the function $F(x)$ is produced by the action of a linear combination of infinitesimal operators upon the function. It can be shown (Smirnov, 1964; Racah, 1965) that the commutator of two infinitesimal operators can always be expressed as a linear combination

$$I_\rho I_\sigma - I_\sigma I_\rho \equiv [I_\rho, I_\sigma] = \sum_{\tau=1}^{r} c_{\rho\sigma}^{\tau} I_\tau. \tag{3.17}$$

The coefficients $c_{\rho\sigma}^{\tau}$ are known as *the structure constants of the group.*

As an example, we consider the group of rotations in three-dimensional space. According to the classification in the previous section, the group $\mathbf{R}_3$ is the group of orthogonal transformations in which the determinants of all the transformations are equal to plus one. The number of independent parameters is equal to $n(n-1)/2 = 3$. We may take as those parameters either the Euler angles or the coordinates of a vector directed along the axis of rotation and whose length is equal to the angle of rotation. It is clear that the specification of such a vector uniquely defines a rotation.

We denote the Cartesian coordinates of a point in the three-dimensional space by x, y, and z, and consider a transformation in which one of the parameters is varied, for instance, by a rotation through an angle α_x about the x axis. In this case Eqs. (3.3) assume the following form:

$$\begin{aligned} x' &= x, \\ y' &= y \cos \alpha_x - z \sin \alpha_x, \\ z' &= y \sin \alpha_x + z \cos \alpha_x. \end{aligned} \tag{3.18}$$

From formula (3.13) we find‡

$$u_{xx} = 0, \qquad u_{yx} = -z, \qquad u_{zx} = y. \tag{3.19}$$

Consequently,

$$I_x = y(\partial/\partial z) - z(\partial/\partial y). \tag{3.20}$$

The two other infinitesimal operators are obtained in a similar manner:

$$I_y = z(\partial/\partial x) - x(\partial/\partial z), \qquad I_z = x(\partial/\partial y) - y(\partial/\partial x). \tag{3.21}$$

A direct calculation shows that the operators (3.20) and (3.21) satisfy the following commutation relations:

$$[I_x, I_y] = -I_z, \qquad [I_y, I_z] = -I_x, \qquad [I_z, I_x] = -I_y. \tag{3.22}$$

PART 2. The Three-Dimensional Rotation Group

3.4. Rotation Operators and Angular Momentum Operators

In the preceding section it was shown that the change in a function due to an infinitely small rotation is determined by the infinitesimal operators (3.20) and (3.21). Thus for an infinitely small rotation about an axis x_ρ through an angle $\delta\alpha$

$$dF(x) = (\delta\alpha)I_\rho F(x). \tag{3.23}$$

In quantum mechanics the change in a wave function due to infinitely small rotations is expressed in terms of the *orbital angular momentum operator* **J** (Landau and Lifshitz, 1965). The operator **J** is a vector quantity and its three components J_ρ are related to the infinitesimal operators I_ρ by the simple equation§

$$J_\rho = -iI_\rho. \tag{3.24}$$

The commutation relations for the J_ρ follow from (3.22):

$$[J_x, J_y] = iJ_z \qquad [J_y, J_z] = iJ_x, \qquad [J_z, J_x] = iJ_y. \tag{3.25}$$

In general, angular momentum operators **J** are defined in quantum mechanics as operators whose components obey the permutational relations (3.25). Spin angular momentum operators are also covered by this definition, the components of which cannot be written in the form (3.20) and (3.21).

We now show that the angular momentum operators determine the change in a function of the coordinates due to an arbitrary finite rotation. According

‡In (3.19) and subsequent equations the same notation will be used for both the coordinates and the parameters of a transformation.

§The eigenvalues of J_ρ are expressed in units of $\hbar = h/2\pi$, where h is Planck's constant The effect of multiplying I_ρ by i is to make the operator Hermitian, so that the matrices of J_ρ satisfy the Hermitian condition $\tilde{J}_\rho{}^* = J^\rho$.

to Eqs. (3.23) and (3.24), an arbitrary function $F(x)$ receives an increment

$$dF(x) = i(\delta\alpha)J_\rho F(x) \tag{3.26}$$

due to an infinitely small rotation about the x_ρ axis through an angle $\delta\alpha$. The function $F(x)$ becomes

$$F'(x) = (1 + i\,\delta\alpha\,J_\rho)F(x). \tag{3.27}$$

A rotation through a finite angle α can be carried out by k consecutive rotations through an angle α/k. If k is a sufficiently large number, Eq. (3.27) holds for each rotation. Hence for a rotation through an angle α the function $F(x)$ becomes

$$F'(x) = \lim_{k\to\infty}\,[1 + i(\alpha/k)J_\rho]^k\,F(x) = e^{i\alpha J_\rho}\,F(x). \tag{3.28}$$

The obvious generalization of this equation to a rotation about an axis in the direction of an arbitrary unit vector $\mathbf{n}$ is

$$F'(x) = [\exp i\alpha(\mathbf{n}\cdot\mathbf{J})]F(x). \tag{3.29}$$

From this equation it follows that

$$R_{\mathbf{n},\alpha} = \exp i\alpha(\mathbf{n}\cdot\mathbf{J}) \tag{3.30}$$

can be regarded as an operator for a finite rotation about an axis $\mathbf{n}$ through an angle α. In order to find the effect of (3.30) upon a function of the coordinates, one must write the operator in the form of a power series and determine the effect of each term of the series upon the function.

A rotation through an angle α about an axis $\mathbf{n}$ can be transformed into a rotation through the same angle about some other axis $\mathbf{n}'$. For this purpose the axis $\mathbf{n}$ is brought into coincidence with $\mathbf{n}'$ by means of a rotation R. The rotation through the angle α is then carried out, and finally $\mathbf{n}$ is returned to its original position by the rotation R^{-1}, i.e.,

$$R_{\mathbf{n},\alpha} = R^{-1}R_{\mathbf{n}',\alpha}R. \tag{3.31}$$

All rotations through a given angle therefore belong to the same class of the group $\mathbf{R}_3$. Although in order to specify a particular rotation operation it is necessary to give its axis and angle of rotation, the class of $\mathbf{R}_3$ to which the rotation belongs is characterized by just the magnitude of the angle of rotation.

3.5. Irreducible Representations‡

It can be shown that a knowledge of the commutation relations for the infinitesimal operators of a continuous group is sufficient to determine all the irreducible representations of the group. The infinitesimal operators of the

‡The reader will find detailed treatments of the representations of the three-dimensional rotation group in the works of Helfand *et al.* (1963), Wigner (1959), and Rose (1957).

three-dimensional rotation group are just the angular momentum operators whose commutation relations are given by Eq. (3.25). In order to find the possible irreducible representations, we make use of the set of eigenfunctions of one of the operators J_p—for example, the eigenfunctions of J_z. We impose upon these eigenfunctions the requirement that they be simultaneously eigenfunctions of the operator for the square of the angular momentum

$$J^2 = J_x{}^2 + J_y{}^2 + J_z{}^2. \tag{3.32}$$

This requirement can always be satisfied, since the operators J_z and J^2 commute.‡ The eigenfunctions $\psi_m^{(j)}$ are therefore characterized by two indices the values of which determine the eigenvalues of J^2 and J_z. In addition, these indices, by means of the commutation relations (3.25), determine the effect of applying J_x, J_y, and J^2 to the functions $\psi_m^{(j)}$. Thus one finds that [see Landau and Lifshitz (1965) and Helfand *et al.* (1963)]

$$\begin{aligned} J^2\psi_m^{(j)} &= j(j+1)\psi_m^{(j)}, \\ J_z\psi_m^{(j)} &= m\psi_m^{(j)}, \\ (J_x + iJ_y)\psi_m^{(j)} &= [(j-m)(j+m+1)]^{1/2}\psi_{m+1}^{(j)}, \\ (J_x - iJ_y)\psi_m^{(j)} &= [(j+m)(j-m+1)]^{1/2}\psi_{m-1}^{(j)}, \end{aligned} \tag{3.33}$$

where j can assume integer and half-integer values only, and, for given j, m can assume the $2j+1$ values within the bounds $|m| \leq j$:

$$m = j, \quad j-1, \ldots, \quad -j. \tag{3.34}$$

It follows from Eqs. (3.33) that functions $\psi_m^{(j)}$ with fixed j transform into one another under infinitely small rotations. Since, according to Eq. (3.30), the operator for a finite rotation can be expressed in terms of angular momentum operators, the $\psi_m^{(j)}$ also transform among themselves under finite rotations, i.e.,

$$R_{\mathbf{n},\alpha}\psi_m^{(j)} = \sum_{m'} D_{m'm}^{(j)}(R_{\mathbf{n},\alpha})\psi_{m'}^{(j)}, \tag{3.35}$$

where the coefficients $D_{m'm}^{(j)}(R_{n,\alpha})$ form a matrix which corresponds to a rotation through an angle α about an axis $\mathbf{n}$. The $2j+1$ functions $\psi_m^{(j)}$ therefore form a $(2j+1)$-dimensional basis for a representation of the rotation group $\mathbf{R}_3$. It can be shown that this representation, which is usually denoted by $D^{(j)}$, is irreducible (Helfand *et al.*, 1963). Its matrix elements satisfy orthogonality relations which form a particular case of relations (3.7). The volume element for $\mathbf{R}_3$ can be written as $d\tau_R = \sin\beta \, d\beta \, d\alpha \, d\gamma$, where α, β, and γ are the Euler angles (Helfand *et al.*, 1963).

‡We recall that the necessary and sufficient condition for two operators to possess a common set of eigenfunctions is that the operators commute [see Landau and Lifshitz (1965)].

As a result, we obtain, instead of (3.7),

$$\int D^{(j)}_{m\mu}(R)^* D^{(j')}_{m'\mu'}(R)\, d\tau_R = [8\pi^2/(2j+1)]\delta_{jj'}\delta_{mm'}\delta_{\mu\mu'}. \tag{3.36}$$

An explicit form for the representation matrices $D^{(j)}(\alpha, \beta, \gamma)$, expressed in terms of the Euler angles has been obtained by Wigner (1959).

It is not difficult to find the character of the class which corresponds to a rotation through an angle α. Since the classes of $\mathbf{R}_3$ are typified by the angle of rotation only and are independent of the direction of the axis of rotation, we take the z axis to be the axis of rotation. The rotation operator (3.30) therefore assumes the form

$$R_{z,\alpha} = e^{i\alpha J_z}. \tag{3.37}$$

In order to find the effect of this operator on a function $\psi^{(j)}_m$, we expand the exponent and note that

$$J_z{}^k \psi^{(j)}_m = m^k \psi^{(j)}_m,$$

i.e., that in every term in the series the operator is replaced by its eigenvalue. We therefore obtain

$$R_{z,\alpha}\psi^{(j)}_m = e^{im\alpha}\psi^{(j)}_m. \tag{3.38}$$

The matrix of the rotation $R_{z,\alpha}$ is thus diagonal and its character in the representation $D^{(j)}$ is given by

$$\chi^{(j)}(\alpha) = \sum_{m=-j}^{j} e^{im\alpha} = \frac{e^{i(j+1)\alpha} - e^{-ij\alpha}}{e^{i\alpha} - 1}. \tag{3.39}$$

Upon dividing the numerator and denominator of (3.39) by $e^{i\alpha/2}$, we obtain an expression which is more convenient in computations,

$$\chi^{(j)}(\alpha) = \frac{\sin(j + 1/2)\alpha}{\sin(\alpha/2)}. \tag{3.40}$$

From this formula it follows that

$$\chi^{(j)}(-\alpha) = \chi^{(j)}(\alpha).$$

Two rotations through identical angles but in opposite directions thus belong to the same class; i.e., a class is characterized by the absolute value of the angle of rotation.

The set of rotations about a fixed axis forms the two-dimensional rotation group $\mathbf{R}_2$. This group is Abelian, and all of its irreducible representations are one dimensional. It follows from Eq. (3.38) that the $2j+1$ basis functions for an irreducible representation $D^{(j)}$ of the group $\mathbf{R}_3$ belong to different irreducible representations of $\mathbf{R}_2$, each of which is characterized by a particular value of the number m.

According to Eq. (3.38), when j, and consequently m, is half-integral, a basis function $\psi^{(j)}_m$ changes sign on rotation by 2π. The identity transformation corresponds to a rotation by 4π. As a result, two matrices $D^{(j)}(R_{n,\alpha})$ which differ from one another in sign correspond to each rotation $R_{n,\alpha}$. In order to obtain well-defined representations, we introduce a group known as the *double group* of three-dimensional rotations (Bethe, 1939) in which a rotation by 2π is regarded as an element that is distinct from the identity. Rotations by α ($\alpha \leq 2\pi$) and $2\pi + \alpha$ are regarded as different elements, as a result of which the volume of the double group is twice that of the ordinary group $\mathbf{R}_3$.

When j is integral, $j = l$, the explicit form of the basis functions can be found by solving the first two equations of (3.33) directly. However, it is first of all necessary to replace J^2 and J_z by their expressions in spherical polar coordinates. The solutions of the resulting equations are the spherical harmonics $Y_{lm}(\theta, \phi)$, which are very well known in mathematical physics. It can be shown (Helfand *et al.*, 1963) that the matrix elements of the representations $D^{(j)}$ for integral j are proportional to the spherical harmonics:

$$D^{(l)}_{m'0}(\alpha, \beta, 0) = \left(\frac{4\pi}{2l+1}\right)^{1/2} Y_{lm'}(\beta, \alpha),$$

$$D^{(l)}_{0m}(0, \beta, \gamma) = (-1)^m \left(\frac{4\pi}{2l+1}\right)^{1/2} Y_{lm}(\beta, \gamma),$$

where α, β, and γ are the Euler angles. The matrix elements $D^{(j)}_{m'm}(\alpha, \beta, \gamma)$ are therefore known as *generalized spherical harmonics* of the jth order.

The irreducible representations of the three-dimensional rotation group are thus characterized by a number j which can be either integral or half-integral. The dimension of an irreducible representation $D^{(j)}$ is $2j + 1$. The basis functions $\psi^{(j)}_m$ are characterized by two indices, the index j denoting the particular irreducible representation $D^{(j)}$ and the index m enumerating the basis functions within the representation. On the other hand, the functions $\psi^{(j)}_m$ are eigenfunctions of the operators for the square of the angular momentum and the projection of it upon the z axis. The indices j and m thus have an additional interpretation: Index j determines the eigenvalue of the square of the angular momentum, which is equal to $j(j + 1)$, and m the projection of the angular momentum upon the z axis.‡

3.6. Reduction of the Direct Product of Two Irreducible Representations

Let us construct the direct product of two irreducible representations, $D^{(j_1)} \times D^{(j_2)}$. The representation which is obtained is of dimension $(2j_1 + 1)(2j_2 + 1)$ and is reducible. Products of the basis functions for the individual representations, $\psi^{(j_1)}_{m_1}\psi^{(j_2)}_{m_2}$, form a basis for this representation.

‡One usually says that the magnitude of the angular momentum vector $\mathbf{J}$ is j, meaning that the maximum value of the projection of the angular momentum is equal to j.

Since each of the functions $\psi^{(j_i)}_{m_i}$ is an eigenfunction of the operator J_{iz}, it is easy to see that a product of two such functions is an eigenfunction of the operator

$$J_z = J_{1z} + J_{2z} \tag{3.41}$$

with eigenvalue

$$m = m_1 + m_2. \tag{3.42}$$

However, the products $\psi^{(j_1)}_{m_1}\psi^{(j_2)}_{m_2}$ are not eigenfunctions of $J^2 = (J_1 + J_2)^2$. By forming linear combinations that are eigenfunctions of this operator, we carry out the reduction of the direct product into irreducible representations $D^{(j)}$. In order to find which irreducible representations $D^{(j)}$ can occur in the decomposition of this representation, we write out, using rule (3.42), the $(2j_1 + 1)(2j_2 + 1)$ eigenvalues m of the projection of the angular momentum. These eigenvalues are then divided into sets of $2j + 1$ values of the form (3.34), each set corresponding to an irreducible representation $D^{(j)}$. We obtain as a result the desired decomposition‡

$$D^{(j_1)} \times D^{(j_2)} \doteq D^{(j_1+j_2)} + D^{(j_1+j_2-1)} + \cdots D^{(|j_1-j_2|)}. \tag{3.43}$$

Each irreducible representation $D^{(j)}$ occurs once in the decomposition, the values of j lying in the interval

$$|j_1 - j_2| \leq j \leq j_1 + j_2, \qquad j_1 + j_2 + j \text{ integral.} \tag{3.44}$$

If one constructs a triangle with integral perimeter length and sides of lengths j_1, j_2, and j, the condition which the sides must obey is just Eq. (3.44). Equation (3.44) is called the triangle rule with integral perimeter, and is denoted by $\Delta(j_1 j_2 j)$.

The linear transformation from the basis functions of the reducible representation to basis functions for the irreducible representations which occur in the decomposition (3.43) is carried out by means of the matrix of Clebsch–Gordan coefficients (see Section 1.17).

$$\psi^{(j)}_m = \sum_{m_1, m_2} \psi^{(j_1)}_{m_1}\psi^{(j_2)}_{m_2}\langle j_1 m_1, j_2 m_2 | jm\rangle. \tag{3.45}$$

The Clebsch–Gordan coefficients $\langle j_1 m_1, j_2 m_2 | jm\rangle$ are nonzero only if conditions (3.42) and (3.44) are fulfilled. The summation over m_2 in Eq. (3.45) is therefore purely formal, since $m_2 = m - m_1$. According to Eq. (1.78), the $\langle j_1 m_1, j_2 m_2 | jm\rangle$ coefficients satisfy the orthogonality relations

$$\begin{aligned} &\sum_{j,m} \langle j_1 m_1, j_2 m_2 | jm\rangle\langle j_1 m_1', j_2 m_2' | jm\rangle = \delta_{m_1 m_1'}\delta_{m_2 m_2'}, \\ &\sum_{m_1, m_2} \langle j_1 m_1, j_2 m_2 | jm\rangle\langle j_1 m_1, j_2 m_2 | j'm'\rangle = \delta_{jj'}\delta_{mm'}. \end{aligned} \tag{3.46}$$

‡The decomposition of the direct product into irreducible components, Eq. (3.43), is equivalent to finding the possible values j that the total angular momentum vector can assume when the angular momentum vectors of the individual wave functions are coupled together. The decomposition (3.43) is therefore known in quantum mechanics as the rule for coupling angular momenta.

A number of different explicit expressions for the Clebsch–Gordan coefficients in terms of the parameters j_1, j_2, j, m_1, and m_2 which occur in them may be found in the works of Edmonds (1957), Yutsis *et al.* (1962a), and Yutsis and Bandzaitis (1965). References to available tables can also be found in these books. From the explicit expressions for the Clebsch–Gordan coefficients we obtain the following relations:

$$\langle j_1m_1, j_2m_2 | jm\rangle = \begin{cases} (-1)^{j_1+j_2-j}\langle j_1 - m_1, j_2 - m_2 | j - m\rangle, & \text{(3.47a)} \\ (-1)^{j_1+j_2-j}\langle j_2m_2, j_1m_1 | jm\rangle, & \text{(3.47b)} \\ (-1)^{j_1-m_1}\left(\dfrac{2j+1}{2j_2+1}\right)^{1/2}\langle j_1m_1, j - m | j_2 - m_2\rangle. & \text{(3.47c)} \end{cases}$$

If two identical angular momenta are coupled to give a resultant angular momentum of zero, the Clebsch–Gordan coefficients differ only in phase and are given by

$$\langle jm, j - m | 00\rangle = (-1)^{j-m}(2j + 1)^{-1/2}. \tag{3.48}$$

We note that when constructing a function of the form (3.45) the ordering of the functions being coupled is important. If the angular momenta are equal, $j_1 = j_2 = j$, then according to Eq. (3.47b), the function (3.45) becomes multiplied by a phase factor when the positions of $\psi_{m_1}^{(j)}$ and $\psi_{m_2}^{(j)}$ are permuted,

$$P_{12}\psi_M^{(J)}[j(1)j(2)] = \psi_M^{(J)}[j(2)j(1)] = (-1)^{2j-J}\psi_M^{(J)}[j(1)j(2)]. \tag{3.49}$$

Instead of the Clebsch–Gordan coefficients, one often make use of more symmetric coefficients known as 3-j symbols. The 3-j symbols are written in the form of two-rowed matrices,

$$\begin{pmatrix} j_1 & j_2 & j_3 \\ m_1 & m_2 & m_3 \end{pmatrix}.$$

They are related to the Clebsch–Gordan coefficients by the equation

$$\langle j_1m_1, j_2m_2 | jm\rangle = (-1)^{-j_1+j_2-m}(2j + 1)^{1/2}\begin{pmatrix} j_1 & j_2 & j \\ m_1 & m_2 & -m \end{pmatrix}. \tag{3.50}$$

The 3-j symbols are invariant under any even permutation of their columns, and are multiplied by $(-1)^{j_1+j_2+j_3}$ under any odd permutation of their columns.

3.7. Reduction of the Direct Product of k Irreducible Representations. $3n$-j Symbols

The decomposition of the direct product of k irreducible representations

$$D^{(j_1)} \times D^{(j_2)} \times \cdots \times D^{(j_k)} \tag{3.51}$$

into irreducible components is carried out by successively reducing the direct products of pairs of irreducible representations. A reduction such as this can be effected in several different ways. In cases where a precise specification is unnecessary, we shall denote the reduction scheme by a capital letter such as $A, B, \ldots$, etc. Thus for $k = 3$ there are two different reduction schemes:

(a) The direct product of the representations $D^{(j_1)} \times D^{(j_2)}$ is reduced first, followed by the reduction of the direct product of the representations that are obtained in this way, $D^{(j_{12})}$, with the remaining representation $D^{(j_3)}$.

(b) The direct product of the last two representations $D^{(j_2)} \times D^{(j_3)}$ is reduced first, following which the direct products of $D^{(j_1)}$ with the representations obtained in the first reduction, $D^{(j_{23})}$, are reduced.

The order of reduction is conveniently denoted by enclosing the reducible representations in brackets. The two reduction schemes discussed here can then be written in the form

$$\text{(a)} \quad (D^{(j_1)} \times D^{(j_2)}) \times D^{(j_3)}, \qquad \text{(b)} \quad D^{(j_1)} \times (D^{(j_2)} \times D^{(j_3)}). \tag{3.52}$$

The matrix which reduces the direct product (3.51) is a generalization of the matrix of Clebsch–Gordan coefficients. Since the reduction process consists of successively reducing pairs of representations, the coefficients in the linear transformation from the basis functions for the initial, reducible representation to basis functions $\psi_m^{(j)}$ for the resulting, irreducible representations can be expressed as products of the corresponding Clebsch–Gordan coefficients. The reduction process involves $k - 2$ intermediate irreducible representations $D^{(\text{int})}$ and therefore the basis functions $\psi_m^{(j)}$ are characterized by $k - 2$ additional quantum numbers j_{int}. These quantum numbers can be regarded as the eigenvalues of $k - 2$ intermediate angular momentum operators J_{int} which occur in the vector coupling of k angular momenta $\mathbf{J}_1$, $\mathbf{J}_2, \ldots, \mathbf{J}_k$. The representations $D^{(j)}$ which are obtained on reducing the direct product (3.51) are independent of the reduction scheme. The form of the basis functions $\psi_m^{(j)}$, however, does depend upon the reduction scheme, since the set of Clebsch–Gordan coefficients which expresses the $\psi_m^{(j)}$ in terms of the basis functions of the direct product (3.51) depends upon the particular reduction scheme. The basis functions which are obtained from different schemes are usually referred to in quantum mechanics as wavefunctions for states with different *coupling schemes for the angular momenta.* We denote the coupling scheme for the angular momenta by capital letters $A, B, \ldots$, etc., the same as for the reduction scheme. A basis function for a representation $D^{(j)}$ with scheme A for coupling the angular momenta is written

$$\psi_m^{(j)}((j_1 \cdots j_k)(j_{\text{int}})^A), \tag{3.53}$$

in which $(j_{\text{int}})^A$ denotes the set of $k - 2$ intermediate angular momenta which occur in the given scheme for coupling the angular momenta $\mathbf{J}_1$, $\mathbf{J}_2, \ldots, \mathbf{J}_k$.

Two sets of basis functions for a representation $D^{(j)}$ which differ in the coupling scheme for the angular momenta can be transformed into each other by means of some orthogonal transformation,

$$\psi_m^{(j)}((j_1 \cdots j_k)(j_{\text{int}})^A) = \sum_{(j_{\text{int}})^B} \psi_m^{(j)}((j_1 \cdots j_k)(j_{\text{int}})^B)\langle (j_{\text{int}})^B | (j_{\text{int}})^A \rangle^{(j)}. \quad (3.54)$$

The coefficients in this transformation cannot depend on the index of the basis function, i.e., upon m, and due to the orthonormality of the basis functions, are given by the following integral in configuration space:

$$\langle (j_{\text{int}})^B | (j_{\text{int}})^A \rangle^{(j)} = \int \psi_m^{(j)}((j_1 \cdots j_k)(j_{\text{int}})^B)^* \psi_m^{(j)}((j_1 \cdots j_k)(j_{\text{int}})^A)\, dV. \quad (3.55)$$

We shall refer to the matrix for the transformation (3.54) as the *transformation matrix for the three-dimensional rotation group*. This matrix is obviously diagonal in j. We now seek an explicit form of it for the case of three coupled angular momenta.

There are two possible schemes for coupling three angular momenta, corresponding to the two reduction schemes (3.52). The basis functions which correspond to these coupling schemes are found by the successive application of formula (3.45):

$$\begin{aligned}\psi_m^{(j)}((j_1 j_2) j_{12} j_3) &= \sum_{m_{12}, m_3} \psi_{m_{12}}^{(j_{12})} \psi_{m_3}^{(j_3)} \langle j_{12} m_{12}, j_3 m_3 | jm \rangle \\ &= \sum_{\substack{m_1, m_2, \\ m_3, m_{12}}} \psi_{m_1}^{(j_1)} \psi_{m_2}^{(j_2)} \psi_{m_3}^{(j_3)} \langle j_1 m_1, j_2 m_2 | j_{12} m_{12} \rangle \langle j_{12} m_{12}, j_3 m_3 | jm \rangle. \end{aligned} \quad (3.56)$$

Similarly,

$$\begin{aligned}&\psi_m^{(j)}(j_1 (j_2 j_3) j_{23}) \\ &= \sum_{\substack{m_1, m_2, \\ m_3, m_{23}}} \psi_{m_1}^{(j_1)} \psi_{m_2}^{(j_2)} \psi_{m_3}^{(j_3)} \langle j_1 m_1, j_{23} m_{23} | jm \rangle \langle j_2 m_2, j_3 m_3 | j_{23} m_3 \rangle. \end{aligned} \quad (3.57)$$

We find the transformation matrix which connects the basis functions (3.56) and (3.57) from formula (3.55). Making use of the orthonormality of the $\psi_{m_i}^{(j_i)}$ and of the reality of the Clebsch–Gordan coefficients, we obtain

$$\begin{aligned}\langle (j_1 j_2) j_{12} j_3 | j_1 (j_2 j_3) j_{23} \rangle^{(j)} = &\sum_{\substack{m_1, m_2, m_3, \\ m_{12}, m_{23}}} \langle j_1 m_1, j_2 m_2 | j_{12} m_{12} \rangle \\ &\times \langle j_{12} m_{12}, j_3 m_3 | jm \rangle \langle j_2 m_2, j_3 m_3 | j_{23} m_{23} \rangle \\ &\times \langle j_1 m_1, j_{23} m_{23} | jm \rangle. \end{aligned} \quad (3.58)$$

The matrix elements (3.58) do not depend upon the m_i, and are functions of six variables. They are related to coefficients known as *Racah W coefficients*‡ by the following equation:

$$\begin{aligned}&\langle (j_1 j_2) j_{12} j_3 | j_1 (j_2 j_3) j_{23} \rangle^{(j)} \\ &\quad = [(2j_{12} + 1)(2j_{23} + 1)]^{1/2} W(j_1 j_2 j j_3; j_{12} j_{23}). \end{aligned} \quad (3.59)$$

‡These coefficients were introduced by Racah in his classic papers on the theory of complex atomic spectra (Racah, 1942).

In a similar way it can be shown that when the order in which the angular momenta are coupled is changed, the elements of the transformation matrix can be expressed in terms of Racah coefficients. Thus if we interchange angular momenta j_2 and j_3, we have (Racah, 1943)

$$\langle (j_1 j_2) j_{12} j_3 | (j_1 j_3) j_{13} j_2 \rangle^{(j)} = [(2j_{12}+1)(2j_{13}+1)]^{1/2} W(j_{12} j_3 j_2 j_{13}; j j_1). \tag{3.60}$$

From (3.58) it follows that the $W(j_1 j_2 j j_3; j_{12} j_{23})$ coefficients are nonzero only if the four triangle conditions

$$\Delta(j_1 j_2 j_{12}), \qquad \Delta(j_2 j_3 j_{23}), \qquad \Delta(j_{12} j_3 j), \qquad \Delta(j_1 j_{23} j)$$

are satisfied.

Instead of the Racah coefficients, one often uses more symmetric expressions known as 6-j symbols which are written in the form of a two-rowed matrix,

$$\begin{Bmatrix} j_1 & j_2 & j_{12} \\ j_3 & j & j_{23} \end{Bmatrix} = (-1)^{j_1+j_2+j_3+j} W(j_1 j_2 j j_3; j_{12} j_{23}). \tag{3.61}$$

The 6-j symbols are invariant under any permutation of their columns, and under permutations of the upper and lower arguments in any pair of columns. These correspond to the following symmetry relations among the Racah coefficients:

$$W(abcd; ef) = W(badc; ef) = W(cdab; ef) = W(acbd; ef) = (-1)^{e+f-a-d} W(ebcf; ad) = (-1)^{e+f-b-c} W(aefd; bc). \tag{3.62}$$

If one of the arguments is equal to zero, the Racah coefficients assume a very simple form. Thus

$$W(abcd; e0) = (-1)^{a+b-e} \delta_{ac} \delta_{bd} / [(2a+1)(2b+1)]^{1/2}. \tag{3.63}$$

The Racah coefficients and the associated 6-j symbols have found a wide use in a variety of problems in physics [see, for example, Rose (1957), Edmonds (1957), Fano and Racah (1959), Judd (1963), and Sobel'man (1972)]. Tables of Racah coefficients and 6-j symbols have been published which cover a fairly wide range of arguments [see, for example, Rotenberg *et al.* (1959) and Ishidzu *et al.* (1960)].

The 6-j symbols occur in the coupling of three angular momenta. Similarly, 9-j symbols (Wigner, 1965) make their appearance in problems which involve four angular momenta. These can be expressed as a sum of products of three 6-j symbols. In the general case in which $n+1$ angular momenta are coupled the transformation matrix can be expressed as a 3n-j symbol. [A more detailed account of 3n-j symbols is given by Yutsis *et al.* (1962a) and Yutsis and Bandzaitis (1965).]

PART 3. Point Groups

3.8. Symmetry Elements and Symmetry Operations

According to our classification of groups of linear transformations in a vector space (Section 3.2), every subgroup of the group of orthogonal transformations is called a point group. The group of rotations in three-dimensional space which was considered in the previous section is therefore a point group. These groups are referred to as "point" groups because the transformations which occur in them leave fixed at least one point of the space (the origin of coordinates). Now, all orthogonal transformations reduce to combinations of two types of transformation: to rotations, and to reflections in planes which pass through the origin of coordinates. Hence the point in the space corresponding to the origin of coordinates remains fixed under these transformations.

A body is said to possess the symmetry of some point group if it is sent into itself by the transformations of this group. In such a case the transformations of the point group coincide with the symmetry transformations of the body. A necessary condition for the body to be symmetric is that it possess axes and planes such that transformations with respect to these send the body into itself. Such axes and planes are usually referred to as *symmetry elements.* Each symmetry element gives rise to corresponding symmetry transformations which are commonly called *symmetry operations.*

If the body coincides with itself n times under a rotation through an angle 2π about some axis, the axis is called an *n-fold (or nth-order) axis of symmetry* and is denoted by C_n (Fig. 3.1a). In this case the smallest angle of rotation under which the body can be made to coincide with itself is equal to $2\pi/n$. This rotation operation is denoted by the symbol C_n, the same as for the axis of rotation. A succession of k rotations through $2\pi/n$, i.e., a rotation through $2\pi k/n$, is denoted by C_n^k. The operation C_n when carried out n times is obviously equivalent to the identity transformation. This last is usually denoted by the symbol E, i.e.,

$$C_n^n = E.$$

The other possible symmetry element is a *plane of symmetry*, and the corresponding symmetry operation is a reflection of the body in the plane. The equilateral triangle with small tags at its vertices shown in Fig. 3.1a possesses no planes of symmetry apart from the trivial plane of the triangle. The dumbbell in Fig. 3.1b is obviously symmetric with respect to a reflection in the plane perpendicular to its connecting rod and which passes through its midpoint. A plane of symmetry is denoted in general by the symbol σ. A plane which is perpendicular to an axis C_n is denoted by σ_h, and a plane passing through an axis C_n is denoted by σ_v (or by σ_d; see subsequent discus-

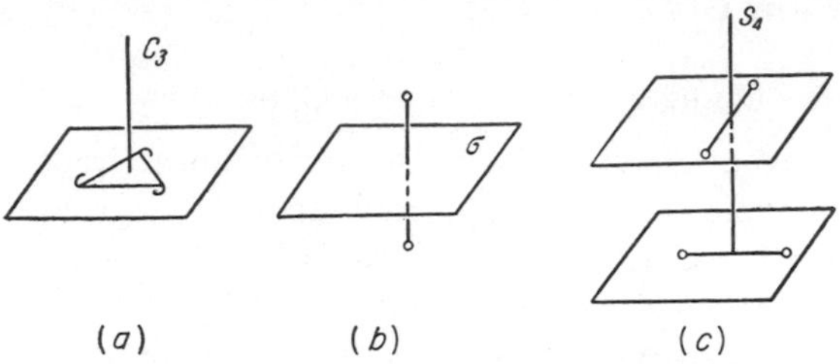

Figure 3.1.

sion). This notation is also used for the corresponding symmetry operations. Clearly,

$$\sigma^2 = E.$$

It may happen that a body can only be made to coincide with itself after the application of two symmetry operations in succession: a rotation through an angle $2\pi/n$ and a reflection in a plane perpendicular to the axis of rotation. This symmetry operation is called a *rotation–reflection*, and is denoted by the symbol S_n. The order in which the rotation and reflection operations are carried out is unimportant:

$$S_n = \sigma_h C_n = C_n \sigma_h. \tag{3.64}$$

In Fig. 3.1c, two dumbbells are shown in a configuration where they are fixed at 90° with respect to each other. This configuration possesses a fourfold rotation–reflection axis.

A rotation–reflection axis only constitutes a new symmetry element if n is even. If n is odd, it follows from (3.64) that

$$S_{2m+1}^2 = C_{2m+1}^2, \qquad S_{2m+1}^{2m} = C_{2m+1}^{2m}, \qquad S_{2m+1}^{2m+1} = \sigma_h, \tag{3.65}$$

i.e., there is an independent $(2m + 1)$-fold axis of symmetry, and an independent plane of symmetry perpendicular to it. We note that if n is even, a rotation–reflection axis S_n is simultaneously a $C_{n/2}$-fold rotation axis.

A twofold rotation–reflection axis is equivalent to the presence of a *center of symmetry* in the body. This lies at the point where the axis S_2 and the plane σ_h intersect. The operation S_2 constitutes an *inversion operation* through the center of symmetry, and is denoted by the symbol I:

$$I \equiv S_2 = C_2 \sigma_h. \tag{3.66}$$

Symmetry operations usually do not commute. The following operations only are an exception to this:

(a) Rotations about a given axis.

(b) Rotations through π about mutually perpendicular axes (the result is equivalent to a rotation by π about an axis perpendicular to the first two).

(c) A rotation, and a reflection in a plane perpendicular to the axis of rotation [see Eq. (3.64)].

(d) Reflections in mutually perpendicular planes (the resulting operation is equivalent to a rotation by π about the axis formed by the line of intersection of the two planes).

From (a) and (b) it follows that the operation of inversion commutes with any symmetry operation.

The set of symmetry operations for a given body forms the point group of the body. In the next section we consider the symmetry point groups of molecules. The symmetry elements possessed by a molecule determine the particular point group to which it belongs. The point groups are therefore conveniently classified by the symmetry elements that generate the operations of a given group. The following rules, the proofs of which are similar to the proof of Eq. (3.31), are useful guides in sorting the operations of a group into classes.

(a) Two rotations through equal angles about different axes belong to the same class if there is an operation in the group which brings the two axes into coincidence.

(b) Two rotations about a single axis through equal angles but in opposite directions belong to the same class either if there is a rotation operation in the group which reverses the direction of the axis, or if there is a reflection in a plane passing through the axis (the axis is then said to be *bilateral*).

(c) Two reflections in different planes belong to the same class if there is an operation in the group which transforms one of the planes into the other.

3.9. Classification of Point Groups

There are several notations for point groups. We shall adhere to the notation due to Schönflies, which is adopted, in particular, by Landau and Lifschitz (1965), and Eyring *et al.* (1944). We begin by classifying the simplest groups, and proceed from these to groups with additional symmetry elements. Finally, we consider continuous point groups, picking out those to which it is possible for molecules to belong.

A. *Discrete Axial Point Groups*

These groups possess a single axis of symmetry, whose order is greater than twofold.‡

$\mathbf{C}_n$ Groups. These possess a single symmetry element, an n-fold axis of symmetry. A $\mathbf{C}_n$ group consists of n operations, these being rotations through

‡The subscript n in the notation for the point groups $\mathbf{C}_n$, $\mathbf{D}_n$, etc , denotes the order of the main symmetry axis. This should not be confused with the notation used for the linear groups $\mathbf{GL}_n$, $\mathbf{U}_n$, etc , where the subscript n denotes the dimension of the vector space of the particular group.

angles $2\pi k/n$ about the axis of symmetry. All the operations of the group commute, i.e., the group is Abelian. Every operation of the group therefore forms a class, and the irreducible representations are all one dimensional.

A necessary condition for a molecule to belong to a group C_n is the absence of a plane of symmetry. As an example, we cite the dichloro-substituted allene molecule (Fig. 3.2a). The molecule $H_2C{=}CCl_2$ belongs to the group C_2 and the molecule $H_3C{-}CCl_3$ to the group C_3 if the terminal groups of atoms are slightly rotated with respect to each other (the equilibrium configurations of both molecules in fact possess higher symmetries).

S_{2n} Groups. These possess a single rotation–reflection axis S_{2n}. An S_{2n} group is Abelian, and consists of $2n$ rotation–reflection operations S_{2n}^k. As was shown in the previous section, a rotation–reflection axis can only be of even order, since for an odd order the axis reduces to an ordinary symmetry axis of odd order and to a plane perpendicular to it. The operation S_2 is the

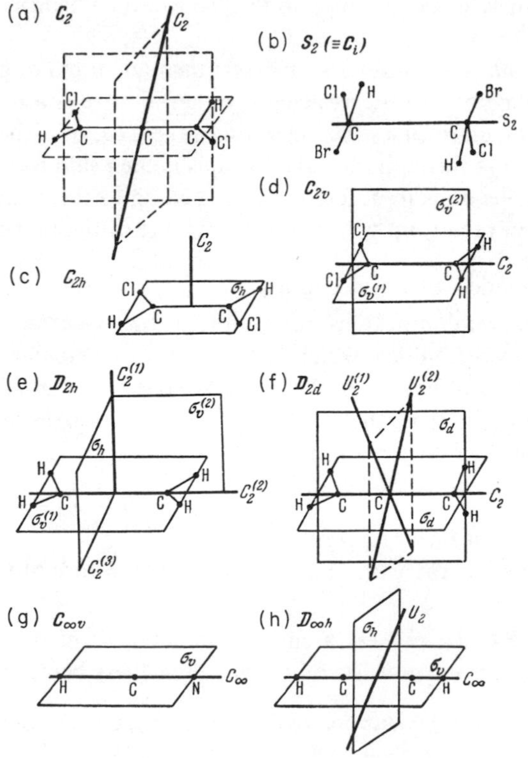

Figure 3.2. Axial point groups.

inversion operation, so that the group $\mathbf{S}_2$ consists of just two elements, E and I, and is often denoted by the symbol $\mathbf{C}_i$. The trans form of the ClBrHC—CHBrCl molecule (Fig. 3.2b) serves as an example of a molecule which belongs to $\mathbf{S}_2$.

$\mathbf{C}_{nh}$ Groups. These possess an n-fold symmetry axis and a plane of symmetry perpendicular to it, σ_h. A C_{nh} group consists of $2n$ operations, n rotations C_n^k, and n rotation–reflection operations $C_n^k\sigma_h$. All the operations of the group commute, and there are therefore $2n$ classes. The group C_{1h} contains just two elements, E and σ_h; this group is usually denoted by $\mathbf{C}_s$. A group $\mathbf{C}_{nh}$ can obviously be written as a direct product of the groups $\mathbf{C}_n$ and $\mathbf{C}_s$, i.e., $\mathbf{C}_{nh} = \mathbf{C}_n \times \mathbf{C}_s$. All groups $\mathbf{C}_{2mh}$ possess a center of symmetry. The planar trans isomers of the molecules $C_2H_2Cl_2$ (Fig. 3.2c) and $C_6H_2Cl_2Br_2$ are examples of molecules which belong to the group $\mathbf{C}_{2h}$.

$\mathbf{C}_{nv}$ Groups. These possess an n-fold symmetry axis and n planes of symmetry which pass through the axis. A $\mathbf{C}_{nv}$ group consists of $2n$ elements, n rotations C_n^k and n reflections σ_v. The rotation and reflection operators do not commute with one another. The symmetry axis is bilateral due to the presence of planes σ_v which pass through the axis, i.e., the rotations all belong to the same class. The n reflections all belong to the same class only if n is odd. If $n = 2m$, the reflection operations fall into two classes with m elements in each. This is because rotations about an axis only bring alternating planes into coincidence with each other; adjacent planes cannot be made to coincide.

Many molecules belong to the $\mathbf{C}_{nv}$ point group. Thus H_2O, H_2S, SO_2, NO_2, *cis*-$C_2H_2Cl_2$ (Fig. 3.2d), H_2CO, etc. belong to $\mathbf{C}_{2v}$, and NH_3, PCl_3, CH_3Cl *etc.* belong to $\mathbf{C}_{3v}$.

$\mathbf{D}_n$ Groups. These possess an n-fold symmetry axis, and n twofold symmetry axes perpendicular to it, the twofold axes intersecting at an angle of π/n. A $\mathbf{D}_n$ group contains $2n$ operations: n rotations C_n^k about the vertical axis and n rotations through π about the horizontal axes, these usually being denoted by U_2. The operations of the group are divided into classes in the same way as for $\mathbf{C}_{nv}$ (the groups $\mathbf{D}_n$ and $\mathbf{C}_{nv}$ are isomorphic). The group $\mathbf{D}_2$ possesses three mutually perpendicular twofold axes, and is denoted by the letter $\mathbf{V}$.

Molecules whose equilibrium configuration is characterized by the symmetry axes of the $\mathbf{D}_n$ group usually possess additional planes of symmetry, and so belong to point groups of higher symmetry than $\mathbf{D}_n$.

$\mathbf{D}_{nh}$ Groups. The axis system of the $\mathbf{D}_n$ group has added to it a horizontal plane of symmetry which passes through the n axes U_2. This leads to the appearance of n vertical planes of symmetry which pass through the C_n axis. *A* $\mathbf{D}_{nh}$ group contains $4n$ operations: n reflections σ_v and n rotation–reflection operations $C_n^k\sigma_h$ added to the $2n$ operations of $\mathbf{D}_n$. The group $\mathbf{D}_{nh}$ can be written as a direct product $\mathbf{D}_n \times \mathbf{C}_s$, since the reflection σ_h commutes with all operations of $\mathbf{D}_n$.

The ethylene molecule in its equilibrium configuration (Fig. 3.2e) serves as an example of a molecule which belongs to the group $\mathbf{D}_{2h}$. All planar XY_3 molecules belong to $\mathbf{D}_{3h}$, e.g., BF_3. In addition, cyclopropane, ethane (eclipsed conformation), 1,3,5-trichlorobenzene, etc., also belong to $\mathbf{D}_{3h}$. The symmetry group of the benzene molecule is $\mathbf{D}_{6h}$.

$\mathbf{D}_{nd}$ Groups. The axis system of the $\mathbf{D}_n$ group has added to it n planes of symmetry which pass through the C_n axis and which bisect the angle between two adjacent twofold axes. A $\mathbf{D}_{nd}$ group contains $4n$ operations: $2n$ operations of $\mathbf{D}_n$, n reflections σ_d, and n rotation–reflection operations $\sigma_d U_2$. The rotation–reflection operations $\sigma_d U_2$ are equivalent to a rotation about the C_n axis followed by a reflection in the plane perpendicular to C_n. The C_n axis consequently becomes a rotation–reflection axis of twice the order, S_{2n}. If n is odd, $\mathbf{D}_{2m+1,d} = \mathbf{D}_{2m+1} \times \mathbf{C}_i$.

The allene molecule, in which the planes of the two CH_2 groups are mutually perpendicular (Fig. 3.2f), forms an example of a molecule possessing the symmetry of $\mathbf{D}_{2d}$. The ethane molecule in its straggered conformation (the angle between the CH_3 groups is 60°) belongs to $\mathbf{D}_{3d}$.

B. *Cubic Point Groups*

Point groups of this kind possess several symmetry axes, the orders of which are greater than twofold.

The Group T. If the group $\mathbf{D}_2 \equiv \mathbf{V}$ has added to it four oblique threefold axes, such that rotations about them carry the twofold axes into each other, we obtain the axis system of a tetrahedron. This system of axes is conveniently represented by drawing the threefold axes as the spatial diagonals of a cube and the twofold axes as passing through the centers of opposite faces of the cube. (Fig. 3.3a). Rotations about these axes constitute a group which we denote by the symbol **T**. The group **T** consists of 12 operations, which are divided into four classes: E, three rotations C_2, four rotations C_3, and four rotations $C_3{}^2$.

There is no molecule whose equilibrium configuration possesses the symmetry of the point group **T**. However, if the methyl groups in the tetramethylmethane molecule $C(CH_3)_4$ (which belongs to a group with higher symmetry, $\mathbf{T}_d$) are slightly rotated, the nonequilibrium configuration which is obtained possesses the symmetry of **T**.

The Group $\mathbf{T}_d$. The axis system of the group **T** has added to it six planes of symmetry, each of which passes through two of the threefold axes and through one of the twofold axes. The twofold axes thereby become fourfold rotation–reflection axes. The 24 operations of the group are divided into five classes:

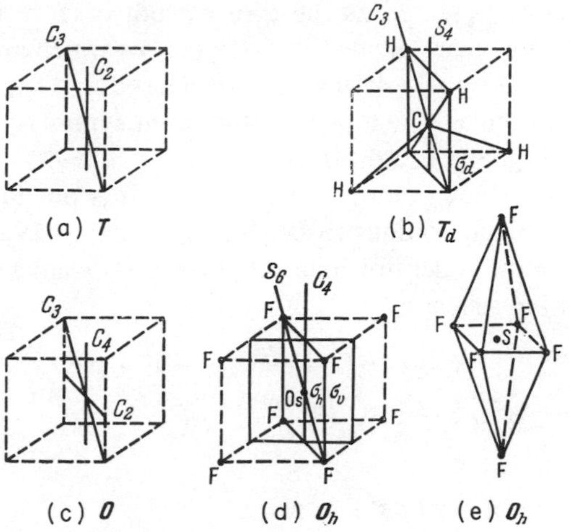

Figure 3.3. Cubic point groups.

eight rotations C_3 and $C_3{}^2$; six reflections σ_d; six rotation–reflection operations S_4; $S_4{}^3$; and three rotations $C_2 \equiv S_4{}^2$.

All molecules which possess the symmetry of a tetrahedron belong to the group $\mathbf{T}_d$, e.g., CH_4 (Fig. 3.3b), CCl_4, P_4, etc.

The Group $\mathbf{T}_h$. The axis system of the group **T** has added to it a center of symmetry. Three mutually perpendicular planes automatically appear as result, each of which passes through two of the twofold axes. Obviously, $\mathbf{T}_h = \mathbf{T} \times \mathbf{C}_i$. The point group $\mathbf{T}_h$ is apparently not realized as the symmetry group of a molecule.

The Group O. The symmetry axes of a cube form the symmetry elements of this group: three fourfold axes and four threefold axes (Fig. 3.3c). The 24 operations of the group are divided into five classes: E, eight rotations C_3 and $C_3{}^2$, six rotations C_4 and $C_4{}^3$, three rotations $C_4{}^2$ about the fourfold axes, and six rotations C_2 about the twofold axes.

Molecules which possess the axis system of the **O** group as a rule have additional planes of symmetry, and so belong to a group with higher symmetry, $\mathbf{O}_h$.

The Group $\mathbf{O}_h$. The symmetry elements of this group are all the symmetry elements of a cube or of a regular octahedron. The symmetry axes of the group **O** have added to them six planes of symmetry σ_h passing through opposite edges of the cube and three planes of symmetry σ_v which pass through the center of the cube, parallel to its faces. The C_3 axes thereby become rotation–reflection axes S_6. The group $\mathbf{O}_h$ contains a center of symmetry and can be represented as the direct product $\mathbf{O}_h = \mathbf{O} \times \mathbf{C}_i$. The 48 operations of the group are divided into ten classes which are obtained from those of the group $\mathbf{O}$ by calculating the direct product $\mathbf{O} \times \mathbf{C}_i$.

Examples of molecules which possess the point symmetry of $\mathbf{O}_h$ are OsF_8 (Fig. 3.3d), SF_6 (Fig. 3.3e), and UF_6.

This exhausts the finite point groups to which it is possible for molecules with rigid nuclear configurations to belong. By placing the groups which we have considered in the order in which new symmetry elements are added, we obtain the following scheme:

$$
\begin{array}{l}
\mathbf{C}_n \longrightarrow \mathbf{D} \longrightarrow (\mathbf{V}) \longrightarrow \mathbf{T} \longrightarrow \mathbf{O}\\
\swarrow\ \downarrow\ \searrow \quad \swarrow\ \searrow \qquad\qquad \swarrow\ \searrow \quad \downarrow\\
\mathbf{S}_{2n}\ \mathbf{C}_{nh}\ \mathbf{C}_{nv}\ \mathbf{D}_{nh}\ \mathbf{D}_{nd} \qquad\qquad \mathbf{T}_h\ \mathbf{T}_d\ \mathbf{O}_h.
\end{array}
$$

C. *Continuous Point Groups*

If in the discrete axial groups the order of the axis C_n becomes infinite, we obtain the continuous axial point groups $\mathbf{C}_\infty$, $\mathbf{C}_{\infty h}$, $\mathbf{C}_{\infty v}$, $\mathbf{D}_\infty$, and $\mathbf{D}_{\infty h}$ (the group $\mathbf{D}_{\infty d}$ cannot be defined). All of these continuous groups are subgroups of the group of orthogonal transformations in three-dimensional space, $\mathbf{O}_3$. The groups $\mathbf{C}_\infty$, $\mathbf{C}_{\infty h}$, and $\mathbf{D}_\infty$ are not realized as symmetry groups for molecules, and we shall not consider them further.

The Group $\mathbf{O}_3$. The transformations of this group are combinations of rotations about any axis which passes through the origin of coordinates, and the inversion. The group $\mathbf{O}_3$ can be represented as the direct product of the three-dimensional rotation group and the group $\mathbf{C}_i$, $\mathbf{O}_3 = \mathbf{R}_3 \times \mathbf{C}_i$.

The classes of the group $\mathbf{O}_3$ can be obtained from those of the group $\mathbf{R}_3$. Besides the two classes with one element in each, E and I, the group $\mathbf{O}_3$ contains two infinite sets of classes with a continuous set of elements in each class. These consist of rotations through an angle with absolute magnitude $|\phi|$ and a combination of rotations with the inversion. The group $\mathbf{O}_3$ is the symmetry point group of an atom.

The Group $\mathbf{C}_{\infty v}$. This group possesses a symmetry axis C_∞ and an infinite set of planes σ_v which pass through the axis. The group consists of a continuous set of classes each of which contains two rotations, $C(\phi)$ and $C(-\phi)$, and of a single class of reflections containing an infinite set of reflection operations σ_v.

All linear molecules which are asymmetric about their midpoints have the symmetry of the group $\mathbf{C}_{\infty v}$, e.g., HD, NO, N_2O, HCN (see Fig. 3.2g).

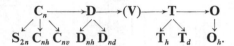

The Group $\mathbf{D}_{\infty h}$. The symmetry elements of the group $\mathbf{C}_{\infty v}$ have added to them a center of symmetry. This leads to the appearance of a plane of symmetry σ_h and of a continuous set of twofold axes U_2. The group $\mathbf{D}_{\infty h}$ can be represented as the direct product $\mathbf{D}_{\infty h} = \mathbf{C}_{\infty v} \times \mathbf{C}_i$. The classes of the group $\mathbf{D}_{\infty h}$ are obtained directly from those of $\mathbf{C}_{\infty v}$.

All linear molecules which are symmetric about their midpoint belong to $\mathbf{D}_{\infty h}$. This includes all diatomic molecules with identical nuclei, the molecules C_2H_2 (see Fig. 3.2h), C_2N_2, CO_2, C_3O_2 etc.

Character tables are given for the point groups in Appendix 1, and the matrices of the irreducible representations for the most important point groups are given in Appendix 2.

D. *The Symmetry Groups of Nonrigid Molecules*

The point groups that have been considered above consist of sets of rotation and reflection operations of the molecule as a whole. The molecule is regarded as a rigid configuration of atoms fixed in their equilibrium positions (*the equilibrium configuration*). However, there are a number of instances in which a molecule can possess several equilibrium configurations, each of which corresponds to an identical energy, and separated by a finite potential barrier. The molecule does not change its energy in making a transition from one configuration to another. In this way new symmetry elements make their appearance which enlarge the original point group. The symmetry groups of such "nonrigid" molecules have been investigated by Longuet-Higgins (1963) [see also the papers by Altmann (1967) and Serre (1968)].

The simplest example is that of the ammonia molecule NH_3, whose point group is $\mathbf{C}_{3v}$. The nitrogen atom can be found either above or below the plane of the three hydrogen atoms (Fig. 3.4a). It has been shown experimentally that the nitrogen atom makes transitions from one position to the other. One adds to the operations of the symmetry group $\mathbf{C}_{3v}$ reflections in the plane σ_h, thereby doubling the total number of symmetry operations. Hence in situations in which the transitions of NH_3 from one configuration to the other must be taken into account the molecule must be characterized by the point group $\mathbf{D}_{3h}$.

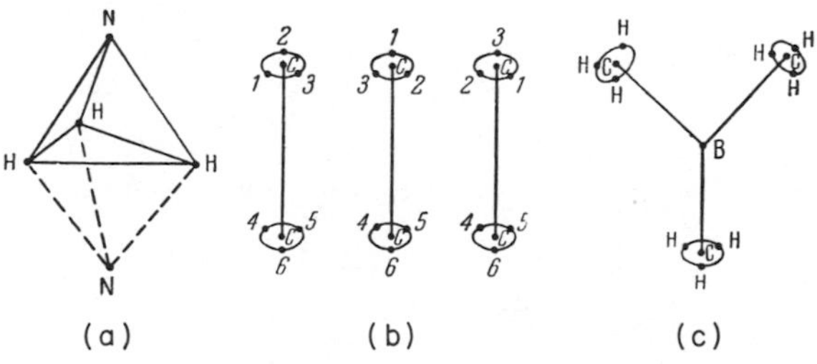

Figure 3.4. Symmetry groups of nonrigid molecules.

Other examples can be found in molecules with an internal rotation, where one part of the molecule can rotate through large angles with respect to another. This usually occurs with methyl groups which are bound to the rest of the molecule by single σ bonds. For example, the two CH_3 groups in ethane can rotate about the C—C bond. There are three equilibrium configurations, corresponding to the rotation of one of the CH_3 groups with respect to the other through angles of 60, 180, and 300° (Fig. 3.4b). These configurations differ only in permutations of the hydrogen atoms. Transitions from one configuration to another are equivalent to the following permutations of the hydrogen atoms of one of the methyl groups: I, P_{123}, and P_{132}. As a result, the symmetry group of the ethane molecule $\mathbf{D}_{3d}$ (or $\mathbf{D}_{3h}$, for the cis isomer), which contains 12 elements, is enlarged, by taking into account the internal rotation, to a symmetry group of 12.3 = 36 elements.

The presence in a molecule of several parts with internal rotations causes a sharp increase in the order of the symmetry group. Thus in the trimethyl boron molecule $B(CH_3)_3$ (Fig. 3.4c) there are internal rotations due to all three methyl groups. The point symmetry of the boron and carbon atom framework is $\mathbf{D}_{3h}$. Each of the 12 symmetry operations of the group $\mathbf{D}_{3h}$ has added to it 3.3.3 = 27 possible orientations of the methyl groups. The order of the group which is then obtained is found to be equal to 12.27 = 324.

We note that according to Cayley's theorem (Section 1.3), all symmetry point groups of nonrigid molecules must be isomorphic with the corresponding subgroup of the permutation group of the nuclei.

CHAPTER IV

Tensor Representations and Tensor Operators

PART 1. The Interconnection between Linear Groups and Permutation Groups

4.1. Construction of a Tensor Representation

Consider a set of n orthonormal functions ψ_i. This set can be regarded as a vector in an n-dimensional vector space. Under a unitary transformation of the space U a vector with components ψ_i becomes a new vector whose components are given by

$$\psi_i' = \sum_{k=1}^{n} u_{ik}\psi_k. \tag{4.1}$$

Let us form products of the components of two such vectors, and apply the transformation U to both vectors. As a result, the products transform according to the following rule:

$$\psi'_{i_1}\psi'_{i_2} = \sum_{k_1,k_2} u_{i_1k_1}u_{i_2k_2}\psi_{k_1}\psi_{k_2}. \tag{4.2}$$

The matrix of this transformation is the direct product of the transformation matrices (4.1). The n^2 quantities $\psi_{i_1}\psi_{i_2}$ constitute a *second rank tensor* defined in the n-dimensional vector space. We denote the components of this tensor by $T_{i_1i_2}$. In a similar way, we may form an Nth *rank tensor*, defined as the set of n^N components

$$T_{i_1i_2\cdots i_N} = \psi_{i_1}\psi_{i_2}\cdots\psi_{i_N}, \tag{4.3}$$

which under transformations of the vector space (4.1) transform according to the rule

$$T'_{i_1i_2\cdots i_N} = \sum_{k_1,k_2,\ldots,k_N} u_{i_1k_1}u_{i_2k_2}\cdots u_{i_Nk_N}T_{k_1k_2\cdots k_N}. \tag{4.4}$$

A transformation of the n-dimensional space U therefore induces a transformation

$$\Pi_N(U) = \underbrace{U \times U \times \cdots \times U}_{N} \tag{4.5}$$

in the n^N-dimensional space of Nth rank tensors. The matrices which correspond to the transformation $\prod_N(U)$ form an Nth rank *tensor representation* of the unitary group $\mathbf{U}_n$. This representation is a direct product of N n-dimensional representations of $\mathbf{U}_n$ and, according to the results of Section 1.16, must be reducible.

If one now introduces the arguments upon which the functions ψ_i depend, the tensor (4.3) can be written as a function of these arguments:

$$T_{i_1 i_2 \cdots i_N}(1, 2, \ldots, N) = \psi_{i_1}(1)\psi_{i_2}(2)\cdots\psi_{i_N}(N), \tag{4.6}$$

where the numbers $1, 2, \ldots, N$ enumerate the sets of arguments for each of the functions ψ_i.‡

4.2. Reduction of a Tensor Representation into Irreducible Components

It follows from the form of the transformation (4.4) that if one carries out a permutation $\bar{P}$ of the indices of the tensor (4.6), the new tensor which is obtained obeys the same transformation law as the original tensor. This can be shown by taking a second rank tensor as an example:

$$(\bar{P}_{12}T_{i_1 i_2})' = T'_{i_2 i_1} = \sum_{k_2, k_1} u_{i_2 k_2} u_{i_1 k_1} T_{k_2 k_1} = \sum_{k_1, k_2} u_{i_1 k_1} u_{i_2 k_2} \bar{P}_{12} T_{k_1 k_2}. \tag{4.7}$$

Hence the two operations of carrying out a unitary transformation and of permuting the indices commute. If we form from the tensor components linear combinations which possess definite symmetries with respect to permutations of the indices, then under unitary transformations these linear combinations transform only among themselves. Such linear combinations can be obtained by applying the Young operators (2.30) to the arguments of the tensor (4.6):

$$\omega_{rt}^{[\lambda]} T_{i_1 i_2 \cdots i_N}(1, 2, \ldots, N). \tag{4.8}$$

The letter t in such tensors characterizes their symmetry with respect to permutations of the indices, and the letter r their symmetry with respect to permutations of the arguments (see Section 2.9). If (4.8) is subjected to a unitary transformation (4.5), then because of the commutation relations

$$\prod_N(U)\omega_{rt}^{[\lambda]} = \omega_{rt}^{[\lambda]}\prod_N(U) \tag{4.9}$$

we obtain

$$\begin{aligned} \prod_N(U)\omega_{rt}^{[\lambda]} T_{i_1 i_2 \cdots i_N}(1, \ldots, N) &= \omega_{rt}^{[\lambda]}\prod_N(U) T_{i_1 i_2 \cdots i_N}(1, \ldots, N) \\ &= \sum_{k_1, \ldots, k_N} u_{i_1 k_1} u_{i_2 k_2} \cdots u_{i_N k_N} \omega_{rt}^{[\lambda]} T_{k_1 k_2 \cdots k_N}(1, \ldots, N). \end{aligned} \tag{4.10}$$

‡If we let $\psi_i(k)$ denote the wavefunction for the kth particle in a state characterized by the set of quantum numbers i (see Chapter VI), then (4.6) is just a wavefunction for a system of N noninteracting particles.

The components of a tensor of the form (4.8) with fixed $[\lambda]$, r, and t and with running indices $i_1, i_2, \ldots, i_N$ therefore transform among themselves under unitary transformations. However, not all of the components are independent. The number of independent components of a symmetrized tensor is determined by the dimension of the irreducible representation of the group $\mathbf{U}_n$ characterized by the symmetry diagram $[\lambda]$. We denote this representation by $U_n^{[\lambda]}$ and its dimension by $\delta_\lambda(n)$. A rule for finding $\delta_\lambda(n)$ can be obtained from the method of determining a basis for a representation $U_n^{[\lambda]}$.

First we note that there are only representations $U_n^{[\lambda]}$ for which the Young diagrams do not have any rows or columns longer than n. This is due to the fact that not more than n of the N indices of a tensor $T_{i_1 i_2 \ldots i_N}$ can be distinct (see Section 2.9, where n corresponds to the quantity $N - m + 1$ in the notation of that section). For a given Young diagram $[\lambda]$ there are $f_\lambda{}^2$ operators $\omega_{rt}^{[\lambda]}$. The symmetry of a tensor (4.8) under permutations of its arguments is characterized by the Young tableau r. For a given r we obtain a set of components that transform into each other under unitary transformations. In all there will be f_λ such sets, differing from one another in the tableau r. A basis for an irreducible representation $U_n^{[\lambda]}$ can be constructed from each set of components. It should be noted that tensor components that are obtained by the action of $\omega_{r_0 t}^{[\lambda]}$ with fixed r_0 upon the $T_{i_1 i_2 \cdots i_N}$, and that differ from one another only in a permutation of the indices, are linearly dependent. Thus, letting

$$T_{i_1 i_2 \cdots i_N} = \bar{P} T^0_{i_1 i_2 \cdots i_N},$$

then, because of relations (2.43) and (2.48),

$$\omega_{r_0 t_0}^{[\lambda]} T_{i_1 \cdots i_N} = \omega_{r_0 t_0}^{[\lambda]} \bar{P} T^0_{i_1 i_2 \cdots i_N} = \bar{P} \omega_{r_0 t_0}^{[\lambda]} T^0_{i_1 i_2 \cdots i_N} = \sum_u \Gamma_{u t_0}^{[\lambda]}(P) \omega_{r_0 u}^{[\lambda]} T^0_{i_1 i_2 \cdots i_N}. \tag{4.11}$$

The following procedure can therefore be used to find the dimension of an irreducible representation $U_n^{[\lambda]}$. We first pick out all the tensor components that do not transform into each other under permutations of the indices. We apply to them in succession the f_λ operators $\omega_{r_0 t}^{[\lambda]}$ with a fixed first index. This may give a zero result in some cases. The dimension $\delta_\lambda(n)$ of the irreducible representation is then given by the number of nonzero components of the symmetrized tensor. Examples of this procedure will be given presently.

The irreducible representations of the unitary group $\mathbf{U}_n$ that occur in the decomposition of an Nth rank tensor representation are thus characterized by Young diagrams consisting of N cells, the number of cells in a column not exceeding n. The number of times that each irreducible representation of the unitary group with symmetry diagram $[\lambda]$ occurs in a decomposition is equal to the dimension of the irreducible representation of the permutation group corresponding to the same symmetry diagram $[\lambda]$.‡ As a result of the

‡These results were first obtained by Weyl (1946). The derivation given here differs somewhat from the original by the use of the Young operators $\omega_{rt}^{[\lambda]}$.

reduction of a tensor representation, the n^N basis functions can be schematically arranged in a plane diagram in the form of a series of rectangles, each rectangle being characterized by a particular Young diagram $[\lambda]$ and containing $f_\lambda \delta_\lambda(n)$ functions. Functions which lie on the same row of a rectangle transform into each other under unitary transformations, and functions lying in the same column transform into each other under permutations of the arguments. It is obvious that

$$\sum_\lambda f_\lambda \delta_\lambda(n) = n^N.$$

A very simple example is provided by the decomposition into irreducible components of a second rank tensor representation in a two-dimensional space. One can form from the four components

$$T_{\alpha\alpha}, \qquad T_{\beta\beta}, \qquad T_{\alpha\beta}, \qquad T_{\beta\alpha},$$

three symmetric combinations $\omega^{[2]}T_{i_1 i_2}$:

$$T_{\alpha\alpha}, \qquad T_{\beta\beta}, \qquad T_{\alpha\beta} + T_{\beta\alpha},$$

and a single antisymmetric combination $\omega^{[1^2]}T_{i_1 i_2}$:

$$T_{\alpha\beta} - T_{\beta\alpha}.$$

As a more complicated example we consider a third rank tensor representation in a three-dimensional space. The 27 components can be divided into three types: (a) those with all indices identical, (b) those with two of the indices identical, and (c) those with all indices different. We write these out, placing in the same column the components that transform into each other under permutations:

$$\begin{array}{llllllllll}
\text{(a)} & T_{xxx} \quad T_{yyy} \quad T_{zzz} & \text{(b)} & T_{xxy} & T_{xxz} & T_{yyx} & T_{yyz} & T_{zzx} & T_{zzy} \\
 & & & T_{xyx} & T_{xzx} & T_{yxy} & T_{yzy} & T_{zxz} & T_{zyz} \\
 & & & T_{yxx} & T_{zxx} & T_{xyy} & T_{zyy} & T_{xzz} & T_{yzz} \\
 & & \text{(c)} & T_{xyz} & & & & & \\
 & & & T_{yxz} & & & & & \\
 & & & T_{xzy} & & & & & \\
 & & & T_{zyx} & & & & & \\
 & & & T_{yzx} & & & & & \\
 & & & T_{zxy} & & & & &
\end{array} \tag{4.12}$$

The three Young diagrams with three cells, [3], [21], and [1³], are all allowed, and consequently there will be three nonequivalent irreducible representations in the decomposition of the tensor representation. We now determine their dimensions.

When $[\lambda] = [3]$ there is just a single Young operator $\omega^{[3]}$. The independent linear combinations can be obtained by applying $\omega^{[3]}$ to the components in the first row of (4.12). We obtain, in all, ten such combinations, and therefore $\delta_{[3]}(3) = 10$.

When $[\lambda] = [21]$ there are four Young operators $\omega_{rt}^{[21]}$; these are explicitly written out on p. 43. Young operators which differ in the index r produce tensors which belong to different bases for an irreducible representation of the unitary group. The independent tensor components that belong to a single basis are found by applying operators $\omega_{r1}^{[21]}$ and $\omega_{r2}^{[21]}$ to the components in the first row of (4.12). The application of $\omega_{rt}^{[21]}$ to the first type of components gives zero (see Section 2.9). By applying $\omega_{r1}^{[21]}$ to the six components in the first row of (b) in (4.12), we obtain six linearly independent basis functions (the application of $\omega_{r2}^{[21]}$ to these components gives zero, since the components are symmetric in the first two indices). We obtain a further two basis functions by applying $\omega_{r1}^{[21]}$ and $\omega_{r2}^{[21]}$ to T_{xyz}. The dimension of the representation is therefore $\delta_{[21]}(3) = 8$. In all, two such representations occur in the desired decomposition, corresponding to the two tableaux r.

Finally, there is only one way of forming an antisymmetric tensor. This is achieved by applying the antisymmetrizing operator $\omega^{[1^3]}$ to a component in which all three indices are different, for example, to T_{xyz}. The dimension of the representation is therefore given by $\delta_{[1^3]}(3) = 1$.

The 27-dimensional space of a third rank tensor can thus be decomposed into one ten-dimensional, two eight-dimensional, and a single one-dimensional irreducible subspace. By representing each basis function of an irreducible subspace as a point on a plane diagram, we obtain three rectangles. Functions which lie on the same row of a rectangle transform into each other under unitary transformations, and functions in the same column transform into each other under permutations of their arguments (Fig. 4.1).

A general formula can be derived which gives the dimension of an irreducible representation of $\mathbf{U}_n$ in terms of the characteristic parameters of the associated Young diagram $[\lambda]$ (Murnaghan, 1938; Hamermesh, 1962):

$$\delta_\lambda(n) = \{\prod_{i<j} (h_i - h_j)\}/\{(n-1)!\,(n-2)! \cdots 1!\},$$
$$h_i = \lambda^{(i)} + n - i. \tag{4.13}$$

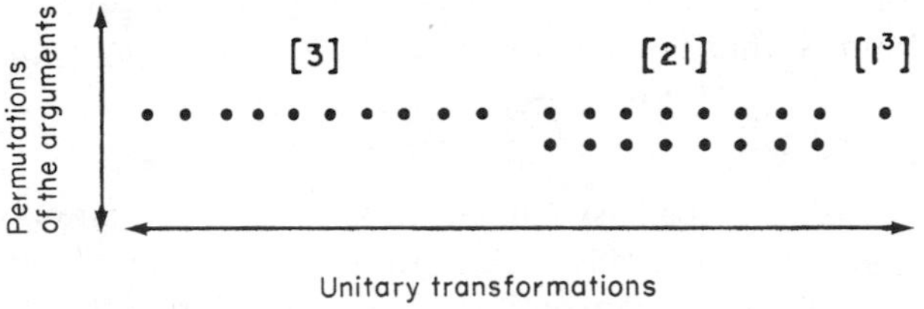

Figure 4.1.

In this equation i and j run from one to n, and $\lambda^{(i)}$ denotes the number of cells on the ith row of the Young diagram. If the number of rows in the diagram m is less than n, then $\lambda^{(i)}$ must be put equal to zero in the formula for h_i when $i > m$.

For the symmetric representation $[\lambda] = [N]$, formula (4.13) assumes the form

$$\delta_{[N]}(n) = (n + N - 1)!/[N!\,(n - 1)!] \tag{4.14}$$

and for the antisymmetric representation $[\lambda] = [1^N]$, the formula becomes

$$\delta_{[1^N]}(n) = n!/[N!\,(n - N)!]. \tag{4.15}$$

Expression (4.14) coincides with the formula for the number of ways of distributing N particles among n single particle states in Bose–Einstein statistics, and expression (4.15) with the analogous formula in Fermi–Dirac statistics. This concurrence is not accidental, but arises from the fact that the total wavefunction characterizing the state of a system of N particles must be symmetric under permutations of the particle in the case of Bose–Einstein statistics, and antisymmetric in the case of Fermi–Dirac statistics.

4.3. Formulae for the Characters of Symmetrized Powers of Representations

In the previous section it was shown that the use of the Young operators $\omega_{rt}^{[\lambda]}$ to symmetrize a given Nth rank tensor in an n-dimensional vector space leads to the decomposition of the initial n^N-dimensional tensor representation into irreducible components. The characters $\chi^{[\lambda]}(U)$ of the irreducible representations $U_n^{[\lambda]}$ which are obtained can be expressed in terms of the characters $\chi(U)$ of an n-dimensional vector representation of the unitary group $\mathbf{U}_n$. This relation is stated without proof [cf. Lyubarskii (1960, §25)]:

$$\chi^{[\lambda]}(U) = (f_\lambda/N!) \sum_{\mathbf{C}_\nu} g_{\mathbf{C}_\nu} \chi^{[\lambda]}(\mathbf{C}_\nu) \chi^{\nu_1}(U) \chi^{\nu_2}(U^2) \cdots \chi^{\nu_m}(U^m). \tag{4.16}$$

The summation is taken over all classes $\mathbf{C}_\nu$ of the permutation group $\boldsymbol{\pi}_N$, in which each class consists of $g_{\mathbf{C}_\nu}$ permutations with the identical cyclic structure of ν_1 unit cycles, ν_2 transpositions, . . . , and ν_m cycles of length m. $\chi^{[\lambda]}$ $(\mathbf{C}_\nu)$ is the character of $\mathbf{C}_\nu$ in the $\boldsymbol{\Gamma}^{[\lambda]}$ irreducible representation of $\boldsymbol{\pi}_N$.

In the case of an arbitrary group $\mathbf{G}$ one can associate with the tensor representation of Section 4.1 the direct product of a representation $\boldsymbol{\Gamma}$ of $\mathbf{G}$ taken with itself N times, thus forming the Nth power of the representation:

$$\boldsymbol{\Gamma}^N = \underbrace{\boldsymbol{\Gamma} \times \boldsymbol{\Gamma} \times \cdots \times \boldsymbol{\Gamma}}_{N}. \tag{4.17}$$

The symmetrization of the basis functions for this representation with respect to Young diagrams with N cells leads to the decomposition of the representation into ones with smaller dimensions. These are referred to as *symmetrized powers of* the *representation* Γ. The representation whose basis is symmetric, corresponding to $[\lambda] = [N]$, occurs very frequently in physical problems. This representation is known as the symmetric Nth power of Γ, and is denoted by $[\Gamma]^N$. Thus, for example, the wavefunctions corresponding to vibrational overtone states of molecules which have been excited by many quanta in the infrared belong to such a representation (Herzberg, 1947). The characters of $[\Gamma]^N$ are denoted by $[\chi]^N(R)$ (R denotes an arbitrary operation of $\mathbf{G}$), and are obtained as a special case of formula (4.16) with $[\lambda] = [N]$:

$$[\chi]^N(R) = (1/N!) \sum_{\mathbf{C}_\nu} g_{\mathbf{C}_\nu} \chi^{\nu_1}(R)\chi^{\nu_2}(R^2) \cdots \chi^{\nu_m}(R^m). \tag{4.18}$$

By taking the cyclic structure of the classes $\mathbf{C}_\nu$ and the values of the $g_{\mathbf{C}_\nu}$ for the groups $\boldsymbol{\pi}_2$, $\boldsymbol{\pi}_3$, and $\boldsymbol{\pi}_4$ from character tables (Appendix 4), we derive the following formulae:

$$\begin{aligned} [\chi]^2(R) &= \tfrac{1}{2}\chi^2(R) + \tfrac{1}{2}\chi(R^2), \\ [\chi]^3(R) &= \tfrac{1}{6}\chi^3(R) + \tfrac{1}{2}\chi(R)\chi(R^2) + \tfrac{1}{3}\chi(R^3), \\ [\chi]^4(R) &= \tfrac{1}{24}\chi^4(R) + \tfrac{1}{4}\chi^2(R)\chi(R^2) + \tfrac{1}{3}\chi(R)\chi(R^3) + \tfrac{1}{8}\chi^2(R^2) + \tfrac{1}{4}(R^4). \end{aligned} \tag{4.19}$$

The calculation of the characters by means of formula (4.18) becomes very tedious for large N. There exist, however, more compact formulae by means of which the $[\chi]^N(R)$ are easily found for any N. The form of these equations depends upon the dimension of the initial representation Γ.

For one-dimensional representations

$$\chi(R^k) = \chi^k(R),$$

and hence one obtains from (4.18) the very simple result

$$[\chi]^N(R) = \chi^N(R). \tag{4.20}$$

The formulae for two- and three-dimensional representations have been derived by Kompaneyets (1940). We consider here the derivation for the two-dimensional case.

We denote the basis functions for such a representation Γ by the letters x and y. These can be regarded as components of a two-dimensional vector. For any transformation R of a point group $\mathbf{G}$ the transformed functions Rx and Ry can be expressed in terms of x and y by means of an orthogonal matrix which can be interpreted as a rotation or reflection in the space xy. Thus the matrix

$$\begin{bmatrix} \cos\alpha_R & -\sin\alpha_R \\ \sin\alpha_R & \cos\alpha_R \end{bmatrix}, \tag{4.21}$$

whose sum of the diagonal elements equals $2\cos\alpha_R$, corresponds to a rotation through an angle α_R. Equating the character of an operation $\chi(R)$ to $2\cos\alpha_R$ yields the angle of rotation α_R. It is important to note that if the

Cartesian coordinates x and y do not transform according to the particular representation under consideration, then the rotation has no connection with an actual symmetry transformation.

The matrix

$$\begin{bmatrix} 1 & \\ & -1 \end{bmatrix} \tag{4.22}$$

corresponds to a reflection in two-dimensional space. In order to ascertain whether a particular operation R is a reflection in the xy space or a rotation through an angle $\alpha_R = \pi/2$ (the character of which is also equal to zero), it is sufficient to examine $\chi(R^2)$. If $\chi(R^2) = 2$, the transformation is a reflection, but if $\chi(R^2) = -2$, it is a rotation.

The symmetric product of a two-dimensional representation with itself N times is a representation whose basis consists of the $N + 1$ functions

$$x^N, x^{N-1}y, \ldots, y^N. \tag{4.23}$$

The character of a reflection defined by the matrix (4.22) is obviously given in this basis by

$$[\chi]^N(R) = [1 + (-1)^N]/2. \tag{4.24}$$

In order to find the character in this basis of an operation R which corresponds to a rotation, it is convenient to go over to the spherical components, $x \pm iy$, of the two-dimensional vector. The basis (4.23) now becomes

$$(x + iy)^N, \qquad (x + iy)^{N-1}(x - iy), \ldots, \quad (x - iy)^N. \tag{4.25}$$

The rotation matrix (4.21) is then diagonal in the spherical basis, and is given by

$$\begin{bmatrix} e^{-i\alpha_R} & \\ & e^{i\alpha_R} \end{bmatrix}.$$

The rotation matrix generated by (4.25) is now also diagonal, the sum of its elements forming a geometric progression

$$e^{iN\alpha_R} + e^{i(N-2)\alpha_R} + \cdots + e^{-iN\alpha_R}.$$

From this it follows that

$$[\chi]^N(R) = \{\sin[(N + 1)\alpha_R]\}/\sin \alpha_R. \tag{4.26}$$

The following formulae for three-dimensional representations can be derived in a similar manner (Kompaneyets, 1940): For an operation R that is equivalent to a rotation in the xyz space

$$[\chi]^N(R) = \frac{\sin[\frac{1}{2}(N + 2)\alpha_R] \sin[\frac{1}{2}(N + 1)\alpha_R]}{\sin \alpha_R \sin(\alpha_R/2)}; \tag{4.27}$$

and for an operation R that is equivalent to a transformation in the xyz space containing a reflection

$$[\chi]^N(R) = \frac{\sin[\frac{1}{2}(N+2)\bar{\alpha}_R]\sin[\frac{1}{2}(N+1)\bar{\alpha}_R]}{\sin\alpha_R \sin(\bar{\alpha}_R/2)}, \qquad \bar{\alpha}_R = \alpha_R + \pi. \tag{4.28}$$

If the x, y, and z coordinates do not transform according to the three-dimensional representation under consideration, the form of the transformation in the xyz space corresponding to an operation R is determined by an operation R' of an isomorphic group in which the coordinates x, y, and z do possess the desired transformation properties. A similar association can be made for two-dimensional representations. This can easily be established by means of character tables for the point groups (Appendix 1).

As an example, we give the characters of some of the symmetric powers of the E representation of the group $\mathbf{C}_{3v}$. In this particular case the group elements do correspond to the analogous operations in the xy space. The characters of the symmetric powers are easily calculated from formulae (4.24) and (4.26):

$\mathbf{C}_{3v}$	E	$2C_3$	$3\sigma_v$
α_R	0	$2\pi/3$	
$\chi(R)$	2	-1	0
$[\chi]^4(R)$	5	-1	1
$[\chi]^7(R)$	8	-1	0
$[\chi]^9(R)$	10	1	0

In calculations with formulae (4.26)–(4.28) it is convenient to make use of the identities

$$(\sin m\pi)/\sin\pi = (-1)^{m-1}m, \qquad (\sin 2m\pi)/\sin 2\pi = m.$$

4.4. Littlewood's Theorem

Littlewood (1940) has proved a theorem that enables one to determine which irreducible representations $\Gamma^{[\lambda]}$ of a group π_N can be formed from the direct product of two irreducible representations $\Gamma^{[\lambda_1]}$ and $\Gamma^{[\lambda_2]}$ of the groups π_{n_1} and π_{n_2} ($n_1 + n_2 = N$), i.e., on enlarging the subgroup $\pi_{n_1} \times \pi_{n_2}$ to the group π_N. In contrast to the formation of the more usual direct product, this process is known as forming the *outer product*, and is denoted by the symbol $\otimes$. The decomposition of an outer product into irreducible components is easily carried out with the aid of Littlewood's theorem:

$$\Gamma^{[\lambda_1]} \otimes \Gamma^{[\lambda_2]} \doteq \sum_\lambda a(\lambda_1, \lambda_2, \lambda)\Gamma^{[\lambda]}. \tag{4.29}$$

Since an Nth rank tensor which has been symmetrized with respect to a Young diagram $[\lambda]$ can simultaneously serve as a basis for both permutation and unitary groups (see the previous section), Littlewood's theorem can also be used to find which irreducible representations $U_n^{[\lambda]}$ occur in the decomposition of the direct product

$$U_n^{[\lambda_1]} \times U_n^{[\lambda_2]} \doteq \sum_{\lambda} a(\lambda_1, \lambda_2, \lambda) U_n^{[\lambda]}. \tag{4.30}$$

The summation on the right of this equation includes only those Young diagrams of (4.29) in which the length of the columns does not exceed n. This restriction automatically ceases to arise when $n \geq N$.

Before stating Littlewood's theorem, we introduce the concept of a *lattice permutation*. Consider an expression of the form $x_1^{n_1} x_2^{n_2} x_3^{n_3} \cdots (n_1 \geq n_2 \geq n_3 \geq \cdots)$. A reordering of the elements $x_1, x_2, x_3, \ldots,$ is called a lattice permutation if among the first m terms of such a permutation, the numbers of times x_1 occurs is equal to or greater than the number of times x_2 occurs, which is equal to or greater than the number of times x_3 occurs, etc.

As an example, the six possible lattice permutations of the expression ${x_1}^3 x_2 x_3$ are

$$\begin{array}{lll} {x_1}^3 x_2 x_3, & {x_1}^2 x_2 x_1 x_3, & {x_1}^2 x_2 x_3 x_1, \\ x_1 x_2 {x_1}^2 x_3, & x_1 x_2 x_1 x_3 x_1, & x_1 x_2 x_3 {x_1}^2. \end{array}$$

We now proceed to state the theorem:

The possible Young diagrams which can be constructed from the diagrams $[\lambda_1] \equiv [\lambda_1^{(1)} \lambda_1^{(2)} \cdots]$ and $[\lambda_2] \equiv [\lambda_2^{(1)} \lambda_2^{(2)} \cdots]$ are found by consecutively adding to the diagram $[\lambda_1]$ in all possible ways $\lambda_2^{(1)}$ cells containing the index α, $\lambda_2^{(2)}$ cells containing the index β, etc., such that in the augmented diagram the indices which have been added satisfy two conditions: (a) two identical indices may not occur in the same column, and (b) when the total of all the indices which have been added is read from right to left in consecutive rows, we obtain a lattice permutation of the expression $\alpha^{\lambda_2^{(1)}}, \beta^{\lambda_2^{(2)}}, \cdots$.

EXAMPLE 1. $[1^3] \otimes [1^2]$:

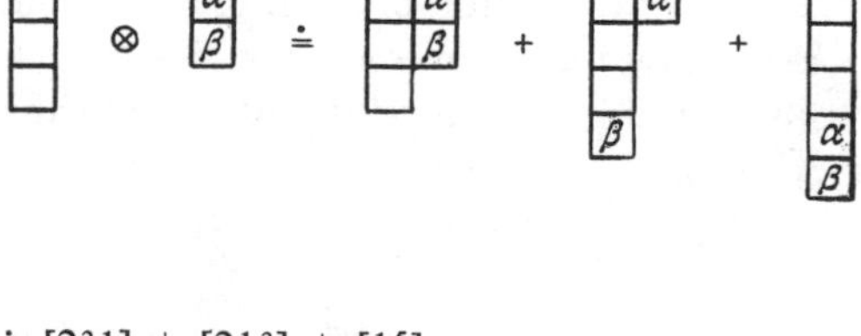

i.e., $[1^3] \otimes [1^2] \doteq [2^2 1] + [21^3] + [1^5]$.

EXAMPLE 2. $[21]\otimes[21]$:

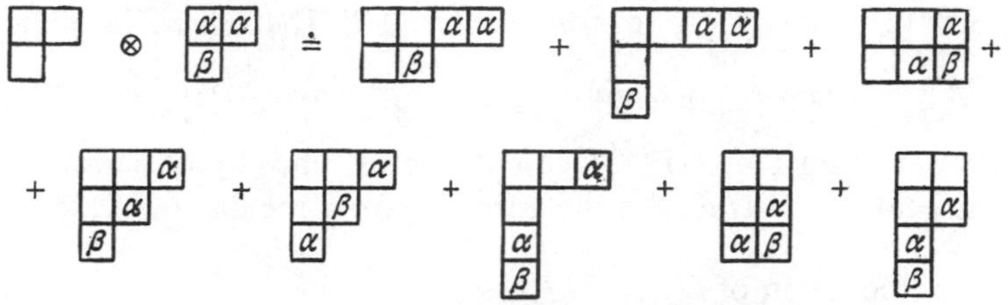

On collecting similar terms together, we obtain

$$[21]\otimes[21] \doteq [42]+[41^2]+[3^2]+2[321]+[31^3]+[2^3]+[2^21^2].$$

The decomposition (4.30) must coincide with this for any $n \geq N$. This circumstance can be used to check a decomposition carried out by using Littlewood's theorem. Namely, the condition

$$\delta_{\lambda_1}(n)\delta_{\lambda_2}(n) = \sum_{\lambda} a(\lambda_1, \lambda_2, \lambda)\delta_\lambda(n) \tag{4.31}$$

must be fulfilled. The dimensions of the representations $\delta_\lambda(n)$ are calculated by formula (4.13). We can verify that Eq. (4.31) is satisfied for Example 1. From formula (4.13) we find

$$\begin{array}{llllll} [\lambda]: & [1^3] & [1^2] & [2^21] & [21^3] & [1^5] \\ \delta_\lambda(5): & 10 & 10 & 75 & 24 & 1. \end{array}$$

which gives

$$10.10 = 75 + 24 + 1,$$

in complete agreement with (4.31). Equation (4.31) is also satisfied for large n. Indeed for $n = 7$ we obtain

$$35.21 = 490 + 224 + 21.$$

Littlewood's theorem can also be used to find the representations $\Gamma^{[\lambda_1]} \times \Gamma^{[\lambda_2]}$ into which $\Gamma^{[\lambda]}$ decomposes on reduction to the subgroup $\pi_{n_1} \times \pi_{n_2}$. For this purpose it is necessary to apply the theorem to every possible term in the decomposition‡

$$\Gamma^{[\lambda]} \doteq \sum_{\lambda_1, \lambda_2} a(\lambda, \lambda_1, \lambda_2)\Gamma^{[\lambda_1]} \times \Gamma^{[\lambda_2]} \tag{4.32}$$

and to reject those terms that do not yield the given $[\lambda]$. This method is quicker than a decomposition carried out with the aid of character tables for π_N and $\pi_{n_1} \times \pi_{n_2}$ if n_2 is not too large.

‡By Frobenius' theorem (see p. 138) it follows that the coefficients $a(\lambda, \lambda_1, \lambda_2)$ in Eq. (4.32) are equal to the coefficients $a(\lambda_1, \lambda_2, \lambda)$ in Eq. (4.29).

For example, let us find the representations that arise on reducing $\Gamma^{[\lambda]} = \Gamma^{[32]}$ with respect to the subgroup $\pi_3 \times \pi_2$. The diagram $[\lambda_1]$ is restricted to being [3] or [21], since $[1^3]$ is not contained in [3 2]. The desired decomposition is

$$[32] \doteq [3] \times [2] + [21] \times [2] + [21] \times [1^2].$$

The representation $\Gamma^{[3]} \times \Gamma^{[1^2]}$ does not appear, since by Littlewood's theorem, one cannot construct the diagram [32] from the diagrams [3] and $[1^2]$.

4.5. The Reduction of $U_{2j+1} \to R_3$

Let us consider the $2j + 1$ functions $\psi_m^{(j)}$, which form a basis for an irreducible representation $D^{(j)}$ of the rotation group in three dimensions $\mathbf{R}_3$. The basis functions transform into each other under rotations:

$$R\psi_m^{(j)} = \sum_{m'} D_{m'm}^{(j)}(R)\psi_{m'}^{(j)}. \tag{4.33}$$

The set of $2j + 1$ functions $\psi_m^{(j)}$ can be regarded as a vector in a $(2j + 1)$-dimensional space. Every rotation in three-dimensional space thus generates some unitary transformation in the $(2j + 1)$-dimensional space which is implemented by the matrices of the representation $D^{(j)}$. The group of such transformations forms a subgroup of the group of all unitary transformations $\mathbf{U}_n$.

Under rotations of the three-dimensional space the product of functions

$$\psi_{m_1}^{(j)}\psi_{m_2}^{(j)} \cdots \psi_{m_N}^{(j)} \tag{4.34}$$

transforms according to the $m_1 m_2 \cdots m_N$ column of the direct product

$$\underbrace{D^{(j)} \times D^{(j)} \times \cdots \times D^{(j)}}_{N}.$$

The irreducible representations $D^{(J)}$ which occur in the decomposition of this direct product are found by successive applications of the decomposition (3.43). On the other hand, the product (4.34) transforms under unitary transformations as an Nth rank tensor representation of the group $\mathbf{U}_{2j+1}$. The irreducible representations which occur in its decomposition $U_{2j+1}^{[\lambda]}$ are characterized by Young diagrams with N cells, the columns of which do not exceed $2j + 1$ cells in length (see Section 4.2). Since the group $\mathbf{R}_3$ is a subgroup of $\mathbf{U}_{2j+1}$, the irreducible representations $U_{2j+1}^{[\lambda]}$ in general become reducible upon restricting the operations to those of $\mathbf{R}_3$ and hence the $U_{2j+1}^{[\lambda]}$ split into irreducible representations $D^{(J)}$. In classifying the states of a system of N identical particles, it is often important to know which $D^{(J)}$ occur in the decomposition

$$U_{2j+1}^{[\lambda]} \doteq \sum_J a^{(J)} D^{(J)}, \tag{4.35}$$

and in order to discover this, we make use of a recursive method proposed by Jahn (1950).

We note first of all that the one-dimensional representation $U_{2j+1}^{[1^{2j+1}]}$ must correspond to the representation $D^{(0)}$. This is because all the indices m_i in the nonzero components of the tensor $T^{[1^{2j+1}]}_{m_1 \cdots m_{2j+1}}$ must be different, and hence there is in fact only one nonzero component. The tensor therefore corresponds to an angular momentum $J = 0$. It can be shown that as far as the decomposition (4.35) is concerned, the representations $U^{[\lambda]}_{2j+1}$ are equivalent to the Young diagrams

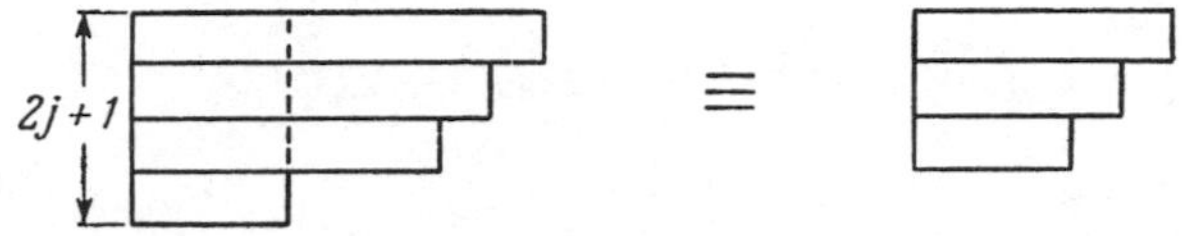

(4.36a)

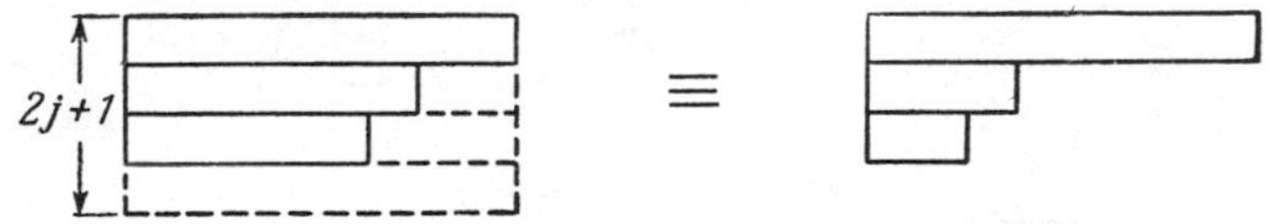

(4.36b)

Relation (4.36b) depends upon the fact that a Young diagram whose columns are all of length $2j + 1$ corresponds to an angular momentum $J = 0$. Hence the two Young diagrams which make up the rectangle in (4.36b) must correspond to the same values of J, otherwise one would not obtain $J = 0$ on coupling the two corresponding vectors.

We consider the method of determining the form of the decomposition (4.35) for the group $\mathbf{U}_3$ ($j = 1$). The representation $U^{[1]}_3$ is generated by the three functions $\psi^{(1)}_1$, $\psi^{(1)}_0$, and $\psi^{(1)}_{-1}$, which under rotations transform according to the irreducible representation $D^{(1)}$. Hence we have

$$U^{[1]}_3 \doteq D^{(1)}. \tag{4.37}$$

When $N = 2$ two irreducible representations $U^{[2]}_3$ and $U^{[1^2]}_3$ occur in the decomposition of the tensor representation. However, when the direct product $D^{(1)} \times D^{(1)}$ is reduced we obtain three irreducible representations $D^{(2)}$, $D^{(1)}$, and $D^{(0)}$. In order to find the correspondence between these representations, we make use of the symmetry of a basis function of a representation $D^{(J)}$, $\psi^{(J)}_M(j(1)j(2))$, with respect to a permutation of the order in which the angular momenta $j(i)$ are coupled. According to Eq. (3.49), the function is multiplied by $(-1)^{2j-J} = (-1)^J$ (for j integral) under such a permutation; i.e., the function is symmetric for J even and antisymmetric for J odd. Since the basis functions for $U^{[2]}_3$ are symmetric with respect to a permutation of the indices whereas those for $U^{[1^2]}_3$ are antisymmetric, we obtain

$$U^{[2]}_3 \doteq D^{(2)} + D^{(0)}, \qquad U^{[1^2]}_3 \doteq D^{(1)}. \tag{4.38}$$

Proceeding further with this decomposition, we form from the Young diagram [2] all the possible diagrams with three cells. We find the representations $D^{(J)}$ that correspond to these diagrams by means of the triangle rule (3.44). The diagram $[1^2]$ is dealt with similarly:

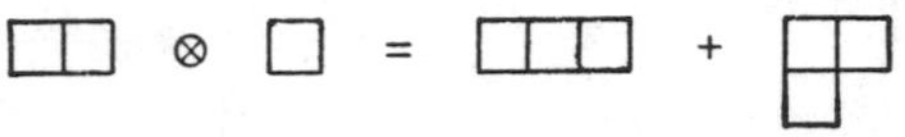

$$(D^{(2)} + D^{(0)}) \times D^{(1)} \doteq D^{(3)} + D^{(2)} + 2D^{(1)},$$

(4.39)

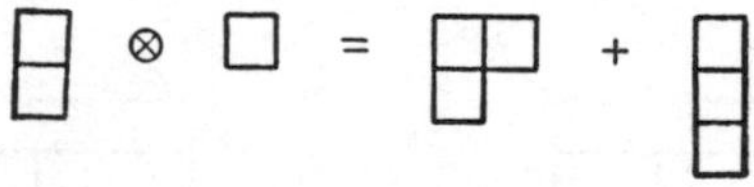

$$D^{(1)} \times D^{(1)} \doteq D^{(2)} + D^{(1)} + D^{(0)}.$$

As a result we obtain the two equations

$$\begin{aligned} U_3^{[3]} + U_3^{[21]} &\doteq D^{(3)} + D^{(2)} + 2D^{(1)}, \\ U_3^{[21]} + U_3^{[1^3]} &\doteq D^{(2)} + D^{(1)} + D^{(0)}. \end{aligned} \qquad (4.40)$$

Since only $D^{(0)}$ may correspond to $U^{[1^3]}$, we arrive, with the aid of Eq. (4.40), at the desired decomposition:

$$U_3^{[3]} \doteq D^{(3)} + D^{(1)}, \qquad U_3^{[21]} \doteq D^{(2)} + D^{(1)}, \qquad U_3^{[1^3]} \doteq D^{(0)}. \qquad (4.41)$$

For $N = 4$ we have, similarly to (4.39),

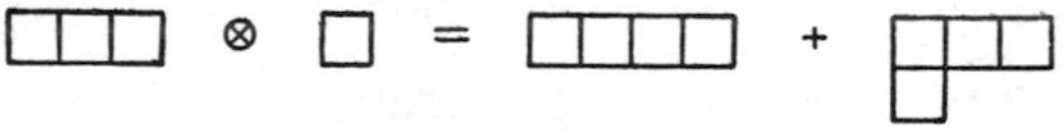

$$(D^{(3)} + D^{(1)}) \times D^{(1)} \doteq D^{(4)} + D^{(3)} + 2D^{(2)} + D^{(1)} + D^{(0)},$$

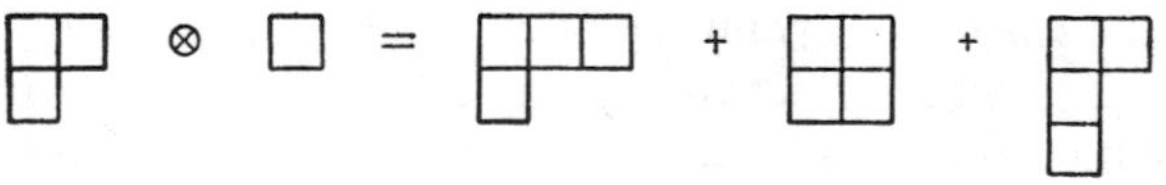

(4.42)

$$(D^{(2)} + D^{(1)}) \times D^{(1)} \doteq D^{(3)} + 2D^{(2)} + 2D^{(1)} + D^{(0)},$$

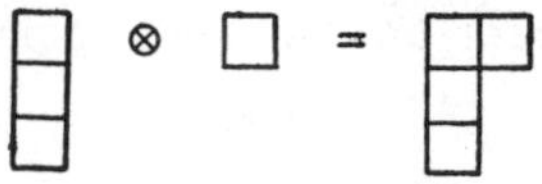

$$D^{(0)} \times D^{(1)} \doteq D^{(1)}.$$

According to (4.36b), the representation $U_3^{[2^2]}$ possesses the same J structure as $U_3^{[2]}$, and $U_3^{[21^2]}$ the same J structure as $U_3^{[1^2]}$. Taking this into account, we obtain from (4.42)

$$U_3^{[4]} \doteq D^{(4)} + D^{(2)} + D^{(0)}, \qquad U_3^{[31]} \doteq D^{(3)} + D^{(2)} + D^{(1)},$$
$$U_3^{[2^2]} \doteq D^{(2)} + D^{(0)}, \qquad U_3^{[21^2]} \doteq D^{(1)}. \tag{4.43}$$

The J structure of any representation $U_{2j+1}^{[\lambda]}$ can be found similarly. It is only necessary to bear in mind that if j is half-integral, an odd value of J corresponds to the representation $U_{2j+1}^{[2]}$ and an even value to $U_{2j+1}^{[1^2]}$. In Appendix 3 we give the results of reducing a number of representations $U_{2j+1}^{[\lambda]}$ for $j = 1$ to $j = 3$.

The irreducible representations of the group $\mathbf{U}_{2j+1}$ remain irreducible on passing to the subgroup $\mathbf{SU}_{2j+1}$ (Hamermesh, 1962). However, some of the irreducible representations may as a result become equivalent. It turns out that the conditions for this to occur are the same as the conditions (4.36). The reduction of the irreducible representations $\mathbf{SU}_{2j+1}$ to those of $\mathbf{R}_3$ of course proceeds similarly to that of $\mathbf{U}_{2j+1} \to \mathbf{R}_3$. However in the case of $\mathbf{SU}_{2j+1}$ there is now a unique correspondence between the various irreducible representations and their J structure.

The reduction of the unitary group in two-dimensional space, $\mathbf{U}_2 \longrightarrow \mathbf{R}_3$, does not cause any splitting of the representations $U_2^{[\lambda]}$. This is connected with the fact that the groups $\mathbf{SU}_2$ and $\mathbf{R}_3$ are isomorphic, and hence there is a one-to-one correspondence between the irreducible representations of these two groups.‡ This can easily be verified by carrying out the procedure of successively adding cells to the Young diagrams. The diagrams for the representations $U_2^{[\lambda]}$ may not have more than two cells in a column, i.e., they consist of two rows. The angular momentum J that corresponds to a diagram $[\lambda]$ is determined by the lengths of the rows $\lambda^{(1)}$ and $\lambda^{(2)}$ as follows:

$$J = \tfrac{1}{2}(\lambda^{(1)} - \lambda^{(2)}). \tag{4.44}$$

Hence

$$U_2^{[41]} \longleftrightarrow D^{(3/2)}, \qquad U_2^{[31]} \longleftrightarrow D^{(1)}, \qquad U_2^{[2^2]} \longleftrightarrow D^{(0)}.$$

PART 2. Irreducible Tensor Operators

4.6. Definition

In the foregoing section it was shown that the n^N components of an Nth rank tensor transform under a unitary transformation of the n-dimensional space as a reducible representation which decomposes into irreducible components when the tensor is symmetrized according to a Young diagram. The

‡This circumstance is of great importance in the classification of the states of systems of particles with spin $\frac{1}{2}$, since one can associate with every value of the resultant spin S a particular Young diagram $[\lambda]$ (see Chapter VI).

tensors which are symmetrized in this way are irreducible with respect to the operations of the unitary group $\mathbf{U}_n$, i.e., under unitary transformations the components of a symmetrized tensor transform only among themselves. In general, however, these tensors are not irreducible with respect to the operations of the three-dimensional rotation group $\mathbf{R}_3$. For example, according to Eq. (4.38), one can form from the six components of a symmetrized second rank tensor which transforms according to the irreducible representation $U_3^{[2]}$ one scalar (the irreducible representation $D^{(0)}$) and five linear combinations which transform according to the irreducible representation $D^{(2)}$.

In general, a set of f_α quantities $T_i^{(\alpha)}$ is called an *irreducible tensor* of a group of linear transformations if under the operations of the group the $T_i^{(\alpha)}$ transform according to an irreducible representation $\mathbf{\Gamma}^{(\alpha)}$ of this group:

$$RT_i^{(\alpha)} = \sum_{k=1}^{f_\alpha} \Gamma_{ki}^{(\alpha)}(R) T_k^{(\alpha)}. \tag{4.45}$$

From this definition it follows that any set of basis functions for an irreducible representation can be regarded as an irreducible tensor. Thus the set of $2J+1$ spherical harmonic functions‡ $Y_{JM}(\theta, \phi)$ form an example of an irreducible tensor which belongs to the representation $D^{(J)}$ of the group $\mathbf{R}_3$. The Cartesian components of an arbitrary vector $\mathbf{A}$ form a first rank tensor. However, their transformational properties under three-dimensional rotations are more complicated than those of the spherical components

$$A_0 = A_z, \qquad A_{\pm 1} = \mp(\sqrt{\tfrac{1}{2}})(A_x \pm iA_y),$$

which constitute an irreducible tensor of the representation $D^{(1)}$. When calculating matrix elements of vector quantities with basis functions of the group $\mathbf{R}_3$ it is therefore more convenient to use the spherical components of the vectors.

Three types of operation can be defined for ordinary (i.e., Cartesian) tensors T^N (N denotes the rank of a tensor):

(1) Addition of tensors of identical rank,

$$V_{ik}^{(2)} = T_{ik}^{(2)} + U_{ik}^{(2)}. \tag{4.46}$$

(2) Tensor multiplication,

$$V_{iklmn}^{(5)} = T_{ik}^{(2)} V_{lmn}^{(3)}, \tag{4.47}$$

which leads to a tensor whose rank is equal to the sum of the ranks of the tensors forming the product.

(3) Contraction over a pair of indices, which leads to a lowering of the rank of the tensor,

$$T_i^{(1)} = \sum_k T_{ikk}^{(3)}. \tag{4.48}$$

‡Irreducible tensors of the group $\mathbf{R}_3$ are called *spherical tensors*.

Successive contractions of a tensor of even rank $\frac{1}{2}N$ lead to a scalar. For example,

$$T^{(0)} = \sum_{i,k} T^{(4)}_{iikk}. \tag{4.49}$$

The scalar product of two tensors is defined as a contraction over all indices:

$$V^{(0)} = \sum_{i,k} T^{(2)*}_{ik} V^{(2)}_{ik}. \tag{4.49a}$$

The addition of irreducible tensors is similar to the addition of Cartesian tensors

$$V_i^{(\alpha)} = T_i^{(\alpha)} + U_i^{(\alpha)}. \tag{4.50}$$

However, instead of tensor multiplication or contraction, the following operation is defined for irreducible tensors:

$$V_t^{(\tau)} = \sum_{i,k} T_i^{(\alpha)} U_k^{(\beta)} \langle \alpha i, \beta k \mid \tau t \rangle. \tag{4.51}$$

As a result one obtains a tensor that transforms according to an irreducible representation $\mathbf{\Gamma}^{(\tau)}$ which occurs in the decomposition of the direct product $\mathbf{\Gamma}^{(\alpha)} \times \mathbf{\Gamma}^{(\beta)}$. The coefficients in Eq. (4.51) are Clebsch–Gordan coefficients (see Section 1.17).

A tensor which belongs to the unit (totally symmetric) representation of a group behaves as a scalar with respect to the operations of the particular group. According to (1.73), the necessary and sufficient condition for the unit representation to occur in the decomposition of the direct product of two representations is that the representations are complex conjugates of each other. Consequently, one can never form a scalar from a product of two irreducible tensors that belong to two different representations. In the case of the group $\mathbf{R}_3$ a scalar is characterized by a value of zero for the angular momentum, $J = 0$, and is constructed from two spherical tensors with the same J. Substituting the values of the Clebsch–Gordan coefficients (3.48) into (4.51), we obtain the following scalar from two spherical tensors:

$$V^{(0)} = [(-1)^J/(2J+1)^{1/2}] \sum_m (-1)^m T_m^{(J)} U_{-m}^{(J)}. \tag{4.52}$$

Expression (4.52) without the multiplying factor in front of the sum is usually called the scalar product of two spherical tensors and is denoted by

$$(T^{(J)} \cdot U^{(J)}) = \sum_m (-1)^m T_m^{(J)} U_{-m}^{(J)}. \tag{4.53}$$

When $J = 1$ this expression coincides with the usual scalar product of two vectors expressed in spherical coordinates.

In calculating the matrix elements of any operator it is important to know according to which irreducible representations the factors in the integrand transform. This knowledge enables one, for example, immediately to distinguish nonzero matrix elements and to obtain the group-theoretical selection rules (see Chapter V). It is therefore convenient to write the operator which occurs in the matrix element in the form of a sum of operators each of which transforms according to definite irreducible representations of the group. *An irreducible tensor operator* is defined (by analogy with an irreducible tensor) as a set of f_α quantities $T_i^{(\alpha)}$ whose transformation law is‡

$$R^{-1}T_i^{(\alpha)}R = \sum_k \Gamma_{ki}^{(\alpha)}(R)T_k^{(\alpha)}. \tag{4.54}$$

The use of irreducible tensor operators facilitates the calculation of matrix elements considerably, since one can employ a whole series of useful relations first obtained by Wigner (1959) and Racah [see Fano and Racah (1959), Judd (1963), and Sobel'man (1972)]. These relations rest upon the Wigner–Eckart theorem.

4.7. The Wigner–Eckart Theorem

Let us consider a matrix element of an irreducible tensor operator $T_t^{(\tau)}$,

$$\langle \alpha i \,|\, T_t^{(\tau)} \,|\, \beta k \rangle. \tag{4.55}$$

defined in terms of functions which transform according to the irreducible representations $\Gamma^{(\alpha)}$ and $\Gamma^{(\beta)}$ of the same group as the representation $\Gamma^{(\tau)}$. We examine the transformational properties with respect to the operations of the group of the function which results by allowing $T_t^{(\tau)}$ to operate upon the function $\psi_k^{(\beta)}$:

$$R(T_t^{(\tau)}\psi_k^{(\beta)}) = (R^{-1}T_t^{(\tau)}R)(R\psi_k^{(\beta)}) = \sum_{t',k'} \Gamma_{t't}^{(\tau)}(R)\Gamma_{k'k}^{(\beta)}(R)(T_{t'}^{(\tau)}\psi_{k'}^{(\beta)}).$$

The function $T_t^{(\tau)}\psi_k^{(\beta)}$ consequently transforms according to the direct product $\Gamma^{(\tau)} \times \Gamma^{(\beta)}$, and by using Clebsch–Gordan coefficients, it can be written in the form of a decomposition into basis functions for irreducible representations of the group [see Eq. (1.79)]:

$$T_t^{(\tau)}\psi_k^{(\beta)} = \sum_{a,\mu,m} \Phi_m^{(a\mu)}(\tau\beta)\langle a\mu m \,|\, \tau t, \beta k \rangle. \tag{4.56}$$

In this equation the index a distinguishes irreducible representations that are repeated, and the symbols τ and β in the argument of the function Φ indicate that the form of Φ depends upon the basis functions in the direct product. Substituting (4.56) into (4.55), we obtain

$$\langle \alpha i \,|\, T_t^{(\tau)} \,|\, \beta k \rangle = \sum_{a,\mu,m} \langle \alpha\mu m \,|\, \tau t, \beta k \rangle \int \psi_i^{(\alpha)*}\Phi_m^{(a\mu)}(\tau\beta)\, dV. \tag{4.57}$$

According to the orthogonality conditions (1.97) for basis functions of irreducible representations,

$$\int \psi_i^{(\alpha)*}\Phi_m^{(a\mu)}(\tau\beta)\, dV = \delta_{\alpha\mu}\delta_{im}A_a(\alpha, \tau, \beta), \tag{4.58}$$

‡Irreducible tensor operators transform under the operations of a group in the same way as irreducible tensors. The difference between Eqs. (4.54) and (4.45) is due to the fact that operators in the new and old bases are related by Eq. (1.35).

where the quantity $A_a(\alpha, \tau, \beta)$ is determined by the form of the functions $\psi_i^{(\alpha)}$ and $\Phi_m^{(a\alpha)}(\tau\beta)$, but does not depend upon i and m. On substituting (4.58) into (4.57), we obtain an analytical expression of the Wigner–Eckart theorem‡:

$$\langle\alpha i \,|\, T_t^{(\tau)} \,|\, \beta k\rangle = \sum_a \langle a\alpha i \,|\, \tau t, \beta k\rangle A_a(\alpha, \tau, \beta). \qquad (4.59)$$

It follows from this equation that the Clebsch–Gordan coefficients completely determine the dependence of the matrix element upon the column numbers of the irreducible representations according to which the factors in the integrand transform. The presence of these coefficients allows one immediately to obtain the conditions under which a matrix element reduces to zero. These are as follows:

The matrix element (4.59) is zero whenever the decomposition of the direct product $\Gamma^{(\tau)} \times \Gamma^{(\beta)}$ into irreducible components does not contain the representation $\Gamma^{(\alpha)}$.

Systems with differing physical structures may possess identical symmetries. However, the matrix elements which occur in quantal calculations on such systems differ only in the factor A_a, the Clebsch–Gordan coefficients being identical. Consequently, the Wigner–Eckart theorem enables one to "separate off" the symmetry properties of a system which is being studied from the detailed physical structure.

The operators which are usually employed in quantum mechanics are symmetric with respect to all particles, i.e., they transform according to the unit representation $\Gamma^{[N]}$ of the permutation group (see Chapter II). Since $\Gamma^{[N]} \times \Gamma^{[\lambda]} = \Gamma^{[\lambda]}$, the sum over a in (4.59) reduces to a single term. As a result, the Wigner–Eckart theorem assumes the following form:

$$\begin{aligned}\langle[\lambda_2]r_2 \,|\, T^{[N]} \,|\, [\lambda_1]r_1\rangle &= \langle[\lambda_2]r_2 \,|\, [N], [\lambda_1]r_2\rangle\langle[\lambda_2] \,||\, T^{[N]} \,||\, [\lambda_1]\rangle \\ &= \delta_{\lambda_1\lambda_2}\delta_{r_1r_2}\langle[\lambda_1] \,||\, T^{[N]} \,||\, [\lambda_1]\rangle, \qquad (4.60)\end{aligned}$$

where the double bars in the matrix element denote its independence of the Young tableaux r which enumerate the basis functions. Equation (4.60) is the basis of the well-known quantal selection rule according to which a perturbation described by a symmetric operator only induces transitions in the system between states with the same permutational symmetry.

4.8. Matrix Elements of Spherical Tensors

In the case of the group $\mathbf{R}_3$ the decomposition of a direct product into irreducible components does not give rise to repeated representations [see formula (3.43)], and therefore the sum over a in (4.59) reduces to a single

‡This theorem was proved by Wigner (1927) and by Eckart (1930a) for the three-dimensional rotational group. By writing it in the form (4.59), Koster (1958) subsequently extended the theorem to include an arbitrary finite group.

term,

$$\langle jm|T_k^{(\kappa)}|j'm'\rangle = \langle jm|\kappa k, j'm'\rangle A(j,\kappa,j'). \tag{4.61}$$

In place of the quantities $A(j,\kappa,j')$, one usually introduces the so-called reduced matrix elements $\langle j||T^{(\kappa)}||j'\rangle$, which are related to the $A(j,\kappa,j')$, by the equation

$$A(j,\kappa,j') = \{(-1)^{j+\kappa-j'}/[(2j+1)]^{1/2}\}\langle j||T^{(\kappa)}||j'\rangle. \tag{4.62}$$

This leads to the following two equivalent statements of the Wigner–Eckart theorem:

$$\begin{aligned}\langle jm|T_k^{(\kappa)}|j'm'\rangle &= (-1)^{j+\kappa-j'}\frac{\langle jm|\kappa k, j'm'\rangle}{(2j+1)^{1/2}}\langle j||T^{(\kappa)}||j'\rangle \\ &= (-1)^{j-m}\begin{pmatrix} j & \kappa & j' \\ -m & k & m'\end{pmatrix}\langle j||T^{(\kappa)}||j'\rangle.\end{aligned} \tag{4.63}$$

Since a reduced matrix element does not depend upon the values of the projections of the angular momenta m, k, and m', it is sufficient from the calculational point of view to find the simplest matrix element $\langle jm|T_k^{(\kappa)}|j'm'\rangle$ and to determine $\langle j||T^{(\kappa)}||j'\rangle$ from formula (4.63). For example, we might calculate the reduced matrix element of the angular momentum operator **J**. The spherical components of the angular momentum vector,

$$J_0 = J_z, \qquad J_{\pm 1} = \mp\sqrt{\tfrac{1}{2}}(J_x \pm iJ_y),$$

constitute a first rank irreducible spherical tensor. Since the $\psi_m^{(j)}$ are eigenfunctions of the operator J_z,

$$\langle jm|J_z|j'm'\rangle = \delta_{jj'}\delta_{mm'}m. \tag{4.64}$$

On the other hand, $J_z \equiv T_0^{(1)}$, and hence, according to (4.63),

$$\langle jm|J_z|jm\rangle = (-1)^{j-m}\begin{pmatrix} j & 1 & j \\ -m & 0 & m\end{pmatrix}\langle j||J||j\rangle. \tag{4.65}$$

The 3-j symbol which occurs in this equation is given by (Landau and Lifshitz, 1965)

$$\begin{pmatrix} j & 1 & j \\ -m & 0 & m\end{pmatrix} = \frac{(-1)^{j-m}m}{[j(j+1)(2j+1)]^{1/2}}. \tag{4.66}$$

Upon equating (4.64) and (4.65), we obtain

$$\langle j||J||j'\rangle = \delta_{jj'}[j(j+1)(2j+1)]^{1/2}. \tag{4.67}$$

For a scalar operator we have the equation

$$\langle jm|T^{(0)}|j'm'\rangle = [\delta_{jj'}\delta_{mm'}/(2j+1)^{1/2}]\langle j||T^{(0)}||j\rangle. \tag{4.68}$$

Let us determine the matrix element of the scalar product of two spherical tensors. According to (4.53),

$$\begin{aligned}&\langle jm|(T^{(\kappa)}\cdot U^{(\kappa)})|j'm'\rangle \\ &= \sum_k \sum_{j'',m''}(-1)^k\langle jm|T_k^{(\kappa)}|j''m''\rangle\langle j''m''|U_{-k}^{(\kappa)}|j'm'\rangle.\end{aligned} \tag{4.69}$$

We apply the Wigner–Eckart theorem to each term on the right-hand side of the equation. As a result, we obtain the following expression:

$$\sum_{j''}(-1)^{j+\kappa-j''}\frac{\langle j\|T^{(\kappa)}\|j''\rangle}{(2j+1)^{1/2}}(-1)^{j''+\kappa-j'}\frac{\langle j''\|U^{(\kappa)}\|j'\rangle}{(2j''+1)^{1/2}}$$
$$\times\sum_{k,m''}(-1)^{k}\langle jm\,|\,\kappa k, j''m''\rangle\langle j''m''\,|\,\kappa -k, j'm'\rangle. \quad (4.70)$$

By Eq. (3.47a) and (3.47b),

$$\langle j''m''\,|\,\kappa-k, j'm'\rangle$$
$$=(-1)^{2\kappa+j'-j''-k}\left(\frac{2j''+1}{2j'+1}\right)^{1/2}\langle j'm'\,|\,\kappa k, j''m''\rangle. \quad (4.71)$$

We substitute Eq. (4.71) into (4.70) and carry out the summation over the product of Clebsch–Gordan coefficients, making use of the orthogonality relations (3.46). We finally obtain

$$\langle jm\,|(T^{(\kappa)}\cdot U^{(\kappa)})|\,j'm'\rangle$$
$$=[\delta_{jj'}\delta_{mm'}/(2j+1)]\sum_{j''}(-1)^{j-j''}\langle j\|T^{(\kappa)}\|j''\rangle\langle j''\|U^{(\kappa)}\|j'\rangle. \quad (4.72)$$

Since the scalar product of two tensors is a zeroth rank tensor, equating Eqs. (4.68) and (4.72) leads to the following formula for the reduced matrix element of a scalar product:

$$\langle j\|(T^{(\kappa)}\cdot U^{(\kappa)})\|j'\rangle$$
$$=[\delta_{jj'}/(2j+1)]\sum_{j''}(-1)^{j-j''}\langle j\|T^{(\kappa)}\|j''\rangle\langle j''\|U^{(\kappa)}\|j'\rangle. \quad (4.73)$$

The summation is taken over all j'' that satisfy the triangle rule $\Delta(j\kappa j'')$.

The operators $T^{(\kappa)}$ and $U^{(\kappa)}$ in formulae (4.72) and (4.73) operate upon the coordinates of the whole system. Let the system being studied consist of two subsystems, characterized by angular momenta j_1 and j_2, respectively. Let $T^{(\kappa)}$ operate upon the coordinates of the first subsystem and $U^{(\kappa)}$ upon the coordinates of the second subsystem. We quote without proof several formulae which are useful in applications (Edmonds, 1957; Yutsis *et al.*, 1962a; Yutsis and Bandzaitis, 1965).

The reduced matrix element of a scalar product of operators which act upon different subsystems is equal to

$$\langle j_1j_2j\|(T^{(\kappa)}\cdot U^{(\kappa)})\|j_1'j_2'j'\rangle=\delta_{jj'}(-1)^{j_1'+j_2+j}(2j+1)^{1/2}\begin{Bmatrix} j_1 & j_2 & j \\ j_2' & j_1' & \kappa \end{Bmatrix}$$
$$\times\langle j_1\|T^{(\kappa)}\|j_1'\rangle\langle j_2\|U^{(\kappa)}\|j_2'\rangle. \quad (4.74)$$

The following two equations hold for operators which act upon the coordinates of only one subsystem:

$$\langle j_1j_2j\|T^{(\kappa)}\|j_1'j_2'j'\rangle=\delta_{j_2j_2'}(-1)^{j_1+j_2+j'+\kappa}[(2j+1)(2j'+1)]^{1/2}$$
$$\times\begin{Bmatrix} j_1 & j & j_2 \\ j' & j_1' & \kappa \end{Bmatrix}\langle j_1\|T^{(\kappa)}\|j_1'\rangle. \quad (4.75)$$

$$\langle j_1 j_2 j \| U^{(\kappa)} \| j_1' j_2' j' \rangle = \delta_{j_1 j_1'} (-1)^{j_1 + j_2' + j + \kappa} [(2j+1)(2j'+1)]^{1/2} \cdot \begin{Bmatrix} j_2 & j & j_1 \\ j' & j_2' & \kappa \end{Bmatrix} \langle j_2 \| U^{(\kappa)} \| j_2' \rangle. \tag{4.76}$$

Equations (4.75) and (4.76) assume a simple form in the case of a scalar operator. From Eqs. (3.61) and (3.63) it follows that

$$\begin{Bmatrix} a & b & e \\ d & c & 0 \end{Bmatrix} = \frac{(-1)^{a+b+e} \delta_{ab} \delta_{bd}}{[(2a+1)(2b+1)]^{1/2}}. \tag{4.77}$$

Substituting (4.77) for the 6-j symbol into (4.75) and (4.76), we obtain

$$\langle j_1 j_2 j \| T^{(0)} \| j_1' j_2' j' \rangle = \delta_{j_1 j_1'} \delta_{j_2 j_2'} \delta_{jj'} \left(\frac{2j+1}{2j_1+1} \right)^{1/2} \langle j_1 \| T^{(0)} \| j_1 \rangle, \tag{4.78}$$

$$\langle j_1 j_2 j \| U^{(0)} \| j_1' j_2' j' \rangle = \delta_{j_1 j_1'} \delta_{j_2 j_2'} \delta_{jj'} \left(\frac{2j+1}{2j_2+1} \right)^{1/2} \langle j_2 \| U^{(0)} \| j_2 \rangle. \tag{4.79}$$

Equations for the corresponding total matrix elements are easily found from expressions (4.78) and (4.79) for the reduced matrix elements if one makes use of Eq. (4.68). As a result, we arrive at two fairly obvious equations:

$$\langle j_1 j_2 jm | T^{(0)} | j_1' j_2' j' m' \rangle = \delta_{j_1 j_1'} \delta_{j_2 j_2'} \delta_{jj'} \delta_{mm'} \langle j_1 | T^{(0)} | j_1 \rangle, \tag{4.80}$$

$$\langle j_1 j_2 jm | U^{(0)} | j_1' j_2' j' m' \rangle = \delta_{j_1 j_1'} \delta_{j_2 j_2'} \delta_{jj'} \delta_{mm'} \langle j_2 | U^{(0)} | j_2 \rangle. \tag{4.81}$$

The matrix elements on the right-hand sides of Eqs. (4.80) and (4.81) do not depend upon the projections of the angular momenta. These matrix elements should not be confused with reduced matrix elements, since they are related to these last by Eq. (4.68).

SYMMETRY AND QUANTAL CALCULATIONS

A useful theory of complex systems is
like a well-drawn caricature, in which
the most typical properties of such systems
are exaggerated and all the other
nonessential properties deliberately ignored.

Ya. I. Frenkel

CHAPTER V

Principles of the Application of Group Theory to Quantum Mechanics

5.1. The Symmetry of the Schrödinger Equation and the Classification of States

The application of group theory to quantum mechanics is based upon the invariance of the Schrödinger equation with respect to transformations of a number of groups. The Schrödinger equation for an isolated quantal system of particles is invariant with respect to transformations of the following groups:

(a) The group of bodily translations of the system through space.

(b) The group of orthogonal transformations $\mathbf{O}_3$, consisting of bodily rotations of the system about any axis passing through the center of mass of the system (this point being taken as the origin of coordinates), and inversions through the origin.

(c) The permutation group of identical particles.

If the system is placed in an external field of force, the Schrödinger equation describing the motion of the system is then invariant with respect to (a) the symmetry group of the potential field in which the particles are moving, and (b) the permutation group of identical particles.

When the electronic and nuclear motions are separated (by the Born–Oppenheimer approximation (Born and Oppenheimer, 1927; Slater, 1963), one can regard the electrons as moving in an external field created by the nuclei fixed in their equilibrium configuration. The potential field then possesses the symmetry of the molecular point group.

Let the time-independent Schrödinger equation for a system of particles

$$\mathcal{H}\psi(x) = E\psi(x) \tag{5.1}$$

be invariant with respect to the transformations R of some group $\mathbf{G}$. This means that the Hamiltonian $\mathcal{H}$ does not alter its form under the transformations of $\mathbf{G}$. In this equation $\psi(x)$ denotes a stationary-state wavefunction of the system in a state of energy E, and x denotes the set of coordinates for the particles (including spin). We operate on both sides of this equation with an operation $R \in \mathbf{G}$. In accordance with our assumption, the Hamiltonian $\mathcal{H}$ remains invariant with respect to such an operation. Equation (5.1) therefore becomes

$$\mathcal{H} R\psi(x) = ER\psi(x). \tag{5.2}$$

From this it follows that $R\psi(x)$ is also a solution of the Schrödinger equation with the eigenvalue E.‡ One of two cases may occur:

(1) The energy level E is nondegenerate, so that

$$R\psi(x) = c\psi(x). \tag{5.3}$$

The normalization condition for the wavefunction requires $|c|^2 = 1$.

(2) The energy level E is f-fold degenerate, i.e., there are f linearly independent wavefunctions $\psi_i(x)$ which are solutions of Eq. (5.1) with the particular eigenvalue E. Since the function $R\psi_i(x)$ is also a solution of (5.1) with the same eigenvalue E, it can be expressed as a linear combination of the f linearly independent functions $\psi_i(x)$, i.e.,

$$R\psi_i(x) = \sum_{k=1}^{f} \Gamma_{ki}(R)\psi_k(x) \tag{5.4}$$

for any operation $R \in \mathbf{G}$. The system of eigenfunctions $\psi_i(x)$ therefore forms a basis for an f-dimensional representation of $\mathbf{G}$.

It thus follows from the symmetry properties of the Schrödinger equation that wavefunctions which belong to a given energy level transform according to a representation of the symmetry group of this equation. In general, this representation is reducible and can be decomposed into irreducible components by a suitable transformation. As a result of this, the wavefunctions are split up into irreducible sets. All wavefunctions which are members of the same set belong to the same energy level. A concurrence in the values of the energy for different sets not caused by the symmetry of the Schrödinger equation is called an *accidental depeneracy*. For example, in the case of a system of particles subjected to a magnetic field two energy levels which correspond to two different irreducible representations may coincide at a certain field strength. At this particular magnetic field strength, therefore, there is an accidental degeneracy.

Special mention must be made of accidental degeneracies which occur as a result of an approximate solution of a problem. For example in the Born–Oppenheimer approximation, which is used in quantal studies of the properties of molecules and solids, the electrons are regarded as moving in the potential field of fixed nuclei. At certain configurations of the nuclei different

‡This result, which is basic for all applications of group theory to quantum mechanics, was first obtained by Wigner (1927). It is sometimes referred to as *Wigner's theorem*.

electronic terms may cross, i.e., there is an accidental degeneracy. The energies of the electronic terms are found approximately. In solving this problem in the zeroth order of perturbation theory, one neglects the interactions between the electrons. The solutions of the approximate Schrödinger equation which is obtained must be classified, just as for the exact nonrelativistic equation, according to the irreducible representations of the permutation group of the electrons. However in the zeroth approximation, wavefunctions that transform according to different irreducible representations of the permutation group describe states that possess identical energies. This additional degeneracy is caused by the factorization in this approximation of the Schrödinger equation for the system of electrons into Schrödinger equations for the individual electrons.

The occurrence of this kind of accidental degeneracy can usually be detected since as a rule degenerate energy levels which correspond to solutions of the exact Schrödinger equation arise from the symmetry of the equation. The eigenfunctions of every energy level consequently induce an irreducible representation of the symmetry group of the Schrödinger equation, the dimension of this representation being determined by the degree of degeneracy of the particular energy level. As a result, it is possible without solving the Schrödinger equation immediately to determine the permissible degrees of degeneracy of the energy levels and the transformation laws for the wavefunctions under the operations of the symmetry group of the system.

As an example, we consider the molecule H_2O. The equilibrium configuration of the nuclei possesses the symmetry of the point group $\mathbf{C}_{2v}$. With the nuclei in this configuration one classifies the electronic energy levels of the molecule according to the irreducible representations of $\mathbf{C}_{2v}$. This group is Abelian, and all its irreducible representations are one dimensional. From this we conclude immediately that the system cannot possess any degenerate levels (provided there is no accidental degeneracy). The wavefunctions of the system can therefore be divided into four types, corresponding to the four irreducible representations of the group $\mathbf{C}_{2v}$.

In the case of $\mathbf{C}_{3v}$ point symmetry (molecules NH_3, CH_3Cl, etc.), two kinds of nondegenerate states and one kind of doubly degenerate state are possible. No other kinds of states exist in such systems. For example, there are no triply degenerate levels.

The nonrelativistic motion of an electron in a Coulombic field (H atom, He^+ ion, etc.) constitutes an interesting example as regards the classification of states. The Schrödinger equation for this problem is invariant with respect to transformations of the group $\mathbf{O}_3$. This last is the direct product of $\mathbf{R}_3 \times \mathbf{C}_i$. The spatial wavefunctions of the electron are classified according to the $(2l + 1)$-dimensional irreducible representations of the group $\mathbf{R}_3$ (which specify a value for the orbital angular momentum of the electron and a value

m for its projection along the z axis), and according to the one-dimensional irreducible representations of the group $\mathbf{C}_i$ (which specify the parity of a state). One would expect the energy levels of the electron to be degenerate only in m, so that the degree of degeneracy is equal to $2l + 1$, where $l = 0, 1, \ldots$. However, the solution of the Schrödinger equation shows that in addition to the quantum numbers l and m, the electron states are characterized by a principal quantum number n such that the energy levels are designated by this quantum number only. All states with the same n but with different l belong to a single energy level, the degeneracy of which is equal to n^2. It turns out that this degeneracy is not accidential at all. As was shown by Fock (1935a,b) [see also Petrashen and Trifonov (1969)], the Schrödinger equation for the motion of an electron in a Coulombic field, when written in the momentum representation, is invariant with respect to transformations of the four-dimensional rotation group $\mathbf{R}_4$, of which $\mathbf{R}_3$ is a subgroup. The energy levels must therefore be classified according to the irreducible representations of the group $\mathbf{R}_4$. Upon reducing the group $\mathbf{R}_4 \longrightarrow \mathbf{R}_3$, the irreducible representations of $\mathbf{R}_4$ according to which the energy levels have been classified decompose into irreducible representations $D^{(l)}$ which naturally belong to a single energy level.

It should be noted that in the absence of a magnetic field the Hamiltonian is real, and therefore a wavefunction and its complex conjugate must belong to one and the same energy level. In fact $\psi^*(x)$ satisfies Eq. (5.1) with the same eigenvalue as $\psi(x)$:

$$\mathcal{H}\psi^*(x) = E\psi^*(x). \tag{5.1a}$$

An irreducible representation and its complex conjugate must therefore belong to the same energy level, i.e., when classifying states they can be regarded as a single representation with double the dimension.‡ The only molecular point groups that possess complex conjugate representations are the axial point groups $\mathbf{C}_n$ and $\mathbf{C}_{nh}$.

5.2. Conservation Laws

Equation (5.2) can be written in the form

$$\mathcal{H}R\psi(x) = R\mathcal{H}\psi(x), \tag{5.5}$$

from which it follows that

$$\mathcal{H}R = R\mathcal{H}, \tag{5.6}$$

i.e., from the invariance of the Hamiltonian with respect to the transformation

‡Translator's note: From this it follows that a wavefunction belonging to a nondegenerate energy level may always be taken to be real.

R it follows that their commutator

$$[\mathcal{H}, R] \equiv \mathcal{H}R - R\mathcal{H}$$

is equal to zero.

In quantum mechanics it can be proved [see for example, Landau and Lifshitz (1965)] that the necessary condition for a physical quantity which corresponds to a Hermitian operator T to be conserved is that the commutator $[T, \mathcal{H}] = 0$. In such cases one speaks of the existence of a *conservation law.*

Since any operation of the symmetry group of the Schrödinger equation commutes with the Hamiltonian, there is a certain conservation law corresponding to every element of the group. Furthermore, in the case of continuous groups there is a continuum of conservation laws. In order to distinguish the independent conservation laws for such groups,‡ one can make use of the following general theorem (Lyubarskii, 1960):

Let the Schrödinger equation be invariant with respect to the transformations of some r-parameter Lie group. Each of the r infinitesimal operators then corresponds to a conserved physical quantity. Furthermore, any conservation law which is due to symmetry with respect to the transformations of the particular Lie group is a consequence of the r conservation laws corresponding to the infinitesimal operators of the group.

As an example, we consider a system of N particles in the absence of an external field. The potential energy of such a system remains unchanged when the system is subjected to a parallel displacement in space. Consequently, the Hamiltonian of the system is invariant with respect to the group of parallel displacements. This group constitutes a three-parameter Lie group, and we seek the form of its infinitesimal operators.

The transformations of the group are defined in the $3N$-dimensional configuration space of the system which is formed by the set of three coordinates x_i, y_i, and z_i for each particle. Each operation is characterized by specifying three parameters, for which one could choose the three Cartesian components of the displacement vector $\mathbf{a}$. As a result of an operation of the group, the position vector of every particle in the system is increased by the vector $\mathbf{a}$:

$$\mathbf{r}_i' = \mathbf{r}_i + \mathbf{a}. \tag{5.7}$$

From formula (3.13) we find

$$\begin{gathered} u_{x_i x} = u_{y_i y} = u_{z_i z} = 1, \\ u_{y_i x} = u_{z_i x} = u_{x_i y} = u_{z_i y} = u_{x_i z} = u_{y_i z} = 0. \end{gathered} \tag{5.8}$$

Upon substituting (5.8) into the definition (3.16) of infinitesimal operators, we

‡The determination of the independent conservation laws in the case of finite groups has been studied by Sokolov and Shirokovsky (1956).

obtain

$$I_x = \sum_{i=1}^{N} \frac{\partial}{\partial x_i}, \qquad I_y = \sum_{i=1}^{N} \frac{\partial}{\partial y_i}, \qquad I_z = \sum_{i=1}^{N} \frac{\partial}{\partial z_i}. \tag{5.9}$$

These infinitesimal operators are proportional to the operators for the components of the linear momentum of the system:

$$P_x = -i\hbar I_x, \qquad P_y = -i\hbar I_y, \qquad P_z = -i\hbar I_z. \tag{5.10}$$

Hence in the absence of a field there is a conservation law for each of the three components of the linear momentum of the system, or, as a result of this, there is a conservation law for the linear momentum vector of the system.

If the system is placed in a field with central symmetry, the Schrödinger equation is invariant with respect to rotations in three-dimensional space. The infinitesimal operators of the group $\mathbf{R}_3$ are, to within a constant, just the operators for the components of the angular momentum (3.24). In a field with central symmetry, therefore, there is a conservation law for each component of the angular momentum of the system. The angular momentum vector of the system, in short, is conserved.

As a result of the homogeneity and isotropy of space, conservation laws for both the linear momentum and the angular momentum vectors are valid for an isolated system. In the case of an external field with the symmetry of a discrete point group a set of independent conserved quantities can always be picked out. Thus in a potential field with cubic symmetry a certain function of the components of the angular momentum is conserved (Sokolov and Shirokovsky, 1956).

5.3. Perturbation Theory

We consider a physical system, the Hamiltonian of which is invariant with respect to the operations of a group $\mathbf{G}_0$. The solutions of the Schrödinger equation

$$\mathcal{H}_0 \psi^{(0)} = E^{(0)} \psi^{(0)} \tag{5.11}$$

belong to irreducible representations of the group $\mathbf{G}_0$. If the system is placed in an external time-independent field with a potential V which is invariant under the transformations of a group $\mathbf{G}_1$, the states of the system are described by the solutions of the equation

$$(\mathcal{H}_0 + V)\psi = E\psi, \tag{5.12}$$

and are classified according to the irreducible representations of the symmetry group of this equation. We denote this group by $\mathbf{G}$. This group is the "intersection" of the groups $\mathbf{G}_0$ and $\mathbf{G}_1$, and consists of elements which are common to both $\mathbf{G}_0$ and $\mathbf{G}_1$. We distinguish two cases:

(1) The group $\mathbf{G}_1$ is either larger than $\mathbf{G}_0$ or coincides with it. In this case the symmetry group of Eq. (5.12) remains the same as for Eq. (5.11). The

states of the system are classified according to the irreducible representations of $\mathbf{G}_0$, just as before the imposition of the field V.

(2) The group $\mathbf{G}_1$ constitutes a subgroup of the group $\mathbf{G}_0$. The symmetry of Eq. (5.12) is then the same as the symmetry of the field V.‡ The new levels of the system are classified according to the irreducible representations of the group $\mathbf{G}_1$. The dimensions of the irreducible representations of a subgroup are always less than or equal to the corresponding dimensions for the group (reduction with respect to a subgroup). Therefore in this case the imposition of a field V lowers the degree of degeneracy of the energy levels.

If V can be regarded as a small correction to a field which is already acting on the system, a solution of Eq. (5.12) can be obtained by the methods of perturbation theory (Landau and Lifshitz, 1965; Eyring *et al.*, 1944). The solution is sought in the form of an expansion in some small parameter. Let the energy level $E^{(0)}$ of the unperturbed system be f-fold degenerate; we denote the wavefunctions which belong to this level by $\psi_\nu^{(0)}$, $\nu = 1, 2, \ldots, f$. In the first order of perturbation theory, the corrections $E^{(1)}$ to the energy $E^{(0)}$ are found as solutions of a *secular equation*

$$\begin{vmatrix} V_{11} - E^{(1)} & V_{12} & \cdots & V_{1f} \\ V_{21} & V_{22} - E^{(1)} & \cdots & V_{2f} \\ \vdots & \vdots & \vdots & \vdots \\ V_{f1} & V_{f2} & \cdots & V_{ff} - E^{(1)} \end{vmatrix} = 0, \tag{5.13}$$

where the matrix elements of the operator V are defined in terms of the unperturbed wavefunctions:

$$V_{\nu\nu'} = \langle \psi_\nu^{(0)} | V | \psi_{\nu'}^{(0)} \rangle. \tag{5.14}$$

Perturbation theory can be applied as long as the splitting of the level $E^{(0)}$ is small compared to the difference between the levels of the unperturbed problem.

We wish to discover into how may levels an originally degenerate level splits when a perturbation with different symmetry is applied. Before answering this question we demonstrate the following theorem:

Let $T^{(0)}$ be an operator which is invariant with respect to the transformations of a group $\mathbf{G}$. A matrix element of this operator which connects basis functions belonging to different irreducible representations of $\mathbf{G}$ is then diagonal both with respect to the irreducible representations and with respect to the

‡It is obvious that if one adds to an expression which possesses the symmetry of some group an expression which possesses the symmetry of a subgroup of this, their sum will possess the symmetry of the subgroup. The symmetry of a function can thus be completely destroyed by adding to it an asymmetric term.

indices of the individual basis functions. In addition, the matrix element does not depend upon these indices, i.e.,

$$\langle a\alpha i | T^{(0)} | a'\alpha' i' \rangle = \delta_{\alpha\alpha'} \delta_{ii'} \langle a || T^{(0)} || a' \rangle^{(\alpha)}, \tag{5.15}$$

where a distinguishes bases for equivalent irreducible representations $\mathbf{\Gamma}^{(\alpha)}$.

The proof follows immediately from the Wigner–Eckart theorem (4.59) if one takes into account that an operator which is invariant with respect to the transformations of a group belongs to the unit (totally symmetric) representation A_1. Since for any representation

$$\mathbf{\Gamma} \times A_1 = \mathbf{\Gamma},$$

the summation over a in (4.59) reduces to a single term, and the Clebsch–Gordan coefficient is equal to

$$\langle \alpha i | A_1, \alpha' i' \rangle = \delta_{\alpha\alpha'} \delta_{ii'}. \tag{5.16}$$

Let the symmetry of a perturbation V be the same as or higher than that of $\mathcal{H}_0$. The operator V is then invariant with respect to transformations of the group $\mathbf{G}_0$. By theorem (5.15), the off-diagonal matrix elements in the secular equation (5.13) are zero, and the diagonal elements are all equal. Equation (5.13) therefore possesses a single root, repeated f times, whose value gives the displacement of the level $E^{(0)}$. No splitting occurs in this case.

However, the situation is quite different if an accidental degeneracy is present (for example, the degeneracy resulting from the factorization of the Schrödinger equation when the interelectronic interactions are neglected; see Section 5.1). In such a case the representation $\mathbf{\Gamma}$ which is generated by the f functions $\psi_i^{(0)}$ contains m irreducible representations $\mathbf{\Gamma}^{(\beta)}$:

$$\mathbf{\Gamma} \doteq \sum_{\beta=1}^{m} \mathbf{\Gamma}^{(\beta)}, \tag{5.17}$$

which, because of the accidental degeneracy, all belong to a single energy level. Such a degeneracy can also be removed by a perturbation. It is first of all necessary to form from the f functions $\psi_i^{(0)}$ basis functions for the irreducible representations $\mathbf{\Gamma}^{(\beta)}$. If none of the irreducible representations in (5.17) are repeated, then according to (5.15), the secular equation is diagonalized, with the diagonal elements that correspond to a single irreducible representation all identical. If one of the irreducible representations, $\mathbf{\Gamma}^{(\alpha)}$ say, occurs a number of times, then there are nonzero off-diagonal matrix elements which connect functions that transform according to one and the same column of $\mathbf{\Gamma}^{(\alpha)}$. One now has to solve a secular equation whose dimension is equal to the number of times $\mathbf{\Gamma}^{(\alpha)}$ occurs in (5.17). In both cases the level $E^{(0)}$ is split into not more than m levels. This is to be expected since if one assumes that the perturbation does not lower the symmetry, then the degeneracy associated with the symmetry of the Hamiltonian must be preserved.

We now consider the case when the symmetry group of the perturbation is a subgroup of the symmetry group of the unperturbed Hamiltonian, $\mathbf{G}_0$. The operator V is invariant with respect to the transformations of a group $\mathbf{G}_1 \subset \mathbf{G}_0$. An irreducible representation of the group may be reducible as regards a subgroup, and may be decomposed into irreducible representations of the subgroup. In such a case the energy level $E^{(0)}$ is split. The matrix elements will, as usual, conform to theorem (5.15) if one forms from the functions $\psi_i^{(0)}$ basis functions for irreducible representations of $\mathbf{G}_1$. By making use of character tables one can determine, without having to solve the secular equation, the possible splitting of an originally degenerate energy level due to the application of the perturbation.

For example, let the Hamiltonian $\mathcal{H}_0$ be symmetric with respect to the operations of the group **T**. We examine the behavior of a triply degenerate level, belonging to the F_1 irreducible representation, when a perturbing field of symmetry $\mathbf{D}_2$ is applied. We write out the part of the character table for **T** that corresponds to the operations of the group $\mathbf{D}_2$:

	E	C_2^x	C_2^y	C_2^z
$\chi^{(F_1)}$	3	-1	-1	-1

The characters of the irreducible representations of $\mathbf{D}_2$ are given in Appendix 1, and by formula (1.63) we obtain

$$a^{(A_1)} = 0, \qquad a^{(B_1)} = a^{(B_2)} = a^{(B_3)} = 1.$$

The originally triply degenerate F_1 level is split into three nondegenerate energy levels, corresponding to the B_1, B_2, and B_3 irreducible representations of $\mathbf{D}_2$. In this example the perturbation lifts the degeneracy completely.

As another example we take a system whose Hamiltonian is centrally symmetric and study the splitting of its energy levels which occurs when a perturbing field with octahedral symmetry (point group **O**) is applied. The energy levels of the unperturbed problem are classifed according to the irreducible representations $D^{(j)}$ of the group $\mathbf{R}_3$. The energy levels of the perturbed system must be classified according to the irreducible representations of the group **O** with respect to which the representations $D^{(j)}$ are reducible (except for the lowest value of j). In order to find which irreducible representations of **O** are contained in a representation $D^{(j)}$, we determine the characters of $D^{(j)}$ that correspond to the operations of **O** by formula (3.40). Thus for $j = 2$

$$\chi^{(2)}(\alpha) = (\sin \tfrac{5}{2}\alpha)/(\sin \tfrac{1}{2}\alpha). \tag{5.18}$$

Substituting in this equation the angles α that correspond to the rotations C_2, C_3, and C_4, we obtain a reducible representation of the group **O** with the characters:

	E	$8C_3$	$6C_2$	$6C_4$	$3C_4^2$
$\chi^{(2)}(\alpha)$	5	−1	1	−1	1

This decomposes into the following irreducible representations:

$$D^{(2)} \doteq E + F_2.$$

A fivefold degenerate energy level with $j = 2$ thus splits in an octahedral field into a doubly degenerate and a triply degenerate energy level. The decomposition of the representations $D^{(j)}$ for several integral values of j is given in Table 5.1.

Table 5.1

Reduction of $\mathbf{R}_3 \longrightarrow \mathbf{O}$

$D^{(j)}$	$\sum_\beta \Gamma^{(\beta)}$
$j = 0$	A_1
1	F_1
2	$E + F_2$
3	$A_2 + F_1 + F_2$
4	$A_1 + E + F_1 + F_2$
5	$E + 2F_1 + F_2$
6	$A_1 + A_2 + E + F_1 + 2F_2$

5.4. The Variation Method

As is well known, the Schrödinger equation can be obtained from the variation principle

$$\delta \int \psi^*(\mathcal{H} - E)\psi \, dV = 0. \tag{5.19}$$

In an alternative formulation of this principle the energy E occurs as a Lagrange multiplier in the problem of finding the extremum of

$$\delta \int \psi^* \mathcal{H} \psi \, dV = 0, \tag{5.20}$$

subject to the additional condition

$$\int \psi^* \psi \, dV = 1. \tag{5.21}$$

For many-electron systems the variational function ψ is usually sought in the form of a linear combination of trial functions ψ_ν (Slater, 1963; Gombas, 1950):

$$\psi = \sum_{\nu=1}^{n} c_\nu \psi_\nu. \tag{5.22}$$

The functions ψ_ν should be chosen so as to conform as closely as possible to the specific characteristics of the system under study. If (5.22) is substituted into (5.20) and the coefficients c_ν varied, a set of equations is obtained from which the c_ν can be determined. This set of equations only has nonzero solutions for those values of the energy E that satisfy the secular equation

$$\begin{vmatrix} \mathcal{H}_{11} - S_{11}E & \mathcal{H}_{12} - S_{12}E & \cdots & \mathcal{H}_{1n} - S_{1n}E \\ \mathcal{H}_{21} - S_{21}E & \mathcal{H}_{22} - S_{22}E & \cdots & \mathcal{H}_{2n} - S_{2n}E \\ \vdots & \vdots & \vdots & \vdots \\ \mathcal{H}_{n1} - S_{n1}E & \mathcal{H}_{n2} - S_{n2}E & \cdots & \mathcal{H}_{nn} - S_{nn}E \end{vmatrix} = 0. \tag{5.23}$$

The appearance in (5.23) of the overlap integral

$$S_{\nu\nu'} = \int \psi_\nu^* \psi_{\nu'} \, dV$$

is due to the fact that the trial functions do not always form an orthonormal set. The smallest root gives an approximation to the energy of the lowest state, and the other roots give approximations to the energies of higher states. These last, however, are usually worse approximations than that for the lowest state.

It is interesting to note that if in perturbation theory one takes, instead of the perturbation V, the entire Hamiltonian $\mathcal{H} = \mathcal{H}_0 + V$ and for the sake of complete generality assumes that the functions $\psi_\nu^{(0)}$ are not orthonormal, then the secular equation (5.13) so modified becomes identical to the secular equation (5.23). Hence the variation method is formally identical to perturbation theory for degenerate states. The difference between the two methods is that in perturbation theory for degenerate states the functions $\psi_\nu^{(0)}$ in the zeroth approximation all belong to a single energy level, whereas in the variation method there is no such restriction, and the functions ψ_ν in (5.22) may be arbitrary.

The more functions one puts into the sum (5.22), the better are the results given by the variation method. This, however, leads to secular equations of very high order, which are very difficult to solve even with the use of computers. For example, the problem of finding the energy levels of a system of six hydrogen atoms, in which all the configurations that can be constructed from six 1s orbitals are taken into account, leads to a secular equation of order 924 to be solved (see Section 6.11). A solution is possible only if the secular equation is partially diagonalized beforehand. This is achieved by forming from the initial trial functions linear combinations which transform according to irreducible representations of the symmetry group of the Hamiltonian. In some cases this procedure may diagonalize the secular equation completely. We now consider this process in more detail.

Let the Hamiltonian of the problem be invariant under the operations of a group **G**. We divide the n trial functions which occur in the expansion of the variational function (5.22) into separate sets, within each of which the functions transform among themselves under the operations of **G**. Each such set induces a representation of **G**. These representations obviously cannot have dimensions greater than that of the regular representation, and consequently in their decomposition into irreducible components the number of times with which each irreducible representation $\Gamma^{(\alpha)}$ occurs cannot be greater than its dimension f_α. The construction of basis functions for irreducible representations is conveniently effected by means of the operators $\epsilon_{ik}^{(\alpha)}$ (see Section 1.19). The set of f_α functions $\psi_{ik}^{(\alpha)}$

$$\psi_{ik}^{(\alpha)} = \epsilon_{ik}^{(\alpha)} \psi_\nu \tag{5.24}$$

with fixed index k forms a basis for an irreducible representation $\Gamma^{(\alpha)}$. The index k distinguishes linearly independent bases. In cases where the representation $\Gamma^{(\alpha)}$ occurs f_α times in the decomposition of the reducible representation, index k assumes all possible f_α values. If the number of times an irreducible representation occurs is $a^{(\alpha)} < f_\alpha$, then in order to construct $a^{(\alpha)}$ linearly independent bases for the representation $\Gamma^{(\alpha)}$, one arbitrarily picks out $a^{(\alpha)}$ sets of operators $\epsilon_{ik}^{(\alpha)}$ with differing second index. Any one of the functions in the given set may be chosen as the function ψ_ν in (5.24). Should the application of $\epsilon_{ik}^{(\alpha)}$ to a particular ψ_ν give zero, then the next function in the set is chosen, this process being continued until a nonzero result is obtained.

According to theorem (5.15), the secular equation is considerably simplified when one carries out the linear transformation from the functions ψ_ν to the functions $\psi_{ik}^{(\alpha)}$. Diagonal elements which belong to a single irreducible representation become equal. All off-diagonal elements reduce to zero except those that connect basis functions that transform according to one and the same column of equivalent irreducible representations, i.e., the nonzero matrix elements are of the form

$$\langle \psi_{ik}^{(\alpha)} | \mathcal{H} | \psi_{ik'}^{(\alpha)} \rangle. \tag{5.25}$$

All the overlap integrals also reduce to zero except for

$$S_{kk'} = \int \psi_{ik}^{(\alpha)*} \psi_{ik'}^{(\alpha)} \, dV. \tag{5.26}$$

Consequently, the original secular equation factorizes into a set of secular equations, one for each irreducible representation which is repeated. The order of the secular equation corresponding to an irreducible representation $\Gamma^{(\alpha)}$ is equal to the total number of times that $\Gamma^{(\alpha)}$ occurs in the decomposition of all the reducible representations which are generated by the original set of functions ψ_ν. If each irreducible representation occurs just once, the secular equation is completely diagonalized. Examples of this procedure are described in Sections 6.11 and 8.8.

5.5. Selection Rules

In the previous two sections it was shown that in finding the energy levels of a system, either by perturbation theory or by the variation method, one has to calculate the matrix elements of a scalar operator which belongs to the totally symmetric or unit representation of the symmetry group of the Schrödinger equation. In a number of problems one can divide the system into two interacting subsystems; for example, the interaction of atoms and molecules with a photon field, the spin–orbit interaction, etc. In determining the probability that the subsystem undergoes a transition from one state to another, it is necessary to calculate the matrix elements of tensor operators. The application of the Wigner–Eckart theorem substantially simplifies the calculation of such matrix elements (see Sections 4.7 and 4.8). Furthermore, the application of this theorem makes it possible, without resorting to an actual calculation, to determine which matrix elements reduce to zero. This gives rise to rules known as *selection rules*, and which can be formulated as follows:

Consider the matrix element

$$\langle \alpha i | T | \beta k \rangle \equiv \int \psi_i^{(\alpha)*} T \psi_k^{(\beta)} \, dV, \tag{5.27}$$

in which the components of the tensor T transform according to a representation $\Gamma^{(T)}$ of some group, and the functions $\psi_i^{(\alpha)}$ and $\psi_k^{(\beta)}$ according to the representations $\Gamma^{(\alpha)}$ and $\Gamma^{(\beta)}$, respectively, of the same group. Then a necessary condition for this matrix element to be nonzero is that the representation $\Gamma^{(\alpha)}$ occur in the decomposition of the direct product $\Gamma^{(T)} \times \Gamma^{(\beta)}$.

The representation $\Gamma^{(T)}$ is usually reducible and when determining the selection rules it is convenient to decompose it beforehand into irreducible representations:

$$\Gamma^{(T)} \doteq \sum_{\tau} a^{(\tau)} \Gamma^{(\tau)}. \tag{5.28}$$

The integrand in (5.27) belongs to a basis for a representation which is the direct product of the three representations

$$\Gamma^{(\alpha)*} \times \Gamma^{(T)} \times \Gamma^{(\beta)}. \tag{5.29}$$

If the direct product $\Gamma^{(T)} \times \Gamma^{(\beta)}$ contains $\Gamma^{(\alpha)}$, then the direct product (5.29) necessarily contains the unit representation (see Section 1.16). In this way we arrive at the following alternative formulation of selection rules:

A necessary condition for the matrix element (5.27) to be nonzero is that the unit representation occur in the decomposition of the direct product (5.29).

We consider some examples of the application of selection rules. Let the operator T be a scalar with respect to the operations of a group $\mathbf{G}$. The selection rules for this case have, in fact, already been given in Section 5.3 [Eq. (5.15)]. The quantal transition probabilities can be stated as follows:

A scalar operator induces transitions between states of the same type of symmetry only, i.e., between states which transform according to equivalent irreducible representations. Transitions between states of differing symmetry type are forbidden.

The probability of a radiative transition is proportional, in the dipole approximation, to the square of the matrix element of the dipole moment vector, which in the case of electric radiation is a polar vector and in the case of magnetic radiation is an axial vector.‡ We determine the selection rules for the point group $\mathbf{D}_{2d}$. The three components of the dipole moment vector form a basis for a reducible representation with respect to the operations of $\mathbf{D}_{2d}$. In the case of the electric dipole moment this representation can be decomposed into the irreducible representations B_2 and E. The character table of $\mathbf{D}_{2d}$ (Appendix 1) shows the irreducible representations according to which the components of a polar vector transform. The z component belongs to the B_2 representation, and the x and y components form a basis for the E representation. In order to determine the allowed transitions, one must decompose the direct products $(B_2 + E) \times \mathbf{\Gamma}^{(\beta)}$, where $\mathbf{\Gamma}^{(\beta)}$ runs through all the irreducible representations of $\mathbf{D}_{2d}$. We find, by formulae (1.63) and (1.65),

$$\begin{aligned} &B_2 \times A_1 \doteq B_2, \qquad && E \times A_1 \doteq E, \\ &B_2 \times A_2 \doteq B_1, && E \times A_2 \doteq E, \\ &B_2 \times B_1 \doteq A_2, && E \times B_1 \doteq E, \\ &B_2 \times B_2 \doteq A_1, && E \times B_2 \doteq E, \\ &B_2 \times E \doteq E, && E \times E \doteq A_1 + A_2 + B_1 + B_2. \end{aligned} \tag{5.30}$$

From this equation it follows that for electric dipole radiation along the z axis the allowed transitions are

$$A_1 \longleftrightarrow B_2, \qquad A_2 \longleftrightarrow B_1, \qquad E \longleftrightarrow E, \tag{5.31}$$

and for radiation polarized in the xy plane

$$A_1, A_2, \quad B_1, B_2 \longleftrightarrow E. \tag{5.32}$$

All other transitions are forbidden.

According to the character table for $\mathbf{D}_{2d}$ (Appendix 1), the z component of an axial vector belongs to the A_2 representation, and the x and y components

‡An ordinary, or polar, vector changes sign on inversion, and remains unchanged under a reflection in a plane σ_v passing through the vector. An axial vector remains unchanged on inversion, but changes sign under a reflection σ_v. The vector product of two polar vectors is an example of an axial vector.

to the E representation. The selection rules for magnetic dipole radiation when polarized in the xy plane therefore coincide with the selection rules (5.32). However, the selection rules for magnetic dipole radiation polarized along the z axis will differ from (5.31). A similar procedure in this case leads to the following allowed transitions:

$$A_1 \longleftrightarrow A_2, \qquad B_1 \longleftrightarrow B_2, \qquad E \longleftrightarrow E. \tag{5.33}$$

The operator in the transition matrix element in the case of electric quadrupole radiation is a second rank spherical tensor. Its five components transform according to the representation $D^{(2)}$ of the group $\mathbf{R}_3$. The selection rules for transitions between states with the symmetry of irreducible representations of the group $\mathbf{R}_3$ follow from the triangle rule $\Delta(j2j')$ [see Eq. (3.44)]. In the case of discrete point groups it is necessary to decompose the tensor $T^{(2)}$ into irreducible tensors corresponding to the particular point group, i.e., it is necessary actually to carry out the reduction of $D^{(2)}$ with respect to the point group. For example, for the group $\mathbf{O}$, according to Table 5.1 $D^{(2)} = E + F_2$. The direct products of the E and F_2 representations with all the irreducible representations of $\mathbf{O}$ are then decomposed. As a result, we obtain for the allowed transitions the following:

$$E \longleftrightarrow A_1, A_2, E; \quad F_1 \longleftrightarrow A_2, E, F_1, F_2; \quad F_2 \longleftrightarrow A_1, E, F_1, F_2. \tag{5.34}$$

The selection rules for 2^l-pole radiation of any order are similarly determined. The representation $D^{(l)}$ is first decomposed into irreducible representations of the particular point group, after which the usual procedure is followed.

In determining the selection rules for diagonal matrix elements, one should note that if the wavefunctions are real, their product in a diagonal matrix element forms a basis, not for the direct product of the representation, but for the symmetric product of the representation with itself (see Section 1.16). This situation arises in the absence of a magnetic field since all the wavefunctions can be chosen to real, as can also the representations according to which they transform. The condition for a diagonal matrix element to be nonzero is that one of the irreducible representations according to which the operator transforms occurs in the decomposition of the symmetric product $[\Gamma^{(\alpha)}]^2$.

For example, in the case of the group $\mathbf{D}_{2d}$ the electric dipole moment operator belongs to the representation $B_2 + E$. We find the characters of the symmetric products of all the irreducible representations from formula (1.69). Their decomposition into irreducible components is as follows:

$$[A_1]^2 \doteq A_1, \qquad [A_2]^2 \doteq A_1, \qquad [B_1]^2 \doteq A_1,$$
$$[B_2]^2 \doteq A_1, \qquad [E]^2 \doteq A_1 + B_1 + B_2.$$

Since, with the exception of $[E]^2$ none of the symmetric products contains the representations B_2 or E, we conclude that the only nonvanishing diagonal matrix element is that with the E basis functions.

The bands in molecular spectra in the infrared region show, in addition to the fundamental lines, other lines of lower intensity known as *overtones*, which are due to the absorption of N quanta of the fundamental frequency. The selection rules for overtones are the same as those for ordinary dipole transitions. It is only necessary to recall that the representation to which the wavefunction of a bound state belongs is the symmetric Nth power of the representation according to which the corresponding normal coordinate transforms (Landau and Lifshitz, 1965). In Section 4.3 formulae were given from which the characters of the symmetric Nth power of a representation can easily be found for any values of N.

This procedure for determining selection rules is equivalent to calculating by means of first-order perturbation theory the probability that a quantal system undergoes a transition from state 1 to state 2. This is because in this approximation the transition probability is proportional to the squared modulus of the matrix element of the transition operator (we denote this matrix element by V_{12}). It may happen that $V_{12} = 0$ for the transition under consideration. In second-order perturbation theory a quantum transition is regarded as arising from intermediate states. This occurs, for example, in Mandelshtam–Raman spectra (Herzberg, 1947). The transition probability is proportional to the squared modulus of the sum over all intermediate states ν (Landau and Lifshitz, 1965):

$$\sum_{\nu} V_{1\nu} V_{\nu 2}/(E_1 - E_\nu). \tag{5.35}$$

The selection rules for both the matrix elements $V_{1\nu}$ and $V_{2\nu}$ must be satisfied if their product in this equation is not to reduce to zero. We denote the representations to which the wavefunctions of states 1, 2, and ν belong by $\Gamma^{(1)}$, $\Gamma^{(2)}$, and $\Gamma^{(\nu)}$, respectively, and the representation according to which the operator in the matrix elements transforms by $\Gamma^{(T)}$. The matrix element $V_{\nu 2}$ is only nonzero for those $\Gamma^{(\nu)}$ that are contained in the direct product $\Gamma^{(T)} \times \Gamma^{(2)}$. Similarly the matrix elements $V_{1\nu}$ are nonzero only if the representation $\Gamma^{(1)}$ is contained in the direct product of the representations $\Gamma^{(T)} \times \Gamma^{(\nu)}$. By combining these two conditions, we obtain as a necessary condition for expression (5.35) to be nonzero that the representation $\Gamma^{(1)}$ occur in the decomposition of the direct product

$$\Gamma^{(T)} \times \Gamma^{(T)} \times \Gamma^{(2)}. \tag{5.36}$$

This result can, of course, be generalized to nth-order perturbation theory. Thus we have the following condition:

A necessary condition for a quantum transition between states 1 and 2 to be allowed is that the representation $\Gamma^{(1)}$ occur in the decomposition of the direct product

$$\underbrace{(\Gamma^{(T)} \times \Gamma^{(T)} \times \cdots \times \Gamma^{(T)})}_{n} \times \Gamma^{(2)}. \tag{5.37}$$

For example, a dipole transition ($\Gamma^{(T)} = D^{(1)}$) in a central field between two states characterized by values of J and $J + k$ for their angular momenta is allowed only in kth-order perturbation theory. This is because the representation $D^{(k)}$ appears in the decomposition of a direct product which contains not less than k representations $D^{(1)}$.

CHAPTER VI

Classification of States

PART 1. Electrons in a Central Field

6.1. Equivalent Electrons. *L–S* Coupling

It is well known that the Schrödinger equation for a many-electron system does not permit an exact solution in finite form. One possible alternative would be to tabulate the solution numerically, but even for very coarse intervals such a table would have to be of colossal dimensions. Hartree (1957) gives the following example: If in tabulating the wavefunction for a stationary state of the Fe atom (26 electrons) one were to give just ten values of each variable, the table would have to contain 10^{78} values of the wavefunction. This exceeds the number of atoms in the entire solar system. This example shows that in calculations of many-particle problems, it is not only desirable to make use of approximations, it is absolutely necessary to do so.

In quantal calculations on many-electron systems the so-called one-particle approximation is widely used. In this approximation each electron is regarded as being in a stationary state in the field of the nuclei and of all the other electrons. For atoms this field can be taken to be centrally symmetric to a very good approximation, and the one-electron states can be classified according to the irreducible representations of the group $\mathbf{O}_3$. Besides the parity, the state of each electron is characterized by a set of four quantum numbers: n, l, m, and σ, where n is the principal quantum number, l is the orbital angular momentum, m is the projection of l upon the z-axis, and σ is the projection of the spin of the electron upon the z axis. Each electron possesses its own set of four quantum numbers. In determining the energy levels of atoms with low atomic weights, one usually neglects relativistic interactions (these are subsequently taken into account as a perturbation). As a result, the spin and orbital states of a system are considered separately. This approximation is known as Russell–Saunders or *L–S* coupling.

In the *L–S* coupling approximation the total spin of the electrons S and the total orbital angular momentum L are conserved. Values of L and S characterize the energy levels of the system. In addition, each energy level is

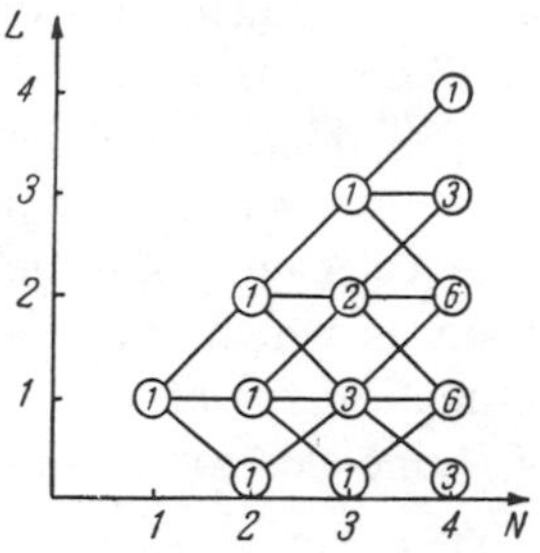

Figure 6.1.

characterized by the configuration of one-electron states from which it arises. In order to specify an electronic configuration, it suffices to give the principal quantum number n and the orbital angular momentum l for each electron. Electrons with the same n and l quantum numbers are said to be *equivalent*. A set of equivalent electrons of a given type constitutes a *shell*. Since, according to the Pauli principle, each electron must be characterized by a distinct set of four quantum numbers n, l, m, and σ, a shell $(nl)^N$ may contain not more than $2(2l + 1)$ electrons.‡ A shell in which all the states are filled is said to be *closed*.

The total orbital and spin angular momenta are obtained by coupling vectorially the angular momenta of the individual electrons. The coupling of angular momenta can be conveniently represented in the form of a diagram, so long as the number of momenta to be coupled is small. Figure 6.1 shows the diagram for $l = 1$. The numbers enclosed in circles give the number of times the given value for the total angular momentum occurs in a system of N electrons.§ We denote this number by a_{LN}. Since the number of states for a system of N angular momenta is unchanged by the coupling,

$$\sum_L a_{LN}(2L + 1) = (2l + 1)^N.$$

This equation is easily verified by Fig. 6.1. Thus for $N = 4$ we have

$$3 \cdot 1 + 6 \cdot 3 + 6 \cdot 5 + 3 \cdot 7 + 1 \cdot 9 = 81.$$

The a_{LN}, however, are by no means equal to the number of different energy levels of a configuration $(nl)^N$ with the given value of L. In fact, a number of states with the same L differ from one another only by a permutation of the

‡This statement of the Pauli principle arises from a more general formulation which asserts that the only states of many-electron systems which occur in nature are those whose total wavefunction is antisymmetric with respect to permutations of the electrons.

§The diagram corresponds in fact to the decomposition of the direct product of N irreducible representations $D^{(l)}$ of the group $\mathbf{R}_3$ into irreducible representations $D^{(L)}$.

electrons, and because of the indistinguishability of electrons, such states must possess the same energy (this is known as the permutational degeneracy). Some of the remaining states are forbidden, since they do not satisfy the Pauli principle. The true number of energy levels for a system of equivalent electrons can be elegantly determined from group-theoretical considerations.

In the zeroth approximation, in which one neglects the electronic interactions, the coordinate wavefunction for an electronic configuration $(nl)^N$ is constructed as a product of N one-electron functions, usually called *orbitals*:

$$\phi_{nlm_1}(1)\phi_{nlm_2}(2) \cdots \phi_{nlm_N}(N), \tag{6.1}$$

in which the number i in the argument of an orbital denotes the set of coordinates for the ith electron. One can form $(2l + 1)^N$ such products altogether, all of which in the absence of interactions between the electrons belong to a single energy level. Orbitals ϕ_{nlm} which differ in the quantum number m can be regarded as basis vectors for a $(2l + 1)$-dimensional vector space. Under unitary transformations of this space the products (6.1) transform as tensors and generate a basis for a $(2l + 1)^N$-dimensional tensor representation of the group U_{2l+1}. The decomposition of such a representation was discussed in Section 4.2. From the results of that section it follows that the number of times with which each irreducible $U_{2l+1}^{[\lambda]}$ occurs in the decomposition of the tensor representation is equal to the dimension f_λ of the irreducible representation of the permutation group characterized by the same Young diagram $[\lambda]$ in which, moreover, the number of cells in any column does not exceed $2l + 1$. This means that one can always find a linear transformation which takes one from the initial $(2l + 1)^N$ functions (6.1) to a set of functions each of which is characterized by a specified Young diagram of N cells whose columns do not exceed $2l + 1$ in length. Such a set contains $f_\lambda \delta_\lambda(2l + 1)$ functions, where $\delta_\lambda(2l + 1)$ is the dimension of the representation $U_{2l+1}^{[\lambda]}$. The functions in each set can be represented as points on a plane diagram placed in the form of a rectangle:

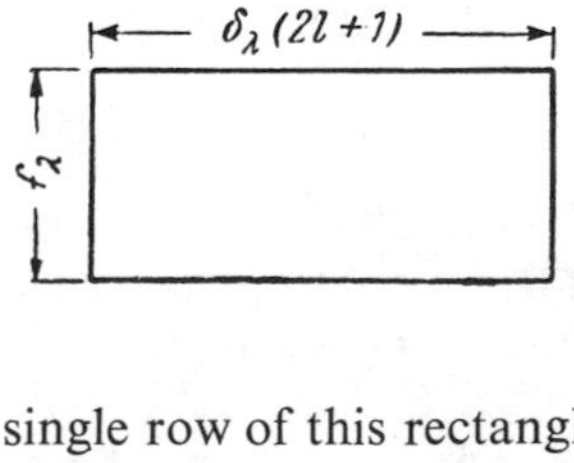

Functions which lie on a single row of this rectangle transform according to the irreducible representation $U_{2l+1}^{[\lambda]}$; functions lying on a single column transform according to the irreducible representation $\Gamma^{[\lambda]}$ of the group π_N and are enumerated by the tableaux r.

For example, for $l = 1$, $N = 4$ (the configuration p^4), the 81 functions (6.1) divide into four sets corresponding to $[\lambda] = [4], [31], [2^2]$, and $[21^2]$ ($[\lambda] = [1^4]$ is not allowed). From formulae (2.18) and (4.13) we find the dimensions

$[\lambda]$:	[4]	[31]	$[2^2]$	$[21^2]$
$\delta_\lambda(3)$:	15	15	6	3
f_λ:	1	3	2	3

It is easily verified that $\sum_\lambda f_\lambda \delta_\lambda = 81$. We thus arrive at a diagram similar to that given in Section 4.2 for the case $l = 1$, $N = 3$ (see Fig. 6.2).

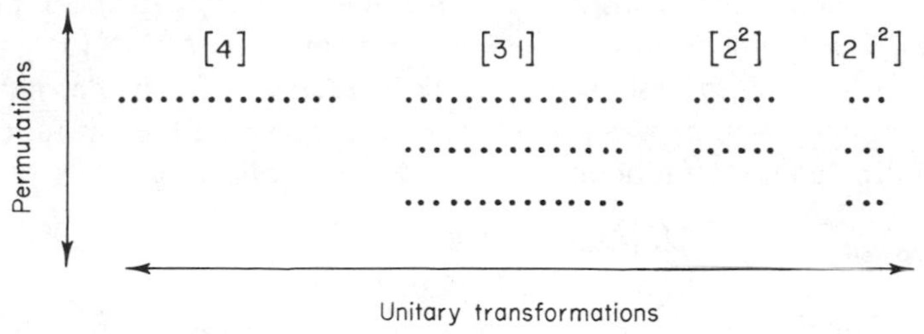

Figure 6.2.

The total wavefunction describing the electron states must be antisymmetric with respect to permutations of the electrons. In the *L–S* coupling approximation the total wavefunction Ψ can be written in the form of a linear combination of products of coordinate wavefunctions Φ with spin wavefunctions Ω. By applying the mathematical apparatus of group theory, one easily writes down the desired linear combination. As was shown in Section 2.7, in order to construct an antisymmetric function from binary products, the functions making up the products must belong to conjugate representations of the group π_N. According to (2.29),

$$\Psi^{[1^N]} = (f_\lambda)^{-1/2} \sum_r \Phi_r^{[\lambda]} \Omega_{\tilde{r}}^{[\tilde{\lambda}]}. \tag{6.2}$$

Every electron possesses a spin angular momentum of $\frac{1}{2}$. The symmetrization of a spin function according to a Young diagram $[\tilde{\lambda}]$ automatically leads to the result that functions $\Omega_{\tilde{r}}^{[\tilde{\lambda}]}$ with differing values of the projection of the spin angular momentum transform according to irreducible representations $U_2^{[\lambda]}$ of the unitary group $\mathbf{U}_2$. The irreducible representations $U_2^{[\tilde{\lambda}]}$ are characterized by Young diagrams which contain up to two rows. Since the Young diagrams $[\lambda]$ and $[\tilde{\lambda}]$ are dual to each other, the coordinate diagrams may not contain more than two columns.

The only orbital states which are realized in the electron configuration just considered are those that correspond to the rectangles $[\lambda] = [2^2]$ and $[\lambda] = [21^2]$ in Fig. 6.2. In addition, it should be noted that all functions that lie on a single column of a rectangle enter into one antisymmetric combination (6.2), because each column corresponds to a single physical state. The number of physically distinct orbital states is determined by the number of columns, i.e., is equal to $\sum_\lambda \delta_\lambda(2l+1)$. We have in this case altogether $\delta_{[2^3]}(3) + \delta_{[21^2]}(3) = 9$ distinct orbital states.

In order to determine the values of the total orbital angular momentum L that correspond to the permitted representations $U^{[\lambda]}_{2l+1}$, it is necessary to carry out the reduction $\mathbf{U}_{2l+1} \longrightarrow \mathbf{R}_3$. On doing so, the irreducible representations $U^{[\lambda]}_{2l+1}$ decompose into irreducible representations $D^{(L)}$ of the group $\mathbf{R}_3$. This process has been described in detail in Section 4.5. The decomposition for the configuration p^4 was given by Eqs. (4.43). From these it follows that the coordinate representations correspond to the following values of L:

$$[\lambda]_{\text{coord}}: \quad \begin{matrix} [2^2] & \qquad [21^2] \\ L = 2,0 & \qquad L = 1. \end{matrix} \tag{6.3}$$

The associated representations of the spin wavefunctions each correspond uniquely to a single value of the total spin

$$[\lambda]_{\text{spin}}: \quad \begin{matrix} [2^2] & \qquad [31] \\ S = 0, & \qquad S = 1. \end{matrix}$$

Consequently, a configuration of four equivalent electrons, p^4, gives rise to the following permitted multiplets‡ ${}^{2S+1}L$: ^{1}S, ^{3}P, and ^{1}D. The configuration p^2, which is the complementary configuration to a closed shell, gives rise to the same multiplets. This result follows from the equivalence relations (4.36).

Thus in order to determine the allowed multiplets of a configuration of equivalent electrons $(nl)^N$, we propound the following general procedure: All the allowed Young diagrams with N cells that consist of two columns whose lengths do not exceed $2l + 1$ are first written out, and put into correspondence with the appropriate dual spin diagrams. The permitted values of L are then found from the reduction $\mathbf{U}_{2l+1} \longrightarrow \mathbf{R}_3$ (which in practice can be taken from the tables of Appendix 3), and the multiplicity derived from the form of the Young spin diagram. This procedure can be represented schematically as

‡Values of the total orbital angular momentum L for atomic terms are denoted by capital letters:

	S	P	D	F	G	H	I	K
for $L =$	0	1	2	3	4	5	6	7.

The letter S for $L = 0$ should not be confused with the notation S for the total spin of a system follows‡:

$$\begin{array}{ccc} U_{2l+1}^{[\lambda]} & \longleftrightarrow & U_2^{[\tilde{\lambda}]} \\ \downarrow & & \downarrow \\ D^{(L)} & \longleftrightarrow & D^{(S)} \end{array} \quad . \tag{6.4}$$

The simplest example is that of the two-electron configuration $(nl)^2$. According to the results of Section 4.5, only representations $D^{(L)}$ with L even may occur in the decomposition of $U_{2l+1}^{[1^2]}$. Since the spin representation $U_2^{[1^2]}$ corresponds to $S = 0$ and $U_2^{[2]}$ to $S = 1$, application of scheme (6.4) allows one to arrive at a general conclusion: In the electron configuration $(nl)^2$ terms with even values of L are singlets and terms with odd values of L are triplets.

As another example of the application of scheme (6.4), we determine the allowed multiplets for the configuration d^3: It is convenient to present the results of this investigation in the form of a table (the values of L in the first column of Table 6.1 are taken from Table 4 of Appendix 3).

Table 6.1

Allowed Terms for the Configuration d^3

$U_5^{[\lambda]} \longrightarrow D^{(L)}$	$U_2^{[\tilde{\lambda}]} \longleftrightarrow D^{(S)}$	Multiplets
[21] $L = 1, 2^2, 3, 4, 5$	[21] $S = \frac{1}{2}$	^2P, $(^2D)^2$, ^2F, ^2G, ^2H
$[1^3]$ $L = 1, 3$	[3] $S = \frac{3}{2}$	^4P, ^4F

6.2. Additional Quantum Numbers. The Seniority Number

From the fact that a single representation $D^{(S)}$ corresponds to each representation $U_2^{[\lambda]}$ and that, because of the Pauli principle, there is a one-to-one correspondence between the coordinate and spin Young diagrams, the specification of the coordinate Young diagram is equivalent to specifying the total spin S. Atomic terms can therefore be enumerated by the quantum numbers L and $[\lambda]$ instead of by the quantum numbers L and S.§ The assign-

‡Tables of the permitted multiplets already exist for equivalent electrons. These have been derived by a different method which does not require the application of group theory (Landau and Lifshitz, 1965). However, the group-theoretical method given here is more transparent than the conventional method, and furthermore, it allows a more detailed classification in cases where terms occur more than once. The scheme is also of methodological interest because it can also be applied in a number of more complicated situations [for example, in finding the allowed molecular multiplets (see Section 6.5)].

§We recall that since we do not take into account relativistic interactions, each term is $(2S + 1)(2L + 1)$-fold degenerate.

ment of a Young diagram $[\lambda]$ and orbital angular momentum L to each term is equivalent to classifying the orbital states by the irreducible representations of the groups $\mathbf{U}_{2l+1}$ and $\mathbf{R}_3$. The specification of the projection of the orbital angular momentum along the z axis, M, signifies an additional classification of the orbital states by the irreducible representations of the two-dimensional rotation group $\mathbf{R}_2$.

A classification of this kind for a configuration p^N uniquely determines a state. However, in the case of d^N and f^N configurations several values of L may correspond to a single symmetry diagram $[\lambda]$. Thus for the configuration d^3 two states with $L = 2$ possess the identical permutation symmetry $[\lambda] = [21]$ (see Table 6.1). In order to distinguish between repeated terms, it is useful to introduce additional quantum numbers.‡ For this purpose one would naturally make use of some subgroup of $\mathbf{U}_{2l+1}$ which itself contains $\mathbf{R}_3$ as a subgroup. Racah (1949) showed that the $(2l+1)$-dimensional rotation group $\mathbf{R}_{2l+1}$ is such an additional group. The coordinate wavefunctions of a configuration are then constructed so as to be simultaneously basis functions for irreducible representations of the following chain of subgroups:

$$\mathbf{U}_{2l+1} \supset \mathbf{R}_{2l+1} \supset \mathbf{R}_3 \supset \mathbf{R}_2. \tag{6.5}$$

According to the definition of the group $\mathbf{R}_{2l+1}$, the operations in it leave invariant the scalar product of two vectors defined in the $(2l+1)$-dimensional space. The only scalar product that is invariant with respect to the operations of both the groups $\mathbf{R}_{2l+1}$ and $\mathbf{R}_3$ is the wavefunction of the configuration l^2 for the state $L = 0$ [see (4.52)]:

$$\psi_0^{(0)} = \frac{(-1)^l}{(2l+1)^{1/2}} \sum_m (-1)^m \psi_m^{(l)}(1)\psi_{-m}^{(l)}(2). \tag{6.6}$$

A system of two electrons with $L = 0$ of necessity has a value of $S = 0$ (see p. 129). Such electrons are said to form a closed pair. If a closed pair is added to a configuration l^{N-2}, then because of the invariance with respect to operations of the group $\mathbf{R}_{2l+1}$, the configuration $l^{N-2}l^2\,(^1S)$ which is formed belongs to the same irreducible representation of $\mathbf{R}_{2l+1}$ as the configuration l^{N-2}. This circumstance allows each irreducible representation of the group $\mathbf{R}_{2l+1}$ to be characterized by the minimum number of electrons v of a configuration l^v for which the particular representation first makes its appearance. In addition, it is obvious that terms whose L values match the L structure of the irreducible representation of $\mathbf{R}_{2l+1}$ corresponding to a given v, make their first appearance in the configuration l^v. Each term can therefore be characterized by the additional quantum number v. This number is called the *seniority number*, and

‡As a rule such additional quantum numbers are not "good" quantum numbers, i.e., the matrix of the Hamiltonian is not diagonal in them. The δ forces (de-Shalit and Talmi, 1963) are a special example of interactions for which the seniority number (to be introduced) is, in fact, a good quantum number.

is usually written as a left suffix to the multiplet symbol: ${}^{2S+1}_{\ v}L$. Thus the two ^{2}D terms of the configuration d^3 are distinguished by the seniority number, and are written ${}^2_1$D and ${}^2_3$D.

In this way repeated multiplets are classified by values of the seniority number v signifying the configuration in which the given term makes its first appearance. The presence of a state $[\lambda]vL$ in a configuration l^N means that there are $(N - v)/2$ closed pairs in the given configuration. We emphasize that the Hamiltonian matrix is not diagonal in the number v. This is due to the fact that the Hamiltonian is not, in general, invariant with respect to the operations of $\mathbf{R}_{2l+1}$. Nevertheless, such a classification substantially simplifies both the construction of wavefunctions for repeated terms and the calculation of matrix elements [see Judd (1963)].

The set of quantum numbers $[\lambda]$, v, and L uniquely characterizes the terms of a configuration d^N. For a configuration f^N it turns out that this classification is insufficient, terms that possess identical sets $[\lambda]$, v, and L occurring more than once. However, for the f shell—and only for this case—an extra group can be added to the chain of groups (6.5). This is the group of orthogonal transformations in seven-dimensional space which leaves a trilinear antisymmetric form invariant (Judd, 1963; Racah, 1949). This group is usually denoted by $\mathbf{G}_2$.‡ As a result, the states which arise from the configuration f^N are uniquely classified by means of the irreducible representations of the extended chain of groups:

$$\mathbf{U}_7 \supset \mathbf{R}_7 \supset \mathbf{G}_2 \supset \mathbf{R}_3 \supset \mathbf{R}_2. \tag{6.7}$$

6.3. Equivalent Electrons. *j–j* Coupling

In heavy atoms it is necessary to take into account the spin–orbit interaction even in the zeroth order of perturbation theory. The one-electron energy levels are characterized by values of the total angular momentum of an electron, j, obtained by vector coupling the orbital and spin angular momenta:

$$\mathbf{j} = \mathbf{l} + \mathbf{s}. \tag{6.8}$$

Since $S = \frac{1}{2}$, j takes one of two values for a given value of l: $j = l \pm \frac{1}{2}$. The wavefunctions describing the states of a single electron in this case form bases for the irreducible representations $D^{(j)}$ of the group $\mathbf{R}_3$ that occur in the decomposition of the direct product $D^{(l)} \times D^{(1/2)}$. According to the general

‡The classification of states according to the irreducible representations of the group $\mathbf{G}_2$ is equivalent to the diagonalization of a certain scalar operator which simultaneously belongs to irreducible representations of the groups $\mathbf{R}_7$ and $\mathbf{G}_2$ (Judd, 1963). An analogous scalar operator for the group $\mathbf{R}_9$ has been introduced by Judd (1968). The eigenvalues of this operator facilitate the classification of the electronic terms of the configuration g^N (Judd, 1968).

formula (3.45),

$$\psi_{nj\mu} = \sum_{m,\sigma} \phi_{nlm_s}\chi_\sigma \langle lm, s\sigma | j\mu \rangle. \tag{6.9}$$

The energy levels of an N-electron system are characterized by a resultant angular momentum J which is obtained by coupling vectorially the angular momenta j of the individual electrons. This scheme for constructing the energy levels is called the j–j coupling scheme. In this, electrons with identical values of n and j are equivalent. Because of the Pauli principle, it is obvious that a configuration of equivalent electrons $(nj)^N$ cannot contain more than $2j + 1$ electrons.

In the zeroth approximation, in which the electronic interactions are neglected, the total wavefunction of a configuration $(nj)^N$ is constructed as a product of one-electron orbitals (6.9):

$$\psi_{nj\mu_1}(1)\psi_{nj\mu_2}(2) \cdots \psi_{nj\mu_N}(N). \tag{6.10}$$

In contrast with the L–S coupling case in which one constructs from the product (6.1) a coordinate wavefunction $\Phi_r^{[\lambda]}$ with the permutational symmetry of a Young diagram $[\lambda]$, in j–j coupling one constructs from (6.10) a total wavefunction whose permutational symmetry is confined to just the antisymmetric representation $[\lambda] = [1^N]$.

The values of the total angular momentum J that are allowed for a configuration $(nj)^N$ are just those for which the coupled function is antisymmetric. For the configuration $(nj)^2$ it follows from relations (3.49) that

$$P_{12}\psi_M^{(J)}(j(1)j(2)) = -(-1)^J\psi_M^{(J)}(j(1)(2)), \tag{6.11}$$

since $(-1)^{2j} = -1$. Hence J can only take even values. In the case of an arbitrary configuration $(nj)^N$ the allowed values of J are found from the decomposition

$$U_{2j+1}^{[1^N]} \doteq \sum_J a^{(J)} D^{(J)}. \tag{6.12}$$

The procedure for carrying out such decompositions was described in Section 4.5. Table 6.2 gives the allowed values of J for configurations $(nj)^N$ from $j = \frac{1}{2}$ to $j = \frac{7}{2}$. Only those configurations are given for which the j shell is not more than half-filled, since the J values for the configurations $(nj)^N$ and $(nj)^{2j+1-N}$ coincide. It can be seen from this table that for the configuration $(\frac{7}{2})^4$ terms with $J = 2$ and $J = 4$ occur twice. Repeated terms in j–j coupling can be distinguished by the seniority number v, just as in the case of L–S coupling. The terms of the configuration $(\frac{7}{2})^4$ that are repeated can be divided into those with $v = 2$, which are obtained from the corresponding terms of the configuration $(\frac{7}{2})^2$, and terms with $v = 4$, which first appear in the configuration $(\frac{7}{2})^4$.

Table 6.2

Allowed Values of the Total Angular Momentum J for Configurations j^N

j	N	J
$\frac{1}{2}$	1	$\frac{1}{2}$
$\frac{3}{2}$	1	$\frac{3}{2}$
	2	0, 2
$\frac{5}{2}$	1	$\frac{5}{2}$
	2	0, 2, 4
	3	$\frac{3}{2}, \frac{5}{2}, \frac{9}{2}$
$\frac{7}{2}$	1	$\frac{7}{2}$
	2	0, 2, 4, 6
	3	$\frac{3}{2}, \frac{5}{2}, \frac{7}{2}, \frac{9}{2}, \frac{11}{2}, \frac{15}{2}$
	4	0, 2 (twice), 4 (twice), 5, 6, 8

6.4. Configurations of Several Groups of Equivalent Electrons

We consider first a configuration of two nonequivalent electrons $n_1l_1n_2l_2$ in the L–S coupling scheme. In this case the Pauli principle does not impose any restrictions upon the possible values of the total orbital angular momentum L and total spin S of the system. The requirement that the total wavefunction be antisymmetric can be met for any set of values of L and S. This becomes obvious if one recalls that the one-electron orbitals in a central field are of the form of a product of a radial wavefunction $R_{nl}(r)$ and an angular function $Y_{lm}(\theta, \phi)$ (Landau and Lifshitz, 1965):

$$\phi_{nlm}(r, \theta, \phi) = R_{nl}(r)Y_{lm}(\theta, \phi). \tag{6.13}$$

For equivalent electrons the radial wavefunctions are all identical. An antisymmetric total wavefunction is obtained by forming the appropriate linear combinations of coordinate and spin wavefunctions. The radial parts of such combinations are always symmetric with respect to permutations of the electrons, and this imposes restrictions on the possible values of L and S. For nonequivalent electrons the radial wavefunctions for the electrons are different. The total wavefunction can always be antisymmetrized by antisymmetrizing the radial wavefunctions.

Hence in a system of nonequivalent electrons one obtains all the possible L and S values which can be obtained by vectorially coupling the orbital and spin angular momenta of the individual electrons.

Configurations with three or more nonequivalent electrons are characterized by a large number of repeated terms. In order to distinguish such multiple terms, one can use as additional quantum numbers the terms of the configuration with $N - 1$ electrons from which the given term is derived. In such cases one speaks of specifying the genealogy of a term. For example, the terms which arise from a configuration of three nonequivalent p electrons, $npn'pn''p$, are conveniently characterized by specifying the terms of the configuration $npn'p$:

$$\begin{array}{llll} (npn'p): & {}^1Sn''p, & {}^2P; & \\ (npn'p): & {}^3Sn''p, & {}^2P, & {}^4P; \\ (npn'p): & {}^1Pn''p, & {}^2SPD; & \\ (npn'p): & {}^3Pn''p, & {}^2SPD, & {}^4SPD; \\ (npn'p): & {}^1Dn''p, & {}^2PDF; & \\ (npn'p): & {}^3Dn''p, & {}^2PDF, & {}^4PDF. \end{array}$$

Thus the term ^2P occurs six times in this configuration but each of these six terms possesses a distinct genealogy with respect to the configuration $npn'p$. This configuration gives rise to 21 terms altogether, consisting of $\sum_{S,L}(2S+1)(2L+1) = 216$ states. In fact all the $(2s+1)^3(2l+1)^3 = 216$ states of the configuration $npn'pn''p$ are allowed. For three equivalent p electrons three terms in all are allowed: ^1S, ^1D, and ^3P, or 15 states.

The extension to the general case of a configuration of several groups of electrons‡

$$l_1^{n_1} l_2^{n_2} \cdots l_k^{n_k} \tag{6.14}$$

is obvious. In the approximation of L–S coupling each shell $l_a^{n_a}$ is characterized by a Young diagram $[\lambda_a]$ (which corresponds uniquely to a value of the spin for the shell, S_a), by an orbital angular momentum L_a, and by any additional quantum numbers which we denote by α_a. The orbital states for the whole configuration are characterized by a total symmetry diagram $[\lambda]$, by a resultant orbital angular momentum L, and by its projection M. In order to classify the orbital states uniquely, we need in addition to specify $k-2$ partial resultant orbital angular momenta L_{int} (see Section 3.7), and also the mode of coupling the shells, which we denote by A. The specification of the total symmetry diagram for an orbital state $[\lambda]$ is equivalent to specifying the total spin of the system, S. In addition, there are $2S+1$ spin states for each orbital state of the system.

As a result, the energy terms which arise from the configuration (6.14) are characterized by the following set of quantum numbers:

$$l_1^{n_1}[\lambda_1]\alpha_1 L_1, \ldots, l_k^{n_k}[\lambda_k]\alpha_k L_k, \qquad [\lambda_{\text{int}}]^A (L_{\text{int}})^A, \qquad [\lambda]LM. \tag{6.15}$$

Once the L_a are specified, the possible values for the total orbital angular

‡We omit the principal quantum numbers of the shells in the notation for a configuration.

momentum L and also the set of $k - 2$ partial resultant angular momenta [which we denote by $(L_{int})^A$], are determined by the rules for coupling k angular momenta. The possible Young diagrams $[\lambda]$ and the set of $k - 2$ intermediate diagrams $[\lambda_{int}]^A$ are determined either by successive applications of Littlewood's theorem (see Section 4.4) to the product

$$[\lambda_1] \otimes [\lambda_2] \otimes \cdots \otimes [\lambda_k] \tag{6.16}$$

(with the condition that the Young diagrams which are obtained in this way contain no more than two columns), or by the rules for coupling k angular momenta $\mathbf{S}_a$ (in view of the one-to-one correspondence $[\lambda_a] \longleftrightarrow \mathbf{S}_a$).

In the case of j–j coupling the states which arise from a configuration of k groups of equivalent electrons are characterized by a value of the total angular momentum J, and by $k - 2$ partial resultant angular momenta J_{int}, which are determined by the rules for coupling the angular momenta J_a of the individual shells. The allowed values for the J_a can be found in Table 6.2.

PART 2. The Connection between Molecular Terms and Nuclear Spin

6.5. The Classification of Molecular Terms and the Total Nuclear Spin

When the direct influence of a nuclear spin upon the molecular terms is neglected (the hyperfine interaction) there is still a substantial indirect effect which manifests itself in the kinds of symmetries that are allowed for the terms. When hyperfine interactions are ignored the total wavefunction for a molecule can be written in the form of a linear combination of products of coordinate wavefunctions $\Phi(x, X)$ (where x denotes the electronic and X the nuclear coordinates) with spin functions for the nuclei $\Omega(\sigma_i)$, where σ_i denotes the nuclear spin coordinates. The permutational symmetry of the total wavefunction is determined by the exclusion principle, which can be stated as follows:

The only states of a system of identical particles with spin i that are found in nature are those for which the total wavefunction is multiplied by $(-1)^{2i}$ when any pair of particles is permuted, i.e., the total wavefunction is symmetric for integral values of i (Bose statistics), and antisymmetric for half-integral values (Fermi statistics).

The exclusion principle imposes a specific relationship between the permutational symmetries of coordinate and spin wavefunctions when these are considered separately. The construction of a total wavefunction which is antisymmetric with respect to permutations of identical nuclei is carried out in the same way as the formation of a total wavefunction for a system of electrons in L–S coupling [see Eq. (6.2)]:

$$\Psi^{[1^N]} = (f_\lambda)^{-1/2} \sum_r \Phi_r^{[\lambda]}(x, X)\Omega_r^{[\tilde{\lambda}]}(\sigma_i), \tag{6.17}$$

where the permutational symmetries of the coordinate and spin functions are respectively characterized by dual Young diagrams with N cells.

In order to construct a symmetric total wavefunction, the coordinate and spin functions must possess the identical permutational symmetry with respect to the nuclear coordinates:

$$\Psi^{[N]} = (f_\lambda)^{-1/2} \sum_r \Phi_r^{[\lambda]}(x, X)\Omega_r^{[\lambda]}(\sigma_i). \tag{6.18}$$

A spin wavefunction for a system of identical nuclei $\Omega_r^{[\lambda]}$ transforms according to the irreducible representation $U_{2i+1}^{[\lambda]}$ of the group of unitary transformations in the $(2i + 1)$-dimensional vector space of the spins. A spin Young diagram $[\lambda]$ therefore cannot have more than $2i + 1$ rows (see Section 4.2). This immediately imposes limitations upon the form of the coordinate Young diagram. When i is integral this also cannot have more than $2i + 1$ rows, and when i is half-integral not more than $2i + 1$ columns.

For nuclei with spin $i = \frac{1}{2}$ the values of the resultant spin I correspond uniquely to the form of the Young diagram which defines the permutational symmetry of the spin wavefunction. There is no such unique connection for nuclei with arbitrary spin i (the situation is similar to the coupling of orbital angular momenta of equivalent electrons). The possible values of the resultant spin I are determined from the decomposition

$$U_{2i+1}^{[\lambda]} \doteq \sum_I a_\lambda^{(I)} D^{(I)}, \tag{6.19}$$

the method for which was given in Section 4.5 (see also Appendix 3).

A number of approximations are usually made in determining coordinate wavefunctions for a molecule. Because of the large difference in their masses, the velocities of the electrons greatly exceed those of the nuclei. The electrons can therefore be regarded as moving in a potential field created by stationary nuclei. As a result of this, one obtains electronic wavefunctions and energy terms $E(X)$ which depend upon the positions of the nuclei X as parameters. Furthermore, the nuclei can be regarded as moving in a field of potential energy $E(X)$. This is just the Born–Oppenheimer approximation discussed in Section 5.1, and within its framework the coordinate wavefunction of a molecule can be written in the form of a product of an electronic wavefunction with a nuclear wavefunction‡:

$$\Phi(x, X) = \Phi_e(x, X)\Phi_{nuc}(X). \tag{6.20}$$

‡Approximation (6.20) is frequently called the *adiabatic approximation* [for the conditions for its applicability see Davydov (1965, §116)].

The X in the argument of function (6.20) denotes the set of $3n$ nuclear coordinates corresponding to the $3n$ degrees of freedom of a system of n nuclei. Three of these correspond to the translational motion of the molecule as a whole, and can be excluded by transforming to a system of coordinates which is referred to the center of mass of the molecule. Three further degrees of freedom correspond to the rotation of the molecule as a rigid body.‡ The remaining $3n - 6$ degrees of freedom describe the vibrational motion of the nuclei with respect to each other. One normally takes as coordinates for the vibrational motion the so-called *normal coordinates*, which lead to an expression for the energy of small vibrations in the form of a sum of squares (Landau and Lifshitz, 1965; Davydov, 1965). If it is assumed that the relative displacements of the nuclei from their equilibrium positions are small, one may take for the electronic wavefunction its value at the equilibrium configuration of the nuclei. The coordinate wavefunction for a molecule (6.20) is then of the form of a product of wavefunctions for each type of motion:

$$\Phi(x, X) = \Phi_e(x, X^\circ)\Phi_{\text{vib}}(Q)\Phi_{\text{rot}}(\Theta), \tag{6.21}$$

where x stands for the electronic coordinates, X° the coordinates of the equilibrium configuration of the nuclei, Q the vibrational normal coordinates, and Θ the set of three Euler angles which define the orientation of the molecule in space.

Equation (6.21) assumes that the nuclei execute small vibrations about equilibrium positions corresponding to an equilibrium configuration of the molecule. In this approximation the potential energy for the system of nuclei is invariant only under those permutations of identical nuclei that correspond to operations of the point symmetry of the equilibrium nuclear configuration. The electronic and vibrational energy levels are classified according to the irreducible representations $\Gamma^{(\alpha)}$ of the point symmetry group of the molecule, the wavefunctions forming bases for these representations. The permutations of N identical nuclei are generally not all realized by the operations of the point group, and hence the group is isomorphic with a subgroup of the permutation group for identical nuclei, π_N. In order to construct basis functions for representations of the permutation group from basis functions for its point subgroup, one must proceed from the $N!/g$ equilibrium configurations of the nuclei which differ from one another by the permutations of identical nuclei that do not occur in the operations of the point symmetry (g here denotes the dimension of the point group). The desired basis functions are then obtained by letting the corresponding Young operators act upon basis functions for the representations $\Gamma^{(\alpha)}$.

‡For linear molecules a rotation about the molecular axis is not defined because the moment of inertia about this axis is taken to be zero. For such molecules there are therefore $3n - 5$ independent vibrational degrees of freedom.

One is now faced with the question of which representations $\Gamma^{[\lambda]}$ can be constructed from a basis for a representation $\Gamma^{(\alpha)}$ when the point subgroup is extended to cover the entire permutation group. The answer to this is provided by the Frobenius reciprocity theorem‡:

A reducible representation $\boldsymbol{\Gamma}$ of a group $\mathbf{G}$ which can be constructed from a basis for an irreducible representation $\boldsymbol{\Gamma}^{(h)}$ of a subgroup $\mathbf{H} \in \mathbf{G}$ contains each irreducible representation $\boldsymbol{\Gamma}^{(g)}$ of $\mathbf{G}$ the same number of times as the representation $\boldsymbol{\Gamma}^{(h)}$ occurs in the decomposition of $\boldsymbol{\Gamma}^{(g)}$ on reducing $\mathbf{G} \longrightarrow \mathbf{H}$.

Consequently, the number of independent bases for a representation $\Gamma^{[\lambda]}$ which can be constructed from a basis for a representation $\Gamma^{(\alpha)}$ is given by the coefficients in the decomposition

$$\Gamma^{[\lambda]} \doteq \sum_{\alpha} a_{\lambda}^{(\alpha)} \Gamma^{(\alpha)}. \tag{6.22}$$

The coefficients $a_{\lambda}^{(\alpha)}$ are found by means of formula (1.63) with the aid of character tables for the point and permutation groups. However, it is first of all necessary to place each operation of the point group into correspondence with the appropriate permutation of the nuclei.

The connection between the permutational symmetries of the coordinate and spin functions depends upon the statistics of the nuclei [see relations (6.17) and (6.18)]. The existence of such a connection, together with the decompositions (6.22) and (6.19), makes it possible to associate with each type of point symmetry $\Gamma^{(\alpha)}$ a possible value of the total nuclear spin I. This method of finding the allowed nuclear molecular multiplets (Kaplan, 1959) can be represented schematically as follows:

$$\begin{array}{ccc} \text{(a) Bose statistics} & & \text{(b) Fermi statistics} \\ \begin{array}{ccc} \Gamma^{[\lambda]} & \longleftrightarrow & U_{2i+1}^{[\lambda]} \\ \downarrow & & \downarrow \\ \Gamma^{(\alpha)} & \longleftrightarrow & D^{(I)} \end{array} & & \begin{array}{ccc} \Gamma^{[\lambda]} & \longleftrightarrow & U_{2i+1}^{[\tilde{\lambda}]} \\ \downarrow & & \downarrow \\ \Gamma^{(\alpha)} & \longleftrightarrow & D^{(I)} \end{array} \end{array} . \tag{6.23}$$

We now consider several examples.

EXAMPLE 1.

The methane molecule $^{12}CH_4$,§ $i(^{12}C) = 0$, $i(H) = \frac{1}{2}$; the point group, of symmetry T_d, is isomorphic with the group $\boldsymbol{\pi}_4$. There is therefore a one-to-one correspondence between the classes of these groups. Thus

$$\begin{array}{llllll} \text{Classes of } T_d\text{:} & E & C_3 & C_2 & \sigma_d & S_4 \\ \text{Classes of } \pi_4\text{:} & \{1^4\} & \{13\} & \{2^2\} & \{1^2 2\} & \{4\} \end{array} \tag{6.24}$$

‡For the proof of this theorem see Petrashen and Trifonov (1969, p. 235).

§The superscript on the chemical symbol for an element denotes a particular isotopic species. Nuclei of different isotopes may possess different spins.

There must also be a one-to-one correspondence between the irreducible representations of the two groups. Comparing character tables, we find

$$\begin{array}{llllll} \text{Irreducible representations of } T_d: & A_1 & A_2 & E & F_1 & F_2 \\ \text{Irreducible representations of } \pi_4: & [4] & [1^4] & [2^2] & [21^2] & [31] \end{array} \qquad (6.25)$$

For $i = \frac{1}{2}$ we obtain the spin Young diagrams

$$\begin{array}{rccc} [\lambda]_{\text{spin}}: & [4] & [31] & [2^2] \\ I: & 2 & 1 & 0\,. \end{array}$$

The coordinate Young diagrams which correspond to these are

$$[\lambda]_{\text{coord}}: \quad [1^4] \quad [21^2] \quad [2^2],$$

or, by (6.25), to the irreducible representations

$$A_2, \quad F_1, \quad E.$$

As a result, we obtain the following allowed multiplets.

$$^5A_2, \quad ^3F_1, \quad \text{and} \quad ^1E, \qquad (6.26)$$

in which the multiplicities now, of course, refer to the total nuclear spins.

EXAMPLE 2.

In deuteromethane, $^{12}CD_4$, $i(D) = 1$, the point group is the same as that of methane, CH_4. However, the nuclei now obey Bose statistics. We write out all the possible spin Young diagrams together with the values of the total nuclear spin, these last being taken from Table 2, Appendix 3:

$$\begin{array}{rcccc} [\lambda]_{\text{spin}}: & [4] & [31] & [2^2] & [21^2] \\ I: & 4,2,0 & 3,2,1 & 2,0 & 1\,. \end{array}$$

The coordinate Young diagrams coincide with those for the spins. Taking into account the one-to-one correspondence between the irreducible representations (6.25), we obtain the following allowed multiplets:

$$^9A_1, \quad ^7F_2, \quad ^5A_1EF_2, \quad ^3F_1F_2, \quad ^1A_1E. \qquad (6.27)$$

EXAMPLE 3.

The ethylene molecule, $^{12}C_2H_4$, has the point symmetry $\mathbf{D}_{2h}$. The possible spin Young diagrams are the same as for methane:

$$\begin{array}{rccc} [\lambda]_{\text{spin}}: & [4] & [31] & [2^2] \\ I: & 2 & 1 & 0\,. \end{array}$$

The dual coordinate Young diagrams which correspond to these are

$$[\lambda]_{\text{coord}}: \quad [1^4] \quad [21^2] \quad [2^2].$$

The group $\boldsymbol{\pi}_4$, however, is larger than the point group $\mathbf{D}_{2h}$, and contains eight operations. Table 6.3 gives the correspondence between the classes of

Table 6.3

Characters of the Irreducible Representations of π_4 Corresponding to Operations of the Group $\mathbf{D}_{2h}$

Classes of $\mathbf{D}_{2h}$	E	$C_2{}^x$	$C_2{}^y$	$C_2{}^z$	I	σ_{yz}	σ_{xz}	σ_{xy}
Classes of π_4	$\{1^4\}$	$\{2^2\}$	$\{2^2\}$	$\{2^2\}$	$\{2^2\}$	$\{2^2\}$	$\{2^2\}$	$\{1^4\}$
$\chi^{[1^4]}$	1	1	1	1	1	1	1	1
$\chi^{[21^2]}$	3	−1	−1	−1	−1	−1	−1	3
$\chi^{[2^2]}$	2	2	2	2	2	2	2	2

the two groups and also the characters of the irreducible representations $\Gamma^{[\lambda]}$ corresponding to the operations of the point group (the $C_2{}^x$ axis is chosen to be along the C═C bond, and the $C_2{}^z$ axis perpendicular to the plane of the molecule). The decomposition of the representations $\Gamma^{[\lambda]}$ into irreducible representations of $\mathbf{D}_{2h}$ has the following form:

$$\Gamma^{[1^4]} \doteq A_g, \qquad \Gamma^{[2^2]} \doteq 2A_g, \qquad \Gamma^{[21^2]} \doteq B_{1g} + B_{2u} + B_{3u}. \tag{6.28}$$

One thus obtains the following nuclear multiplets for the ethylene molecule:

$$^5A_g, \qquad ^3B_{1g}B_{2u}B_{3u}, \qquad ^1A_g(2). \tag{6.29}$$

6.6. The Determination of the Nuclear Statistical Weights of Coordinate States

In the previous section it was shown how the methods of group theory allow one to determine the permitted types of coordinate symmetry and the values of the total nuclear spin which correspond to them. In the zeroth approximation in which nuclear spin interactions are neglected all the nuclear multiplets for a given symmetry type of coordinate wavefunction belong to a single energy level. The degree of degeneracy of an energy level $\Gamma^{(\alpha)}$ with respect to the nuclear spin states is called the *nuclear statistical weight* of the level, and we denote this by $\rho_{\text{nuc}}^{(\alpha)}$.

If a molecule consists of nuclei with spin $i = 0$ and of a single group of identical nuclei with nonzero spin, then the nuclear statistical weight of the term $\Gamma^{(\alpha)}$ is equal to the sum of all its multiplicities, multiplied by the number of times each multiplet occurs. We denote this last quantity by $a_{2I+1}^{(\alpha)}$; i.e.,

$$\rho_{\text{nuc}}^{(\alpha)} = \sum_I a_{2I+1}^{(\alpha)}(2I + 1). \tag{6.30}$$

This formula makes the calculation of the nuclear statistical weights of the terms easy if their multiplicities are known. As an example, we write out the nuclear statistical weights of the allowed terms for the molecules which were considered in the previous section. The value of $\rho_{\text{nuc}}^{(\alpha)}$ is given as a factor in front of the term symbol:

$$\begin{array}{lllll} {}^{12}CH_4: & 5A_2, & 1E, & 3F_1; & \\ {}^{12}CD_4: & 15A_1, & 6E, & 3F_1, & 15F_2; \\ {}^{12}C_2H_4: & 7A_g, & 3B_{1g}, & 3B_{2u}, & 3B_{3u}. \end{array} \tag{6.31}$$

If the molecule contains several distinct nuclei with spin $i_a \neq 0$ in addition to identical nuclei with spin i, the statistical weights which are obtained by taking into account just the identical nuclei must be multiplied by the number of spin states of the distinct nuclei. We denote this last quantity by γ. If such nuclei are enumerated by the index a, then

$$\gamma = \prod_a (2i_a + 1). \tag{6.32}$$

The total number of spin states for a molecule is equal to

$$\rho_{\text{nuc}} = \gamma \sum_\alpha \rho_{\text{nuc}}^{(\alpha)} f_\alpha = (2i + 1)^N \prod_a (2i_a + 1), \tag{6.33}$$

where $\rho_{\text{nuc}}^{(\alpha)}$ takes into account the contribution from identical nuclei only, and f_α is the dimension of the irreducible representation $\Gamma^{(\alpha)}$.

The nuclear statistical weights of terms can be found without first determining the allowed values of the total nuclear spin (Landau and Lifshitz, 1965; Wilson, 1935; Godnev, 1945; Trifonov, 1959). The method which will be set out here is based upon the scheme (6.23); however it is unnecessary to carry out the reduction (6.19).

We consider a molecule which contains a group of identical nuclei with spin i and several individual nuclei, the number of whose spin states is equal to γ [see (6.32)]. For a system of identical nuclei one can construct $(2i + 1)^N$ distinct spin functions, which can be resolved by a linear transformation into irreducible sets of functions—just as in the case of a configuration of equivalent electrons l^N (see Section 6.1). Each set of such functions is characterized by a particular Young diagram $[\lambda]$ consisting of N cells whose column lengths do not exceed $2i + 1$, and each set contains $f_\lambda \delta_\lambda (2i + 1)$ functions. These functions can be represented as points arranged in the form of a rectangular matrix:

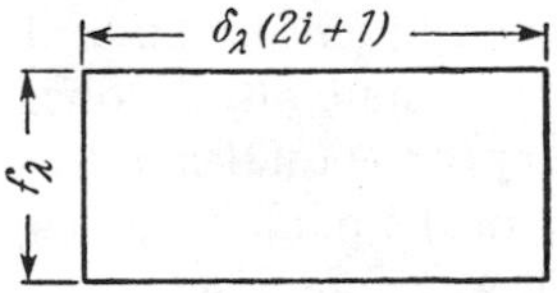

Spin functions which lie on a single column form a basis for an irreducible representation $\Gamma^{[\lambda]}$ of the permutation group, and, on constructing a total wavefunction according to formulas (6.17) and (6.18), lead to a single linear combination. The number of distinct total wavefunctions which can be constructed for a coordinate state $\Gamma^{[\lambda]}$ is equal to $\delta_{\lambda}(2i+1)$ in the case of a symmetric total wavefunction and to $\delta_{\tilde{\lambda}}(2i+1)$ in the case of an antisymmetric function.‡ It is obvious that

$$\delta_{\lambda}(2i+1) = \sum_{I} a_{\lambda}^{(I)}(2I+1), \tag{6.34}$$

where the $a_{\lambda}^{(I)}$ are determined by the decomposition (6.19). At the same time $\delta_{\lambda}(2i+1)$ can be calculated independently by formula (4.13).

The number of coordinate states $\Gamma^{[\lambda]}$ which can be constructed from the coordinate states of point symmetry $\Gamma^{(\alpha)}$ is equal, by Frobenius' theorem, to the coefficients $a_{\lambda}^{(\alpha)}$ in the decomposition (6.22). As a result we arrive at the following formulae for the nuclear statistical weights $\rho_{\text{nuc}}^{(\alpha)}$:

(a) Bose statistics (b) Fermi statistics

$$\rho_{\text{nuc}}^{(\alpha)} = \gamma \sum_{\lambda} a_{\lambda}^{(\alpha)}\delta_{\lambda}(2i+1) \qquad \rho_{\text{nuc}}^{(\alpha)} = \gamma \sum_{\lambda} a_{\lambda}^{(\alpha)}\delta_{\tilde{\lambda}}(2i+1) \tag{6.35}$$

This method is particularly simple for molecules whose point symmetry group is isomorphic with the permutation group for the identical nuclei. In such cases the summation over λ vanishes in formulae (6.35), the coefficients $a_{\lambda}^{(\alpha)}$ being equal to one or zero. In more complicated cases, such as in molecules possessing several groups of identical nuclei, it is simpler to determine the nuclear statistical weights by decomposing the representation generated by the nuclear spins under the operations of the point group directly into irreducible representations of the point group [see Landau and Lifshitz (1965, §105)].

6.7. Statistical Weights of Rotational Levels and Molecular Spin Modifications

For a given symmetry of the total coordinate wavefunction (6.21) the methods developed in sections 6.5 and 6.6 allow one to determine the possible symmetries of the coordinate functions for the separate kinds of motion. For example, we consider the rotational motion of molecules in their ground electronic and vibrational states. For most molecules the electronic wavefunction for the ground state is totally symmetric, i.e., belongs to the unit representation of the symmetry point group of the molecule. The wavefunction for the lowest vibrational state also belongs to the unit representation. Consequently, the symmetry of the total coordinate wavefunction (6.21) with respect to transformations of the point group coincides in this case with the symmetry of the rotational wavefunction.

The determination of the allowed rotational terms and their associated nuclear statistical weights is simplest for diatomic molecules. The point

‡We note that since all the f_{λ} coordinate wavefunctions which belong to a basis for a representation $\Gamma^{[\lambda]}$ occur in a single linear combination when constructing a total wavefunction, these functions describe a single physical state.

symmetry group for a diatomic molecule with identical nuclei is $\mathbf{D}_{\infty h} = \mathbf{C}_{\infty v} \times \mathbf{C}_i$. The irreducible representations of the group $\mathbf{C}_i$ characterize the behavior of the coordinate wavefunction (6.21) with respect to an inversion. Since in practically all molecules in their ground states the electronic wavefunction Φ_e remains unchanged under an inversion, and Φ_{vib} is always totally symmetric for diatomic molecules, the operations of $\mathbf{C}_i$ affect Φ_{rot} only. The behavior of rotational wavefunctions under inversions is given by the factor $(-1)^K$, where K denotes the orbital angular momentum of the molecule. [The rotational energy levels are related to the value of K by the simple equation

$$\epsilon_K = (\hbar^2/2I_0)K(K+1), \tag{6.36}$$

where I_0 is the moment of inertia of the molecule.] States which are symmetric with respect to an inversion, A_g, occur with even values of K, and states which are antisymmetric with respect to an inversion, A_u, occur with odd values of K. The inversion operation corresponds to a permutation of the nuclei of the molecule because the group $\mathbf{C}_i$ is isomorphic with the permutation group π_2. There is therefore a unique correspondence between the irreducible representations of these two groups:

π_2:	[2]	[1²]
$\mathbf{C}_i$:	A_g	A_u
K:	Even	Odd

Once the correspondence between the permutational symmetry of the coordinate wavefunction and the rotational states has been established, the nuclear statistical weights of the rotational levels are easily found by the rules of the previous section.

Let us determine the nuclear statistical weights of the rotational levels of the molecules H_2 and D_2. For clarity, we treat the two molecules together.

(a) H_2 ($i = \frac{1}{2}$, Fermi statistics):

$[\lambda]_{spin}$:	[2]	[1²]
I:	1	0
$[\lambda]_{coord}$:	[1²]	[2]
K:	1, 3, 5, ...	0, 2, 4, ...
ρ_{nuc}:	3	1

(b) D_2 ($i = 1$, Bose statistics):

$[\lambda]_{spin}$:	[2]	[1²]
I:	0,2	1
$[\lambda]_{coord}$:	[2]	[1²]
K:	0, 2, 4, ...	1, 3, 5, ...
ρ_{nuc}:	6	3

Because of the smallness of the interaction between electrons and nuclear spins, transitions are extremely improbable between states whose nuclear spin functions are of different symmetry. Hence molecules in states with differing symmetry $[\lambda]_{\text{spin}}$ behave as practically different modifications of matter. These are usually called *spin modifications*. Diatomic molecules in states with high nuclear statistical weights are called *ortho modifications*. For H_2, $\rho_{\text{ortho}} : \rho_{\text{para}} = 3:1$, and for D_2, $\rho_{\text{ortho}} : \rho_{\text{para}} = 2:1$.

As shown earlier, states with even values of the rotational angular momentum K possess different nuclear statistical weights from those with odd values. In molecules in which the nuclei have zero spin some values of K may be totally forbidden.

As an example we consider the molecule $^{16}O_2$. The nuclei of the ^{16}O isotope of oxygen have $i = 0$ and obey Bose statistics. Hence the coordinate wavefunction, which in this case coincides with the total wavefunction, can only be symmetric with respect to a permutation of the nuclei, i.e., it must belong to the representation $\Gamma^{[2]}$. The ground electronic state of the oxygen molecule, $^3\Sigma_g^-$, is symmetric with respect to inversions, and changes sign under a reflection σ_v in a plane passing through the molecular axis. Since for diatomic molecules an inversion is equivalent to a permutation of the nuclei, the electronic wavefunction for the ground state of O_2 is symmetric with respect to a permutation of the nuclei. From this it follows that the rotational wavefunction must also be symmetric with respect to permutations of the nuclei. Hence odd values of rotational angular momentum are forbidden for this molecule in its ground electronic state.

In a state of thermal equilibrium the rotational partition function for homonuclear diatomic molecules with nonzero nuclear spin must be formed from the partition functions for the ortho and para modifications by multiplying each partition function by its corresponding nuclear statistical weight and adding:

$$\begin{aligned} H_2: &\quad Z_{\text{rot}} = 3Z_1 + Z_2, \\ D_2: &\quad Z_{\text{rot}} = 3Z_1 + 6Z_2, \end{aligned} \tag{6.37}$$

where‡

$$Z_1 = \sum_{K=1,3,5,\ldots} (2K+1)e^{-\epsilon_K/T}, \qquad Z_2 = \sum_{K=0,2,4,\ldots} (2K+1)e^{-\epsilon_K/T}. \tag{6.38}$$

However, thermal equilibrium is reached extremely slowly in a mixture of the two modifications. Therefore in the absence of special catalysts mixtures of the two modifications are usually not at equilibrium.

Polyatomic molecules can also exist in different spin modifications which practically do not interconvert. For this to occur it is necessary for states

‡The value of the moment of inertia I_0 which occurs in the expression for the energies ϵ_K in the sums of (6.38) is different for H_2 and D_2. Note that ϵ_K is given in units of $k = 1$.

whose spin functions are of different symmetry to belong to energy levels the intervals between which are large compared to the nuclear spin interaction energy. Otherwise the internal nuclear spin interactions with the motion of the electrons lead to interconversions between spin modifications. The spin modifications of polyatomic molecules are classified according to the irreducible representations of the point group for the rotation of the molecule.

In determining the rotational energy levels of a polyatomic molecule regarded as a rigid rotator, one distinguishes among three cases.

(1) *An asymmetric top.* All three principal moments of inertia of the molecule are different. Molecules of this kind either possess no axes of symmetry at all or possess a twofold axis.

(2) *A symmetric top.* Two of the principal moments of inertia are identical. This occurs in all molecules which possess an axis of symmetry whose order is greater than two.

(3) *A spherical top.* All three principal moments of inertia coincide. Molecules which possess the symmetry of one of the cubic point groups belong to this class.

For an asymmetric top $2J + 1$ nondegenerate energy levels correspond to each value of the angular momentum J.‡ The Hamiltonian for the rotation of a body of arbitrary shape is invariant under the operations of the point group $\mathbf{D}_2$. The energy levels of an asymmetric top can therefore be classified according to the irreducible representations of this group. As can be proved by a direct calculation, one has for each value of J the following distribution of energy levels with symmetry as indicated (Landau and Lifshitz, 1965):

	A	B_1, B_2, B_3
Even J	$(J/2) + 1$	$J/2$
Odd J	$(J - 1)/2$	$(J + 1)/2$

(6.39)

The ethylene molecule $^{12}C_2H_4$ forms an example of an asymmetric top molecule. It possesses a center of symmetry and hence its point symmetry is $\mathbf{D}_{2h}$. The allowed symmetries and the nuclear statistical weights for the total coordinate function were found in the previous section [see (6.31)]. If we call states which are even with respect to inversion *positive* in conformity with the usual terminology, and those which are odd, *negative*, then the rotational energy levels of ethylene are classified as follows:

$$\overset{(+)}{7A,\ 3B_1;} \qquad \overset{(-)}{3B_2,\ 3B_3.} \tag{6.40}$$

‡Here and further on when speaking of the degeneracy of rotational levels we mean the degeneracy connected with different orientations of the angular momentum with respect to the *molecule*. In addition to this, there is always a $(2J + 1)$-fold degeneracy in the orientations of the angular momentum with respect to an external fixed coordinate system.

Each of the irreducible representations (6.40) corresponds to a particular energy level. An ethylene molecule can occur in one of four spin modifications, corresponding to the four irreducible representations of $\mathbf{D}_{2h}$, the nuclear statistical weights of which are in the ratio 7: 3: 3: 3.

Degenerate rotational levels‡ occur in symmetric and spherical top molecules. The representation generated by the rotational wavefunctions which belong to a single energy level may be reducible with respect to the operations of the point group of the molecule. In such a case the nuclear statistical weight of a rotational level is equal to the sum of the statistical weights for those irreducible representations of the point group that occur in the decomposition of the original reducible representation. In order to calculate the rotational partition function, it is necessary in addition to take into account the degeneracy associated with inversions of the molecule; i.e., it is necessary to sum over both positive and negative energy levels.

We carry out this procedure for a spherical top molecule such as $^{12}CH_4$, for example, the symmetry group of which is $\mathbf{T}_d$. The rotational levels of a spherical top are degenerate in the $2J + 1$ orientations of the angular momentum J with respect to the molecule. The nuclear statistical weights of the coordinate states of $^{12}CH_4$ were found in Section 6.6. Three types of terms are allowed: $5A_2$, $3F_1$, and $1E$. Besides rotations, the group $\mathbf{T}_d$ contains reflection operations σ_d. The behavior of the rotational functions can only be determined with respect to rotations about an axis. However, if the behavior of a rotational function under a rotation C_2 about an axis perpendicular to a plane of symmetry is known, then the behavior of the total coordinate function under a reflection σ_d can be determined if its behavior under an inversion I is given, since

$$\sigma_d = IC_2. \tag{6.41}$$

If the symmetry planes of $\mathbf{T}_d$ are replaced by twofold axes perpendicular to them, we obtain the group $\mathbf{O}$, which is isomorphic with $\mathbf{T}_d$. The decomposition of the irreducible representations $D^{(J)}$ on reducing $\mathbf{R}_3 \longrightarrow \mathbf{O}$ is given in Table 5.1. We determine the correspondence between the symmetry types of the rotational functions and the total coordinate function by means of character tables for $\mathbf{T}_d$ and $\mathbf{O}$, taking (6.41) into account.§ The results are conveniently represented in Table 6.4. It is clear from this table that only the E-type rotational states are split by the possibility of inversion. Rotational states of symmetry A_1 and F_2 correspond to negative coordinate functions of symmetries A_2 and F_1, respectively.

‡See footnote on p. 145.

§Since the group $\mathbf{T}_d$ does not contain a center of symmetry, eigenfunctions of the inversion operator are constructed directly from basis functions for irreducible representations of $\mathbf{T}_d$ by applying the appropriate projection operator. Thus $\Phi^{(\alpha)\pm} = \sqrt{\tfrac{1}{2}}(1 \pm I)\Phi^{(\alpha)}$

Table 6.4

Classification of the Rotational States of the Methane Molecule

Value of J	Symmetries of the rotational states	Symmetries and nuclear statistical weights of the total coordinate function (+)	(−)
0	A_1	—	$5A_2$
1	F_1	$3F_1$	—
2	E, F_2	$1E$	$1E, 3F_1$
3	A_2, F_1, F_2	$5A_2, 3F_1$	$3F_1$
4	A_1, E, F_1, F_2	$1E, 3F_1$	$5A_2, 1E, 3F_1$

According to the classification of its energy states by the irreducible representations of $\mathbf{T}_d$, methane exists in the form of three spin modifications. Each spin modification is characterized by its own rotational partition function. In determining the statistical weight of a coordinate state of symmetry $\Gamma^{(\alpha)}$, one must remember when constructing a total wavefunction that the f_α coordinate functions belonging to this state yield a single linear combination when combined with the spin functions. Each irreducible representation $\Gamma^{(\alpha)}$ is therefore counted as one state in determining the statistical weight. Furthermore, one must also take into account the $(2J+1)$-fold degeneracy in the orientations of the angular momentum in space. As a result, the rotational partition function for a particular spin modification is given by the following formula:

$$Z_{\text{rot}}^{(\alpha)} = \rho_{\text{nuc}}^{(\alpha)} \sum_J (a_J^{(\alpha)+} + a_J^{(\alpha)-})(2J+1) \exp[-\epsilon_J^{(\alpha)}/T] \qquad (6.42)$$

where $a_J^{(\alpha)+}$ is the number of times the positive coordinate state with symmetry $\Gamma^{(\alpha)}$ occurs in the reducible representation formed by the $2J+1$ rotational functions, $\rho_{\text{nuc}}^{(\alpha)}$ is the nuclear statistical weight of the state $\Gamma^{(\alpha)}$, and $\epsilon_J^{(\alpha)}$ is a rotational energy level of a spherical top, the additional symbol α taking account of the possibility that the level with the given J may split in a crystalline field of symmetry $\mathbf{T}_d$.

The symmetries and statistical weights of the coordinate states of methane which occur for the first few values of J are given in Table 6.4. Upon substituting the values which are given there into formula (6.42), we obtain the rotational partition function for each of the three spin modifications of methane:

$$\begin{aligned} Z_{\text{rot}}^{(A_2)} &= 5(1 + 7e^{-\epsilon_3/T} + 9e^{-\epsilon_4/T} + \cdots), \\ Z_{\text{rot}}^{(F_1)} &= 3(3e^{-\epsilon_1/T} + 5e^{-\epsilon_2/T} + 14e^{-\epsilon_3/T} + 18e^{-\epsilon_4/T} + \cdots), \\ Z_{\text{rot}}^{(E)} &= 2(5e^{-\epsilon_2/T} + 9e^{-\epsilon_4/T} + \cdots). \end{aligned} \qquad (6.43)$$

In gaseous methane the rotational levels are classified by the value of the angular momentum J only. Since for $J \geq 2$ degenerate irreducible representations of the point group occur, the intramolecular spin interactions effectively lead to interconversions among the spin modifications (Cure *et al.*, 1966). In a state of thermal equilibrium

$$Z_{\text{rot}} = \sum_{\alpha} Z_{\text{rot}}^{(\alpha)} = 5 + 9e^{-\epsilon_1/T} + 25e^{-\epsilon_2/T} + 77e^{-\epsilon_3/T} + 117e^{-\epsilon_4/T} + \cdots. \tag{6.44}$$

It is interesting to note that a small deviation from the equilibrium configuration of a symmetric molecule in which the energy is almost unchanged leads to a sudden change in the rotational partition function. This is due to the fact that the nuclear statistical weights of rotational levels increase when the symmetry of the molecule is lowered. For example, let methane become an asymmetric top with point symmetry $\mathbf{D}_{2h}$ as a result of a distortion from its equilibrium configuration. The rotational energy levels of the molecule are then classified according to the irreducible representation of the group $\mathbf{D}_2$, their correspondence with the values of the angular momentum J having been given earlier [see (6.39)]. We write out the result of the reduction $\mathbf{R}_3 \rightarrow \mathbf{D}_2$ for the first few values of J:

J	$7A_1$	$3B_1$	$3B_2$	$3B_3$
0	1	—	—	—
1	—	1	1	1
2	2	1	1	1
3	1	2	2	2
4	3	2	2	2

On neglecting the splitting of the energy levels of a given J (distortion of the spherical top small), the rotational partition function assumes the following form:

$$Z_{\text{rot}} = 7 + 27e^{-\epsilon_1/T} + 115e^{-\epsilon_2/T} + 175e^{-\epsilon_3/T} + 351e^{-\epsilon_4/T} + \cdots. \tag{6.45}$$

The values of Z_{rot} calculated by this equation are considerably larger than those calculated by formula (6.44). As a result, a jump in the value of the rotational partition function occurs at the moment when the symmetry of the molecule alters. The reason for this is that when the molecular configuration becomes more symmetric, the number of independent rotational states diminishes. This effect also occurs in classical statistical mechanics (see the following section).

6.8. Transition from the Rotational Partition Function to an Integral over States. The Symmetry Number

We now carry out the limiting transition from a quantal rotational partition function to a classical integral over states for a spherical top molecule. The rotational partition function for a molecule of symmetry $\mathbf{T}_d$ when there is equilibrium among the spin modifications is given by expression (6.42) summed over α (we regard the rotational energy levels as depending only upon J),

$$Z_{\text{rot}} = \sum_{\alpha} \sum_{J} \rho_{\text{nuc}}^{(\alpha)} (a_J^{(\alpha)+} + a_J^{(\alpha)-})(2J+1) e^{-\epsilon_J/T}. \tag{6.46}$$

At high temperatures when $T \gg \epsilon_J$ the main contribution to the sum in (6.46) comes from the terms with high values of J. At high J values, however, the rotation is semiclassical and this allows one to replace the sum over J by the corresponding integral. In addition, $a_J^{(\alpha)+}$ and $a_J^{(\alpha)-}$ can be equated because when J is large a representation $D^{(J)}$ contains sets of regular representations so that positive and negative terms occur an equal number of times. We determine the form of $a_J^{(\alpha)}$ at high J values. According to Eq. (1.62),

$$a_J^{(\alpha)} = (1/g) \sum_R \chi^{(J)}(R) \chi^{(\alpha)}(R)^*, \tag{6.47}$$

where R runs over the g operations of the point group. If we replace the group $\mathbf{T}_d$ by the isomorphic group $\mathbf{O}$, then all the operations R become equivalent to rotations about symmetry axes, the characters of which depend only upon the angle of rotation α and are given by formula (3.40). For all operations except the identity E, $\chi^{(J)}(R)$ is of order unity, and for the identity operation $\chi^{(J)}(E) = 2J+1$. Hence as $J \longrightarrow \infty$ the main contribution to the sum (6.47) is given by the operation E, from which it follows that

$$a_{J\to\infty}^{(\alpha)} = (f_\alpha/g)(2J+1). \tag{6.48}$$

As a result, the partition function (6.46) becomes

$$\begin{aligned} Z_{\text{rot}}(J \longrightarrow \infty) &= (2/g) \sum_{\alpha} f_\alpha \rho_{\text{nuc}}^{(\alpha)} \int 4J^2 \exp[-(\hbar^2 J^2/2I_0 T)]\, dJ \\ &= (1/\sigma) \rho_{\text{nuc}} Z_{\text{rot}}^{\text{class}}, \end{aligned} \tag{6.49}$$

where $\sigma = g/2$ is equal to the number of different rotations about symmetry axes which send the molecule into itself. This number is called the *symmetry number* of the molecule. The factor ρ_{nuc} is the total number of nuclear spin states [see (6.33)]. Integration over J in (6.49) leads to a classical value for the rotational partition function:

$$Z_{\text{rot}}^{\text{class}} = (\sqrt{\pi}/\hbar^3)(2I_0 T)^{3/2}. \tag{6.50}$$

Thus at high temperatures the rotational partition function or sum-over-states goes over to a classical integral over rotational states, multiplied by the number of nuclear spin states and divided by the symmetry number of the molecule. In the derivation given here σ arises naturally in the classical expression (6.49). The symmetry number was introduced by Ehrenfest and Trkal (1921) before the advent of quantum mechanics. In order to take each physical state into account once, they divided the function of states obtained by integrating over all the angles of rotation by the number of physically indistinguishable orientations of the molecule in space.

It follows from the classical approximation (6.49), just as from quantal considerations, that when the symmetry of the molecule is altered the accompanying change in the symmetry number σ leads to a jump in the value of the rotational partition function.

PART 3. Classification of States in Approximate Quantal Calculations

6.9. The Configuration Interaction Method and Partial Diagonalization of the Secular Equation

As is well known, the nature of the chemical bond was only elucidated after the advent of quantum mechanics, and after the introduction of the concept of electron spin. However, the spins of the electrons take no direct part in the formation of molecules, the interactions that lead to chemical bonding being entirely electrostatic in nature. The dependence of the energy of a system of electrons upon the value of its total spin arises from a correlation between the positions of the electrons in space and the directions of their spins. If one neglects spin interactions, the total wavefunction for a system of N electrons is expressed, according to Eq. (6.2), as a sum of products of coordinate wavefunctions $\Phi_r^{[\lambda]}$ with spin functions $\Omega_r^{[\tilde{\lambda}]}$ respectively symmetrized according to mutually dual Young diagrams. The energy levels of a system of interacting electrons are characterized by the permutational symmetry $[\lambda]$ of the coordinate wavefunction. Since every spin Young diagram $[\tilde{\lambda}]$ is uniquely associated with a value of the spin S, one can assign to each energy level a specific value of the total spin. All properties of particular systems that do not involve spin interactions are completely determined by the coordinate wavefunction $\Phi_r^{[\lambda]}$.

In quantal calculations one normally starts from a one-particle approximation. A configuration of k one-electron spatial orbitals ϕ_a is selected in which N electrons are accommodated in a particular way:

$$\phi_1^{n_1}\phi_2^{n_2}\cdots\phi_k^{n_k}, \qquad \sum_{a=1}^{k} n_a = N. \tag{6.51}$$

In accordance with the Pauli principle, no orbital may contain more than two electrons, i.e., $n_a \leq 2$, $k \geq N/2$. When $n_a = 2$ the sets of quantum numbers for the two electrons occupying the ath orbital differ in the values for the projections of the spin. One says that the spins of such electrons are paired. The one-electron orbitals should be solutions of the Hartree–Fock equations. However, in cases where such solutions are unknown, one takes for the ϕ_a either orbitals which are localized on the individual atoms (*the valence bond method*), or molecular orbitals in the form of linear combinations of atomic orbitals (*the LCAO–MO method*).

The number of configurations of k orbitals in which m orbitals are doubly filled and $N-2m$ are singly occupied is obviously equal to the number of ways of selecting m orbitals from k (number of combinations of m from k), multiplied by the number of ways of distributing the remaining $N-2m$ electrons among $k-m$ orbitals, i.e., is equal to

$$n(k, m, N) = \binom{k}{m}\binom{k-m}{N-2m} = \frac{k!}{m!\,(N-2m)!\,(k+m-N)!}. \tag{6.52}$$

For each configuration one can construct 2^{N-2m} different antisymmetric wavefunctions because each of the $N-2m$ unpaired electrons can possess one of two different values of the projection of the spin. The total number of states for a set of k orbitals with N electrons is obtained by multiplying the number of states of one configuration by the number of configurations of that particular type, and summing over all possible configurations‡:

$$n(k, N) = \sum_{m=0}^{E(N/2)} n(k, m, N)2^{N-2m} = \sum_{m=0}^{E(N/2)} \frac{k!\,2^{N-2m}}{m!\,(N-2m)!\,(k+m-N)!}. \tag{6.53}$$

Instead of summing expression (6.53), the number of states can be found directly from the following simple considerations. If we associate a particular value of the projection of the spin with an orbital ϕ_a, we obtain a *spin orbital* which can accommodate not more than one electron. The number of ways in which N electrons can be distributed among $2k$ spin orbitals is equal to the number of combinations of $2k$ from N, i.e.,

$$n(k, N) = \binom{2k}{N} = \frac{(2k)!}{(2k-N)!\,N!}. \tag{6.54}$$

The sum over m in formula (6.53) leads to a result which is equivalent to this formula (Kaplan, 1966).

The most general wavefunction for a system which can be constructed from k one-electron orbitals is of the form of a linear combination of $\binom{2k}{N}$ antisymmetric functions, each function consisting of products of orbitals ϕ_a with the corresponding spin functions. The coefficients in this linear combination are found by minimizing the energy of the system (see Section 5.4). A variation method in which one takes into account different possible configurations of one-electron orbitals is called a *configuration interaction* method. In this approximation the system under study is not described by any one particular configuration of one-electron orbitals, and therefore the method constitutes an improvement upon the one-particle approximation. The use of a large enough number of configurations often leads to good agreement with experiment (Slater, 1963).

If all the configurations which arise from the chosen set of one-electron orbitals are taken into account, the resulting secular equation is of very high

‡$E(N/2)$ denotes the integral part of the number $N/2$.

order even for small systems. Thus for four electrons and four orbitals we have $n(k, N) = 8!/(4! \times 4!) = 70$ states. However, in the present approximation in which the total wavefunction is factorized into coordinate and spin functions it is possible partially to diagonalize the secular equation. Since the Hamiltonian of the molecule is invariant under permutations of the electrons, the secular equation breaks up into blocks when the wavefunctions for the states are constructed according to formula (6.2), each block corresponding to a particular irreducible representation $\Gamma^{[\lambda]}$ for the coordinate wavefunction. Since a coordinate Young diagram $[\lambda]$ corresponds uniquely to a value of the total spin S of the system of electrons, there will be a secular equation for each value of S, the dimension of which is equal to the number of ways in which the particular value of S can be realized in the system.

For symmetric molecules it is possible to diagonalize the secular equation further. For this purpose it is necessary to construct from the wavefunctions corresponding to a particular spin S, basis functions for irreducible representations of the point symmetry group of the molecule. According to Eq. (5.15), all diagonal elements of the energy matrix that correspond to a single irreducible representation are equal, and all off-diagonal elements reduce to zero, except for matrix elements between basis functions that transform according to one and the same column of equivalent irreducible representations.

We shall refer to the states of an N-electron system of point symmetry $\Gamma^{(\alpha)}$ and whose total electron spin is S as *molecular electronic multiplets* and denote them by ${}^{2S+1}\Gamma^{(\alpha)}$. If one constructs linear combinations of the initial set of $\binom{2k}{N}$ functions so as to form eigenfunctions of the multiplets, then the original secular equation breaks up into a number of secular equations, each of which corresponds to some multiplet ${}^{2S+1}\Gamma^{(\alpha)}$. The orders of these secular equations are equal to the number of times a particular multiplet occurs in the decomposition of the representation induced by the original set of basis functions. If all $\binom{2k}{N}$ configurations are taken into account, calculations by the valence bond and the molecular orbital methods are equivalent (Slater, 1963). This is because in order to pass from the one-electron orbitals in the valence bond method to the molecular orbitals in the LCAO-MO method, it is necessary to carry out a linear transformation of the basis functions in the secular equation. Now the roots of a secular equation are quite independent of a particular choice of linear combination of basis functions. Hence the multiplet structures obtained by the two methods of calculation are identical.

It is of some interest to determine, before carrying out a direct calculation, which multiplets may arise, and also the maximum order of the secular equations which will occur in solving the problem. In the next section a method is set out for finding the possible multiplets and the number of times they occur in calculations by the valence bond method.

6.10. Determination of the Multiplets Which May Arise in Calculations by the Valence Bond Method[‡]

We confine ourselves to the case where there is one nondegenerate one-electron orbital per atom, the number of valence electrons being equal to the number of orbitals.[§] Each orbital is then singly occupied in a configuration of neutral atoms. The interactions of the valence electrons in such a configuration lead to the formation of *covalent chemical bonds*. If one or several electrons move from their "own" orbitals to those on neighboring atoms, an ionic configuration of the atom is formed, which is responsible for the formation of an *ionic chemical bond*. In real molecules contributions to a chemical bond are made by both ionic and covalent configurations. The more configurations one takes into account, the more accurate calculations on molecular electronic states become.

A. Covalent Structures

In the present case a configuration of singly occupied orbitals, characterized by a certain mode of coupling the spins of the individual electrons to form a total spin S, is called a covalent structure. Generally speaking, there are several independent ways of coupling the electron spins to give a particular value of the total spin. The number of such ways determines the number of independent covalent structures with spin S that can be constructed from N valence electrons. We denote this number by $n(N, S)$.

In order to find the independent covalent structures, one makes use in quantum chemistry of what is known as *Rumer's rule* (Eyring *et al.*, 1944; Hellmann, 1937; Kauzmann, 1957). According to this, we place symbols representing the orbitals of the valence electrons around the circumference of a circle and join pairs of orbitals by bonding lines in all possible ways without any bonds intersecting one another. A bond joining two orbitals signifies that the spins of the participating electrons are paired. For example, for four orbitals there are two independent covalent structures with $S = 0$:

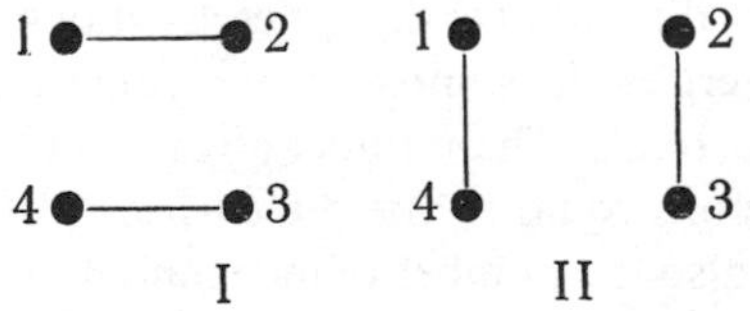

$$
\begin{array}{cc}
\begin{array}{c}
1\!\bullet\!\!-\!\!-\!\!-\!\!\bullet 2 \\[18pt]
4\!\bullet\!\!-\!\!-\!\!-\!\!\bullet 3
\end{array}
&
\begin{array}{c}
1\!\bullet\quad\quad\bullet 2 \\
\Big|\quad\quad\Big| \\
4\!\bullet\quad\quad\bullet 3
\end{array}
\\[6pt]
\mathrm{I} & \mathrm{II}
\end{array}
\qquad (6.55)
$$

[‡]See Kaplan (1966).
[§]This situation occurs in calculations on the π-electron systems of conjugated and aromatic hydrocarbons, and also in problems involving interactions between s electrons (see Section 6.11). The general case is treated by Kaplan and Rodimova (1968).

Usually in order to construct the wavefunction corresponding to a covalent structure, one forms the particular linear combination of determinants of spin orbitals which corresponds to the given type of coupling of electron spins (Eyring *et al.*, 1944; Kauzmann, 1957). However, the use of the theory of the permutation group enables one to write the total wavefunction for a system of electrons in terms of products of appropriately symmetrized coordinate wavefunctions with spin functions. This makes a different approach to the description of covalent structures possible, based upon the use of coordinate wavefunctions. Such an approach was developed by the author (1963, 1965a) and, independently, by Matsen (1964a, b).‡ The author's method has been applied to configurations in which the orbitals are arbitrarily occupied, and together with the technique of coefficients of fractional parentage (see Chapters VII and VIII), leads to compact expressions for the matrix elements of the Hamiltonian.

In the zeroth approximation the coordinate wavefunction for a system of valence electrons is written as a product of N one-electron orbitals ϕ_a, each of which is localized on a particular atom,

$$\Phi_0 = \phi_1(1)\phi_2(2) \cdots \phi_N(N). \tag{6.56}$$

When each orbital is singly occupied, i.e., when all the ϕ_a in (6.56) are different, the configuration corresponds to one of neutral atoms. We assume that the orbitals constitute an orthonormal set. The construction of basis functions for irreducible representations $\Gamma^{[\lambda]}$ of the permutation group π_N is carried out by applying the Young operators $\omega_{rt}^{[\lambda]}$ to the function (6.56) (see Section 2.9). The $f_\lambda{}^2$ functions which are obtained in this way, $\Phi_{rt}^{[\lambda]}$, are divided into f_λ sets, the sets being enumerated by the index t. Each set consists of f_λ functions $\Phi_{rt}^{[\lambda]}$ with a fixed index t, and forms a basis for an irreducible representation $\Gamma^{[\lambda]}$. All functions which belong to a single basis hence correspond to a single physical state, because when the total wavefunction is constructed they all occur in a single antisymmetric combination:

$$\Psi_t^{[\lambda]} = \sum_r \Phi_{rt}^{[\lambda]}\Omega_{\tilde{r}}^{[\tilde{\lambda}]}. \tag{6.57}$$

For a given symmetry diagram $[\lambda]$ the different states are enumerated by the index t which characterizes the symmetry of a function (6.57) with respect to permutations of the orbitals. The number of distinct states of permutational symmetry $[\lambda]$ is therefore equal to the dimension f_λ of the irreducible representation $\Gamma^{[\lambda]}$. This is also the number of independent covalent structures with

‡Coordinate wavefunctions were first used in molecular problems by Kotani and Siga (1937). The method was applied to configuration interaction calculations on diatomic molecules by Kotani and his co-workers (Kotani *et al.*, 1957, Ishiguro *et al.*, 1957). The theory of the permutation group has been used to construct molecular wavefunctions in recent papers by Goddard (1967a), and Gallup (1968, 1969).

spin S, because the functions (6.57) describe a configuration of neutral atoms in a state with a particular value of the spin S.

We now express f_λ in terms of the number of valence electrons N and the value of the spin S. For this purpose we note that the dimensions of the representations $\Gamma^{[\lambda]}$ and $\Gamma^{[\tilde{\lambda}]}$ are the same. According to Eq. (4.44),

$$S = \tfrac{1}{2}(\tilde{\lambda}^{(1)} - \tilde{\lambda}^{(2)}), \tag{6.58}$$

and, in addition,

$$N = \tilde{\lambda}^{(1)} + \tilde{\lambda}^{(2)}. \tag{6.59}$$

Relations (6.58) and (6.59) enable one to express the row lengths of a spin Young diagram in terms of the two parameters N and S:

$$\tilde{\lambda}^{(1)} = \tfrac{1}{2}N + S, \qquad \tilde{\lambda}^{(2)} = \tfrac{1}{2}N - S. \tag{6.60}$$

Substituting these values into the formula for the dimension of an irreducible representation, Eq. (2.18), we obtain the desired expression:

$$n(N, S) \equiv f_{\tilde{\lambda}} = \frac{N!\,(2S+1)}{(N/2+S+1)!\,(N/2-S)!}. \tag{6.61}$$

For $N = 4$, $S = 0$, this formula gives $n(4, 0) = 2$, in agreement with (6.55).

In calculations on the lower electronic states of the benzene molecule, one often takes into account just the six π electrons of the benzene ring. The number of covalent structures with $S = 0$ is equal to $n(6, 0) = 5$. When the five independent covalent structures are written out by means of Rumer's rule, one sees that they are just the two Kekulé and three Dewar structures so well known in organic chemistry (Kauzmann, 1957). The number of covalent structures with $S = 1$ is equal to $n(6, 1) = 9$, the number of independent covalent structures for higher multiplicities being found similarly.

The wavefunctions for the covalent structures are linear combinations of the functions (6.57). Since the roots of a secular equation are invariant with respect to a linear transformation of the basis functions, calculations with the functions (6.57) are completely equivalent to calculations with covalent structure functions. However, use of the functions $\Phi_{rt}^{[\lambda]}$ makes the calculation of matrix elements considerably simpler and more systematic (see Chapter VIII).

If it should be necessary for some reason actually to write out the structure functions, then the coordinate wavefunctions which correspond to these are easily expressed in terms of the functions $\Phi_{rt}^{[\lambda]}$. For this purpose it is necessary to use the fact that the symmetry of the coordinate wavefunctions with respect to permutations of the orbitals is determined by specifying the Young tableaux t of the standard representation of the permutation group. The coordinate wavefunctions for the covalent structures must obviously be symmetric with respect to permutations of any two atomic orbitals whose electrons are paired and the functions therefore belong to a nonstandard representation of the permutation group. A transition from basis functions for a standard representation to those for a nonstandard representation is carried out by means of the appropriate set of Young operators, and was described in Part 3 of Chapter II.

Thus for $N = 4$ and $[\lambda] = [2^2]$ we have two standard Young tableaux:

$$t_1: \begin{array}{|c|c|}\hline 1 & 2 \\ \hline 3 & 4 \\ \hline\end{array} \qquad t_2: \begin{array}{|c|c|}\hline 1 & 3 \\ \hline 2 & 4 \\ \hline\end{array} \tag{6.62}$$

By its construction, the function $\Phi_{rt_1}^{[2^2]}$ is symmetric with respect to permutations of the orbitals $\bar{P}_{12}$ and $\bar{P}_{34}$ (in accordance with the convention explained in Section 2.9 permutations of the orbitals are distinguished from those of electronic coordinates by placing a bar over the permutation symbol for the former). Consequently,

$$\Phi_{\rm I} \equiv \Phi_{rt_1}^{[2^2]}. \tag{6.63}$$

In order to obtain a function which is symmetric with respect to the permutations $\bar{P}_{14}$ and $\bar{P}_{23}$, we apply the following symmetrizing operator to $\Phi_{rt_1}^{[2^2]}$:

$$\Phi_{\rm II} = c(I + \bar{P}_{14})(I + \bar{P}_{23})\Phi_{rt_1}^{[2^2]}, \tag{6.64}$$

where c is a normalization constant. The successive application of formula (2.48) with matrix elements for the transpositions taken from Table 2, Appendix 5 leads to the following normalized coordinate function for structure II:

$$\Phi_{\rm II} = \tfrac{1}{2}\Phi_{rt_1}^{[2^2]} + \tfrac{1}{2}\sqrt{3}\,\Phi_{rt_2}^{[2^2]}. \tag{6.65}$$

Additional degeneracy occurs in symmetric molecules due to the symmetry of the molecular potential field in which the electrons are regarded as moving. All the $f^{(\alpha)}$ states which belong to a single irreducible representation $\Gamma^{(\alpha)}$ of the point symmetry group of the molecule belong to a single energy level. In order to find which irreducible representations $\Gamma^{(\alpha)}$ can occur with a particular permutational symmetry $\Gamma^{[\lambda]}$, we use the fact that a discrete point symmetry group for a molecule is isomorphic with a subgroup of the permutation group of the orbitals. As a result, any operation of the point symmetry can be represented as some permutation of the orbitals, since each orbital is localized on a particular atom. On passing from the complete permutation group to its point subgroup, a representation $\Gamma^{[\lambda]}$ breaks up into irreducible representations $\Gamma^{(\alpha)}$ which are found by decomposing the characters. Thus with each $\Gamma^{(\alpha)}$ there is associated a value of the spin S which corresponds uniquely to the permutational symmetry $[\tilde{\lambda}]$ of the spin wavefunction.

The method of finding the possible electronic multiplets for covalent structures is thus similar to the method of determining the nuclear molecular multiplets (6.23b), and can be represented diagramatically as follows:

$$\begin{array}{ccc} \Gamma^{[\lambda]} & \longleftrightarrow & U_2^{[\tilde{\lambda}]} \\ \downarrow & & \updownarrow \\ \Gamma^{(\alpha)} & \longleftrightarrow & D^{(S)} \end{array} \cdot \qquad (6.66)$$

Examples in which this scheme is applied are given in the next section.

The scheme (6.66) is based upon the assumption that the particular orbitals on the atoms are of a single type which remain invariant under the operations of the point group, so that an arbitrary operation of the group is equivalent to a permutation of the orbitals. A connection is thus established between the permutational symmetry of the coordinate wavefunction and its point symmetry. If the orbitals on the atoms are of different types, the procedure for finding the allowed multiplets becomes a little more complicated. The problem of determining which molecular multiplets may arise from arbitrary atomic terms was first treated by Kotani (1937). The same problem is also considered by Kaplan and Rodimova (1968), where the permutational symmetry of the coordinate wavefunction and the possibility of mixing by different configurations is taken into account.

The effect of an operation R of the point group upon a configuration of degenerate localized orbitals corresponding to a definite value of the angular momentum j, is given by two successive operations: a permutation of the orbitals $\bar{P}$ and an operation of the point symmetry in the space of each orbital. The permutation $\bar{P}$ can always be broken down into independent cycles, and consequently the atoms can be divided into sets which do not transform into each other under R. The character is nonzero only if the atoms within a set are all in the same state. It can be shown for a state with permutational symmetry $[\lambda]$ that the character of an operation R corresponding to a permutation $\bar{P}$ which is made up of cycles of length $n_1, n_2, \ldots, n_k$ is equal to (Kaplan and Rodimova, 1968)

$$\chi^{[\lambda]}(R) = \chi^{[\lambda]}(\bar{P})\chi^{(j_1)}(R^{n_1})\chi^{(j_2)}(R^{n_2}) \cdots \chi^{(j_k)}(R^{n_k})\tau(R), \qquad (6.67)$$

where $\tau(R)$ denotes the number of configurations which remain invariant under R. Some of the angular momenta j_i in (6.67) may coincide.‡ For nondegenerate orbitals all of the same type so that scheme (6.66) is applicable, formula (6.67) reduces to

$$\chi^{[\lambda]}(R) = \chi^{[\lambda]}(\bar{P}). \qquad (6.67a)$$

B. *Ionic Structures*

Just as in the case of covalent structures, we take the number of orbitals to be equal to the number of interacting electrons. We consider nondegenerate

‡Formulae for the characters in the case of a single atomic shell in a crystal field, γ^N (or j^N in a central field), have been obtained by Kotani (1964); see also Goscinski and Öhrn (1968).

orbitals of a single type. An electronic configuration in which some of the orbitals are doubly occupied hence is associated with an ionic configuration. According to formula (6.52), the number of ionic configurations which contain m doubly filled orbitals and hence $N - 2m$ singly occupied orbitals is equal to

$$n(N, m, N) = N!/(m!)^2(N - 2m)!. \tag{6.68}$$

An ionic configuration characterized by a certain coupling of the unpaired spins to form a total spin S is called an *ionic structure*. Each doubly filled orbital $\phi_a{}^2$ has the coordinate Young diagram [2]. For a number of doubly filled orbitals in an ionic configuration

$$\phi_1{}^2\phi_2{}^2 \cdots \phi_m{}^2\phi_{m+1} \cdots \phi_{N-2m} \tag{6.69}$$

the coordinate wavefunction for the doubly filled orbitals is obviously characterized by the following Young diagram with $2m$ cells:

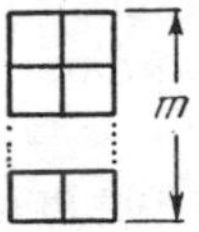

corresponding to a spin $S = 0$. The Young diagrams [2] for each orbital pair are uniquely connected with the diagram $[2^m]$, and therefore there is only one state for doubly filled orbitals. The number of independent states of the configuration (6.69) with spin S is equal to the dimension of the representation characterized by the Young diagram for the singly occupied orbitals. This diagram, which we denote by $[\lambda^{(2m)}]$, is obtained by removing the m upper rows from the full diagram $[\lambda]$. In Fig. 6.3, $[\lambda^{(2m)}]$ is shaded. The diagram $[\lambda^{(2m)}]$ corresponds to a state for which the value of the total spin S is obtained by vector coupling $N - 2m$ unpaired spins. According to (6.61), the number of independent states is given in this case by

$$n(N - 2m, S) \equiv f_{\lambda^{(2m)}} = \frac{(N - 2m)!\,(2S + 1)}{(\frac{1}{2}N - m + S + 1)!\,(\frac{1}{2}N - m - S)!}. \tag{6.70}$$

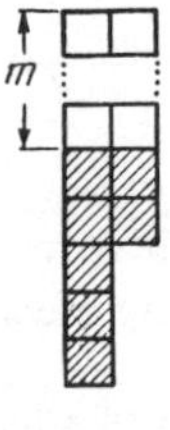

Figure 6.3.

The total number of ionic structures of the form (6.69) with spin S is equal to

$$n(N, m, S) = \frac{N!\,f_{\lambda^{(2m)}}}{(m!)^2(N-2m)!}$$
$$= \frac{N!\,(2S+1)}{(m!)^2(\frac{1}{2}N - m + S + 1)!\,(\frac{1}{2}N - m - S)!}. \tag{6.71}$$

The secular equation for asymmetric molecules can be partially diagonalized only with respect to values of the total spin S. Its order is equal to the total number of states with the given spin S that arise from all possible ionic configurations (6.69) and from the covalent configuration ($m = 0$); i.e., its order is given by summing expression (6.71) over all values of m for which the particular S can be realized. As a result of such a summation [see Mulder (1966)], we obtain the following expression:

$$\sum_{m=0}^{E(N/2-S)} n(N, m, S) = \frac{2S+1}{N+1}\binom{N+1}{\frac{1}{2}N - S}^2 \equiv \frac{N+1}{2S+1}[n(N, S)]^2, \tag{6.72}$$

where $n(N, S)$ is defined by (6.61), and $E(\frac{1}{2}N - S)$ is the integral part of $\frac{1}{2}N - S$.

For symmetric molecules a further reduction can be made in the order of the secular equation by constructing from the structure functions basis functions for irreducible representations $\mathbf{\Gamma}^{(\alpha)}$ of the point symmetry group of the molecule. The $\mathbf{\Gamma}^{(\alpha)}$ which occur in the decomposition of the reducible representation of the point group generated by the ionic structures can be easily found if the characters are known. We now consider the process of finding the characters of such a representation.

The ionic configurations must first of all be divided up into sets whose members transform into each other under the operations of the point group **G**. For this purpose an ionic configuration is identified by the corresponding number of positively and negatively charged atoms placed in the equilibrium nuclear configuration of the molecule. One then proceeds according to the following rules:

(a) If an ionic configuration does not possess any elements of symmetry, the application to it of the g operations of the molecular point group leads to g distinct configurations which generate the regular representation.

(b) Let there be some subgroup **H** (whose order we denote by h) of the molecular point group with respect to which a particular ionic configuration is invariant. Then the representation that is generated by the configuration is of smaller dimension than that of the regular representation, and is equal to g/h.

In order to prove this last statement, we expand all the elements of G in left cosets of the subgroup **H** : **H**, R_1**H**, R_2**H**, . . . , R_m**H** (where $m = g/h$; see

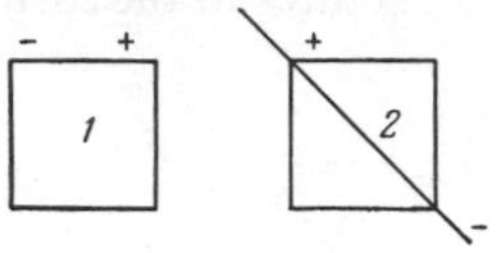

Figure 6.4.

Section 1.4). It is clear that all the operations within a single coset have the same effect upon the particular ionic configuration. Since there are g/h cosets altogether, the number of distinct configurations which are generated by a configuration whose symmetry group is **H** is equal to g/h. As an example, we consider a hypothetical molecule consisting of four atoms, and which possesses the symmetry of the group $\mathbf{D}_4$ ($g = 8$). According to formula (6.68) there are $4!/2! = 12$ configurations with one pair of ions. These 12 configurations can be divided into two sets which do not mix under the operations of the point group (Fig. 6.4). Configuration 1 has no elements of symmetry, and so set 1 consists of eight configurations. Configuration 2 is invariant with respect to operations of a group **H** whose elements are E and U_2. Set 2 therefore contains $8/2 = 4$ configurations.

One can construct from a given ionic configuration $f_{\lambda^{(2m)}}$ ionic structures with spin S [see (6.70)]. These transform according to the irreducible representation $\Gamma^{[\lambda^{(2m)}]}$ of the permutation group of the orbitals with unpaired spins, π_{N-2m}. At the same time the structures form a basis, which in general is reducible, for a representation of a subgroup of π_{N-2m} that is isomorphic with the point symmetry group of the molecule. The characters of this representation $\chi^{[\lambda^{(2m)}]}(\bar{P})$, are found from character tables after putting each operation R of the point group into correspondence with a permutation $\bar{P}$ of the unpaired orbitals. Since the representations of the point group that are generated in this way are independent of which ionic configuration is picked from a set whose members transform into each other, the following theorem can be proved‡:

The characters of a representation of a point group which is generated by an ionic structure with spin S, and which is taken from a set of configurations whose members transform into each other, are equal to the characters $\chi^{[\lambda^{(2m)}]}(\bar{P})$ multiplied by the number of configurations in the given set that remain unchanged under an operation $R \equiv \bar{P}$.

When the characters of the reducible representation have been found they are decomposed into irreducible components and the possible multiplets

‡We recall that we are considering the case in which there is a single type of nondegenerate orbital upon the atoms. For the general case see Kaplan and Rodimova (1968).

${}^{2S+1}\Gamma^{(\alpha)}$ which can be obtained from the ionic structures are determined. The same process is carried through for each set of configurations. The order of the secular equation corresponding to a particular multiplet ${}^{2S+1}\Gamma^{(\alpha)}$ is equal to the sum of the number of times the multiplet occurs in the decompositions of the reducible representation formed by both the ionic and covalent structures.

In the next section this method is applied to the determination of the possible multiplets that may arise from a ring of six hydrogen atoms.

6.11. Determination of All the Multiplets Which Arise from a Ring of Six s Orbitals with Full Configuration Interaction

We consider six hydrogen atoms placed at the vertices of a regular hexagon. We wish to determine the interaction energy of such a system, taking into account all the possible configurations which can be constructed from the atomic s orbitals. According to Eq. (6.54), the initial secular equation for this variational problem is of order $\binom{12}{6} = 924$. The system under consideration possesses the point symmetry $\mathbf{D}_{6h}$. Since the 1s orbitals are symmetric with respect to a reflection in the plane of the ring, we may confine our considerations to just the symmetry of the point group $\mathbf{D}_6$. By constructing eigenfunctions of the various multiplets, it is possible to break up the initial 924×924 secular equation into secular equations of much smaller orders.

Calculations on a model of this kind have been carried out by Mattheiss (1961), and are discussed in detail by Slater (1963). The method that he used for finding the multiplets is based upon the electron spins being distributed among the atoms in all possible ways. A determinantal function is placed into correspondence with each such distribution, and subsequently one seeks the result of operating upon these determinantal functions with all the operations of the point group. The representations that are found in this way possess a definite value of the projection of the total spin M_S. This method is quite cumbersome even for the relatively simple problem of determining the 14 multiplets which arise from the covalent structures [see Slater (1963, pp. 381–388)]. By using the method which was propounded in the previous section the 268 multiplets which arise in this variational problem can be found without any particular difficulty. We consider the process of determining these multiplets consecutively.

A. *Covalent Structures*

The electronic configuration is here one in which all the orbitals are singly occupied. Since the number of orbitals is equal to the number of electrons, there is only one configuration of this type. The possible spin Young diagrams are

$$\begin{array}{rcccc} [\lambda]_{\text{spin}}: & [6] & [51] & [42] & [3^2] \\ S: & 3 & 2 & 1 & 0. \end{array} \qquad (6.73)$$

The coordinate Young diagrams which correspond to these are

$$[\lambda]_{\text{coord}}: \quad [1^6] \quad [21^4] \quad [2^2 1^2] \quad [2^3]. \tag{6.74}$$

The number of independent covalent structures corresponding to a particular value of S is equal to the dimension of the irreducible representation $\Gamma^{[\lambda]}$ [see (6.61)]. These are as follows:

Value of S	3	2	1	0	(6.75)
Number of structures	1	5	9	5	

The group $\mathbf{D}_6$ is isomorphic with a subgroup of the permutation group $\boldsymbol{\pi}_6$. In Table 6.5 the correspondence between the classes of these groups is given, together with the characters of $\Gamma^{[\lambda]}$ which correspond to the operations of the point group.

Table 6.5

Characters of the Irreducible Representations of $\boldsymbol{\pi}_6$ That Correspond to Operations of the Group $\mathbf{D}_6$

Classes of $\mathbf{D}_6$	E	C_2	$2C_3$	$2C_6$	$3U_2$	$3\bar{U}_2$
Classes of $\boldsymbol{\pi}_6$	$\{1^6\}$	$\{2^3\}$	$\{3^2\}$	$\{6\}$	$\{2^3\}$	$\{1^2 2^2\}$
$\chi^{[1^6]}$	1	−1	1	−1	−1	1
$\chi^{[21^4]}$	5	1	−1	1	1	1
$\chi^{[2^2 1^2]}$	9	−3	0	0	−3	1
$\chi^{[2^3]}$	5	3	2	0	3	1

The decomposition of these representations into irreducible representations of $\mathbf{D}_6$ gives the following:

$$\begin{aligned} &\Gamma^{[1^6]} \doteq B_1, && \Gamma^{[21^4]} \doteq A_1 + E_1 + E_2, \\ &\Gamma^{[2^2 1^2]} \doteq A_2 + 2B_1 + 2E_1 + E_2, && \Gamma^{[2^3]} \doteq 2A_1 + B_2 + E_2. \end{aligned} \tag{6.76}$$

Hence 14 multiplets can be constructed from 20 covalent structures [see (6.75)]

$$^7B_1, \quad ^5A_1E_1E_2, \quad ^3A_2B_1(2)\,E_1(2)\,E_2, \quad ^1A_1(2)\,B_2E_2 \tag{6.77}$$

(the number in parentheses following a multiplet denotes the number of times the multiplet occurs).

B. $H_4H^+H^-$ *Ionic Configurations*

According to Eq. (6.68), there are 30 configurations of this type altogether. These are divided into three sets whose members do not mix under operations of the point group (Fig. 6.5). Configurations of type 1 and 2 do not possess any elements of symmetry and form regular sets with 12 configurations in each. Configurations of type 3 are invariant with respect to a rotation $\bar{U}_2$ (the notation for the symmetry axes is shown in Fig. 6.6), and consequently their

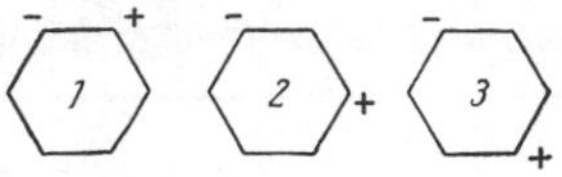

Figure 6.5.

symmetry group consists of the two elements E and $\bar{U}_2$. The set of configurations 3 is smaller than a regular set and, according to the rule given on p. 159, contains $g/h = 6$ configurations.

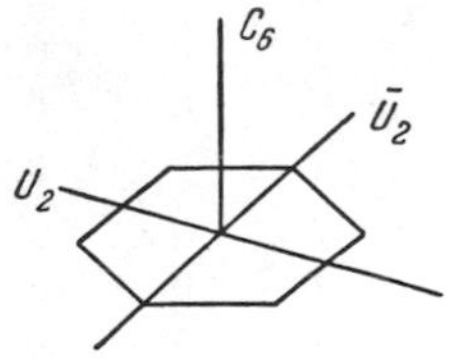

Figure 6.6.

Each configuration in the sets enumerated gives rise to $f_{\lambda^{(2)}}$ ionic structures with spin S [see (6.70)]. The permutational symmetry of the spin wavefunctions for the unpaired electrons is characterized by three Young diagrams $[\lambda^{(2)}]$:

$$\begin{array}{lccc} [\lambda^{(2)}]: & [4] & [31] & [2^2] \\ S: & 2 & 1 & 0. \end{array} \tag{6.78}$$

The coordinate Young diagrams $[\tilde{\lambda}^{(2)}]$ which correspond to these are

$$\begin{array}{lccc} [\tilde{\lambda}^{(2)}]: & [1^4] & [21^2] & [2^2] \\ f_{\lambda^{(2)}}: & 1 & 3 & 2. \end{array} \tag{6.79}$$

The characters of the representation of the point group generated by the ionic structures are found using the rules given in Section 6.10B. The characters of representations generated by regular sets are all zero except that of the identity operation E. This last is equal to the dimension of the representation, i.e., $\chi^{[\lambda^{(2)}]}(E) = 12f_{\lambda^{(2)}}$. However, in the representations generated by set 3, the character of the operation $\bar{U}_2$ is also nonzero. Application of any of the three $\bar{U}_2$ operations leaves two of the configurations of set 3 unchanged. According to the theorem stated on p. 160, $\chi_3^{[\lambda^{(2)}]}(\bar{U}_2) = 2\chi^{[\lambda^{(2)}]}(\bar{P})$ for the representations which are generated by the ionic structures of set 3. The symbol $\bar{P}$ stands for the permutation corresponding to the operation $\bar{U}_2$ of the four orbitals with unpaired spins. In the present case $\bar{P}$ belongs to the class $\{2^2\}$ of the group π_4. Table 6.6 gives the characters of the representations which are generated by sets 1 and 3 (the representations generated by sets 1

Table 6.6

Characters of the Nonequivalent Representations Generated by the Ionic Structures

	E	C_2	$2C_3$	$2C_6$	$3U_2$	$3\bar{U}_2$
$\chi_1^{[1^4]}$	12	0	0	0	0	0
$\chi_1^{[21^2]}$	36	0	0	0	0	0
$\chi_1^{[2^2]}$	24	0	0	0	0	0
$\chi_3^{[1^4]}$	6	0	0	0	0	2
$\chi_3^{[21^2]}$	18	0	0	0	0	−2
$\chi_3^{[2^2]}$	12	0	0	0	0	4
$\chi_6^{[1^2]}$	6	0	0	0	−2	0
$\chi_6^{[2]}$	6	0	0	0	2	0
$\chi_9^{[1^2]}$	6	0	0	0	0	−2
$\chi_9^{[2]}$	6	0	0	0	0	2
$\chi_{12}^{[1^2]}$	6	−6	0	0	0	0
$\chi_{12}^{[2]}$	6	6	0	0	0	0
$\chi_{14}^{[1^2]}$	6	0	0	0	0	2
$\chi_{14}^{[2]}$	6	0	0	0	0	2
χ_{15}	2	0	2	0	0	2

Table 6.7

Multiplets Generated by Different Sets of Configurations in the Problem of a Ring of Six H Atoms

	$^7B_1$	$^5A_1$	$^5A_2$	$^5B_1$	$^5B_2$	$^5E_1$	$^5E_2$	$^3A_1$	$^3A_2$	$^3B_1$	$^3B_2$	$^3E_1$	$^3E_2$	$^1A_1$	$^1A_2$	$^1B_1$	$^1B_2$	$^1E_1$	$^1E_2$
K_0	1	1				1	1		1	2		2	1	2			1		1
K_1		1	1	1	1	2	2	3	3	3	3	6	6	2	2	2	2	4	4
K_2		1	1	1	1	2	2	3	3	3	3	6	6	2	2	2	2	4	4
K_3		1		1		1	1	1	2	1	2	3	3	2		2		2	2
K_4								1	1	1	1	2	2	1	1	1	1	2	2
K_5								1	1	1	1	2	2	1	1	1	1	2	2
K_6									1	1		1	1	1			1	1	1
K_7									1	1		1	1	1			1	1	1
K_8								1	1	1	1	2	2	1	1	1	1	2	2
K_9									1		1	1	1	1		1		1	1
K_{10}									1		1	1	1	1		1		1	1
K_{11}								1	1	1	1	2	2	1	1	1	1	2	2
K_{12}										1	1	2		1	1				2
K_{13}									1	1		1	1	1			1	1	1
K_{14}								1		1		1	1	1		1		1	1
K_{15}														1		1			
K_{16}														1	1	1	1	2	2
K_{17}														1		1		1	1
Total[a]	1	4	2	3	2	6	6	12	18	18	15	33	30	22	10	16	13	27	30

[a]Total number of times given multiplet occurs.

and 2 are equivalent). The decomposition of these representations into irreducible representations of the group $\mathbf{D}_6$ gives us 120 multiplets arising from configurations of the form $H_4H^+H^-$. These are given in the complete Table 6.7.

C. *Ionic Configurations* $H_2(H^+)_2(H^-)_2$

There are $(6!/2!)^3 = 90$ configurations of this form. They fall into 11 sets which do not mix under the operations of the point group. Figure 6.7 shows one configuration from each set. Since there are just two unpaired spins, only triplet and singlet states are possible, corresponding to coordinate Young diagrams $[\lambda^{(4)}] = [1^2]$ and $[2]$, respectively. In both cases $f_\lambda^{(4)} = 1$, and consequently the dimensions of the representations which are generated by these ionic structures coincide with the dimension of the representations generated by the sets of ionic configurations.

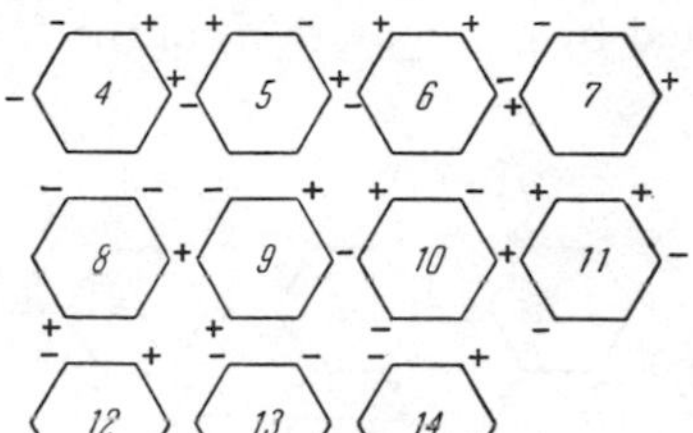

Figure 6.7.

Configurations 4, 5, 8, and 11 form regular sets. Since $f_{[2]} = f_{[1^2]} = 1$, the corresponding ionic structures form regular representations. The number of times each irreducible representation occurs in the decomposition of the regular representation is equal to the dimension of the corresponding irreducible representation. Consequently each of the sets gives rise to 16 multiplets:

$$^{3,1}A_1A_2B_1B_2E_1(2)\,E_2(2). \tag{6.80}$$

The configurations in sets 6, 7, and 13 possess an axis of symmetry U_2 (see Fig. 6.6 for designation of symmetry axes). They generate equivalent six-dimensional representations. The character of the operation U_2 is equal to $2\chi^{[\lambda^{(4)}]}(P) = \pm 2$, the upper sign referring to $[\lambda^{(4)}] = 2$ and the lower sign to $[\lambda^{(4)}] = [1^2]$.

The symmetry group for configurations of type 12 consists of the elements E and C_2. The operation C_2 leaves all six configurations in set 12 invariant. Hence $\chi^{[\lambda^{(4)}]}(C_2) = \pm 6$, the sign depending upon $[\lambda^{(4)}]$.

The configurations in sets 9, 10, and 14 possess an axis of symmetry $\bar{U}_2$ and also generate six-dimensional representations. However, the characters of the representations generated by the ionic structures of sets 9 and 10 are different from those of set 14. This is because in sets 9 and 10 the operation $\bar{U}_2$ corresponds to a permutation of the orbitals with unpaired spins, whereas in the configurations of set 14 the atoms with unpaired spins remain unaffected by this operation since the $\bar{U}_2$ axis passes through these atoms. The characters of the nonequivalent representations are given in Table 6.6. Thus configurations of the form $H_2(H^+)_2(H^-)_2$ generate 120 multiplets (see Table 6.7).

D. *Ionic Configurations* $(H^+)_3(H^-)_3$

It is possible to construct $6!/(3!)^2 = 20$ configurations of this form, and they fall into three sets (Fig. 6.8). Only singlet states are possible since all the spins are paired. The configurations in the set 15 are invariant with respect to a group **H** with elements E, C_3, $C_3{}^2$, and $3\bar{U}_2$. Consequently, they generate a two-dimensional representation, the characters of which are given in Table 6.6. Set 16 is regular. The characters of the representation generated by set 17 coincide with the characters $\chi_9^{[2]}$ of Table 6.6. Decomposition of the reducible representations generated by configurations of the form $(H^+)_3(H^-)_3$ leads to 14 multiplets.

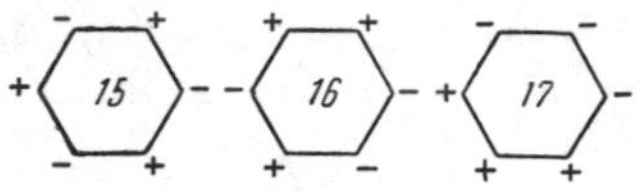

Figure 6.8.

The frequency with which each multiplet occurs on taking into account all the configurations of a system of six s orbitals is given in Table 6.7. K_0 denotes the covalent configuration and K_1–K_{17} the ionic structures illustrated in Figs. 6.5, 6.7, and 6.8. The bottom line of the table gives the total number of times each multiplet occurs, and equals the order of the secular equation for the particular multiplet. The original secular equation of order 924 breaks up into 19 secular equations. The highest order for a secular equation is 33. Solution of these equations leads to 268 energy levels, some of which are degenerate. This degeneracy depends upon the point symmetry of the problem and can only be removed by distorting the ring. The calculations of Mattheiss (1961) show that the lowest energy level of the system has the symmetry ${}^1A_1$. The potential energy curve for this state possesses a minimum, i.e., the energy of a ring of six H atoms for some equilibrium separation is lower than that of the atoms separated to infinity. However, the energy of the ring is higher than the energy of three H_2 molecules calculated by an equivalent method. Consequently, a ring of this kind will decompose spontaneously into three H_2 molecules, in agreement with experiment.

CHAPTER VII

The Method of Coefficients of Fractional Parentage

PART 1. Equivalent Electrons

7.1. Definition of Coefficients of Fractional Parentage

The total wavefunction for a system of electrons in a field with central symmetry must satisfy two conditions: (a) It must be antisymmetric with respect to permutations of the electrons, and (b) it must be an eigenfunction of the operator for the square of the angular momentum J^2 and of the operator for the z component J_z. In the L–S-coupling case the "true" functions in the zeroth order of perturbation theory must also be eigenfunctions of the two pairs of operators: L^2 and L_z, and S^2 and S_z, where L and S, respectively, denote the orbital and spin angular momenta of the system of electrons.

Antisymmetry of the total wavefunction can be easily achieved in the one-electron approximation by constructing the function in the form of a linear combination of determinants (Slater, 1923, 1960). Each determinant corresponds to a definite value for the projection of the orbital angular momentum of the system of electrons M_L and for the projection of the spin angular momentum M_S, but it is not an eigenfunction of the operators L^2 and S^2. Calculations are therefore carried out in the $m\sigma M_L M_S$ scheme, but should any terms occur more than once, a transformation to the LSM_LM_S scheme is necessary. The coefficients of the linear combination of determinants in the initial variational function are found from the requirement that the average energy be a minimum.

The determinantal method for calculating the electrostatic interaction energy of electrons was developed by Slater in 1929 in a classic paper (Slater, 1923, 1960), and was subsequently applied to the calculation of atomic terms in a number of simple electronic configurations (Condon and Shortley, 1964). This remained the basic method for calculating atomic terms for some years until the appearance of Racah's work (1942, 1943). However, attempts to apply Slater's method to complicated electronic configurations came up against a number of computational difficulties, which are due essentially to the cumbrous determination of the matrix elements in the secular equation. The absence of analytical formulae for calculating the matrix elements is also a considerable drawback to the determinantal method, since one cannot obtain recurrence formulae which connect an N-electron problem with a problem involving a smaller number of electrons.

An alternative way of constructing a wavefunction consists in first forming an eigenfunction with coupled angular momenta and then antisymmetrizing this function. For example, in a configuration of three nonequivalent electrons with angular momenta j_1, j_2, and j_3 in j–j coupling, successive coupling of the j_i leads to a wavefunction

$$\psi[(j_1(1)j_2(2)J_{12}j_3(3);JM], \tag{7.1}$$

whose detailed form is given by expression (3.56). Function (7.1) does not satisfy the Pauli principle since it is not antisymmetric. In order to antisymmetrize it, we apply the Young operator $\omega^{[1^3]}$. As a result we obtain a normalized antisymmetric function

$$\begin{aligned}\omega^{[1^3]}\psi[(j_1(1)j_2(2))J_{12}j_3(3);JM]\\ = (1/\sqrt{3!})\{\psi[(j_1(1)j_2(2))J_{12}j_3(3);JM]\\ - \psi[(j_1(2)j_2(1))J_{12}j_3(3);JM] - \psi[(j_1(1)j_2(3))J_{12}j_3(2);JM]\\ - \psi[(j_1(3)j_2(2))J_{12}j_3(1);JM] + \psi[(j_1(2)j_2(3))J_{12}j_3(1);JM]\\ + \psi[(j_1(3)j_2(1))J_{12}j_3(2);JM]\}.\end{aligned} \tag{7.2}$$

However, if the electrons are equivalent, i.e., $j_1 = j_2 = j_3 = j$ (in which case it is assumed that the radial wavefunctions for the electrons are also identical), the functions which appear in the sum (7.2) cease to be orthogonal to one another, and in general a number of them become identical. In addition, function (7.2) will no longer be normalized.

Now according to (3.49), the application of the permutation P_{12} to a function with coupled angular momenta gives, for j half-integral,

$$P_{12}\psi[j(1)j(2);JM] = (-1)^{J+1}\psi[j(1)j(2);JM], \tag{7.3}$$

i.e., the function is only antisymmetric for even values of the resultant angular momentum. Consequently, expression (7.2) becomes zero for odd values of J_{12}, and for even values it becomes an unnormalized sum of three terms:

$$\begin{aligned}\psi[j(1)j(2))J'j(3);JM] - \psi[(j(1)j(3))J'j(2);JM]\\ + \psi[(j(2)j(3))J'j(1);JM].\end{aligned} \tag{7.4}$$

It is obvious that the functions which appear in this sum are not orthogonal to each other since they differ only in the order of coupling the angular momenta, and are therefore connected by the transformation matrix for the group $\mathbf{R}_3$ (see Section 3.7). However, this very circumstance makes it possible to put them in the form of a linear combination of mutually orthogonal functions. For this purpose we make use of the fact that the functions

$$\psi[(j(1)j(2))J'j(3);JM] \qquad \text{and} \qquad \psi[(j(1)j(2)\bar{J}'j(3);JM] \tag{7.5}$$

are orthogonal for $J' \neq \bar{J}'$. The last two terms of (7.4) can be expanded in the functions (7.5) with the aid of the transformation matrix; the order in which the angular momenta j are coupled is specified by their arguments since all the j_i are identical:

$$\psi[(j(1)j(3))\bar{J}'j(2);JM] = \sum_{J'} \psi[(j(1)j(2))J'j(3);JM] \times \langle(j(1)j(2))J'j(3)\,|\,(j(1)j(3))\bar{J}'j(2)\rangle^{(J)}. \tag{7.6}$$

The elements of the tranformation matrix which appear in this equation are given by formula (3.60):

$$\langle(j(1)j(2)J'j(3)\,|\,(j(1)j(3)\bar{J}'j(2)\rangle^{(J)} = [(2J'+1)(2\bar{J}'+1)]^{1/2}W(J'jj\bar{J}';Jj) \tag{7.7}$$

(the suffixes 1, 2, and 3 are omitted in the Racah coefficient). Similarly

$$\psi[(j(2)j(3)\bar{J}'j(1);JM] = \sum_{J'} \psi[(j(1)j(2))J'j(3);JM] \times \langle(j(1)j(2))J'j(3)\,|\,(j(2)j(3))\bar{J}'j(1)\rangle^{(J)}. \tag{7.8}$$

The function on the left-hand side of this equation differs from its analog in Eq. (7.6) by a permutation of the electrons P_{12}. Expression (7.8) is therefore obtained from (7.6) by applying to this last expression the permutation P_{12}. As a result, the functions in the sum (7.6) are multiplied by $(-1)^{J'+1}$. Thus the transformation matrix in expression (7.8) is equal to that in (7.6) multiplied by $(-1)^{J'+1}$,

$$\langle(j(1)j(2))J'j(3)\,|\,(j(2)j(3))\bar{J}'j(1)\rangle^{(J)} = (-1)^{J'+1}[(2J'+1)(2\bar{J}'+1)]^{1/2}W(J'jj\bar{J}';Jj). \tag{7.9}$$

In this way it is possible by taking into account Eq. (7.6)–(7.9) to write a normalized antisymmetric function of three equivalent electrons in the following form:

$$\psi(j^3\alpha JM) = C_{jJ'J} \sum_{J'\text{even}} \{\delta_{J'\bar{J}'} - 2[(2J'+1)(2\bar{J}'+1)]^{1/2} \times W(J'jj\bar{J}';Jj\}\psi(j^2J',j(3);JM), \tag{7.10}$$

where α denotes any additional quantum numbers necessary to characterize a state with a given value of J should it occur several times in the configuration j^3, and C_{jJJ} is a normalizing constant.

Expression (7.10) can formally be written as

$$\psi(j^3\alpha JM) = \sum_{J'\text{even}} \psi(j^2J',j(3);JM)\langle j^2J',j;J|\}j^3\alpha J\rangle. \tag{7.11}$$

The explicit form of the coefficients in this linear combination is given by expression (7.10), from which it follows that they are in fact independent of the quantum number M,

$$\langle j^2J', j; J |\} j^3\alpha J\rangle = C_{jJ'\bar{J}'}\{\delta_{J'\bar{J}'} - 2[(2J'+1)(2\bar{J}'+1)]^{1/2} W(J'jj\bar{J}'; Jj)\}. \qquad (7.12)$$

In cases where an antisymmetric state with a given value of J occurs only once in a configuration j^3 (this is always so for $j \leq 7/2$; see Table 6.2), one can calculate the coefficients (7.12) by starting from any even value of $\bar{J}'$ occurring in the interval $0 \leq \bar{J}' \leq 2j$.

Equation (7.11) shows that a wavefunction for a given state of three electrons can be put into the form of a linear superposition of wavefunctions which correspond to allowed states of two electrons with a value of J' for the angular momentum. The squared moduli of the coefficients in this linear combination express the extent to which a state j^2J' participates in the desired state $j^3\alpha J$. The coefficients (7.12) are therefore called coefficients of fractional parentage.‡

The coefficients of fractional parentage (7.12) may be regarded as elements of a transformation matrix which brings about a transformation from functions with coupled angular momenta, antisymmetric only in the first two electrons, to totally antisymmetric functions. This matrix is rectangular, since the number of states of the configuration j^2 which are permitted by the Pauli principle is not in general equal to the number of states of the configuration j^3. A curly bracket is therefore introduced into the symbol for the coefficients of fractional parentage in order to distinguish them from square matrices. The coefficients of fractional parentage are real quantities, as are the transformation matrices of the group $\mathbf{R}_3$.

Coefficients of fractional parentage for a configuration j^N are defined similarly to (7.11):

$$\psi(j^N\alpha JM) = \sum_{\alpha', J'} \psi(j^{N-1}\alpha'J', j(N); JM)\langle j^{N-1}\alpha'J', j; J |\} j^N\alpha J\rangle. \qquad (7.13)$$

The functions in this sum correspond to states which are antisymmetric in the permutations of the electrons in a configuration j^{N-1}, the summation over $\alpha'J'$ running over all the antisymmetric states of this configuration. In cases where terms are repeated, orthogonality of the wavefunctions is ensured if the coefficients of fractional parentage satisfy the orthogonality conditions

$$\sum_{\alpha', J'} \langle j^{N-1}\alpha'J', j; J |\} j^N\alpha J\rangle\langle j^{N-1}\alpha'J', j; J |\} j^N\bar{\alpha} J\rangle = \delta_{\alpha\bar{\alpha}}. \qquad (7.14)$$

The reality of the fractional parentage coefficients has been taken into account in writing down this formula. Wavefunctions with different J are automatically orthogonal to each other by construction, and therefore this orthogonality does not impose any additional conditions upon the coefficients of fractional parentage.

‡Translator's note: Or "genealogical coefficients" in the Russian literature.

An antisymmetric function of a configuration j^N can always be put into the form of a finite series (7.13). This is because the functions $\psi(j^{N-1}\alpha'J', j(N); JM)$ constitute a complete set of functions which are antisymmetrized with respect to permutations of the first $N - 1$ electrons. Therefore any function which possesses this property, and in particular a function which is antisymmetric with respect to all N electrons, can be expanded in basis functions from this set.

In order to calculate the coefficients of fractional parentage for a configuration with more than three electrons ($N > 3$), one may use the recurrence formula (de-Shalit and Talmi, 1963)

$$\begin{aligned} & N\langle j^{N-1}\alpha'J', j; J \,|\!\}\, j^N[\alpha'J']J\rangle\langle j^{N-1}\bar{\alpha}'\bar{J}', j; J \,|\!\}\, j^N[\alpha'J']J\rangle \\ &\quad = \delta_{\alpha'\bar{\alpha}'}\delta_{J'\bar{J}'} - (N-1)\sum_{\alpha'', J''}(-1)^{J'+\bar{J}'+J''+J}W(J''jjJ; \bar{J}'J') \\ &\quad\quad \times \langle j^{N-2}\alpha''J'', j; J' \,|\!\}\, j^{N-1}\alpha'J'\rangle\langle j^{N-2}\alpha''J'', j; \bar{J}' \,|\!\}\, j^{N-1}\bar{\alpha}'\bar{J}'\rangle. \end{aligned} \tag{7.15}$$

The square brackets on the right-hand side of the coefficients of fractional parentage mean that a function $\psi(j^N[\alpha'J']J)$ is obtained by antisymmetrizing a function $\psi(j^{N-1}\alpha'J', j(N); JM)$. Difficulties in forming independent functions arise in cases where terms are repeated [see de-Shalit and Talmi (1963) for details].

The method of finding the coefficients of fractional parentage just described depends upon equating the result of directly antisymmetrizing a wavefunction with the fractional parentage form. It was proposed by Schwartz and de-Shalit (1954), and by Redmond (1954), for the *j–j*-coupling case, but it can also be used in the *L–S*-coupling scheme. The coefficients of fractional parentage for the *L–S*-coupling case were found earlier by Racah (1943, 1949). In his papers, which have provided the foundation for the subsequent extensive application of the fractional parentage method in atomic and nuclear spectroscopy, Racah developed and perfected the idea of constructing electronic states genealogically. This idea had been contained in a forgotten paper by Bacher and Goudsmit (1934). The main point of Racah's method for finding the coefficients of fractional parentage for the p and d shells consists in the following:

A wavefunction for a configuration l^2 which describes a state with quantum numbers SLM_SM_L is constructed with the aid of vector coupling coefficients

$$\begin{aligned} & \psi(l^2SLM_SM_L) \\ &\quad = \sum_{m_1,m_2}\sum_{\sigma_1,\sigma_2}\psi_{lm_1\sigma_1}(1)\psi_{lm_2\sigma_2}(2)\langle lm_1, lm_2 \,|\, LM_L\rangle\langle \tfrac{1}{2}\sigma_1, \tfrac{1}{2}\sigma_2 \,|\, SM_S\rangle. \end{aligned} \tag{7.16}$$

This function already possesses a definite permutational symmetry by construction. Under a permutation of the electrons, applied to the orbital and spin angular momenta separately, function (7.16) is multiplied according to relation (3.49) by

$$(-1)^{2l+1-L-S}.$$

An antisymmetric state is therefore characterized by an even value of $L + S$. Addition of a third electron leads to a function

$$\psi(l^2S'L', l(3); SLM_SM_L) = \sum_{M'_L, m_3} \sum_{M'_S, \sigma_3} \psi(l^2S'L'M_S'M_L')\psi_{lm_3\sigma_3}(3) \times \langle L'M_L', lm_3 | LM_L\rangle\langle S'M_S', \tfrac{1}{2}\sigma_3 |SM_S\rangle, \tag{7.17}$$

which in general is only antisymmetric with respect to permutations of the first two electrons.

A function which is antisymmetric with respect to all permutations is constructed in the form of a linear combination of functions (7.17)‡

$$\psi(l^3\alpha SL) = \sum_{S',L'} \psi(l^2S'L', l(3); SL)\langle l^2S'L', l; SL|\}l^3\alpha SL\rangle. \tag{7.18}$$

The coefficients of fractional parentage which occur in this formula have the same physical meaning as the coefficients in the linear combination (7.11). The summation over $S'L'$ covers only the allowed terms of the configuration l^2. By changing the order in which the angular momenta are coupled, we obtain

$$\psi(l^2S'L', l(3); SL) = \sum_{S'',L''} \psi(l, (ll)S''L''; SL)\langle l, (ll)S''L''; SL| l^2S'L', l; SL\rangle. \tag{7.19}$$

The transformation matrix which appears in this equation is a product of the transformation matrices for the orbital and spin momenta, and, according to (3.59), is given by

$$\langle l, (ll)S''L''; SL | l^2S'L', l; SL\rangle = [(2S' + 1)(2S'' + 1)(2L' + 1)(2L'' + 1)]^{1/2} \times W(\tfrac{1}{2}\tfrac{1}{2}S\tfrac{1}{2}; S'S'')W(llLl; L'L''). \tag{7.20}$$

The summation in Eq. (7.19) is over all the pairs of values of S'' and L'' that are obtained by the vector coupling and includes terms in which $S'' + L''$ is equal to an odd number. Hence if expression (7.19) is substituted into formula (7.18), the requirement that this function be antisymmetric can be met only if all the coefficients of $\psi(l, (ll)S''L''; SL)$ for $S'' + L''$ odd reduce to zero. This leads to the following system of equations for the coefficients of fractional parentage:

$$\sum_{S',L'} \langle l, (ll)S''L''; SL | l^2S'L', l; SL\rangle \times \langle l^2S'L', l; SL|\}l^3\alpha SL\rangle = 0, \qquad S'' + L'' \text{ odd}. \tag{7.21}$$

The number of independent solutions of this system for a fixed SL gives the number of allowed terms of the given type for the configuration l^3. If this is

‡The quantum numbers M_S and M_L are unimportant for the derivations which follow and are omitted for brevity.

greater than unity, the terms are distinguished by an additional quantum number α.

In a similar way one can derive a system of equations for the fractional parentage coefficients of a configuration l^N if those for the configuration l^{N-1} are known (Racah, 1943):

$$\sum_{\alpha',S',L'} \langle S''L'', (ll)S'''L'''; SL | (S''L'', l)S'L', l; SL\rangle$$
$$\times \langle l^{N-2}\alpha''S''L'', l; S'L' |\} l^{N-1}\alpha'S'L'\rangle$$
$$\times \langle l^{N-1}\alpha'S'L', l; SL |\} l^N\alpha SL\rangle = 0, \qquad S''' + L''' \quad \text{odd}. \tag{7.22}$$

The transformation matrix is defined as in (7.20). The systems of equations which have been derived allow one to find the coefficients of fractional parentage for configurations p^N and d^N fairly easily. They have been tabulated by Racah (1943). In the case of f^N configurations many of the terms occur several times. Racah has devised a group-theoretical method for this case, based upon the factorization of the coefficients of fractional parentage when the states are classified according to the corresponding continuous groups [see Racah (1949) and Judd (1963) for more details].‡

7.2. Calculation of Matrix Elements of Symmetric Operators

In physical applications it is often necessary to calculate matrix elements of operators which are symmetric with respect to all the electrons in the system under consideration. Such operators can be divided into two types:

$$F = \sum_{i=1}^{N} f_i \qquad \text{and} \qquad G = \sum_{i<k}^{N} g_{ik}. \tag{7.23}$$

The operator F is a sum of one-electron operators f_i, each of which acts upon the coordinates of one electron only. The interaction of the system of electrons under consideration with an external field and the interaction of the electrons with nuclei are examples of operators of type F. The operator G is a sum of two-electron operators g_{ik}. The energy of the electrostatic interaction between electrons is an example of such an operator.

A feature of the fractional parentage expansion (7.13) or (7.18) is that the original function, which is antisymmetric in all electrons, is expanded in a set of functions in which the last electron is decoupled and is characterized by definite quantum numbers. This circumstance makes it possible to express matrix elements of the operators F and G in a many-electron problem in terms of one- and two-electron matrix elements and coefficients of fractional parentage. Furthermore, it turns out that in order to calculate the matrix

‡Tables of coefficients of fractional parentage for all allowed terms of the p and d shells, and also for terms of maximum multiplicity for a number of configurations f^N, have been given by Sobel'man (1972). Tables of one- and two-particle coefficients of fractional parentage for configurations $s^\lambda s'^\mu p^q$ have also been given by Chisholm *et al.* (1969).

elements, a knowledge of the specific form of the wavefunction is unnecessary, knowledge of the coefficients of fractional parentage being sufficient. The formulae for the matrix elements to be derived apply to the case of j–j coupling. All the formulae remain correct for L–S coupling also, it being only necessary to replace j by l, and J by L,S.

Since all the electrons in states which are described by antisymmetric functions are on the same footing, the matrix elements of all the operators f_i are identical. One may therefore choose any one operator, for example, the operator f_N, which acts upon the coordinates of the last electron, and write

$$\langle j^N\alpha JM|F|j^N\bar{\alpha}\bar{J}\bar{M}\rangle = N\langle j^N\alpha JM|f_N|j^N\bar{\alpha}\bar{J}\bar{M}\rangle. \tag{7.24}$$

We substitute into this matrix element the wavefunction for the configuration in the form of the expansion (7.13):

$$\begin{aligned}\langle j^N\alpha JM|F|j^N\bar{\alpha}\bar{J}\bar{M}\rangle \\ &= N\sum_{\alpha',J'}\sum_{\bar{\alpha}',\bar{J}'}\langle j^{N-1}\alpha'J',j;J|\}j^N\alpha J\rangle\langle j^{N-1}\bar{\alpha}'\bar{J}',j;\bar{J}|\}j^N\bar{\alpha}\bar{J}\rangle \\ &\quad\times\langle j^{N-1}\alpha'J',j(N);JM|f_N|j^{N-1}\bar{\alpha}'\bar{J}',j(N);\bar{J}\bar{M}\rangle.\end{aligned} \tag{7.25}$$

The matrix element of the operator f_N which occurs in this expression is easily calculated with the aid of the techniques for irreducible tensor operators. For example, let the operator f_N be the kth component of a κth-rank spherical tensor, and denote it by $f_k^{(\kappa)}(N)$. According to the Wigner–Eckart theorem (4.63),

$$\begin{aligned}&\langle j^{N-1}\alpha'J',j(N);JM|f_N|j^{N-1}\bar{\alpha}'\bar{J}',j(N);\bar{J}\bar{M}\rangle \\ &= (-1)^{J-M}\begin{pmatrix} J & \kappa & \bar{J} \\ -M & k & \bar{M}\end{pmatrix}\langle\alpha'J',j(N);J\|f^{(\kappa)}(N)\|\bar{\alpha}'\bar{J}',j(N);\bar{J}\rangle.\end{aligned} \tag{7.26}$$

The operator $f^{(\kappa)}(N)$ in this matrix element acts only upon the coordinates of the Nth electron. We apply Eq. (4.76) in which the role of the second subsystem is now played by the Nth electron:

$$\begin{aligned}&\langle\alpha'J',j(N);J\|f^{(\kappa)}(N)\|\bar{\alpha}'\bar{J}',j(N);\bar{J}\rangle \\ &= \delta_{\alpha'\bar{\alpha}'}\delta_{J'\bar{J}'}(-1)^{J'+j+J+\kappa}[(2J+1)(2\bar{J}+1)]^{1/2}\begin{Bmatrix} j & J & J' \\ \bar{J} & j & \kappa\end{Bmatrix}\langle j\|f^{(\kappa)}\|j\rangle.\end{aligned} \tag{7.27}$$

Substituting (7.26) and (7.27) into (7.25), we obtain the final form for the matrix elements of a tensor operator $F_k^{(\kappa)} = \sum_{i=1}^N f_k^{(\kappa)}(i)$:

$$\begin{aligned}&\langle j^N\alpha JM|F_k^{(\kappa)}|j^N\bar{\alpha}\bar{J}\bar{M}\rangle \\ &= (-1)^{J-M}[(2J+1)(2\bar{J}+1)]^{1/2}\begin{pmatrix} J & \kappa & \bar{J} \\ -M & k & \bar{M}\end{pmatrix}N\langle j\|f^{(\kappa)}\|j\rangle \\ &\quad\times\sum_{\alpha',J'}(-1)^{J'+j+J+\kappa}\begin{Bmatrix} j & J & J' \\ \bar{J} & j & \kappa\end{Bmatrix}\langle j^{N-1}\alpha'J',j;J|\}j^N\alpha J\rangle \\ &\quad\times\langle j^{N-1}\alpha'J',j;\bar{J}|\}j^N\bar{\alpha}\bar{J}\rangle.\end{aligned} \tag{7.28}$$

The satisfaction of the triangle conditions $\Delta(J\kappa\bar{J})$ and $\Delta(j\kappa j)$ are necessary conditions for the matrix element to be nonzero.

Formula (7.28) acquires a very simple form in the case of a scalar operator. For this purpose it is necessary to substitute into (7.28) the value of the 6-j symbol which is given by (4.77), and for the 3-j symbol the value

$$\begin{pmatrix} J & 0 & \bar{J} \\ -M & 0 & \bar{M} \end{pmatrix} = \delta_{J\bar{J}}\delta_{M\bar{M}}\frac{(-1)^{J-M}}{(2J+1)^{1/2}}. \tag{7.29}$$

In addition, by substituting the matrix element of the operator $f^{(0)}$ according to formula (4.68) and making use of the orthogonality relations among the coefficients of fractional parentage, we finally obtain

$$\langle j^N\alpha JM|\,F^{(0)}\,|j^N\bar{\alpha}\bar{J}\bar{M}\rangle = \delta_{\alpha\bar{\alpha}}\delta_{J\bar{J}}\delta_{M\bar{M}}N\langle jm|f^{(0)}|jm\rangle. \tag{7.30}$$

We now consider the calculation of the matrix elements of the scalar operator G. These are diagonal in J and are independent of M [see (4.68)]; we shall therefore drop M from the notation for the matrix element. We shall obtain a recurrence formula which expresses the matrix elements of the operator G in terms of matrix elements of the operator

$$G' = \sum_{i<k}^{N-1} g_{ik},$$

which acts upon the coordinates of the first $N-1$ electrons only. Because of the indistinguishability of the electrons, the matrix elements of the operators g_{ik} do not depend on the particular electrons i and k. The operator G consists of $\frac{1}{2}N(N-1)$ terms and G' of $\frac{1}{2}(N-1)(N-2)$ terms; therefore

$$\langle j^N\alpha J|\,G\,|j^N\bar{\alpha}J\rangle = [N/(N-2)]\langle j^N\alpha J|\,G'\,|j^N\bar{\alpha}J\rangle. \tag{7.31}$$

We substitute expression (7.13) for the wavefunction into (7.31),

$$\begin{aligned}\langle j^N\alpha J|\,G\,|j^N\bar{\alpha}J\rangle = [N/(N-2)]&\sum_{\alpha',J'}\sum_{\bar{\alpha}',\bar{J}'}\langle j^{N-1}\alpha'J', j; J|\}j^N\alpha J\rangle\\ &\times\langle j^{N-1}\bar{\alpha}'\bar{J}', j; J|\}j^N\bar{\alpha}J\rangle\\ &\times\langle j^{N-1}\alpha'J', j(N); J|\,G'\,|j^{N-1}\bar{\alpha}'\bar{J}', j(N); J\rangle.\end{aligned}$$

The operator G' does not act upon the coordinates of the Nth electron. Equation (4.80) is applicable to a scalar operator which acts upon the first subsystem only, and we finally obtain

$$\begin{aligned}\langle j^N\alpha J|\,G\,|j^N\bar{\alpha}J\rangle = [N/(N-2)]&\sum_{\alpha',\bar{\alpha}',J'}\langle j^{N-1}\alpha'J', j; J|\}j^N\alpha J\rangle\\ &\times\langle j^{N-1}\bar{\alpha}'J', j; J|\}j^N\bar{\alpha}J\rangle\langle j^{N-1}\alpha'J'|\,G'\,|j^{N-1}\bar{\alpha}'J'\rangle.\end{aligned} \tag{7.32}$$

A matrix element of the operator G can also be expressed in terms of two-electron matrix elements. For this purpose it is necessary to introduce the fractional parentage coefficients which correspond to the detachment of two electrons simultaneously. These are defined by the relation

$$\begin{aligned}\psi(j^N\alpha JM) = &\sum_{\alpha',J',J''}\psi(j^{N-2}\alpha'J', j^2(N-1, N)J''; JM)\\ &\times\langle j^{N-2}\alpha'J', j^2J''; J|\}j^N\alpha J\rangle,\end{aligned} \tag{7.33}$$

the wavefunctions on the right-hand side being antisymmetric with respect to permutations of the first $N-2$ electrons, and with respect to permutations of the $(N-1)$th and Nth electrons. The summation in (7.33) is taken over all allowed terms of the configurations j^{N-2} and j^2.

It is not difficult to express the coefficients of fractional parentage in expansion (7.33) in terms of coefficients of fractional parentage which correspond to the detachment of a single electron. For this purpose we remove an electron from the configuration j^{N-1} in expression (7.13), using the coefficients of fractional parentage which are appropriate for this configuration [$\alpha' J'$ in expression (7.13) is first changed to $\bar{\alpha}'\bar{J}'$]:

$$\psi(j^N \alpha JM) = \sum_{\bar{\alpha}', \bar{J}'} \sum_{\alpha', J'} \psi((j^{N-2}\alpha' J', j(N-1))\bar{J}', j(N); JM) \times \langle j^{N-2}\alpha' J', j; \bar{J}' | \} j^{N-1}\bar{\alpha}'\bar{J}' \rangle \langle j^{N-1}\bar{\alpha}'\bar{J}', j; J | \} j^N \alpha J \rangle. \tag{7.34}$$

We alter the order in which the angular momenta are coupled, and couple $j(N-1)$ to $j(N)$,

$$\psi((j^{N-2}\alpha' J', j(N-1))\bar{J}', j(N); JM) = \sum_{J''} \psi(j^{N-2}\alpha' J', j^2(N-1, N)J''; JM) \langle J'(jj)J'' | (J'j)\bar{J}'j \rangle^{(J)}. \tag{7.35}$$

The transformation matrix connects the two different coupling schemes for the three angular momenta J', j, and j; it can therefore be expressed in terms of Racah coefficients similarly to (3.59):

$$\langle J'(jj)J'' | (J'j)\bar{J}'j \rangle^{(J)} = [(2J''+1)(2\bar{J}'+1)]^{1/2} W(J'jJj; \bar{J}'J''). \tag{7.36}$$

When (7.35) is substituted into (7.34) we obtain the desired formula:

$$\langle j^{N-2}\alpha' J', j^2 J''; J | \} j^N \alpha J \rangle = \sum_{\bar{\alpha}', \bar{J}'} \langle j^{N-2}\alpha' J', j; \bar{J}' | \} j^{N-1}\bar{\alpha}'\bar{J}' \rangle \langle j^{N-1}\bar{\alpha}'\bar{J}', j; J | \} j^N \alpha J \rangle \times \langle J'(jj)J'' | (J'j)\bar{J}'j \rangle^{(J)}. \tag{7.37}$$

With the aid of expansion (7.33) the matrix elements of a scalar operator are easily expressed in terms of two-electron matrix elements. For this purpose we replace G by the operator $g_{N-1\,N}$ which acts only upon the coordinates of the $(N-1)$th and Nth electrons:

$$\langle j^N \alpha J | G | j^N \bar{\alpha} J \rangle = \tfrac{1}{2} N(N-1) \langle j^N \alpha J | g_{N-1\,N} | j^N \bar{\alpha} J \rangle. \tag{7.38}$$

We substitute into this the expansion (7.33). Because of the orthogonality of the basis functions for the configuration j^{N-2} and the scalar property of the operator $g_{N-1\,N}$, we obtain

$$\langle j^N \alpha J | G | j^N \bar{\alpha} J \rangle = \tfrac{1}{2} N(N-1) \sum_{\alpha', J', J''} \langle j^{N-2}\alpha' J', j^2 J''; J | \} j^N \alpha J \rangle \times \langle j^{N-2}\alpha' J', j^2 J''; J | \} j^N \bar{\alpha} J \rangle \langle j^2 J'' | g_{N-1\,N} | j^2 J'' \rangle. \tag{7.39}$$

Methods of calculating matrix elements of the electrostatic interaction in two-electron configurations are discussed by Judd (1963) and Sobel'man (1972).

Calculations using formula (7.39) become extremely cumbersome with increasing numbers of electrons in a configuration j^N. An alternative method consists in expressing the interaction operator in terms of a sum of scalar products of tensor operators. The problem then reduces to the calculation of matrix elements of operators of type F. A method such as this has been devised by Racah (1943) [see also Judd (1963) and Sobel'man (1972)].

PART 2. Configurations of Several Groups of Equivalent Electrons. A State with Arbitrary Permutational Symmetry

7.3. A Single Shell

In the previous section we considered the coefficients of fractional parentage for antisymmetric states of a configuration of equivalent electrons. When a total wavefunction for a system of electrons is constructed from products of coordinate wavefunctions with spin functions [formula (6.2)] the permutational symmetries of the two kinds of function are defined by mutually dual Young diagrams. If the operator of a particular interaction either acts only upon the spatial coordinates or only upon the spin coordinates, then in order to describe the interaction, it is sufficient to specify the actual form only of either the coordinate wavefunction or only the spin wavefunction. Matrix elements in this case are conveniently calculated if beforehand one carries out a fractional parentage decomposition which connects a state of N electrons whose permutation symmetry is that of a Young diagram $[\lambda]$ with a state in which $N-1$ (or $N-2$) electrons are characterized by the permutational symmetry $[\lambda']$.

In this section we obtain formulae for the coefficients of fractional parentage for an arbitrary configuration of several groups of equivalent electrons in a state with permutational symmetry $[\lambda]$. We denote the angular momentum of an electron by j, this standing for either the orbital angular momentum l or the spin angular momentum s. The representation of a wavefunction in the form (6.2) is incorrect for j–j coupling. However, formulae for this case can be obtained from those to be derived here if one puts $[\lambda] = [1^N]$. We begin our investigation with a single group of equivalent electrons.

First of all, we recall that linearly independent basis functions which belong to an irreducible representation $\Gamma^{[\lambda]}$ of the permutation group π_N are enumerated by Young tableaux r (see Section 2.4). The specification of a particular Young tableau r corresponds to the simultaneous specification of $N-2$ irreducible representations $\Gamma^{[\lambda']}, \Gamma^{[\lambda'']}, \ldots$, to which the given basis function belongs on passing to the groups $\pi_{N-1}, \pi_{N-2}, \ldots$. In fact the permutational symmetry of a function $\Phi_r^{[\lambda]}$ with respect to permutations of its first $N-1$ arguments is characterized by the Young diagram $[\lambda']$ which is obtained by removing the cell containing the number N from the Young tableau r.

The coefficients of fractional parentage are defined as the matrix elements in the transformation

$$\Phi(j^N[\lambda]r\alpha JM) = \sum_{\alpha',J'} \Phi(j^{N-1}[\lambda']r'\alpha'J', j(N); JM) \times \langle j^{N-1}[\lambda']\alpha'J', j; J|\} j^N[\lambda]\alpha J\rangle. \tag{7.40}$$

They are independent of the Young tableaux and of the projection of the angular momentum. The inverse transformation has the following form:

$$\Phi(j^{N-1}[\lambda']r'\alpha'J', j(N); JM) = \sum_{\lambda} \Phi(j^N[\lambda]r\alpha JM)\langle j^N[\lambda]\alpha J\{| j^{N-1}[\lambda']\alpha'J', j; J\rangle. \tag{7.41}$$

The summation in this equation is over all the allowed Young diagrams with N cells which are obtained by adding a single cell to the Young diagram $[\lambda']$. From Eqs. (7.40) and (7.41) it follows that the coefficients of fractional parentage form a rectangular matrix, the rows of which are enumerated by values of the angular momentum $\alpha'J'$ which can occur for the particular $[\lambda']$ and J, and the columns by the possible Young diagrams $[\lambda]$ which can be formed from a given diagram $[\lambda']$ by adding one cell.

The basis functions in the expansion (7.40) possess permutational symmetry $[\lambda']r'$ with respect to permutations of the first $N - 1$ electrons. It is necessary to choose the coefficients of fractional parentage such that the functions (7.40) behave correspondingly under permutations of the Nth electron with the remaining $N - 1$ electrons. Since any permutation can be written as a product of transpositions of the type $P_{i\,i-1}$ and function (7.40) already has the correct transformation properties with respect to the transpositions $P_{12}, P_{23}, \ldots, P_{N-2\,N-1}$, it is sufficient to determine the behavior of (7.40) with respect to the transposition $P_{N-1\,N}$. This is the basis of the method developed by Jahn (1951) and Jahn and van Wieringen (1951) for finding the coefficients of fractional parentage, which will be set out here.

We write out a basis function occurring in the expansion (7.40) with the aid of the coefficients of fractional parentage for the configuration j^{N-1}:

$$\Phi(j^{N-1}[\lambda']r'\alpha'J', j(N); JM) = \sum_{\alpha'',J''} \Phi((j^{N-2}[\lambda'']r''\alpha''J'', j(N-1))J', j(N); JM) \times \langle j^{N-2}[\lambda'']\alpha''J'', j; J'|\} j^{N-1}[\lambda']\alpha'J'\rangle. \tag{7.42}$$

The action of the permutation $P_{N-1\,N}$ upon a function with coupled angular momenta on the right-hand side of (7.42) is to give a function with a different coupling scheme for the angular momenta J'', j, and j:

$$P_{N-1\,N}\Phi((j^{N-2}[\lambda'']r''\alpha''J'', j(N-1))J', j(N); JM) = \sum_{\bar{J}'} \Phi((j^{N-2}[\lambda'']r''\alpha''J'', j(N-1))\bar{J}', j(N); JM) \times \langle (J''j(N-1))\bar{J}'j(N)|(J''j(N))J'j(N-1)\rangle^{(J)}. \tag{7.43}$$

The transformation matrix in (7.43) is expressed in terms of a Racah coefficient, similarly to (7.7):

$$\langle (J''j(N-1))\bar{J}'j(N) \,|\, (J''j(N))J'j(N-1)\rangle^{(J)} = [(2\bar{J}'+1)(2J'+1)]^{1/2} W(J'jj\bar{J}';JJ''). \qquad (7.44)$$

From Eq. (7.43) it follows that the matrix of the operator $P_{N-1\,N}$ in a representation of coupled angular momentum functions which is symmetrized with respect to only the first $N-2$ electrons coincides with the transformation matrix (7.44). We introduce for this the following notation:

$$\langle J''\bar{J}'J \,|\, P_{N-1\,N} \,|\, J''J'J\rangle \equiv \langle (J''j(N-1))\bar{J}'j(N) \,|\, (J''j(N))J'j(N-1)\rangle^{(J)}. \qquad (7.45)$$

The coefficients of fractional parentage in the expansion (7.42) can be regarded as elements of a transformation matrix which takes one from coupled angular momentum functions symmetrized with respect to permutations of the first $N-2$ electrons to functions which are symmetrized with respect to permutations of the first $N-1$ electrons. Hence the matrix of the operator $P_{N-1\,N}$ in the representation (7.42) can be expressed in terms of the matrix (7.45) by a standard formula of matrix algebra [see (1.35)]:

$$\begin{aligned}\langle j^{N-1}[\lambda']r'\alpha'J', j(N); J \,|\, P_{N-1\,N} \,|\, j^{N-1}[\bar{\lambda}']\bar{r}'\bar{\alpha}'\bar{J}', j(N); J\rangle &= \sum_{\alpha'', J''} \langle j^{N-1}[\lambda']\alpha'J' \{| j^{N-2}[\lambda'']\alpha''J'', j; J'\rangle \\ &\times \langle J''J'J \,|\, P_{N-1\,N} \,|\, J''\bar{J}'J\rangle \langle j^{N-2}[\lambda'']\alpha''J'', j; \bar{J}' |\} j^{N-1}[\bar{\lambda}']\bar{\alpha}'\bar{J}'\rangle. \qquad (7.46)\end{aligned}$$

The orthogonality of the matrix of the operator $P_{N-1\,N}$ with respect to the quantum numbers of the configuration j^{N-2} has been taken into account in deriving this equation.

The fractional parentage coefficients in expression (7.40) enable one to carry out a transformation from functions with a definite permutational symmetry with respect to just the first $N-1$ electrons to basis functions for an irreducible representation $\Gamma^{[\lambda]}$ of the full permutation group π_N. We require this representation to coincide with the standard Young–Yamanouchi representation. Hence the matrix of the operator $P_{N-1\,N}$ must coincide with the matrix in the standard representation, $\Gamma^{[\lambda]}(P_{N-1\,N})$, which is obtained by means of the rules given in Section 2.5. We obtain, similarly to (7.46),

$$\begin{aligned}\langle j^N[\lambda]r\alpha J \,|\, P_{N-1\,N} \,|\, j^N[\lambda]\bar{r}\alpha J\rangle &\equiv \Gamma^{[\lambda]}_{r\bar{r}}(P_{N-1\,N}) \\ &= \sum_{\alpha', J'} \sum_{\bar{\alpha}', \bar{J}'} \langle j^N[\lambda]\alpha J \{| j^{N-1}[\lambda']\alpha'J', j; J\rangle \\ &\times \langle j^{N-1}[\lambda']r'\alpha'J', j(N); J \,|\, P_{N-1\,N} \,|\, j^{N-1}[\bar{\lambda}']\bar{r}'\bar{\alpha}'\bar{J}', j(N); J\rangle \\ &\times \langle j^{N-1}[\bar{\lambda}']\bar{\alpha}'\bar{J}', j; J |\} j^N[\lambda]\alpha J\rangle. \qquad (7.47)\end{aligned}$$

These relations allow one to find the coefficients of fractional parentage for the configuration j^N if those for the configuration j^{N-1} are known. A recurrence method such as this could start from the configuration j^3 since the fractional parentage coefficients for the configuration j^2 are equal to unity.

As an example, we consider the process of finding the coefficients of fractional parentage for the configuration p^3 in a state with quantum numbers $[\lambda] = [21]$, $J = 1$. The fractional parentage coefficients for the configuration p^2 are equal to one and therefore the matrix (7.46) coincides with (7.45). The coefficients of fractional parentage are easily found in this case by equating the result of applying the permutation P_{23} to (7.40) [taking (7.43) into account] with the result of applying P_{23} to basis functions for the standard representation.

The matrices of the standard representation $\Gamma^{[21]}$ have been given in Section 2.5. We denote the Young tableaux $r^{(i)}$ by a subscript added to the Young diagram [21]. The matrix of the operator P_{23} is given by

$$\begin{matrix} & [21]_1 & [21]_2 \end{matrix}$$
$$\Gamma^{[21]}(P_{23}) = \begin{bmatrix} -\frac{1}{2} & \frac{1}{2}\sqrt{3} \\ \frac{1}{2}\sqrt{3} & \frac{1}{2} \end{bmatrix},$$

i. e.,

$$P_{23}\Phi(p^3[2\,1]_1P) = -\tfrac{1}{2}\Phi(p^3[2\,1]_1P) + \tfrac{1}{2}\sqrt{3}\,\Phi(p^3[2\,1]_2P). \tag{7.48}$$

The configuration p^2 gives rise to three allowed states: [2]S, D, and $[1^2]$P. Since the Young diagram $[\lambda'] = [2]$ corresponds to the tableau $[21]_1$ and the diagram $[\lambda'] = [1^2]$ to $[21]_2$, the state $[21]_1$P arises from two states of the configuration p^2 (we denote the coefficients of fractional parentage of these states by x and y for short), and the state $[21]_2$P from one state of p^2 (the coefficient of fractional parentage is thus equal to unity):

$$\begin{aligned} \Phi(p^3[21]_1P) &= x\Phi(p^2[2]S, p; P) + y\Phi(p^2[2]D, p; P) \\ \Phi(p^3[21]_2P) &= \Phi(p^2[1^2]P, p; P). \end{aligned} \tag{7.49}$$

We apply the permutation P_{23} to the function $\Phi(p^3[21]_1P)$. According to Eq. (7.43) and (7.44), the transformation matrix which occurs as a result of this is expressed in terms of Racah coefficients. The necessary Racah coefficients are given by

$$\begin{aligned} W(0110; 11) &= W(1110; 11) = W(2110; 11) \\ &= W(0112; 11) = \tfrac{1}{3}, \\ W(1112; 11) &= -1/6, \qquad W(2112; 11) = 1/30. \end{aligned}$$

As a result, we find that

$$\begin{aligned} P_{23}\Phi(p^3[21]_1P) = {} & (\tfrac{1}{3}x + \tfrac{1}{3}\sqrt{5}\,y)\Phi(p^2[2]S, p; P) \\ & + (\tfrac{1}{3}\sqrt{5}\,x + \tfrac{1}{6}y)\Phi(p^2[2]D, p; P) \\ & + (\tfrac{1}{3}\sqrt{3}\,x - \tfrac{1}{6}\sqrt{15}\,y)\Phi(p^2[1^2]P, p; P). \end{aligned} \tag{7.50}$$

On equating the coefficients of identical terms in expressions (7.50) and (7.48) [it is necessary to substitute Eq. (7.49) into (7.48) beforehand], we obtain a system of equations for x and y:

$$\tfrac{1}{2}\sqrt{5}\,x + y = 0, \qquad x - \tfrac{1}{2}\sqrt{5}\,y = \tfrac{3}{2},$$

from which it follows that

$$x = \tfrac{2}{3}, \qquad y = -\tfrac{1}{3}\sqrt{5}$$

or

$$\Phi(p^3[2\,1]_1 P) = \tfrac{2}{3}\Phi(p^2[2]S, p; P) - \tfrac{1}{3}\sqrt{5}\,\Phi(p^2[2]D, p; P).$$

In this way Jahn was able to calculate the coefficients of the states of nucleons for the configurations p^N (Jahn and van Wieringen, 1951) and d^N (Jahn, 1951). Table 7.1 gives the coefficients of fractional parentage for the atomic

Table 7.1

Coefficients of Fractional Parentage

$\langle p^N[\lambda]L\{|p^{N-1}[\lambda']L', p; L\rangle$

$p^3[\lambda]L$	$p^2[\lambda']$: L':	[2] S	[2] D	$[1^2]$ P
[2 1]P		$\tfrac{2}{3}$	$-\tfrac{1}{3}\sqrt{5}$	1
D		—	1	1
$[1^3]$S		—	—	1
$p^4[\lambda]L$	$p^3[\lambda']$: L':	[2 1] P	[2 1] D	$[1^3]$ S
$[2^2]$S		-1	—	—
D		$+\tfrac{1}{2}$	$-\sqrt{\tfrac{3}{2}}$	—
$[2\,1^2]$P		$\sqrt{\tfrac{3}{8}}$	$\sqrt{\tfrac{5}{8}}$	1
$p^5[\lambda]L$	$p^4[\lambda']$: L':	$[2^2]$ S	$[2^2]$ D	$[2\,1^2]$ P
$[2^2\,1]$P		$-\sqrt{\tfrac{1}{6}}$	$-\sqrt{\tfrac{5}{6}}$	1

p^N configuration. When using the table it should be remembered that only fractional parentage coefficients with the same Young diagram for the configuration p^{N-1} occur in a single linear combination. These coefficients are separated into groups in the table.

The calculation of the fractional parentage coefficients by means of the relations (7.47) is substantially simplified when, on the basis of the group-theoretical classification of states, the coefficients are factorized beforehand

(Edmonds and Flowers, 1952). By constructing wavefunctions with the aid of the Young operators $\omega_{rt}^{[\lambda]}$, Hassitt (1955) obtained a relation which expresses the coefficients of fractional parentage in (7.40) directly in terms of the fractional parentage coefficients for the configuration j^{N-1} and Racah coefficients. In the special case of antisymmetric states Hassitt's formula reduces to an equation similar to (7.15).

In order to calculate matrix elements of operators of type G [see (7.23)], it is convenient to put the wavefunction for a configuration j^N into the form of a fractional parentage expansion in wavefunctions of the vector-coupled configurations j^{N-2} and j^2. The coefficients of fractional parentage which arise as a result of this are defined by the equation

$$\begin{aligned}\Phi(j^N[\lambda]r'r''\alpha JM) &= \sum_{\alpha',J',J''} \Phi(j^{N-2}[\lambda']r'\alpha'J', j^2(N-1,N)[\lambda'']J''; JM) \\ &\quad\times \langle j^{N-2}[\lambda']\alpha'J', j^2[\lambda'']J'', J | \} j^N[\lambda]\alpha J\rangle. \qquad (7.51)\end{aligned}$$

These can be expressed in terms of fractional parentage coefficients for the detachment of a single particle. For this purpose we apply the decomposition (7.40) twice to a wavefunction for a configuration j^N characterized by the Young tableau r, and couple the detached electrons to each other:

$$\begin{aligned}\Phi(j^N[\lambda]r\alpha JM) &= \sum_{\alpha',J'} \sum_{\bar{\alpha}',\bar{J}',\bar{J}''} \Phi(j^{N-2}[\lambda']r'\alpha'J', j^2(N-1, N)\bar{J}''; JM) \\ &\quad\times \langle j^{N-2}[\lambda']\alpha'J', j; \bar{J}' | \} j^{N-1}[\bar{\lambda}']\bar{\alpha}'\bar{J}'\rangle \\ &\quad\times \langle j^{N-1}[\bar{\lambda}']\bar{\alpha}'\bar{J}', j; J | \} j^N[\lambda]\alpha J\rangle \\ &\quad\times \langle J'(jj)\bar{J}'' | (J'j)\bar{J}'j\rangle^{(J)}. \qquad (7.52)\end{aligned}$$

The summation over $\bar{J}''$ in this equation includes both even and odd values, i.e., both symmetric and antisymmetric states of electrons $N-1$ and N occur in (7.52). With the aid of the transformation matrix for the permutation group, function (7.52) can be written in the form of a linear combination of functions (7.51):

$$\Phi(j^N[\lambda]r\alpha JM) = \sum_{r''} \Phi(j^N[\lambda]r'r''\alpha JM)\langle r'r'' | r\rangle^{[\lambda]}. \qquad (7.53)$$

The transformation matrix appearing in this equation is diagonal in the Young diagram for the first $N-2$ electrons, $[\lambda']$, and is independent of the Young tableau r'. It can be rewritten in the form

$$\langle \lambda'\lambda'' | (\lambda'1)\bar{\lambda}'1\rangle^{[\lambda]},$$

where $[\bar{\lambda}']$ denotes the Young diagram for the first $N-1$ electrons, and r'' is denoted by $[\lambda'']$ in view of their one-to-one correspondence. The explicit form of a matrix such as this was given in Chapter II [see (2.65)].

If the functions on the right hand side of (7.53) are expanded by formula (7.51) and are then equated with expansion (7.52), we obtain the desired relation:

$$\langle \lambda'\lambda'' \,|\, (\lambda'1)\bar{\lambda}'1\rangle^{[\lambda]} \langle j^{N-2}[\lambda']\alpha'J', j^2[\lambda'']J''; J \,|\}\, j^N[\lambda]\alpha J\rangle$$
$$= \sum_{\bar{\alpha}', \bar{J}'} \langle j^{N-2}[\lambda']\alpha'J', j; \bar{J}' \,|\}\, j^{N-1}[\bar{\lambda}']\bar{\alpha}'\bar{J}'\rangle$$
$$\times \langle j^{N-1}[\bar{\lambda}']\bar{\alpha}'\bar{J}', j; J \,|\, j^N[\lambda]\alpha J\rangle \langle J'(jj)J'' \,|\, (J'j)\bar{J}'j\rangle^{(J)}. \qquad (7.54)$$

By using this formula, coefficients of fractional parentage $\langle p^{N-2}, p^2 |\} p^N\rangle$ for all the possible states of a nuclear shell have been obtained by Elliott *et al.* (1953). Table 7.2 gives the coefficients of fractional parentage for the

Table 7.2

Coefficients of Fractional Parentage $\langle p^N[\lambda]L\{| p^{N-2}[\lambda']L', p^2[\lambda'']L''; L\rangle$

$p^3[\lambda]L$	$p[\lambda'], p^2[\lambda'']$:	[1], [2]		[1], [1²]
	$L'L''$:	PS	PD	PP
[2 1]P		$-\frac{2}{3}$	$\frac{1}{3}\sqrt{5}$	1
D		—	1	−1
[1³]S		—	—	1

$p^4[\lambda]L$	$p^2[\lambda'], p^2[\lambda'']$:	[2], [2]				[2], [1²]		[1²], [2]		[1²], [1²]
	$L'L''$:	SS	DS	SD	DD	SP	DP	PS	PD	PP
[2²]S		$-\frac{2}{3}$	—	—	$\frac{1}{3}\sqrt{5}$	—	—	—	—	−1
D		—	$\frac{1}{3}$	$\frac{1}{3}$	$-\frac{1}{3}\sqrt{7}$	—	—	—	—	−1
[2 1²]P		—	—	—	—	$\sqrt{\frac{1}{6}}$	$\sqrt{\frac{5}{6}}$	$\sqrt{\frac{1}{6}}$	$\sqrt{\frac{5}{6}}$	−1

$p^5[\lambda]L$	$p^3[\lambda'], p^2[\lambda'']$:	[21], [2]			[21], [1²]		[1³], [1²]
	$L'L''$:	PS	PD	DD	PP	DP	SP
[2² 1]P		$-\sqrt{\frac{1}{6}}$	$\sqrt{\frac{5}{24}}$	$\sqrt{\frac{5}{8}}$	$\sqrt{\frac{3}{8}}$	$-\sqrt{\frac{5}{8}}$	1

allowed states which arise from configurations of equivalent p electrons. Coefficients which occur in different linear combinations are grouped separately. Although L'' uniquely determines $[\lambda'']$, we give values of both of these quantities for convenience.

7.4. A Configuration of Two Shells

We now consider a configuration of two groups of equivalent electrons in a state with permutational symmetry of a Young diagram $[\lambda]$ and total angular momentum J. Each group of electrons possesses its own permutational symmetry $[\lambda_a]$ and angular momentum J_a. A wavefunction for this configuration which corresponds to a total angular momentum J and to a projection M of J is constructed by means of Clebsch–Gordan coefficients:

$$\Phi(j_1^{n_1}[\lambda_1]r_1\alpha_1J_1, j_2^{n_2}[\lambda_2]r_2\alpha_2J_2; JM) = \sum_{M_1,M_2} \Phi(j_1^{n_1}[\lambda_1]r_1\alpha_1J_1M_1)\Phi(j_2^{n_2}[\lambda_2]r_2\alpha_2J_2M_2) \times \langle J_1M_1, J_2M_2|JM\rangle. \tag{7.55}$$

This function possesses a definite permutational symmetry with respect to permutations of the first n_1 electrons and the last n_2 electrons. In order to symmetrize it with respect to a Young diagram of $n_1 + n_2$ cells, we make use of formula (2.85):

$$\Phi(j_1^{n_1}[\lambda_1]\alpha_1J_1, j_2^{n_2}[\lambda_2]\alpha_2J_2; [\lambda](r)^A JM) = \left\{\frac{f_\lambda}{f_{\lambda_1}f_{\lambda_2}}\frac{n_1!\,n_2!}{N!}\right\}^{1/2} \sum_{r_1,r_2}\sum_{Q}\langle(r)^A|Q|r_1r_2\rangle^{[\lambda]}Q \times \Phi(j_1^{n_1}[\lambda_1]r_1\alpha_1J_1, j_2^{n_2}[\lambda_2]r_2\alpha_2J_2; JM). \tag{7.56}$$

Any set of $N!/(n_1!\,n_2!)$ permutations which is obtained by choosing one permutation from each of the cosets of the subgroup $\pi_{n_1} \times \pi_{n_2}$ can be taken as the Q in this equation. We shall choose permutations that preserve within each shell the ascending order in which the electrons are numbered. As a result of the permutations Q, therefore, electron N is either unaffected or exchanges places with electron n_1. All the $N!/(n_1!\,n_2!)$ functions in the summation over Q in (7.56) are orthogonal to one another due to the orthogonality of one-electron functions with differing j_a. Function (7.56) is therefore normalized.

In order to calculate matrix elements of the operators F and G, (7.23), it is convenient to write (7.56) in the form of an expansion in wavefunctions which are symmetrized with respect to the first $N - 1$ or first $N - 2$ electrons only. We call the coefficients in such an expansion coefficients of fractional parentage, just as in the case of a single group of equivalent electrons.

a. Fractional parentage coefficients of the type $\langle N-1, 1|\}N\rangle$ are formally defined by the expansion

$$\begin{aligned}\Phi(j_1^{n_1}[\lambda_1]\alpha_1J_1, j_2^{n_2}[\lambda_2]\alpha_2J_2; [\lambda]rJM) &= \sum_{\lambda_1'}\sum_{\alpha_1',J_1',J'} \Phi((j_1^{n_1-1}[\lambda_1']\alpha_1'J_1', j_2^{n_2}[\lambda_2]\alpha_2J_2)[\lambda']r'J', j_1(N); JM) \\ &\times \langle(j_1^{n_1-1}[\lambda_1']\alpha_1'J_1', j_2^{n_2}[\lambda_2]\alpha_2J_2)[\lambda']J', j_1; J|\}j_1^{n_1}[\lambda_1]\alpha_1J_1, j_2^{n_2}[\lambda_2]\alpha_2J_2; [\lambda]J\rangle \\ &+ \sum_{\lambda_2'}\sum_{\alpha_2',J_2',J'} \Phi((j_1^{n_1}[\lambda_1]\alpha_1J_1, j_2^{n_2-1}[\lambda_2']\alpha_2'J_2')[\lambda']r'J', j_2(N); JM) \\ &\times \langle j_1^{n_1}[\lambda_1]\alpha_1J_1, j_2^{n_2-1}[\lambda_2']\alpha_2'J_2')[\lambda']J', j_2; J|\}j_1^{n_1}[\lambda_1]\alpha_1J_1, j_2^{n_2}[\lambda_2]\alpha_2J_2; [\lambda]J\rangle.\end{aligned} \tag{7.57}$$

There is no summation over $[\lambda']r'$ since the specification of a Young tableau r uniquely defines the permutational symmetry of the first $N - 1$ electrons, i.e., $[\lambda]r$ uniquely defines $[\lambda']$ and r'.

In order to find an explicit expression for the coefficients in this expansion, we take expression (7.56) as our starting point. The task before us is to transform this into the form (7.57). We choose $(r)^{A}$ in (7.56) to be a standard tableau r.

We divide the sum over Q in (7.56) into two sums. The first summation includes all permutations Q_1 that carry the Nth electron into the place of electron n_1. This yields the coefficients $\langle j_1^{n_1-1} j_2^{n_2}, j_1 | \} j_1^{n_1} j_2^{n_2} \rangle$. The second summation includes the remaining permutations Q_2 that leave the Nth electron in place. This sum determines the coefficients $\langle j_1^{n_1} j_2^{n_2-1}, j_2 | \} j_1^{n_1} j_2^{n_2} \rangle$. The number of permutations in each sum is clearly equal to

$$n(Q_1) = \frac{(N-1)!}{(n_1-1)!\, n_2!},$$

$$n(Q_2) = \frac{(N-1)!}{n_1!\,(n_2-1)!},$$

$$n(Q) = n(Q_1) + n(Q_2) = \frac{N!}{(n_1!\, n_2!)}. \tag{7.58}$$

Let us consider the first sum. The permutations Q_1 can be written in the form of a product $Q(N-1)P_1^{(N)}$ in which $P_1^{(N)} = P_{N\,N-1\ldots n_1}$ brings N into the n_1th place and $Q(N-1)$ does not act upon the coordinates of the Nth electron, i.e., $Q(N-1) \in \boldsymbol{\pi}_{N-1}$. After applying the permutations $P_1^{(N)}$ to the function (7.56) we have

$$\begin{aligned}\Sigma_1 = c \sum_{r_1, r_2} \sum_{Q(N-1)} &\langle r | Q(N-1) P_1^{(N)} | r_1 r_2 \rangle^{[\lambda]} Q(N-1) \\ &\times \Phi'(j_1^{n_1}[\lambda_1] r_1 \alpha_1 J_1, j_2^{n_2}[\lambda_2] r_2 \alpha_2 J_2; JM | 12 \cdots \\ &n_1 - 1\, N n_1\, n_1 + 1 \cdots N - 1),\end{aligned} \tag{7.59}$$

$$c = \left\{ \frac{f_\lambda}{f_{\lambda_1} f_{\lambda_2}} \frac{n_1!\, n_2!}{N!} \right\}^{1/2}. \tag{7.60}$$

The ordering of the electron coordinates in the argument of the function Φ' in (7.59) corresponds to the distribution of the electrons between the shells. We detach the Nth electron from the $j_1^{n_1}$ shell with the aid of the single-shell coefficients of fractional parentage, and recouple the angular momenta so that the angular momentum of the remaining $n_1 - 1$ electrons in the first shell is coupled to the angular momentum J_2 of the second shell,

$$\begin{aligned}\Phi' = \sum_{\alpha_1', J_1', J'} &\Phi((j_1^{n_1-1}[\lambda_1'] r_1' \alpha_1' J_1', j_2^{n_2}[\lambda_2] r_2 \alpha_2 J_2) J', j_1(N); JM) \\ &\times \langle j_1^{n_1-1}[\lambda_1'] \alpha_1' J_1', j_1; J_1 | \} j_1^{n_1}[\lambda_1] \alpha_1 J_1 \rangle \\ &\times \langle (J_1' J_2) J' j_1 | (J_1' j_1) J_1 J_2 \rangle^{(J)}.\end{aligned} \tag{7.61}$$

Let us write the matrix element of $Q(N-1)P_1^{(N)}$ in (7.59) as a sum of products of matrix elements of $Q(N-1)$ and $P_1^{(N)}$, choosing as the symmetry of the intermediate states the nonstandard representation which is reduced with respect to $(\boldsymbol{\pi}_{n_1-1} \times \boldsymbol{\pi}_{n_2}) \times \boldsymbol{\pi}_1$,

$$\begin{aligned}&\langle r | Q(N-1) P_1^{(N)} | r_1 r_2 \rangle^{[\lambda]} \\ &= \sum_{\bar r_1', \bar r_2, \bar\lambda'} \langle r | Q(N-1) | (\bar r_1' \bar r_2) \bar\lambda' 1 \rangle^{[\lambda]} \langle (\bar r_1' \bar r_2) \bar\lambda' 1 | P_1^{(N)} | r_1 r_2 \rangle^{[\lambda]}.\end{aligned} \tag{7.62}$$

Now $Q(N-1) \in \pi_{N-1}$, and since the representation in terms of which the matrix element of $Q(N-1)$ is defined is already reduced with respect to permutations of π_{N-1} [cf. (2.22)], we can write

$$\langle r | Q(N-1) | (\bar{r}_1{}'\bar{r}_2)\bar{\lambda}'1\rangle^{[\lambda]} = \delta_{\bar{\lambda}'\lambda'}\langle r' | Q(N-1) | \bar{r}_1{}'\bar{r}_2\rangle^{[\lambda']}. \tag{7.63}$$

In addition we can show that the matrix of $P_1^{(N)}$ in (7.62) is diagonal with respect to the Young tableaux for the first $n_1 - 1$ numbers and with respect to the last n_2 numbers. For this purpose, we use the fact that when $P_1^{(N)}$ is applied to a basis function $|[\lambda]r_1r_2\rangle$ the result is a function whose arguments are divided into the same subgroups as those of a function $|[\lambda](\bar{r}_1{}'\bar{r}_2)\bar{\lambda}'1\rangle$, i.e., a matrix element of a permutation $P_1^{(N)}$ can be written in the following form:

$$\int \Phi^{*[\lambda]}(\underbrace{12\cdots n_1-1}_{\bar{r}_1{}'}\,\underbrace{n_1\, n_1+1\cdots N-1\, N}_{\bar{r}_2})$$
$$\times\, \Phi^{[\lambda]}\,\underbrace{12\cdots n_1-1}_{r_1{}'}\,\underbrace{Nn_1\, n_1+1\cdots N-1}_{r_2})\, dV. \tag{7.64}$$

The operator in this equation can be taken to be unity. As far as the permutation group is concerned, this is simply an irreducible tensor operator which transforms according to the unit representation. We may therefore apply the Wigner–Eckart theorem in the form (4.60) to each group of numbers in (7.64), from which it follows that

$$\langle(\bar{r}_1{}'\bar{r}_2)\bar{\lambda}'1 | P_1^{(N)} | r_1r_2\rangle^{[\lambda]} = \delta_{\bar{r}_1{}'r_1{}'}\delta_{\bar{r}_2r_2}\langle(\lambda_1{}'\lambda_2)\bar{\lambda}'1 | P_1^{(N)} | (\lambda_1{}'1)\lambda_1\lambda_2\rangle^{[\lambda]}. \tag{7.65}$$

Consequently, the matrix element is diagonal in the Young diagrams $[\lambda_1{}']$ and $[\lambda_2]$ and is independent of the Young tableaux.

Taking Eqs. (7.63) and (7.65) into account, Eq. (7.62) becomes

$$\langle r | Q(N-1)P_1^{(N)} | r_1r_2\rangle^{[\lambda]}$$
$$= \langle r' | Q(N-1) | r_1{}'r_2\rangle^{[\lambda']}\langle(\lambda_1{}'\lambda_2)\lambda'1 | P_1^{(N)} | (\lambda_1{}'1)\lambda_1\lambda_2\rangle^{[\lambda]}. \tag{7.66}$$

We substitute (7.61) and (7.66) into (7.59) and replace the summation in this last equation over r_1 and r_2 by the equivalent summation over $r_1{}'\lambda_1$ and r_2. Since

$$\sum_{r_1{}',r_2}\sum_{Q(N-1)}\langle r' | Q(N-1) | r_1{}'r_2\rangle^{[\lambda']}Q(N-1)\Phi(j_1^{n_1-1}[\lambda_1{}']r_1{}'\alpha_1{}'J_1{}', j_2^{n_2}[\lambda_2]r_2\alpha_2J_2; J')$$
$$= \left\{\frac{f_{\lambda'}}{f_{\lambda_1{}'}f_{\lambda_2}}\frac{(n_1-1)!\, n_2!}{(N-1)!}\right\}^{-1/2}\Phi(j_1^{n_1-1}[\lambda_1{}']\alpha_1{}'J_1{}', j_2^{n_2}[\lambda_2]\alpha_2J_2; [\lambda']r'J'), \tag{7.67}$$

similarly to (7.56), Σ_1 finally assumes the form:

$$\Sigma_1 = \left\{\frac{f_\lambda f_{\lambda_1{}'}}{f_{\lambda'}f_{\lambda_1}}\frac{n_1}{N}\right\}^{1/2}\sum_{\lambda_1{}'}\sum_{\alpha_1{}',J_1{}',J'}\Phi((j_1^{n_1-1}[\lambda_1{}']\alpha_1{}'J_1{}', j_2^{n_2}[\lambda_2]\alpha_2J_2)[\lambda']r'J', j_1(N); JM)$$
$$\times \langle j_1^{n_1-1}[\lambda_1{}']\alpha_1{}'J_1{}', j_1; J_1 | \} j_1^{n_1}[\lambda_1]\alpha_1J_1\rangle$$
$$\times \langle(J_1{}'J_2)J'j_1 | (J_1{}'j_1)J_1J_2\rangle^{(J)}\langle(\lambda_1{}'\lambda_2)\lambda'1 | P_1^{(N)} | (\lambda_1{}'1)\lambda_1\lambda_2\rangle^{(J)}. \tag{7.68}$$

On equating this sum with the first half of expansion (7.57), we obtain the desired expression for the coefficients of fractional parentage:

$$\begin{aligned}
&\langle (j_1^{n_1-1}[\lambda_1']\alpha_1'J_1', j_2^{n_2}[\lambda_2]\alpha_2J_2)[\lambda']J', j_1; J \,|\!\}\, j_1^{n_1}[\lambda_1]\alpha_1J_1, j_2^{n_2}[\lambda_2]\alpha_2J_2; [\lambda]J\rangle \\
&\quad = \left\{\frac{f_\lambda f_{\lambda_1'}}{f_{\lambda'} f_{\lambda_1}} \frac{n_1}{N}\right\}^{1/2} \langle j_1^{n_1-1}[\lambda_1']\alpha_1'J_1', j_1; J_1 \,|\!\}\, j_1^{n_1}[\lambda_1]\alpha_1J_1\rangle \\
&\quad\quad \times \langle (J_1'J_2)J'j_1 \,|\, (J_1'j_1)J_1J_2\rangle^{(J)} \langle (\lambda_1'\lambda_2)\lambda'1 \,|\, P_1^{(N)} \,|\, (\lambda_1'1)\lambda_1\lambda_2\rangle^{[\lambda]}. \qquad (7.69)
\end{aligned}$$

The sum Σ_2 contains the summation over Q_2, and by means of a similar transformation we obtain an expression for the coefficients of fractional parentage that correspond to the detachment of an electron from the second shell. Since the permutations Q_2 by definition do not affect the coordinates of the Nth electron, the matrix of $P_1^{(N)}$ no longer occurs in this expression, and instead we have the corresponding transformation matrix for the permutation group. The final expression has the form

$$\begin{aligned}
&\langle (j_1^{n_1}[\lambda_1]\alpha_1J_1, j_2^{n_2-1}[\lambda_2']\alpha_2'J_2')[\lambda']J', j_2; J \,|\!\}\, j_1^{n_1}[\lambda_1]\alpha_1J_1, j_2^{n_2}[\lambda_2]\alpha_2J_2; [\lambda]J\rangle \\
&\quad = \left\{\frac{f_\lambda f_{\lambda_2'}}{f_{\lambda'} f_{\lambda_2}} \frac{n_2}{N}\right\}^{1/2} \langle j_2^{n_2-1}[\lambda_2']\alpha_2'J_2', j_2; J_2 \,|\!\}\, j_2^{n_2}[\lambda_2]\alpha_2J_2\rangle \\
&\quad\quad \times \langle (J_1J_2')J'j_2 \,|\, J_1(J_2'j_2)J_2\rangle^{(J)} \langle (\lambda_1\lambda_2')\lambda'1 \,|\, \lambda_1(\lambda_2'1)\lambda_2\rangle^{[\lambda]} \qquad (7.70)
\end{aligned}$$

Expressions (7.69) and (7.70) for the fractional parentage coefficients are diagonal in the quantum numbers for the shells from which no electron is detached; the transformation matrices which occur in them repeat the coupling scheme for the two parts of a coefficient of fractional parentage.

b. Coefficients of fractional parentage of the type $\langle N-2, 2 \,|\!\}\, N\rangle$ are defined as the coefficients in the expansion‡

$$\begin{aligned}
&\Phi(j_1^{n_1}[\lambda_1]\alpha_1J_1, j_2^{n_2}[\lambda_2]\alpha_2J_2; [\lambda]r'r''JM) \\
&= \sum_{\lambda_1'} \sum_{\alpha_1', J_1', J_1''J'} \Phi((j_1^{n_1-2}[\lambda_1']\alpha_1'J_1', j_2^{n_2}[\lambda_2]\alpha_2J_2)[\lambda']r'J', j_1^2(N-1, N)[\lambda'']J_1''; JM) \\
&\quad \times \langle (j_1^{n_1-2}[\lambda_1']\alpha_1'J_1', j_2^{n_2}[\lambda_2]\alpha_2J_2)[\lambda']J', j_1^2[\lambda'']J_1''; J \,|\!\}\, j_1^{n_1}[\lambda_1]\alpha_1J_1, j_2^{n_2}[\lambda_2]\alpha_2J_2; [\lambda]J\rangle \\
&\quad + \sum_{\lambda_2'} \sum_{\alpha_2', J_2', J_2'', J'} \Phi((j_1^{n_1}[\lambda_1]\alpha_1J_1, j_2^{n_2-2}[\lambda_2']\alpha_2'J_2')[\lambda']r'J', j_2^2(N-1, N)[\lambda'']J_2''; JM) \\
&\quad \times \langle (j_1^{n_1}[\lambda_1]\alpha_1J_1, j_2^{n_2-2}[\lambda_2']\alpha_2'J_2')[\lambda']J', j_2^2[\lambda'']J_2''; J \,|\!\}\, j_1^{n_1}[\lambda_1]\alpha_1J_1, j_2^{n_2}[\lambda_2]\alpha_2J_2; [\lambda]J\rangle \\
&\quad + \sum_{\lambda_1', \lambda_2'} \sum_{\substack{\alpha_1', J_1', \alpha_2', \\ J_2', J', J''}} \Phi((j_1^{n_1-1}[\lambda_1']\alpha_1'J_1', j_2^{n_2-1}[\lambda_2']\alpha_2'J_2')[\lambda']r'J', j_1j_2(N-1, N)[\lambda'']J''; JM) \\
&\quad \times \langle (j_1^{n_1-1}[\lambda_1']\alpha_1'J_1', j_2^{n_2-1}[\lambda_2']\alpha_2'J_2')[\lambda']J', j_1j_2[\lambda'']J''; J \,|\!\}\, j_1^{n_1}[\lambda_1]\alpha_1J_1, j_2^{n_2}[\lambda_2]\alpha_2J_2; [\lambda]J\rangle \\
&\qquad (7.71)
\end{aligned}$$

By choosing as a basis function for the representation a function which is reduced with respect to the subgroup $\boldsymbol{\pi}_{N-2} \times \boldsymbol{\pi}_2$, one specifies the permuta-

‡We continue to adhere to the convention employed hitherto according to which we place a prime on the quantum numbers of a configuration that remains after one electron has been detached. Quantum numbers that characterize states of two detached electrons are distinguished by a double prime.

tional symmetry of the first $N-2$ and last two electrons. There is therefore no summation over $[\lambda']r'$ and $[\lambda'']$ in this equation.

In order to obtain an explicit expression for the coefficients of fractional parentage, we take expression (7.56) for the wavefunction as our starting point, and choose $r'r''$ for $(r)^A$ in this, where r' is a Young tableau for the first $N-2$ electrons and r'' is a tableau for the last two electrons. There are only two possibilities for r'', corresponding to $[\lambda''] = [2]$ and $[\lambda''] = [1^2]$. We divide the sum over Q into four sums Σ_i according to the effect the permutations have upon the coordinates of the $(N-1)$th and Nth electrons. Thus we have the following permutations:

Q_1: Permutations which carry electrons $N-1$ and N into the $(n_1 - 1)$th and n_1th positions.

Q_2: Permutations which carry electron $N-1$ into the n_1th position.

Q_3: Permutations which carry electron N into the n_1th position.

Q_4: Permutations which leave electrons $N-1$ and N in place.

Correspondingly, the first summation determines the coefficients $\langle j_1^{n_1-2}j_2^{n_2}, j_1^2 | \} j_1^{n_1}j_2^{n_2}\rangle$, and from the fourth summation one obtains an expression for the coefficients $\langle j_1^{n_1}j_2^{n_2-2}, j_2^2 | \} j_1^{n_1}j_2^{n_2}\rangle$. It is necessary to combine the second and third sums in order to derive an expression for the coefficients $\langle j_1^{n_1-1}j_2^{n_2-1}, j_1 j_2 | \} j_1^{n_1}j_2^{n_2}\rangle$.

A detailed derivation of the expressions for the three types of coefficient of fractional parentage just enumerated is given by Kaplan (1961b). We present here only the final formulae:

$$\begin{aligned}
&\langle (j_1^{n_1-2}[\lambda_1']\alpha_1'J_1', j_2^{n_2}[\lambda_2]\alpha_2J_2)[\lambda']J', j_1^2[\lambda'']J_1''; J | \} j_1^{n_1}[\lambda_1]\alpha_1J_1, j_2^{n_2}[\lambda_2]\alpha_2J_2; [\lambda]J\rangle \\
&\quad = \left\{\frac{f_\lambda f'_{\lambda_1}}{f_{\lambda'}f_{\lambda_1}}\frac{n_1(n_1-1)}{N(N-1)}\right\}^{1/2} \langle j_1^{n_1-2}[\lambda_1']\alpha_1'J_1', j_1^2[\lambda'']J_1''; J_1 | \} j_1^{n_1}[\lambda_1]\alpha_1J_1\rangle \\
&\quad\quad \times \langle (J_1'J_2)J'J_1'' | (J_1'J_1'')J_1J_2\rangle^{(J)} \langle (\lambda_1'\lambda_2)\lambda'\lambda'' | P_1^{N-1\,N} | (\lambda_1'\lambda'')\lambda_1\lambda_2\rangle^{[\lambda]}. \qquad (7.72)
\end{aligned}$$

$$\begin{aligned}
&\langle (j_1^{n_1}[\lambda_1]\alpha_1J_1, j_2^{n_2-2}[\lambda_2']\alpha_2'J_2'[\lambda']J', j_2^2[\lambda'']J''; J | \} j_1^{n_1}[\lambda_1]\alpha_1J_1, j_2^{n_2}[\lambda_2]\alpha_2J_2; [\lambda]J\rangle \\
&\quad = \left\{\frac{f_\lambda f_{\lambda_2'}}{f_{\lambda'}f_{\lambda_2}}\frac{n_2(n_2-1)}{N(N-1)}\right\}^{1/2} \langle j_2^{n_2-2}[\lambda_2']\alpha_2'J_2', j_2^2[\lambda'']J_2''; J_2 | \} j_2^{n_2}[\lambda_2]\alpha_2J_2\rangle \\
&\quad\quad \times \langle (J_1J_2')J'J_2'' | J_1(J_2'J_2'')J_2\rangle^{(J)} \langle (\lambda_1\lambda_2')\lambda'\lambda'' | \lambda_1(\lambda_2'\lambda_2'')\lambda_2\rangle^{[\lambda]}. \qquad (7.73)
\end{aligned}$$

$$\begin{aligned}
&\langle (j_1^{n_1-1}[\lambda_1']\alpha_1'J_1', j_2^{n_2-1}[\lambda_2']\alpha_2'J_2')[\lambda']J', j_1j_2[\lambda'']J''; J | \} j_1^{n_1}[\lambda_1]\alpha_1J_1, j_2^{n_2}[\lambda_2]\alpha_2J_2; [\lambda]J\rangle \\
&\quad = \left\{\frac{f_\lambda f_{\lambda_1'}f_{\lambda_2'}}{f_{\lambda'}f_{\lambda_1}f_{\lambda_2}}\frac{2n_1n_2}{N(N-1)}\right\}^{1/2} \langle j_1^{n_1-1}[\lambda_1']\alpha_1'J_1', j_1; J_1 | \} j_1^{n_1}[\lambda_1]\alpha_1J_1\rangle \\
&\quad\quad \times \langle j_2^{n_1-1}[\lambda_2']\alpha_2'J_2', j_2; J_2 | \} j_2^{n_2}[\lambda_2]\alpha_2J_2\rangle \langle (J_1'J_2')J'(j_1j_2)J'' | (J_1'j_1)J_1(J_2'j_2)J_2\rangle^{(J)} \\
&\quad\quad \times \langle (\lambda_1'\lambda_2')\lambda'\lambda'' | P_1^{(N-1)} | (\lambda_1'1)\lambda_1, (\lambda_2'1)\lambda_2\rangle^{[\lambda]}, \qquad (7.74)
\end{aligned}$$

where $P_1^{N-1\,N}$ is a permutation which transfers electrons $N-1$ and N to the first shell, and $P_1^{(N-1)}$ a permutation which transfers just the $(N-1)$th electron to the first shell. The effect of both these permutations is to preserve within each shell the ascending order in which the electrons are numbered. It is not difficult to show that

$$P_1^{N-1\,N} = \begin{cases} P_{N\,N-2\cdots n_1-1\,N-1\,N-3\cdots n_1}, & n_2 \text{ odd}, \\ P_{N\,N-2\cdots n_1}P_{N-1\,N-3\cdots n_1-1}, & n_2 \text{ even}, \end{cases} \tag{7.75}$$
$$P_1^{(N-1)} = P_{N-1\,N-2\cdots n_1}.$$

The transformation matrices for the group $\mathbf{R}_3$ which appear in Eqs. (7.72) and (7.73) can be expressed in terms of either Racah coefficients or 6-*j* symbols. The transformation matrix which occurs in (7.74) connects functions with different coupling schemes for four angular momenta and can be expressed in terms of 9-*j* symbols. The matrices of the group $\boldsymbol{\pi}_N$ which appear in Eqs. (7.72)–(7.74) are calculated by the methods described in Part 3 of Chapter II. Kaplan (1962b) gives the matrices of the permutation group which are needed for calculating the coefficients of fractional parentage (7.72) –(7.73) in systems with 3–6 electrons. This reference also gives tables of matrices for the more general type of symmetry which occurs in nuclear states.‡

7.5. An Arbitrary Multishell Configuration

We consider a configuration consisting of k groups of equivalent electrons. We denote this configuration by the letter K,

$$K:\ j_1^{n_1}j_2^{n_2}\cdots j_k^{n_k}, \qquad \sum_{a=1}^{k} n_a = N. \tag{7.76}$$

The occupation numbers n_a are arbitrary within the limits one to $2j_a+1$. Each shell is characterized by the quantum numbers $[\lambda_a]\alpha_a J_a$, and the whole configuration is characterized by a total diagram of symmetry $[\lambda]$, a total angular momentum J, and its projection M. In order to describe a state completely, it is necessary to specify in addition $k-2$ partial Young diagrams $[\lambda_{\text{int}}]$ and the coupling scheme between the shells. We denote this last by the letter A (see Section 3.7). The total set of quantum numbers which characterize the configuration (7.76) will be denoted by

$$\rho[\lambda]\alpha JM$$

for short, where ρ denotes the set of k Young diagrams $[\lambda_a]$ and the $k-2$

‡Horie (1964) has also obtained expressions for the coefficients of fractional parentage of the type $\langle N-m, m|\}N\rangle$ for a two-shell configuration. For $m=1$ or 2 these reduce to formulae (7.69)–(7.70) and (7.72)–(7.74), respectively. Later developments in the fractional parentage technique may be found in the works of Smirnov *et al.* (1968), Kurdyumov *et al.* (1969), Kukulin *et al.* (1967), and Kramer (1967, 1968). Kukulin *et al.* (1967) and Kramer (1968) demonstrate the utility of the algebra of the group of unitary transformations in calculations of coefficients of fractional parentage for multishell configurations. In particular, they demonstrate the use of the 6-*f* and 9-*f* symbols (these reduce to the 6-*j* and 9-*j* symbols for the groups $\mathbf{SU}_2$ and $\mathbf{R}_3$ which were described in Section 3.7).

partial diagrams. The letter α has a similar significance with respect to the angular momenta.

The coefficients of fractional parentage for the configuration (7.76) are defined as the coefficients in an expansion which is a generalization of (7.57) and (7.71) for a two-shell configuration.

(a) $\langle N-1, 1 | \} N \rangle$,

$$\Phi(K\rho[\lambda]r\alpha JM) = \sum_{a=1}^{k} \sum_{\rho'\alpha'J'} \Phi(K_a\rho'[\lambda']r'\alpha'J', j_a(N); JM) \times \langle K_a\rho'[\lambda']\alpha'J', j_a; J | \} K\rho[\lambda]\alpha J\rangle. \quad (7.77)$$

(b) $\langle N-2, 2 | \} N \rangle$,

$$\begin{aligned} \Phi(K\rho[\lambda]r'r''\alpha JM) &= \sum_{a=1}^{k} \sum_{\rho',\alpha',J',J_a''} \Phi(K_{aa}\rho'[\lambda']r'\alpha'J', j_a^2(N-1, N)[\lambda'']J_a''; JM) \\ &\quad \times \langle K_{aa}\rho'[\lambda']\alpha'J', j_a^2[\lambda'']J_a''; J | \} K\rho[\lambda]\alpha J\rangle \\ &\quad + \sum_{a<b} \sum_{\rho',\alpha',J',J''} \Phi(K_{ab}\rho'[\lambda']r'\alpha'J', j_a j_b(N-1, N)[\lambda'']J''; JM) \\ &\quad \times \langle K_{ab}\rho'[\lambda']\alpha'J', j_a j_b[\lambda'']J''; J | \} K\rho[\lambda]\alpha J\rangle. \end{aligned} \quad (7.78)$$

The summations in these last two equations are taken over just those symmetry schemes and angular momenta in ρ' and α' that are affected by the decoupling of an electron. There is no summation over $[\lambda']$ or $[\lambda'']$ because they are fixed by the Young tableaux r' and r''. The symbol K_{ab} denotes a configuration in which one electron is detached from the shell $j_a^{n_a}$ and the other from the shell $j_b^{n_b}$. K_{aa} and K_a have analogous meanings.

The derivation of the coefficients of fractional parentage in (7.77) and (7.78) is similar to that given for a configuration with two shells, but is more tedious. The results are as follows (Kaplan, 1961b):

(a) $\langle N-1, 1 | \} N \rangle$,

$$\begin{aligned} &\langle K_a\rho'[\lambda']\alpha'J', j_a; J | \} K\rho[\lambda]\alpha J\rangle \\ &= \left\{\frac{f_\lambda f_{\lambda_a'}}{f_{\lambda'} f_{\lambda_a}} \frac{n_a}{N}\right\}^{1/2} \langle j_a^{n_a-1}[\lambda_a']\alpha_a'J_a', j_a; J_a | \} j_a^{n_a}[\lambda_a]\alpha_a J_a\rangle \\ &\quad \times \langle (J_1 \cdots J_a' \cdots J_k)^{A'} J' j_a | (J_1 \cdots (J_a' j_a) J_a \cdots J_k)^{A}\rangle^{(J)} \\ &\quad \times \langle (\lambda_1 \cdots \lambda_a' \cdots \lambda_k)^{A'} \lambda' 1 | P_a^{(N)} | (\lambda_1 \cdots (\lambda_a' 1)\lambda_a \cdots \lambda_k)^{A}\rangle^{[\lambda]}. \end{aligned} \quad (7.79)$$

(b) $\langle N-2, 2 | \} N \rangle$,

$$\begin{aligned} &\langle K_{aa}\rho'[\lambda']\alpha'J', j_a^2[\lambda'']J_a''; J | \} K\rho[\lambda]\alpha J\rangle \\ &= \left\{\frac{f_\lambda f_{\lambda_a'}}{f_{\lambda'} f_{\lambda_a}} \frac{n_a(n_a-1)}{N(N-1)}\right\}^{1/2} \langle j_a^{n_a-2}[\lambda_a']\alpha_a'J_a', j_a^2[\lambda'']J_a''; J_a | \} j_a^{n_a}[\lambda_a]\alpha_a J_a\rangle \\ &\quad \times \langle (J_1 \cdots J_a' \cdots J_k)^{A'} J' J_a'' | (J_1 \cdots (J_a' J_a'') J_a \cdots J_k)^{A}\rangle^{(J)} \\ &\quad \times \langle (\lambda_1 \cdots \lambda_a' \cdots \lambda_k)^{A'} \lambda' \lambda'' | P_{aa}^{N-1\,N} | (\lambda_1 \cdots (\lambda_a' \lambda'')\lambda_a \cdots \lambda_k)^{A}\rangle^{[\lambda]} \end{aligned} \quad (7.80)$$

$$\begin{aligned}
&\langle K_{ab}\rho'[\lambda']\alpha'J', j_a j_b[\lambda'']J''; J|\}K\rho[\lambda]\alpha J\rangle \\
&= \left\{\frac{f_\lambda f_{\lambda_a'} f_{\lambda_b'}}{f_{\lambda'} f_{\lambda_a} f_{\lambda_b}} \frac{2n_a n_b}{N(N-1)}\right\}^{1/2} \\
&\times \langle j_a^{n_a-1}[\lambda_a']\alpha_a'J_a', j_a; J_a|\}j_a^{n_a}[\lambda_a]\alpha_a J_a\rangle \\
&\times \langle j_b^{n_b-1}[\lambda_b']\alpha_b'J_b', j_b; J_b|\}j_b^{n_b}[\lambda_b]\alpha_b J_b\rangle \\
&\times \langle (J_1 \cdots J_a' \cdots J_b' \cdots J_k)^{A'} J'(j_a j_b) J'' \,|\, (J_1 \cdots (J_a' j_a) J_a \cdots (J_b' j_b) J_b \cdots J_k)^{A}\rangle^{(J)} \\
&\times \langle (\lambda_1 \cdots \lambda_a' \cdots \lambda_b' \cdots \lambda_k)^{A'} \lambda' \lambda'' \,|\, P_{ab}^{N-1\,N} \,|\, (\lambda_1 \cdots (\lambda_a' 1)\lambda_a \cdots (\lambda_b' 1)\lambda_b \cdots \lambda_k)^{A}\rangle^{[\lambda]}.
\end{aligned} \tag{7.81}$$

All permutations in these last three equations preserve the ascending order in which the electrons are numbered within each shell. $P_a^{(N)}$ transfers the Nth electron to the last position in shell $j_a^{n_a}$, this position being denoted by $m_a = \sum_{c=1}^{a} n_c$. $P_{aa}^{N-1\,N}$ transfers electrons $N-1$ and N to the $j_a^{n_a}$ shell, while $P_{ab}^{N-1\,N}$ transfers electron $N-1$ to shell $j_a^{n_a}$ and electron N to $j_b^{n_b}$ ($a < b$). These permutations can be expressed in terms of cycles as follows:

$$\begin{aligned}
P_a^{(N)} &= P_{N\,N-1\cdots m_a}, \\
P_{aa}^{N-1\,N} &= \begin{cases} P_{N\,N-2\cdots m_a-1\;N-1\;N-3\cdots m_a}, & N-m_a \quad \text{odd}, \\ P_{N\,N-2\cdots m_a} P_{N-1\;N-3\cdots m_a-1}, & N-m_a \quad \text{even}, \end{cases} \\
P_{ab}^{N-1\,N} &= \begin{cases} P_{N\,N-2\cdots m_b-1\;m_b-2\cdots m_a N-1\;N-3\cdots m_b}, & N-m_b \quad \text{odd}, \\ P_{N\,N-2\cdots m_b} P_{N-1\;N-3\cdots m_b-1\;m_b-2\cdots m_a}, & N-m_b \quad \text{even}. \end{cases}
\end{aligned} \tag{7.82}$$

The permutation matrices which occur in expressions (7.79)–(7.81) for the coefficients of fractional parentage are calculated in the standard representation. The transformation matrices for the permutation group are calculated according to the rules given in Part 3 of Chapter II, while the transformation matrices for the group $\mathbf{R}_3$ are expressed in terms of $3n$-j symbols (see Section 3.7).

When each of the shells is singly occupied (all $n_a = 1$) expressions (7.79)–(7.81) become identical to the formulae derived by Jahn (1954). The permutation matrices for antisymmetric states $[\lambda] = [1^N]$ are one dimentional and equal to ± 1. Equations (7.79)–(7.81) then reduce to corresponding formulae due to Levinson (1957).

7.6. Formulae for the Matrix Elements of Symmetric Spin-Independent Operators

We consider an arbitrary configuration (7.76) in an L–S-coupled state. The angular momenta j_a then become the orbital angular momenta of the electrons l_a. An antisymmetric total wavefunction is constructed similarly to (6.2) from products of a coordinate wavefunction with a spin wavefunction, each function being symmetrized according to mutually dual Young diagrams,

$$\Psi(K\rho[\lambda]\alpha LM) = (f_\lambda)^{-1/2} \sum_r \Phi(K\rho[\lambda]r\alpha LM)\Omega(s^N[\tilde{\lambda}]\tilde{r}). \tag{7.83}$$

In order to calculate matrix elements of operators which do not contain any spin parameters, it is sufficient to know just the coordinate wavefunction. We now derive formulae for matrix elements of the operators F and G [see Eq. (7.23)] under the assumption that they do not depend upon spin coordinates.

a. The operator F. In view of the indistinguishability of electrons, an operator F can be replaced by Nf_N in a matrix element, where f_N acts upon the Nth electron only. Since f_N does not depend upon spin coordinates, we obtain, taking into account the orthonormality of the spin wavefunctions,

$$\begin{aligned}&\langle K\rho[\lambda]\alpha LM \,|\, F \,|\, K\bar{\rho}[\lambda]\bar{\alpha}\bar{L}\bar{M}\rangle \\ &\quad = N(1/f_\lambda) \sum_r \langle K\rho[\lambda]r\alpha LM \,|\, f_N \,|\, K\bar{\rho}[\lambda]r\bar{\alpha}\bar{L}\bar{M}\rangle.\end{aligned} \tag{7.84}$$

We now expand the coordinate wavefunction by formula (7.77). The sum over r in (7.84) is equivalent to a sum over $[\lambda']r'$, where $[\lambda']$ runs over all the possible Young diagrams that can be obtained from $[\lambda]$ by removing one cell. By taking into account the orthogonality of the basis functions in (7.77) and the fact that the fractional parentage coefficients do not depend upon the tableaux r', we obtain the following formula:

$$\begin{aligned}&\langle K\rho[\lambda]\alpha LM \,|\, F \,|\, K\bar{\rho}[\lambda]\bar{\alpha}\bar{L}\bar{M}\rangle \\ &\quad = (N/f_\lambda) \sum_{a=1}^{k} \sum_{\rho',\lambda'} \sum_{\alpha',L'} f_{\lambda'} \langle K_a\rho'[\lambda']\alpha'L', l_a; L | \} K\rho[\lambda]\alpha L\rangle \\ &\qquad \times \langle K_a\rho'[\lambda']\alpha'L', l_a(N); LM \,|\, f_N \,|\, K_a\rho'[\lambda']\alpha'L', l_a(N); \bar{L}\bar{M})\rangle \\ &\qquad \times \langle K_a\rho'[\lambda']\alpha'L', l_a; L | \} K\bar{\rho}[\lambda]\bar{\alpha}\bar{L}\rangle.\end{aligned} \tag{7.85}$$

In a similar way we obtain for the nonzero matrix elements that are nondiagonal in the configuration

$$\begin{aligned}&\langle K\rho[\lambda]\alpha LM \,|\, F \,|\, K_a{}^b\bar{\rho}[\lambda]\bar{\alpha}\bar{L}\bar{M}\rangle \\ &\quad = (N/f_\lambda) \sum_{\rho',\lambda'} \sum_{\alpha',L'} f_{\lambda'} \langle K_a\rho'[\lambda']\alpha'L', l_a; L | \} K\rho[\lambda]\alpha L\rangle \\ &\qquad \times \langle K_a\rho'[\lambda']\alpha'L', l_a(N); LM \,|\, f_N \,|\, K_a\rho'[\lambda']\alpha'L', l_b(N); \bar{L}\bar{M}\rangle \\ &\qquad \times \langle K_a\rho'[\lambda']\alpha'L', l_b; L | \} K_a{}^b\bar{\rho}[\lambda]\bar{\alpha}\bar{L}\rangle,\end{aligned} \tag{7.86}$$

where $K_a{}^b$ denotes a configuration which is obtained from K by removing an electron from the shell $l_a^{n_a}$ and adding it to the shell $l_b^{n_b}$ ($l_b^{n_b}$ need not even occur in the configuration K, i.e., $n_b = 0$).

The matrix elements of f_N in (7.85) and (7.86) are calculated with the aid of tensor operator techniques (see Section 7.2).

b. The operator G. We write the total wavefunction as in (7.83), but choose a representation which is reduced with respect to the subgroup $\pi_{N-2} \times \pi_2$ instead of the standard representation,

$$\Psi(K\rho[\lambda]\alpha LM) = (f_\lambda)^{-1/2} \sum_{\lambda',r',\lambda''} \Phi(K\rho[\lambda]r'r''\alpha LM)\Omega(s^N[\tilde{\lambda}]\tilde{r}'\tilde{r}''). \tag{7.87}$$

The summation over λ', r', and λ'' in this equation is equivalent to a summation over all the tableaux r of the representation $\Gamma^{[\lambda]}$. The summation over the $[\lambda']$ is over all the diagrams which are obtained from $[\lambda]$ by removing two cells, the diagram $[\lambda'']$ being one of two types only: [2] or $[1^2]$.

We take the operator G to be a scalar and consequently its matrix elements must be diagonal in the angular momenta and independent of their projections. When G is replaced by $\frac{1}{2}N(N-1)g_{N-1\,N}$, the orthonormality of the spin wavefunctions leads to the following expression for a matrix element:

$$\begin{aligned}&\langle K\rho[\lambda]\alpha L\,|\,G\,|\,K\bar{\rho}[\lambda]\bar{\alpha}L\rangle\\ &\quad = \tfrac{1}{2}N(N-1)(1/f_\lambda) \sum_{\lambda',r',\lambda''} \langle K\rho[\lambda]r'r''\alpha L\,|\,g_{N-1\,N}\,|\,K\bar{\rho}[\lambda]r'r''\bar{\alpha}L\rangle.\end{aligned} \tag{7.88}$$

We now substitute the expansion of the coordinate wavefunction (7.78) into the right-hand side of this equation. The operator $g_{N-1\,N}$ acts only upon the coordinates of the $(N-1)$th and Nth electrons. Since electrons $N-1$ and N are decoupled in the basis functions of (7.78), the scalarity of $g_{N-1\,N}$ [see Eq. (4.81)] and the fact that the coefficients of fractional parentage are independent of the Young tableaux r' lead to the desired formula:

$$\begin{aligned}&\langle K\rho[\lambda]\alpha L\,|\,G\,|\,K\bar{\rho}[\lambda]\bar{\alpha}L\rangle\\ &\quad = \frac{N(N-1)}{2}\frac{1}{f_\lambda}\sum_{a\le b}^{k}\ \sum_{\rho',\lambda',\lambda''}\ \sum_{\alpha',L',L''} f_{\lambda'}g^{[\lambda'']}_{ab,ab}(L'')\\ &\qquad \times \langle K_{ab}\rho'[\lambda']\alpha'L', l_a l_b[\lambda'']L''; L\,|\!\}\, K\rho[\lambda]\alpha L\rangle\\ &\qquad \times \langle K_{ab}\rho'[\lambda']\alpha'L', l_a l_b[\lambda'']L''; L\,|\!\}\, K\bar{\rho}[\lambda]\bar{\alpha}L\rangle.\end{aligned} \tag{7.89}$$

The coefficients of fractional parentage which occur in this equation are determined by formula (7.80) for $a = b$ and by formula (7.81) when $a \neq b$. The symbol $g^{[\lambda'']}_{ab,ab}(L'')$ denotes a two-electron matrix element of the operator g,

$$g^{[\lambda'']}_{ab,ab}(L'') = \langle l_a l_b[\lambda'']L''\,|\,g\,|\,l_a l_b[\lambda'']L''\rangle. \tag{7.90}$$

Matrix elements that are off-diagonal in the configuration are nonvanishing only if the configurations do not differ from one another by more than two orbitals. In a similar manner to the notation $K_a{}^b$ introduced previously, we denote by K^{cd}_{ab} a configuration which is obtained from the configuration K by transferring two electrons from the shells $l_a^{n_a}$ and $l_b^{n_b}$ to the shells $l_c^{n_c}$ and $l_d^{n_d}$, respectively. By expanding the coordinate wavefunction according to formula (7.78) and making use of the orthogonality of the basis functions in the expansion, we arrive at the following formulae for matrix elements of a scalar operator that are nondiagonal in the configuration:

$$\langle K\rho[\lambda]\alpha L|G|K_a^{\;b}\bar{\rho}[\lambda]\bar{\alpha}L\rangle$$
$$= \frac{N(N-1)}{2}\frac{1}{f_\lambda}\sum_{c=1}^{k}\sum_{\rho',\lambda',\lambda''}\sum_{\alpha',L',L''} f_{\lambda'}g^{[\lambda'']}_{ac,bc}(L'')$$
$$\times \langle K_{ac}\rho'[\lambda']\alpha'L', l_al_c[\lambda'']L''; L|\} K\rho[\lambda]\alpha L\rangle$$
$$\times \langle K_{ac}\rho'[\lambda']\alpha'L', l_bl_c[\lambda'']L''; L|\} K\bar{\rho}[\lambda]\bar{\alpha}L\rangle, \tag{7.91}$$

$$\langle K\rho[\lambda]\alpha L|G|K_{ab}^{cd}\bar{\rho}[\lambda]\bar{\alpha}L\rangle$$
$$= \frac{N(N-1)}{2}\frac{1}{f_\lambda}\sum_{\rho',\lambda',\lambda''}\sum_{\alpha',L',L''} f_{\lambda'}g^{[\lambda'']}_{ab,cd}(L'')$$
$$\times \langle K_{ab}\rho'[\lambda']\alpha'L', l_al_b[\lambda'']L''; L|\} K\rho[\lambda]\alpha L\rangle$$
$$\times \langle K_{ab}\rho'[\lambda']\alpha'L', l_cl_d[\lambda'']L''; L|\} K_{ab}^{cd}\bar{\rho}[\lambda]\bar{\alpha}L\rangle. \tag{7.92}$$

If one makes use of property (3.47b) of the Clebsch–Gordan coefficients, it is not difficult to show that the two-electron matrix elements which occur in these last two equations are equal to

$$g^{[\lambda'']}_{ab\;cd}(L'') = \begin{cases} g_{ab,cd}(L'') + (-1)^{l_c+l_d-L''}g_{ab,dc}(L'') & \text{for } [\lambda''] = [2], \\ g_{ab,cd}(L'') - (-1)^{l_c+l_d-L''}g_{ab,dc}(L'') & \text{for } [\lambda''] = [1^2], \end{cases} \tag{7.93}$$

where $g_{ab,cd}(L'')$ denotes a matrix element evaluated over unsymmetrized functions of coupled angular momenta, these corresponding to a unique ordering of the electrons on both sides of the matrix element, i.e.,

$$g_{ab,cd}(L'') = \langle l_a(N-1)l_b(N)|g_{N-1\,N}|l_c(N-1)l_d(N)\rangle^{(L'')}. \tag{7.94}$$

Methods of calculating such a matrix element for electrostatic interactions between electrons can be found in the books by Judd (1963) and Sobel'man (1972).

PART 3. Non-Vector-Coupled States

7.7. Configuration of Singly Occupied Orbitals

When a system of electrons is placed in a field of arbitrary symmetry the angular momentum vector ceases to be a good quantum number and the concept of coupling together angular momenta loses its significance. The state of each electron, apart from its spin, is described by specifying a one-electron orbital $\phi_a(i)$. We denote by K an arbitrary configuration of k one-electron orbitals,

$$K: \qquad \varphi_1^{n_1}\varphi_2^{n_2}\cdots\varphi_k^{n_k}, \qquad \sum_{a=1}^{k} n_a = N. \tag{7.95}$$

We denote the configuration which is obtained from K by removing orbital ϕ_a by K_a, the notation K_{ab} possessing an analogous meaning. Let the orbitals in (7.95) be nondegenerate, and let them form an orthonormal set. According to the Pauli principle, not more than two electrons can be found in each orbital, i.e., $n_a \leq 2$.‡

We consider the case in which each orbital is singly occupied. A normalized coordinate wavefunction which describes a state of permutational symmetry $[\lambda]$ can then be constructed by applying the operators $\omega_{rt}^{[\lambda]}$ to an unsymmetrized product of one-electron orbitals,

$$\Phi_{rt}^{[\lambda]}(K) = \omega_{rt}^{[\lambda]}\Phi_0(K) = (f_\lambda/N!)^{1/2} \sum_P \Gamma_{rt}^{[\lambda]}(P)P\Phi_0(K), \tag{7.96}$$

where P runs through all the $N!$ permutations of the group π_N and $\Phi_0(K)$ denotes the product of orbitals

$$\Phi_0(K) = \phi_1(1)\phi_2(2)\cdots\phi_N(N). \tag{7.97}$$

The Young tableau r characterizes the symmetry of the function (7.96) with respect to permutations of the electronic coordinates, while the tableau t characterizes its symmetry with respect to permutations of the orbitals. It is possible to construct f_λ distinct states with permutational symmetry $[\lambda]$. These are distinguished by the form of the tableau t, the specification of which is equivalent to specifying $N-2$ partial Young diagrams for the groups $\pi_{N-1}, \pi_{N-2}, \ldots$, and π_2. Consequently, tableau t now plays the role of the symbol ρ in Section 7.5, which there denoted the set of k diagrams $[\lambda_a]$ and $k-2$ partial diagrams.

The coefficients of fractional parentage for singly occupied orbitals are formally defined by the following expansions.

(a) $\langle N-1, 1 |\} N\rangle$,

$$\Phi_{rt}^{[\lambda]}(K) = \sum_{a=1}^{N} \sum_{u'} \Phi_{r'u'}^{[\lambda']}(K_a)\phi_a(N)\langle K_a[\lambda']u', \phi_a |\} K[\lambda]t\rangle. \tag{7.98}$$

(b) $\langle N-2, 2 |\} N\rangle$,

$$\Phi_{r'r'',t}^{[\lambda]}(K) = \sum_{a<b} \sum_{u'} \Phi_{r'u'}^{[\lambda']}(K_{ab})\Phi^{[\lambda'']}(\phi_a\phi_b \mid N-1, N)$$
$$\times \langle K_{ab}[\lambda']u', \phi_a\phi_b[\lambda''] |\} K[\lambda]t\rangle. \tag{7.99}$$

The summation over u' is taken over all tableaux of the representation $\Gamma^{[\lambda']}$. The choice of a representation in (7.99) which is reduced with respect to the subgroup $\pi_{N-2} \times \pi_2$ is equivalent to specifying the Young diagrams $[\lambda']$ and $[\lambda'']$, and hence there is no summation over these.

Expressions for the coefficients in expansions (7.98) and (7.99) can be obtained from the general formulae of Section 7.5 if one discards all the factors there which refer to the coupling of angular momenta. However, in order to gain a clearer understanding of the present situation, we derive explicit

‡Coefficients of fractional parentage for uncoupled nuclear states are considered by Neudachin and Smirnov (1959).

expressions for the coefficients of fractional parentage, after having first of all carried out a suitable transformation of expression (7.96).

In order to derive the $\langle N-1, 1 |\} N\rangle$ coefficients of fractional parentage, we divide the set of $N!$ permutations in (7.96) into N groups according to the position into which they send the Nth electron. Each of these groups consists of (N − 1)! permutations which can always be written in the form $Q(N-1)P_a^{(N)}$, where the permutation $P_a^{(N)}$ transfers the Nth electron to orbital ϕ_a, and $Q(N-1)$ does not act upon the coordinates of the Nth electron, i.e., $Q(N-1) \in \pi_{N-1}$. It is not imperative to choose as the $P_a^{(N)}$ the permutations $P_{N\,N-1\cdots a}$ which preserve the increasing order in which the electrons are numbered (apart from the Nth) as in (7.82), and one might choose the elementary transpositions P_{aN}.

By making use of the properties of the standard representation, a matrix element of a permutation $P \equiv Q(N-1)P_a^{(N)}$ can be written in the form

$$\begin{aligned}\Gamma_{rt}^{[\lambda]}(P) &\equiv \langle r | Q(N-1)P_a^{(N)} | t\rangle^{[\lambda]} \\ &= \sum_u \langle r | Q(N-1) u\rangle^{[\lambda]} \langle u | P_a^{(N)} | t\rangle^{[\lambda]} \\ &= \sum_{u'} \langle r' | Q(N-1) | u'\rangle^{[\lambda']} \langle u'1 | \, P_a^{(N)} | t\rangle^{[\lambda]}.\end{aligned} \tag{7.100}$$

Since

$$\begin{aligned}&\sum_{Q(N-1)} \langle r' | Q(N-1) | u'\rangle^{[\lambda']} Q(N-1)\, P_a^{[N]} \Phi_0(K) \\ &\qquad = [f'_\lambda/(N-1)!]^{-1/2} \Phi_{r'u'}^{[\lambda']}(K_a)\phi_a(N),\end{aligned} \tag{7.101}$$

substitution of (7.100) and (7.101) into (7.96) makes it possible to write this last equation in the form of expansion (7.98), i.e.,

$$\Phi_{rt}^{[\lambda]}(K) = \left(\frac{f_\lambda}{f_{\lambda'}} \frac{1}{N}\right)^{1/2} \sum_{a=1}^{N} \sum_{u'} \langle u'1 | P_a^{(N)} | t\rangle^{[\lambda]} \Phi_{r'u'}^{[\lambda']}(K_a)\varphi_a(N). \tag{7.102}$$

Equating this expression with Eq. (7.98) yields an explicit expression for the coefficients of fractional parentage:

$$\langle K_a[\lambda']u', \phi_a |\} K[\lambda]t\rangle = \left(\frac{f_\lambda}{f_{\lambda'}} \frac{1}{N}\right)^{1/2} \langle u'1 | P_a^{(N)} | t\rangle^{[\lambda]}. \tag{7.103}$$

In order to derive the $\langle N-2, 2 |\} N\rangle$ coefficients of fractional parentage, we take function (7.96) as our starting point where instead of a tableau r we specify a Young tableau r' for the first $N-2$ electrons and a tableau r'' for the last two electrons. The permutations P can be divided into $N(N-1)$ sets according to the positions into which they send electrons $N-1$ and N. Each of these sets consists of $(N-2)!$ permutations. A straightforward derivation leads to the following expression for the coefficients of fractional parentage:

$$\langle K_{ab}[\lambda']u', \phi_a\phi_b[\lambda''] |\} K[\lambda]t\rangle = \left\{\frac{f_\lambda}{f_{\lambda'}} \frac{2}{(N-1)N}\right\}^{1/2} \langle u'\lambda'' | P_{ab}^{N-1\,N} | t\rangle^{[\lambda]}. \tag{7.104}$$

Any permutations which transfer electron $N-1$ to the orbital ϕ_a and electron N to ϕ_b can be taken as the $P_{ab}^{N-1\,N}$. In particular, one might choose $P_{ab}^{N-1\,N} = P_{a\,N-1}P_{bN}$. The matrix of $P_{ab}^{N-1\,N}$ which occurs in (7.104) can be expressed in terms of the corresponding matrix in the standard representation if this last is multiplied on the left by the transformation matrix $\langle u'\lambda'' \mid u'\rho_2\rangle^{[\lambda]}$. The transformation matrix itself contains either ones on the diagonal or blocks of the form (2.65). Appendix 6 contains tables of matrices $\langle r'\lambda'' | P_{ab}^{N-1\,N} | t\rangle^{[\lambda]}$ for all the allowed Young diagrams $[\lambda]$ in configurations consisting of 3–6 electrons.

7.8. Arbitrarily Occupied Orbitals

In the general case the n_a in configuration (7.95) can assume one of two values: one or two. Each doubly occupied orbital is characterized by a Young diagram with two cells [2]. We use the symbol ρ to denote both the set of Young diagrams for the doubly occupied orbitals and the set of partial diagrams necessary to characterize a state uniquely. Since the representations which characterize the electrons in a single orbital are one dimensional, the number of independent sets of ρ for a given $[\lambda]$ is equal to the dimension f_λ of the irreducible representation $\Gamma^{[\lambda]}$. Not all of these sets can occur, however, since doubly filled orbitals may not have $[\lambda] = [1^2]$. The number of independent states is given by formula (6.70).

The coefficients of fractional parentage for the configuration (7.95) are a particular case of the coefficients of fractional parentage which were derived in Section 7.5. They differ by the absence of vector coupling in the states which are now considered. Instead of (7.78) and (7.79), we have the following expansions.

(a) $\langle N-1, 1 |\} N\rangle$,

$$\Phi(K\rho[\lambda]r) = \sum_{a=1}^{k} \sum_{\rho'} \Phi(K_a\rho'[\lambda']r')\phi_a(N)\langle K_a\rho'[\lambda'], \phi_a |\} K\rho[\lambda]\rangle. \qquad (7.105)$$

(b) $\langle N-2, 2 |\} N\rangle$,

$$\begin{aligned}\Phi(K\rho[\lambda]r'r'') = {} & \sum_{a}{}' \sum_{\rho'} \Phi(K_{aa}\rho'[\lambda']r')\phi_a(N-1)\phi_a(N) \\ & \times \langle K_{aa}\rho'[\lambda'], \phi_a{}^2 |\} K\rho[\lambda]\rangle \\ & + \sum_{a<b} \sum_{\rho'} \Phi(K_{ab}\rho'[\lambda']r')\Phi(\phi_a\phi_b[\lambda''] | N-1, N) \\ & \times \langle K_{ab}\rho'[\lambda'], \phi_a\phi_b[\lambda''] |\} K\rho[\lambda]\rangle. \qquad (7.106)\end{aligned}$$

The summation over ρ' is taken over all remaining independent sets of configurations after one or two electrons have been removed. The first term of (7.106) occurs only in expansions of functions with $r'' = [2]$. The prime on the sum over a means that it is taken over doubly occupied orbitals only.

Expressions for the coefficients in expansions (7.105) and (7.106) can be obtained from formulae (7.79)–(7.81) when one notes that there is now no vector coupling, and that all $n_a \leq 2$. As a result, we obtain the following expressions.

(a) $\langle N-1, 1 |\} N\rangle$,

$$\langle K_a\rho'[\lambda'], \phi_a |\} K\rho[\lambda]\rangle = \left\{\frac{f_\lambda}{f_{\lambda'}} \frac{n_a}{N}\right\}^{1/2} \langle\rho'\lambda' 1 | P_a^{(N)} | \rho\rangle^{[\lambda]}. \tag{7.107}$$

(b) $\langle N-2, 2 |\} N\rangle$,

$$\langle K_{aa}\rho'[\lambda'], \phi_a{}^2 |\} K\rho[\lambda]\rangle = \left\{\frac{f_\lambda}{f_{\lambda'}} \frac{2}{N(N-1)}\right\}^{1/2} \times \langle\rho'[\lambda'][2] | P_a^{N-1\,N} | \rho\rangle^{[\lambda]} \tag{7.108}$$

$$\langle K_{ab}\rho'[\lambda'], \phi_a\phi_b[\lambda''] |\} K\rho[\lambda]\rangle = \left\{\frac{f_\lambda}{f_{\lambda'}} \frac{2n_a n_b}{N(N-1)}\right\}^{1/2} \times \langle\rho'\lambda'\lambda'' | P_{ab}^{N-1\,N} | \rho\rangle^{[\lambda]}. \tag{7.109}$$

The matrices which occur in these three equations are defined by functions which belong to representations with differing types of reduction with respect to subgroups. They are determined in terms of matrices in the standard representation with the aid of Eq. (2.74). The form of the permutations in the matrix elements in (7.107)–(7.109) are in general given by formula (7.82).

In the following chapter we derive, with the aid of the fractional parentage expansions (7.105) and (7.106), formulae for the matrix elements of spin-independent operators F and G for arbitrary molecular systems.

CHAPTER VIII

Calculation of Electronic States of Molecular Systems

PART 1. The Hydrogen Molecule. Configuration Interaction

8.1. The Valence Bond Method

In 1927 Heitler and London (1927) carried out the first calculation on the hydrogen molecule, a work which was fundamental to the development of the theory of chemical bonding. The Heitler–London, or valence bond, method, as it is usually called in quantum chemistry, was subsequently applied to a number of more complicated molecules.

In their original calculation Heitler and London took the covalent configuration of H_2 as their starting point, using the 1s orbitals of the two hydrogen atoms as the set of basis functions. This proved to be sufficient to explain the nature of the chemical bond in this particular case. We now show how one carries out a calculation on H_2 taking into account all the configurations which arise from these two orbitals. This and all subsequent calculations in this chapter are carried out without making any use of spin wavefunctions.

In Part 3 of Chapter VI it was mentioned that all properties of molecules which do not involve spin interactions can be described purely in terms of coordinate wavefunctions which possess the permutational symmetry of the appropriate Young diagrams. The permutational symmetry carries, as it were, all the information about the spin functions. Furthermore, when there are no spin interaction operators, calculations that use coordinate wavefunctions are more visual and systematic than methods which employ spin functions.

We denote the 1s orbitals of the H atoms in an H_2 molecule by ϕ_a and ϕ_b. Two covalent structures can be constructed from a configuration of neutral atoms: a singlet and a triplet. The coordinate wavefunctions which correspond to these are obtained by applying the Young operators $\omega^{[2]}$ and $\omega^{[1^2]}$ to the product of orbitals $\phi_a(1)\phi_b(2)$. When the nonorthogonality of the orbitals is taken into account we obtain the following two normalized combinations:

$$\Phi^{[2]} = [2(1 + \rho^2)]^{-1/2}[\phi_a(1)\phi_b(2) + \phi_b(1)\phi_a(2)], \tag{8.1}$$

$$\Phi^{[1^2]} = [2(1 - \rho^2)]^{-1/2}[\phi_a(1)\phi_b(2) - \phi_b(1)\phi_a(2)], \tag{8.2}$$

where ρ denotes an overlap integral

$$\rho = \int \phi_a(\mathbf{r})\phi_b(\mathbf{r})\, dV \tag{8.3}$$

(the 1s orbitals are assumed to be real).

The H_2 molecule belongs to the point group $\mathbf{D}_{\infty h}$. Its electronic states are classified according to the irreducible representations of this group, and in particular, the states must possess a definite symmetry under inversion through the molecular center of gravity. Function (8.1) is even with respect to inversion by construction, whereas function (8.2) is odd. Since it is formed from 1s orbitals, the state described by (8.1) is $^1\Sigma_g$ and the state described by (8.2) is $^3\Sigma_u$.‡

The electronic part of the Hamiltonian for the molecule can be written in the form of a sum of three terms:

$$\mathcal{H} = \mathcal{H}_a(1) + \mathcal{H}_b(2) + W(1, 2), \tag{8.4}$$

in which $\mathcal{H}_a(1)$ and $\mathcal{H}_b(2)$ are the Hamiltonians for the isolated atoms, and $W(1, 2)$ is the operator for the interaction between the H atoms. In atomic units,

$$\mathcal{H}_a(1) = -\frac{1}{2}\nabla_1{}^2 - \frac{1}{R_{a1}}, \qquad \mathcal{H}_b(2) = -\frac{1}{2}\nabla_2{}^2 - \frac{1}{R_{b2}},$$
$$W(1, 2) = -\frac{1}{R_{a2}} - \frac{1}{R_{b1}} + \frac{1}{R_{ab}} + \frac{1}{r_{12}}. \tag{8.5}$$

The distances which occur in Eq. (8.5) are shown in Fig. 8.1.

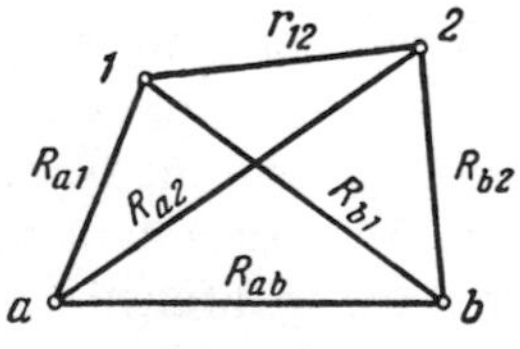

Figure 8.1.

‡The states of molecules which belong to the point groups $\mathbf{D}_{\infty h}$ and $\mathbf{C}_{\infty v}$ are usually classified according to the magnitude of the projection of the total orbital angular momentum of the electrons along the molecular axis. The terms are denoted by different capital Greek letters according to the value of the projection M:

M:	0	1	2
	Σ	Π	Δ

When only covalent configurations are taken into account the problem reduces to solving a second-order secular equation. Now the wavefunctions (8.1) and (8.2) belong to different multiplets and so the secular equation has no nondiagonal elements. The energy of the molecule in the two states (8.1) and (8.2) is then given by the diagonal matrix elements of the Hamiltonian (8.4)‡:

$$E({}^1\Sigma_g) = 2E_{1s}(\mathrm{H}) + \frac{Q + A}{1 + \rho^2}, \tag{8.6}$$

$$E({}^3\Sigma_u) = 2E_{1s}(\mathrm{H}) + \frac{Q - A}{1 - \rho^2}, \tag{8.7}$$

where $E_{1s}(\mathrm{H})$ is the energy of an H atom in a 1s state, and

$$Q = \int \phi_a^2(1)\phi_b^2(2)W(1, 2)\, dV, \tag{8.8}$$

$$A = \int \phi_a(1)\phi_b(1)\phi_a(2)\phi_b(2)W(1, 2)\, dV. \tag{8.9}$$

Q is known as the *Coulomb integral*; it represents the electrostatic interaction energy between two hydrogen atoms. A is called the *exchange integral*, and the energy associated with it the *exchange energy*. The exchange energy has no classical analog; it arises as a consequence of the fact that each electron has an equal probability of being found in one of the states ϕ_a or ϕ_b. Furthermore, the probability of finding the electrons close together is large for the symmetric coordinate function (${}^1\Sigma_g$ state) and small for the antisymmetric function (${}^3\Sigma_u$ state). This causes the energy of a system of electrons to depend upon the total spin. We emphasize that the exchange energy is the part of the ordinary electrostatic interaction energy due to quantal correlation in the electron density distribution.

Heitler and London (1927) evaluated the Coulomb integral, and gave an approximate value for the exchange integral, the exact value of which was

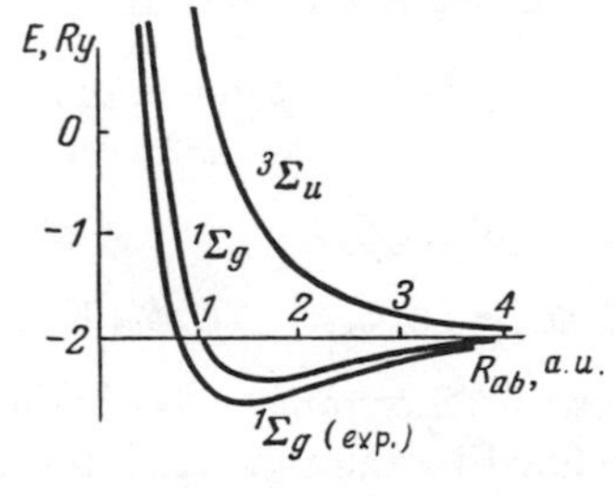

Figure 8.2.

‡We use the relation $\langle\phi_a|\mathcal{H}_a|\phi_b\rangle = E_{1s}(\mathrm{H})\rho$ [see Slater (1963)] in deriving Eqs. (8.6) and (8.7).

obtained later by Sugiura (1927). Values of the integrals Q and A for various separations between the H atoms are tabulated in the book by Slater (1963). In Fig. 8.2 the energies of the ${}^1\Sigma_g$ and ${}^3\Sigma_u$ states are shown as functions of the separation R_{ab} between H atoms. The energies are measured in Rydbergs (Ry). At infinite separation the energy is equal to $2E_{1s}(\mathrm{H})$, or 2 Ry. The singlet-state curve possesses a minimum and corresponds to the ground state of the hydrogen molecule. It agrees qualitatively with the experimental curve, but lies appreciably higher. The binding energy which one obtains from this calculation is 3.14 eV, this being 1.61 eV less than the experimental value. The exchange energy forms the major part of the binding energy, the contribution of the Coulomb energy at the minimum of the potential curve being 5% of the total. This contribution increases to some extent with increasing internuclear distance, but is never greater than 15%.‡

We now consider the ionic configurations. From two 1s orbitals it is possible to construct in all two ionic configurations: $\phi_a{}^2$ and $\phi_b{}^2$. However, in order to obtain functions which have a definite symmetry under an inversion of the molecule, one has to take linear combinations of the two functions. These have the following form:

$$\Phi_2({}^1\Sigma_g) = [2(1+\rho^2)]^{-1/2}[\phi_a(1)\phi_b(2) + \phi_b(1)\phi_b(2)], \tag{8.10}$$

$$\Phi({}^1\Sigma_u) = [2(1-\rho^2)]^{-1/2}[\phi_a(1)\phi_a(2) - \phi_b(1)\phi_b(2)], \tag{8.11}$$

When the interaction between covalent and ionic configurations is considered functions (8.10) and (8.11) are treated on the same footing as functions (8.1) and (8.2). We relabel these last two as $\Phi_1({}^1\Sigma_g) \equiv \Phi^{[2]}$ and $\Phi({}^3\Sigma_u) \equiv \Phi^{[1^2]}$, and we then have four initial variational functions:

$$\Phi_1({}^1\Sigma_g), \qquad \Phi_2({}^1\Sigma_g), \qquad \Phi({}^1\Sigma_u), \qquad \Phi({}^3\Sigma_u). \tag{8.12}$$

The energies of the ${}^1\Sigma_u$ and ${}^3\Sigma_u$ states are determined by diagonal elements of the Hamiltonian. An expression for the diagonal element for the ${}^3\Sigma_u$ state is given by Eq. (8.7). This can be rewritten in the form

$$E({}^1\Sigma_u) = (\gamma - \beta)/(1-\rho^2), \tag{8.13}$$

where

$$\gamma = \langle \phi_a\phi_a | \mathcal{H} | \phi_a\phi_a \rangle, \tag{8.14}$$

$$\beta = \langle \phi_a\phi_a | \mathcal{H} | \phi_b\phi_b \rangle = \langle \phi_a\phi_b | \mathcal{H} | \phi_b\phi_a \rangle. \tag{8.15}$$

The energy of the state with ${}^1\Sigma_g$ symmetry is obtained by solving a second-order secular equation defined by the functions $\Phi_1({}^1\Sigma_g)$ and $\Phi_2({}^1\Sigma_g)$. These

‡The contribution of the Coulomb energy to the total binding energy can be considerably greater than this in other molecules (Fraga and Mulliken, 1960).

functions are not orthogonal to each other, the overlap integral being equal to

$$\int \Phi_1(^1\Sigma_g)\Phi_2(^1\Sigma_g)\,dV = 2\rho/(1-\rho^2). \tag{8.16}$$

In conformity with the notation of (8.14) and (8.15), we write the matrix elements

$$\langle\phi_a\phi_b|\mathcal{H}|\phi_a\phi_b\rangle = \alpha, \tag{8.17}$$

$$\langle\phi_a\phi_b|\mathcal{H}|\phi_a\phi_a\rangle = \langle\phi_a\phi_b|\mathcal{H}|\phi_b\phi_b\rangle = \delta. \tag{8.18}$$

As a result, the secular equation becomes

$$\begin{vmatrix} [(\alpha+\beta)/(1+\rho^2)] - E & [2/(1+\rho^2)](\delta-\rho E) \\ [2/(1+\rho^2)](\delta-\rho E) & [(\gamma+\beta)/(1+\rho^2)] - E \end{vmatrix} = 0.$$

Values of the integrals necessary to calculate the matrix elements α, β, γ, and δ are given in Slater's book (1963).

Figure 8.3 shows the term energies of H_2 which are obtained when configuration interaction is taken into account. The energy of the $^3\Sigma_u$ term is the same as that in the Heitler–London method. There is now a new $^1\Sigma_u$ term,

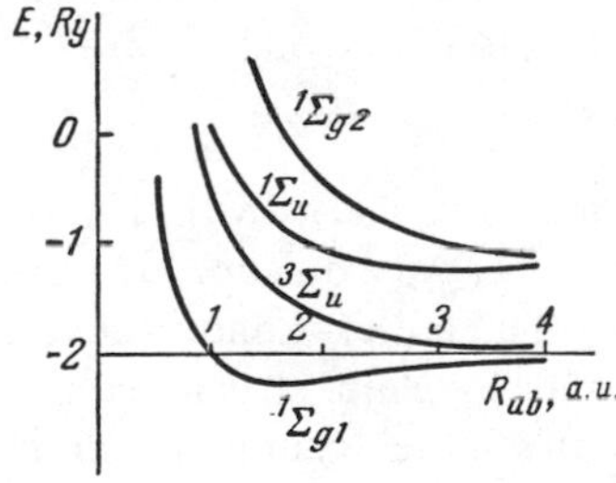

Figure 8.3.

and there are two $^1\Sigma_g$ terms. However, the lowering in the energy of the ground state as a result of taking the ionic structures into account is very small; one obtains a binding energy of 3.21 eV (Weinbaum, 1933). In order to obtain better agreement with experiment, it is imperative to introduce additional variational parameters into the wavefunctions, particularly terms which depend explicitly upon the interelectronic separations. Detailed discussions of these problems can be found in the work of Slater (1963) and Kauzmann (1957).

8.2. The Molecular Orbital Method

This method was proposed by Hund (1927) at the same time as Heitler and London proposed their method. A number of improvements were subsequently made, culminating in the self-consistent-field procedure (Roothaan,

1951; Parr, 1963). In most variants of the molecular orbital method the orbitals are represented in the form of a linear combination of atomic orbitals, the coefficients of which are determined by the requirement that the energy of the system be a minimum. In symmetric molecules the coefficients of such linear combinations can often be obtained for small basis sets by requiring that the molecular orbitals belong to bases for irreducible representations of the molecular symmetry group.

In the case of the H_2 molecule one can construct two molecular orbitals from two atomic 1s orbitals. One of the molecular orbitals is even with respect to inversion (σ_g) and the other is odd (σ_u):

$$\sigma_g = [2(1+\rho)]^{-1/2}(\phi_a + \phi_b), \qquad \sigma_u = [2(1-\rho)]^{-1/2}(\phi_a - \phi_b). \tag{8.19}$$

The energy of a one-electron configuration with a σ_g orbital is considerably lower than that of a configuration with a σ_u orbital. The ground state of H_2 therefore corresponds to the configuration $\sigma_g{}^2$. The coordinate wavefunction for this configuration describes a ${}^1\Sigma_g$ state. It is represented in terms of atomic orbitals as follows:

$$\Phi(\sigma_g{}^2, {}^1\Sigma_g) = \frac{1}{2(1+\rho)}[\phi_a(1)\phi_a(2) + \phi_b(1)\phi_b(2) + \phi_a(1)\phi_b(2) + \phi_b(1)\phi_a(2)]. \tag{8.20}$$

A calculation of the ground-state energy with this wavefunction yields a value of 2.681 eV for the binding energy (Coulson, 1937), which is much worse than the value 3.14 eV given by the Heitler–London covalent structure. This is due to the fact that ionic and covalent structures occur in (8.20) with equal weights. In order to obtain a more accurate result, it is necessary to take into account, in addition to $\sigma_g{}^2$, other configurations which may contribute to the ground state.

From two molecular orbitals one can construct three configurations,

$$\sigma_g{}^2, \qquad \sigma_u{}^2, \qquad \sigma_g\sigma_u. \tag{8.21}$$

The first two configurations correspond to a singlet state. The electrons in the last configuration may occur in either singlet or a triplet states. The coordinate wavefunctions for these last two states are found by applying the operators $\omega^{[2]}$ and $\omega^{[1^2]}$ to the product $\sigma_g(1)\sigma_u(2)$. The coordinate function for the $\sigma_g{}^2$ configuration is given in (8.20). The remaining three coordinate functions have the following form:

$$\begin{aligned}
\Phi(\sigma_u{}^2, {}^1\Sigma_g) &= [2(1-\rho)]^{-1}[\phi_a(1)\phi_a(2) + \phi_b(1)\phi_b(2) \\
&\qquad - \phi_a(1)\phi_b(2) - \phi_b(1)\phi_a(2)], \\
\Phi(\sigma_g\sigma_u, {}^1\Sigma_u) &= [2(1-\rho^2)]^{-1/2}[\phi_a(1)\phi_a(2) - \phi_b(1)\phi_b(2)], \\
\Phi(\sigma_g\sigma_u, {}^3\Sigma_u) &= [2(1-\rho^2)]^{-1/2}[\phi_b(1)\phi_a(2) - \phi_a(1)\phi_b(2)].
\end{aligned} \tag{8.22}$$

When these functions are equated with those obtained by the valence bond method we see that they are linear combinations of one another. Thus,

$$\begin{aligned}\Phi^{\mathrm{MO}}(\sigma_g^{\,2}, {}^1\Sigma_g) &= [(1+\rho^2)/2(1+\rho)^2]^{1/2}[\Phi_2^{\mathrm{VB}}({}^1\Sigma_g) + \Phi_1^{\mathrm{VB}}({}^1\Sigma_g)],\\ \Phi^{\mathrm{MO}}(\sigma_u^{\,2}, {}^1\Sigma_g) &= [(1-\rho^2)/2(1-\rho)^2]^{1/2}[\Phi_2^{\mathrm{VB}}({}^1\Sigma_g) - \Phi_1^{\mathrm{VB}}({}^1\Sigma_g)],\\ \Phi^{\mathrm{MO}}(\sigma_g\sigma_u, {}^1\Sigma_u) &= \Phi^{\mathrm{VB}}({}^1\Sigma_u),\\ \Phi^{\mathrm{MO}}(\sigma_g\sigma_u, {}^3\Sigma_u) &= -\Phi^{\mathrm{VB}}({}^3\Sigma_u).\end{aligned} \tag{8.23}$$

Since the roots of a secular equation are invariant with respect to a linear transformation of the basis functions, a calculation of the electronic states of the hydrogen molecule by the molecular orbital method with configuration interaction must be completely equivalent to a calculation by the valence bond method discussed in the previous section. This result can be given a more general formulation:

For a given atomic basis and with complete configuration interaction, calculations by the valence bond or molecular orbital methods are completely equivalent.

8.3. Orthogonal Localized Orbitals

In many-electron systems the nonorthogonality of the one-electron orbitals seriously complicates the calculation of matrix elements of the Hamiltonian. However, it has been shown by Wannier (1937) that for crystals one can construct orthogonal combinations of atomic orbitals which are also localized at particular points of the lattice. A similar procedure can be applied to molecular systems (Slater, 1963).‡ We consider this in the very simple example of the H_2 molecule.

We take the molecular orbitals (8.19) as our starting point. It is not difficult to show that these are orthogonal to one another. We now form from them a symmetric and an antisymmetric combination:

$$\phi_a^{\,0} = (1/\sqrt{2})(\sigma_g + \sigma_u), \qquad \phi_b^{\,0} = (1/\sqrt{2})(\sigma_g - \sigma_u). \tag{8.24}$$

Figure 8.4.

‡The Wannier functions are constructed from an orthonormal set of delocalized molecular orbitals, these last normally being Bloch functions (Slater, 1963). A method of constructing orthogonal localized orbitals from atomic orbitals was developed by Löwdin (1950).

These orbitals are also orthogonal to each other, but in contrast to the orbitals (8.19), they are localized. In order to show this, we substitute expressions (8.19) into (8.24). As a result, we obtain

$$\phi_a{}^0 = \lambda\phi_a + \mu\phi_b, \qquad \phi_b{}^0 = \mu\phi_a + \lambda\phi_b, \tag{8.25}$$

where

$$\begin{aligned} \lambda &= \tfrac{1}{2}[(1+\rho)^{-1/2} + (1-\rho)^{-1/2}], \\ \mu &= \tfrac{1}{2}[(1+\rho)^{-1/2} - (1-\rho)^{-1/2}]. \end{aligned} \tag{8.26}$$

Orbital $\phi_a{}^0$ is thus mainly localized on nucleus a, and $\phi_b{}^0$ on nucleus b; these are illustrated in Fig. 8.4.

From orbitals (8.25) one constructs a function similar to the coordinate function for the ground state of H_2 in the Heitler–London method:

$$\Phi(^1\Sigma_g) = (1/\sqrt{2})[\phi_a{}^0(1)\phi_b{}^0(2) + \phi_b{}^0(1)\phi_a{}^0(2)]. \tag{8.27}$$

Calculation of the expectation value of the energy in a state described by this function leads to an energy curve which does not possess a minimum and which lies higher than a similar curve for the $^3\Sigma_u$ state (Slater, 1951).

This result is not surprising if one recalls that in the Heitler–London method the main contribution to the bonding is given by the exchange energy, this in turn depending upon the extent to which the orbitals overlap. The orbitals (8.25), however, are constructed precisely so that their overlap is equal to zero. The inclusion of all configurations which can be constructed from the orbitals (8.25) in a configuration interaction calculation leads to the same result as a calculation with the nonorthogonal orbitals ϕ_a and ϕ_b. This is due to the fact that, just as in the molecular orbital method, the coordinate wavefunctions which are constructed from orbitals (8.25) are linear combinations of the coordinate functions of Section 8.1.

PART 2. Calculation of the Energy Matrix for an Arbitrary Molecular System

8.4. Matrix Elements of the Operators *F* and *G*

Let us consider an arbitrary configuration K of k one-electron orbitals in which N electrons are accommodated,

$$K: \qquad \phi_1^{n_1}\phi_2^{n_2}\cdots\phi_k^{n_k}, \qquad \sum_{a=1}^{k} n_a = N. \tag{8.28}$$

In accordance with the Pauli principle, the occupation numbers n_a satisfy the restriction $n_a \leq 2$. We assume that the orbitals ϕ_a constitute an orthonormal set. These can be taken to be either molecular or orthogonalized atomic orbitals.

When spin–orbit coupling is neglected the states of an N-electron system are classified according to the irreducible representations $\Gamma^{[\lambda]}$ of the permutation group (see Chapter VI, Part 3). The number of linearly independent states with permutational symmetry $[\lambda]$ which can be constructed from a particular configuration is given by formula (6.70). The coordinate wavefunctions for such configurations are characterized by their construction from the one-electron orbitals and are distinguished by a subscript ρ which denotes the set of Young diagrams for the doubly occupied orbitals and $k-2$ partial diagrams. An antisymmetric total wavefunction is obtained by forming products of the coordinate wavefunctions with spin wavefunctions, the two sets of functions being symmetrized according to mutually dual Young diagrams:

$$\Psi_{\rho}^{[\lambda]}(K) = (f_{\lambda})^{-1/2} \sum_{r} \Phi_{r\rho}^{[\lambda]}(K)\Omega_{\tilde{r}}^{[\tilde{\lambda}]}(s^N). \tag{8.29}$$

The subscript ρ on the coordinate wavefunctions characterizes their symmetry with respect to permutations of the orbitals, and the subscript r their symmetry with respect to permutations of the electronic coordinates. The spin functions describe a state whose total spin is given by Eq. (6.58). The value of the projection of the spin M_S is not given in (8.29) since it is unimportant for what follows.

We shall use expression (8.29) for the total wavefunction in order to calculate the matrix elements of the symmetric spin-independent operators F and G [see (7.23)]. Explicit expressions for the spin functions Ω are unnecessary for this purpose since they all vanish during the process of calculating the matrix elements due to the spin-independent nature of the operators and the orthonormality of the spin functions.

a. The operator F. In view of the indistinguishability of electrons and the symmetric character of F, this operator may be replaced by Nf_N, where f_N acts upon the Nth electron only. Since f_N does not involve spin coordinates and the spin functions which occur in (8.29) are orthonormal, the matrix element

$$\langle K\rho | F | K\bar{\rho}\rangle^{[\lambda]} \equiv \langle \Psi_{\rho}^{[\lambda]}(K) | F | \Psi_{\bar{\rho}}^{[\lambda]}(K)\rangle \tag{8.30}$$

reduces to

$$(N/f_{\lambda}) \sum_{r} \langle \Phi_{r\rho}^{[\lambda]}(K) | f_N | \Phi_{r\bar{\rho}}^{[\lambda]}(K)\rangle. \tag{8.31}$$

The matrix elements in this equation are easily calculated if the coordinate wavefunctions are expanded according to Eq. (7.105) with coefficients (7.107). We substitute expansion (7.105) into (8.31), replacing the sum over r by an equivalent summation over $[\lambda']r'$. Expression (8.31) then becomes

$$\frac{N}{f_{\lambda}} \sum_{a,b} \sum_{\rho',\bar{\rho}'} \sum_{\lambda',r'} \frac{f_{\lambda}}{f_{\lambda'}} \frac{(n_a n_b)^{1/2}}{N} \langle \Phi_{r'\rho'}^{[\lambda']}(K_a) | \Phi_{r'\bar{\rho}'}^{[\lambda']}(K_b)\rangle$$
$$\times \langle \phi_a | f_N | \phi_b \rangle \langle \rho'\lambda' 1 | P_a^{(N)} | \rho\rangle^{[\lambda]} \langle \bar{\rho}'\lambda' 1 | P_b^{(N)} | \bar{\rho}\rangle^{[\lambda]}. \tag{8.32}$$

From the construction of the $\Phi_{r'\rho'}^{[\lambda']}(K_a)$ functions it follows that

$$\langle \Phi_{r'\rho'}^{[\lambda']}(K_a) | \Phi_{r'\bar{\rho}'}^{[\lambda']}(K_b) \rangle = \delta_{ab} \delta_{\rho'\bar{\rho}'}. \tag{8.33}$$

As a consequence of the orthogonality of the matrices for the permutations,

$$\sum_{\rho', \lambda'} \langle \rho'\lambda'1 | P_a^{(N)} | \rho \rangle^{[\lambda]} \langle \rho'\lambda'1 | P_a^{(N)} | \bar{\rho} \rangle^{[\lambda]} = \delta_{\rho\bar{\rho}}. \tag{8.34}$$

Substituting (8.33) and (8.34) into (8.32) and noting that the sum over r' simply leads to multiplication by $f_{\lambda'}$, we finally obtain the simple result

$$\langle K\rho | F | K\bar{\rho} \rangle^{[\lambda]} = \delta_{\rho\bar{\rho}} \sum_{a=1}^{k} n_a \langle \phi_a | f | \phi_a \rangle. \tag{8.35}$$

From this formula it follows that the value of a matrix element of F which is diagonal in the configuration does not depend upon the permutational symmetry of the state, and hence is independent of the value of the resultant spin.

Matrix elements of F between different configurations are nonzero only if the configurations do not differ by more than one orbital. We write $K_a^{\ b}$ for a configuration which is obtained from K by the transfer of an electron from an orbital ϕ_a to a new orbital ϕ_b:

$$K_a^{\ b}: \qquad \phi_1^{n_1} \phi_2^{n_2} \cdots \phi_a^{n_a - 1} \phi_b \cdots \phi_k^{n_k}. \tag{8.36}$$

On carrying through a computation similar to that used to derive formula (8.35), we arrive at the following:

$$\langle K\rho | F | K_a^{\ b} \bar{\rho} \rangle^{[\lambda]} = \delta_{\rho\bar{\rho}} (n_a)^{1/2} f_{ab}. \tag{8.37}$$

The presence of $\delta_{\rho\bar{\rho}}$ in this equation means that in order for the matrix elements to be nonzero, the orbitals of the configuration $K_a^{\ b}$ must be symmetrized in the same way as those in the configuration K; f_{ab} denotes the one-electron matrix element,

$$f_{ab} = \langle \phi_a | f | \phi_b \rangle. \tag{8.38}$$

In deriving formula (8.37), it was assumed that the ordering of the orbitals which are common to K and $K_a^{\ b}$ is the same in both configurations so that orbital ϕ_b follows ϕ_a. Now let

$$\bar{K}_a^{\ b} = \bar{P} K_a^{\ b},$$

where $\bar{P}$ is a permutation of the orbitals of configuration (8.36) which brings them to the configuration $\bar{K}_a^{\ b}$. According to expressions (2.47) and (2.48),

$$\Phi_{r\bar{\rho}}^{[\lambda]}(\bar{P} K_a^{\ b}) = \bar{P} \Phi_{r\bar{\rho}}^{[\lambda]}(K_a^{\ b}) = \sum_{\bar{\bar{\rho}}} \Gamma_{\bar{\bar{\rho}}\bar{\rho}}^{[\lambda]}(\bar{P}) \Phi_{r\bar{\bar{\rho}}}^{[\lambda]}(K_a^{\ b}). \tag{8.39}$$

On substituting this expression into the right-hand side of the matrix element

$$\langle K\rho | F | \bar{K}_a^{\ b} \bar{\rho} \rangle^{[\lambda]}$$

and applying formula (8.37) to each term in the sum over $\bar{\bar{\rho}}$, we obtain the more general formula

$$\langle K\rho | F | \bar{K}_a{}^b \bar{\rho} \rangle^{[\lambda]} = (n_a)^{1/2} \Gamma_{\rho\bar{\rho}}^{[\lambda]}(\bar{P}) f_{ab}. \tag{8.37a}$$

Suppose that we wish to calculate the matrix element $\langle Kr^{(1)} | F | \bar{K}_a{}^b r^{(2)} \rangle^{[21]}$, where K: $\phi_a\phi_c\phi_d$, $\bar{K}_a{}^b$: $\phi_d\phi_b\phi_c$. In this case $K_a{}^b = \phi_b\phi_c\phi_d$ and $\bar{K}_a{}^b = \bar{P}_{132}K_a{}^b$. From Table 1 of Appendix 5 we find that $\Gamma_{12}^{[21]}(\bar{P}_{132}) = -\sqrt{3}/2$. Hence by formula (8.37a)

$$\langle \phi_a\phi_c\phi_d r^{(1)} | F | \phi_d\phi_b\phi_c r^{(2)} \rangle^{[21]} = -\tfrac{1}{2}\sqrt{3}\, f_{ab}.$$

Both formulae (8.37) and (8.37a) were derived on the assumption that the orbital ϕ_b does not occur in the configuration K, i.e., that $n_b = 0$. If $n_b = 1$, orbital ϕ_b becomes doubly occupied, and configuration $K_a{}^b$ is not of the form (8.36). In this case a formula for the matrix element can be obtained by using the coefficients of fractional parentage to factorize the coordinate wavefunctions as in (8.35). The orbitals on the right-hand side of the matrix element which remain after orbital ϕ_b has been removed are reordered so as to make the configuration coincide with the configuration K_a. We obtain finally‡:

$$\langle K\rho | F | \bar{K}_a{}^b \bar{\rho} \rangle^{[\lambda]} = [n_a(n_b+1)]^{1/2} f_{ab} \sum_{\rho', \bar{\rho}', \lambda'} \Gamma_{\rho'\bar{\rho}'}^{[\lambda']}(\bar{P}') \times \langle \rho'\lambda'1 | P_a^{(N)} | \rho \rangle^{[\lambda]} \langle \bar{\rho}'\lambda'1 | P_b^{(N)} | \bar{\rho} \rangle^{[\lambda]}. \tag{8.37b}$$

The permutations $P_a^{(N)}$ and $P_b^{(N)}$ respectively carry electron N into orbitals ϕ_a and ϕ_b in the configurations K and $\bar{K}_a{}^b$ while preserving the increasing order in which the electrons in the other orbitals are numbered [see Eq. (7.75)]. The permutation $\bar{P}'$ rearranges the orbitals in configuration K_a so as to bring them to the order of the orbitals in $\bar{K}_a$.

Equation (8.37b) is more general than (8.37a), n_b being equal to zero or one. Thus, for example, with $K \equiv \phi_a{}^2\phi_b$ and $K_a{}^b \equiv \phi_b{}^2\phi_a$, one must employ (8.37b) to calculate a matrix element since (8.37a) is inapplicable. We have $\bar{K}_a \equiv \phi_b\phi_a$ and $K_a \equiv \phi_a\phi_b$, so that $\bar{P}' = P_{12}$ and consequently $\Gamma_{\rho'\bar{\rho}'}^{[\lambda']}(\bar{P}') = \pm\delta_{\rho'\bar{\rho}'}$ (with the plus sign for $[\lambda'] = [2]$ and the minus sign for $[\lambda'] = [1^2]$). Here $P_a^{(N)} = P_b^{(N)} = P_{23}$ and taking the matrix $\langle \lambda'1 | P_{23} | r \rangle^{[21]}$ from Table 1 of Appendix 5, we obtain, according to formula (8.37b),

$$\langle \phi_a{}^2\phi_b r^{(1)} | F | \phi_b{}^2\phi_a r^{(1)} \rangle = -f_{ab}.$$

b. The operator G. In constructing the total wavefunction, we choose, instead of the standard representation, one which is reduced with respect to the subgroup $\pi_{N-2} \times \pi_2$. The summation over r is replaced by an equivalent

‡This summation can be rewritten more compactly as

$$\langle K\rho | F | \bar{K}_a{}^b \bar{\rho} \rangle^{[\lambda]} = [n_a(n_b+1)]^{1/2} \Gamma_{\rho\bar{\rho}}^{[\lambda]}(P_a^{(N)-1} \bar{P}' P_b^{(N)}) f_{ab}, \tag{8.37c}$$

in which the orbitals of K_a must be renumbered from one to $N-1$ in order to obtain $\bar{P}'$ correctly.

summation over r', λ', and r'' (r'' is uniquely determined by λ''):

$$\Psi_{\rho}^{[\lambda]}(K) = (f_\lambda)^{-1/2} \sum_{r',\lambda',\lambda''} \Phi_{r'r'',\rho}^{[\lambda]}(K)\Omega_{\tilde{r}'\tilde{r}''}^{[\tilde{\lambda}]}(s^N)$$

Because of the indistinguishability of electrons, or, equivalently, the antisymmetry of the total wavefunction, we may replace the operator G in a matrix element by $\frac{1}{2}N(N-1)g_{N-1\,N}$, where $g_{N-1\,N}$ is the operator for the interaction of electrons N and $N-1$. Since by assumption G does not depend upon any spin coordinates, and the spin functions are orthonormal, we have

$$\begin{aligned}\langle K\rho|G|\bar{K}\bar{\rho}\rangle^{[\lambda]} &\equiv \langle \Psi_{\rho}^{[\lambda]}(K)|G|\Psi_{\bar{\rho}}^{[\lambda]}(\bar{K})\rangle \\ &= \frac{N(N-1)}{2}\frac{1}{f_\lambda}\sum_{r',\lambda',\lambda''}\langle \Phi_{r'r'',\rho}^{[\lambda]}(K)|g_{N-1\,N}|\Phi_{r'r'',\bar{\rho}}^{[\lambda]}(\bar{K})\rangle, \end{aligned} \tag{8.40}$$

in which K and $\bar{K}$ are, in general, different configurations. Into this equation we substitute the factorization of the coordinate wavefunction (7.106) with coefficients of fractional parentage given by (7.108) and (7.109). This leads to the following basic expression, from which more compact formulae can be obtained:

$$\begin{aligned}
&\langle K\rho|G|\bar{K}\bar{\rho}\rangle^{[\lambda]} \\
&= \sum_{a,b}\sum_{r',\lambda'}\sum_{\rho',\bar{\rho}'}\frac{1}{f_{\lambda'}}\langle \rho'\lambda'[2]|P_a^{N-1\,N}|\rho\rangle^{[\lambda]}\langle \bar{\rho}'\lambda'[2]|P_b^{N-1\,N}|\bar{\rho}\rangle^{[\lambda]} \\
&\qquad \times \langle \Phi_{r'\rho'}^{[\lambda']}(K_{aa})|\Phi_{r'\bar{\rho}'}^{[\lambda']}(\bar{K}_{bb})\rangle g_{aa,bb}^{[2]} \\
&+ \sum_{a}\sum_{c<d}\sum_{r',\lambda'}\sum_{\rho',\bar{\rho}}\frac{1}{f_{\lambda'}}\langle \rho'\lambda'[2]|P_a^{N-1\,N}|\rho\rangle^{[\lambda]}\langle \bar{\rho}'\lambda'[2]|P_{cd}^{N-1\,N}|\bar{\rho}\rangle^{[\lambda]} \\
&\qquad \times \langle \Phi_{r'\rho'}^{[\lambda']}(K_{aa})|\Phi_{r'\bar{\rho}'}^{[\lambda']}(\bar{K}_{cd})\rangle(\bar{n}_c\bar{n}_d)^{1/2} g_{aa,cd}^{[2]} \\
&+ \sum_{a<b}\sum_{c}\sum_{r',\lambda'}\sum_{\rho',\bar{\rho}'}\frac{1}{f_{\lambda'}}\langle \rho'\lambda'[2]|P_{ab}^{N-1\,N}|\rho\rangle^{[\lambda]}\langle \bar{\rho}'\lambda'[2]|P_c^{N-1\,N}|\bar{\rho}\rangle^{[\lambda]} \\
&\qquad \times \langle \Phi_{r'\rho'}^{[\lambda']}(K_{ab})|\Phi_{r'\bar{\rho}'}^{[\lambda']}(\bar{K}_{cc})\rangle(n_a n_b)^{1/2} g_{ab,cc}^{[2]} \\
&+ \sum_{a<b}\sum_{c<d}\sum_{r',\lambda',\lambda''}\sum_{\rho',\bar{\rho}'}\frac{1}{f_{\lambda'}}\langle \rho'\lambda'\lambda''|P_{ab}^{N-1\,N}|\rho\rangle^{[\lambda]}\langle \bar{\rho}'\lambda'\lambda''|P_{cd}^{N-1\,N}|\bar{\rho}\rangle^{[\lambda]} \\
&\qquad \times \langle \Phi_{r'\rho'}^{[\lambda']}(K_{ab})|\Phi_{r'\bar{\rho}}^{[\lambda']}(\bar{K}_{cd})\rangle(n_a n_b \bar{n}_c \bar{n}_d)^{1/2} g_{ab,cd}^{[\lambda'']}.
\end{aligned} \tag{8.41}$$

The two-electron matrix elements which occur in this equation are defined by

$$\begin{aligned}
g_{ab,cd}^{[\lambda'']} &= g_{ab,cd} \pm g_{ab,dc} \qquad \begin{cases} + & \text{for } [\lambda''] = [2] \\ - & \text{for } [\lambda''] = [1^2], \end{cases} \\
g_{aa,cd}^{[2]} &= \sqrt{2}\, g_{aa,cd}, \\
g_{aa,bb}^{[2]} &= g_{aa,bb},
\end{aligned} \tag{8.42}$$

where

$$g_{ab,cd} = \langle \phi_a\varphi_b|g|\phi_c\phi_d\rangle = \int \phi_a{}^*(1)\phi_b{}^*(2)g_{12}\phi_c(1)\phi_d(2)\,dV. \tag{8.43}$$

The integral $g_{ab,ab}$ is known as a Coulomb integral, while $g_{ab,ba}$ is called an exchange integral.‡ These are written more compactly as

$$g_{ab,ab} = \alpha_{ab}, \qquad g_{ab,ba} = \beta_{ab}. \tag{8.44}$$

When $a = b$ there is only the Coulomb integral α_{aa}.

Let us consider matrix elements which are diagonal in the configurations $K = \bar{K}$. Because of the orthogonality of the one-electron orbitals,

$$\langle \Phi^{[\lambda']}_{r'\rho'}(K_{ab}) | \Phi^{[\lambda']}_{r'\bar{\rho}'}(K_{cd}) \rangle = \delta_{ac}\delta_{bd}\delta_{\rho'\bar{\rho}'}$$

in Eq. (8.41), so that the only nonzero contributions to this expression are given by the first sum when $a = b$ and the fourth sum when $c = a$ and $d = b$. According to the definitions (8.42) and (8.44), only Coulomb and exchange integrals occur in (8.41). Taking into account the orthogonality of the matrices of the nonstandard representation and the fact that the terms in the sum are independent of r', we obtain the following compact formula (Kaplan, 1967):

$$\langle K\rho | G | K\bar{\rho} \rangle^{[\lambda]} = \delta_{\rho\bar{\rho}} \sum_a{}' \alpha_{aa} + \sum_{a<b} n_a n_b (\delta_{\rho\bar{\rho}} \alpha_{ab} + \tilde{N}^{\rho\bar{\rho}}_{ab} \beta_{ab}), \tag{8.45}$$

in which the prime on the summation over a denotes that it is to be taken over doubly occupied orbitals only. The coefficients $\tilde{N}^{\rho\bar{\rho}}_{ab}$ are found by multiplying together columns ρ and $\bar{\rho}$ of the matrix $\Gamma^{[\lambda]}(P^{N-1\,N}_{ab})$ in the nonstandard representation, and are given by§

$$\tilde{N}^{\rho\bar{\rho}}_{ab} = \sum_{\rho', \lambda', \lambda''} (-1)^{\kappa} \langle \rho'\lambda'\lambda'' | P^{N-1\,N}_{ab} | \rho \rangle^{[\lambda]} \langle \rho'\lambda'\lambda'' | P^{N-1\,N}_{ab} | \bar{\rho} \rangle^{[\lambda]} \tag{8.46}$$

The permutation $P^{N-1\,N}_{ab}$ brings electron $N - 1$ into orbital ϕ_a and electron N into orbital ϕ_b while preserving the increasing order in which the remaining electrons are numbered; κ is equal to zero for all rows with $[\lambda''] = [2]$ and to one for rows with $[\lambda''] = [1^2]$. The matrices $\Gamma^{[\lambda]}(P^{N-1\,N}_{ab})$ are tabulated in Appendix 6 for all allowed symmetries of systems with 3–6 electrons.

Matrix elements which are nondiagonal in the configurations are only nonzero when the configurations differ from one another by not more than two orbitals. On substituting expressions (8.42) for the two-electron matrix elements into (8.41) and taking into account the orthogonality between the coordinate functions which remain after the removal of two orbitals, we

‡Note that a two-electron operator only occurs in the integrals (8.44). This is in contrast to the Coulomb and exchange integrals (8.8) and (8.9) which were defined for nonorthogonal orbitals and whose operators consist of a sum of one- and two-electron operators.

§Note that $(-1)^{\kappa} = \Gamma^{[\lambda'']}(P_{N-1\,N})$, so that this expression is equivalent to the matrix element of the product $(P^{N-1\,N}_{ab})^{-1} P_{N-1\,N} P^{N-1\,N}_{ab} = P_{ab}$, where P_{ab} is a transposition of the orbitals a and b. From this it follows (see Gerratt, 1971) that

$$\tilde{N}^{\rho\bar{\rho}}_{ab} = \Gamma^{[\lambda]}_{\rho\bar{\rho}}(P_{ab}). \tag{8.46a}$$

Matrices of transpositions in the standard representation are given in Appendix 5 for the groups π_3 through π_6.

obtain the following formulae:

$$\langle K\rho|G|K_a^{\ b}\bar{\rho}\rangle^{[\lambda]} = [2(n_a - 1)(n_b + 1)]^{1/2} N^{\rho\bar{\rho}}_{aa,ab}\, g_{aa,ab} + (2n_a n_b)^{1/2} N^{\rho\bar{\rho}}_{ab,bb}\, g_{ab,bb} + [n_a(n_b + 1)]^{1/2} \sum_{c\neq a,b} n_c (N^{\rho\bar{\rho}}_{ca,cb}\, g_{ca,cb} + \tilde{N}^{\rho\bar{\rho}}_{ca,cb}\, g_{ca,bc}), \quad (8.47)$$

$$\langle K\rho|G|K^{cd}_{ab}\bar{\rho}\rangle^{[\lambda]} = [n_a n_b (n_c + 1)(n_d + 1)]^{1/2} (N^{\rho\bar{\rho}}_{ab,cd}\, g_{ab,cd} + \tilde{N}^{\rho\bar{\rho}}_{ab,cd}\, g_{ab,dc}), \quad (8.48)$$

$$\langle K\rho|G|K^{cd}_{aa}\bar{\rho}\rangle^{[\lambda]} = [2(n_c + 1)(n_d + 1)]^{1/2} N^{\rho\bar{\rho}}_{aa,cd}\, g_{aa.cd}, \quad (8.48a)$$

$$\langle K\rho|G|K^{cc}_{ab}\bar{\rho}\rangle^{[\lambda]} = (2n_a n_b)^{1/2} N^{\rho\bar{\rho}}_{ab,cc}\, g_{ab,cc} \quad (8.48b)$$

$$\langle K\rho|G|K^{cc}_{aa}\bar{\rho}\rangle^{[\lambda]} = N^{\rho\bar{\rho}}_{aa}\, g_{aa,cc}. \quad (8.48c)$$

In these formulae K^{cd}_{ab} denotes a configuration which is obtained from configuration K by transferring two electrons, one from orbital ϕ_a to ϕ_c and one from ϕ_b to ϕ_d; n_a, n_b, n_c, and n_d everywhere denote the occupation number in configuration K. We emphasize that the ordering of the indices corresponds to the ordering of the orbitals in the configuration. Thus if orbital ϕ_c is after orbital ϕ_a but before ϕ_b, one must write in (8.47)

$$(N^{\rho\bar{\rho}}_{ac,cb}\, g_{ac,cb} + \tilde{N}^{\rho\bar{\rho}}_{ac,cb}\, g_{ac,bc}).$$

The coefficients in front of the two-electron matrix elements are found by multiplying columns ρ and $\bar{\rho}$ of the corresponding matrices of the nonstandard representation of the permutation group. Thus

$$N^{\rho\bar{\rho}}_{aa,cd} = \sum_{\rho',\lambda'} \langle \rho'\lambda'[2]|P^{N-1\,N}_{a}|\rho\rangle^{[\lambda]} \langle \rho'\lambda'[2]|P^{N-1\,N}_{cd}|\bar{\rho}\rangle^{[\lambda]}, \quad (8.49)$$

$$N^{\rho\bar{\rho}}_{ab,cd} = \sum_{\rho',\lambda',\lambda''} \langle \rho'\lambda'\lambda''|P^{N-1\,N}_{ab}|\rho\rangle^{[\lambda]} \langle \rho'\lambda'\lambda''|P^{N-1\,N}_{cd}|\bar{\rho}\rangle^{[\lambda]}, \quad (8.50a)$$

$$\tilde{N}^{\rho\bar{\rho}}_{ab,cd} = \sum_{\rho',\lambda',\lambda''} (-1)^{\kappa} \langle \rho'\lambda'\lambda''|P^{N-1\,N}_{ab}|\rho\rangle^{[\lambda]} \langle \rho'\lambda'\lambda''|P^{N-1\,N}_{cd}|\bar{\rho}\rangle^{[\lambda]}. \quad (8.50b)$$

The parameter κ and the permutations $P^{N-1\,N}_{ab}$ and $P^{N-1\,N}_{cd}$ have the same meaning as in (8.46), and $P^{N-1\,N}_{a}$ carries electrons $N-1$ and N into the doubly occupied orbital ϕ_a. The matrices which are needed to calculate the coefficients (8.49)–(8.50) are tabulated in Appendix 5 for various types of configurations with 3–6 electrons. Examples of the use of formulae (8.47)–(8.50) are given in later sections, e.g., in Section 8.9.

The formulae given here hold as long as the configurations on both the left- and right-hand sides of the matrix elements possess an identical set of common orbitals so that the coordinate functions for the remaining $N-2$ electrons are orthogonal. Otherwise the configurations in (8.41) which remain after electrons $N-1$ and N have been removed must be rearranged to give the same ordering of the orbitals. This is carried out by means of a relation similar to (8.39). Thus if

$$\bar{K}_{ab} = \bar{P}' K_{ab},$$

then

$$\Phi^{[\lambda']}_{r'\bar{\rho}'}(\bar{P}'K_{ab}) = \bar{P}'\,\Phi^{[\lambda']}_{r'\bar{\rho}'}(K_{ab}) = \sum_{\bar{\bar{\rho}}'} \Gamma^{[\lambda']}_{\bar{\bar{\rho}}'\bar{\rho}'}(\bar{P}')\,\Phi^{[\lambda']}_{r'\bar{\bar{\rho}}'}(K_{ab}), \tag{8.51}$$

where $\Gamma^{[\lambda']}_{\bar{\bar{\rho}}'\bar{\rho}'}(\bar{P}')$ belongs to a nonstandard representation of the group π_{N-2} whose type of reduction is determined by the coupling scheme for the orbitals in the configuration K_{ab}. The formulae for matrix elements nondiagonal in the configurations which are obtained in this way coincide with formulae (8.47)–(8.48). However, the expressions for calculating the coefficients are now somewhat more complicated than expressions (8.49)–(8.50)‡:

$$\begin{aligned} N^{\rho\bar{\rho}}_{aa,cd} &= \sum_{\rho',\bar{\rho}'} \sum_{\lambda'} \Gamma^{[\lambda']}_{\rho'\bar{\rho}'}(\bar{P}') \\ &\quad \times \langle \rho'\lambda'[2] | P^{N-1\,N}_{a} | \rho \rangle^{[\lambda]} \langle \bar{\rho}'\lambda'[2] | P^{N-1\,N}_{cd} | \bar{\rho} \rangle^{[\lambda]}, \\ N^{\rho\bar{\rho}}_{ab,cd} &= \sum_{\rho',\bar{\rho}'} \sum_{\lambda',\lambda''} \Gamma^{[\lambda']}_{\rho'\bar{\rho}'}(\bar{P}') \\ &\quad \times \langle \rho'\lambda'\lambda'' | P^{N-1\,N}_{ab} | \rho \rangle^{[\lambda]} \langle \bar{\rho}'\lambda'\lambda'' | P^{N-1\,N}_{cd} | \bar{\rho} \rangle^{[\lambda]}, \\ \tilde{N}^{\rho\bar{\rho}}_{ab,cd} &= \sum_{\rho',\bar{\rho}'} \sum_{\lambda',\lambda''} (-1)^{\kappa} \Gamma^{[\lambda']}_{\rho'\bar{\rho}'}(\bar{P}') \\ &\quad \times \langle \rho'\lambda'\lambda'' | P^{N-1\,N}_{ab} | \rho \rangle^{[\lambda]} \langle \bar{\rho}'\lambda'\lambda'' | P^{N-1\,N}_{cd} | \bar{\rho} \rangle^{[\lambda]}. \end{aligned} \tag{8.52}$$

An example of a calculation which makes use of these formulae is given on p. 241.

8.5. Expression for the Matrix Elements of the Hamiltonian

The electronic part of the Hamiltonian for a molecule can always be written as a sum of symmetric operators F and G:

$$\mathcal{H} = F + G = \sum_i f_i + \sum_{i<j} g_{ij}. \tag{8.53}$$

In atomic units

$$f_i = -\tfrac{1}{2}\nabla_i^2 - \sum_c (Z_c/R_{ic}), \qquad g_{ij} = 1/r_{ij}, \tag{8.54}$$

where R_{ic} is the distance of the ith electron from nucleus c with charge Z_c, and r_{ij} is the separation between the ith and jth electrons.

In the previous section we derived formulae for the matrix elements of the operators F and G for an arbitrary configuration of orbitals. In this section we consider expressions for matrix elements of the Hamiltonian for various kinds of configurations. However, we observe first that if we take the one-electron operator in (8.54) as f, then a diagonal element (8.38) gives the energy of an electron which is described by a coordinate wave function ϕ_a in the field of the molecular framework. We denote this by ϵ_a:

$$\epsilon_a = \langle \phi_a(i) | -\tfrac{1}{2}\nabla_i^2 - \sum_c (Z_c/R_{ic}) | \phi_a(i) \rangle. \tag{8.55}$$

‡Formulae (8.49), (8.50), and (8.52) can all be put into more compact forms; cf. Eqs. (8.275)–(8.276).

A. *Configurations of Singly Occupied Orbitals*

We denote such a configuration by K_1,

$$K_1: \quad \phi_1\phi_2\cdots\phi_N. \tag{8.56}$$

Instead of being enumerated by the symbol ρ, the different physical states are here distinguished by the Young tableaux r, the specification of which is equivalent to specifying $N-2$ intermediate Young diagrams for the groups $\pi_{N-1}, \pi_{N-2}, \ldots$.‡ For a given configuration the number of states possessing the permutational symmetry of a Young diagram $[\lambda]$ is equal to the dimension of the irreducible representation $\Gamma^{[\lambda]}$ of the group π_N.

With the aid of formulae (8.35) and (8.45) for matrix elements of F and G (with occupation numbers $n_a = n_b = 1$) we now write out expressions for matrix elements of the Hamiltonian (8.53). For matrix elements that are diagonal in the configurations we have

$$\langle K_1 r|\mathcal{H}|K_1\bar{r}\rangle^{[\lambda]} = \delta_{r\bar{r}}\sum_{a=1}^{N}\epsilon_a + \sum_{a<b}(\delta_{r\bar{r}}\alpha_{ab} + \tilde{N}_{ab}^{r\bar{r}}\beta_{ab}). \tag{8.57}$$

A matrix element that is diagonal in r consists of two parts: the energy of the isolated electrons in the field of the molecular framework, and the average value of the interaction energy of the electrons among themselves. The Coulomb integrals occur in diagonal matrix elements only. The dependence upon the permutational symmetry of the system (and, consequently, upon the spin) enters only through the coefficients which precede the exchange integrals.§

Matrix elements that are nondiagonal in the configurations are determined with the aid of formulae (8.37), (8.47), and (8.48) with $n_a = 1$ and are given by¶

$$\begin{aligned}\langle K_1 r|\mathcal{H}|(K_1)_a^b\bar{r}\rangle^{[\lambda]} &= \delta_{r\bar{r}}f_{ab} + (2n_b)^{1/2}N_{ab,bb}^{r\bar{r}}g_{ab,bb}\\ &\quad + (n_b+1)^{1/2}\sum_{c\neq a,b}(N_{ca,cb}^{r\bar{r}}g_{ca,cb} + \tilde{N}_{ca,cb}^{r\bar{r}}g_{ca,bc}),\end{aligned} \tag{8.58}$$

$$\langle K_1 r|\mathcal{H}|(K_1)_{ab}^{cd}\bar{r}\rangle^{[\lambda]} = [(n_c+1)(n_d+1)]^{1/2}(N_{ab,cd}^{r\bar{r}}g_{ab,cd} + \tilde{N}_{ab,cd}^{r\bar{r}}g_{ab,dc}). \tag{8.59}$$

The occupation numbers in these equations are equal to one or zero (the latter value occurring if the electron makes a transition to an unoccupied

‡In all subsequent formulae for matrix elements a Young tableau r characterizes the symmetry with respect to permutations of the *orbitals* (and not permutations of the electrons) because it is just this symmetry which characterizes the states; see Eq. (6.57) or (8.29).

§By using the fractional parentage decomposition of the total wavefunction Gerratt and Lipscomb (1968) have derived expressions for the matrix elements of the molecular Hamiltonian for the case when the orbitals are nonorthogonal. In these formulae the coefficients preceding the molecular integrals consist of linear combinations of overlap integrals. See also Section 8.13.

¶The ordering of the indices of the coefficients N in formula (8.58), and also in (8.47), corresponds to the ordering of the orbitals in the configurations, i.e., in $N_{ca,cb}$ the index c may also be placed after the indices a and b.

orbital). The coefficients which occur in (8.58) and (8.59) are given by Eqs. (8.49) and (8.50) if the symbols ρ in them are replaced by the appropriate Young tableaux. Appendix 6 gives the matrices $\langle r'\lambda'' | P_{ab}^{N-1\, N} | r \rangle^{[\lambda]}$ for all allowable symmetries of systems with three to six electrons. A calculation using formulae (8.46) and (8.50) reduces to a multiplication of the appropriate columns of these matrices.‡

As an example, we consider a configuration of four singly occupied orbitals in a state with spin $S = 0$. The permutational symmetry of the coordinate states is given by the Young diagram $[\lambda] = [2^2]$. There are two linearly independent coordinate functions, the symmetries of which are determined by the Young tableaux

$$r^{(1)} \quad \begin{array}{|c|c|} \hline 1 & 2 \\ \hline 3 & 4 \\ \hline \end{array} \qquad\qquad r^{(2)} \quad \begin{array}{|c|c|} \hline 1 & 3 \\ \hline 2 & 4 \\ \hline \end{array}$$

Since functions which have the symmetry of different tableaux are orthogonal, the secular equation assumes the form

$$\begin{vmatrix} \mathcal{H}_{11} - E & \mathcal{H}_{12} \\ \mathcal{H}_{21} & \mathcal{H}_{22} - E \end{vmatrix} = 0,$$

or,

$$E = \tfrac{1}{2}(\mathcal{H}_{11} + \mathcal{H}_{22}) \pm \tfrac{1}{2}[(\mathcal{H}_{11} - \mathcal{H}_{22})^2 + 4|\mathcal{H}_{12}|^2]^{1/2}.$$

On taking the matrices $\langle | P_{ab}^{34} | \rangle^{[2^2]}$ from Table 2 of Appendix 6, we find, according to formulae (8.57) and (8.46),

$$\begin{aligned} \mathcal{H}_{11} &= \sum_a \epsilon_a + \sum_{a<b} \alpha_{ab} + \beta_{12} + \beta_{34} - \tfrac{1}{2}(\beta_{13} + \beta_{14} + \beta_{23} + \beta_{24}), \\ \mathcal{H}_{22} &= \sum_a \epsilon_a + \sum_{a<b} \alpha_{ab} - \beta_{12} - \beta_{34} + \tfrac{1}{2}(\beta_{13} + \beta_{14} + \beta_{23} + \beta_{24}), \\ \mathcal{H}_{12} &= \mathcal{H}_{21} = \tfrac{1}{2}\sqrt{3}(-\beta_{13} + \beta_{14} + \beta_{23} - \beta_{24}), \end{aligned} \tag{8.60}$$

and hence

$$\begin{aligned} E = {} & \sum_a \epsilon_a + \sum_{a<b} \alpha_{ab} \\ & \pm [(2\beta_{12} + 2\beta_{34} - \beta_{13} - \beta_{14} - \beta_{23})^2 + 3(\beta_{14} + \beta_{23} - \beta_{13} - \beta_{24})]^{1/2}. \end{aligned}$$

If we write [see Eyring *et al.* (1944, p. 246)]

$$\begin{gathered} Q = \sum_a \epsilon_a + \sum_{a<b} \alpha_{ab}, \\ \beta_{12} + \beta_{34} = \alpha, \qquad \beta_{14} + \beta_{23} = \beta, \qquad \beta_{13} + \beta_{24} = \gamma, \end{gathered} \tag{8.61}$$

‡For matrix elements diagonal in the configurations it is more convenient to use formula (8.46a) to calculate the $\tilde{N}_{ab}^{\rho\tilde{\rho}}$.

we obtain the well-known London–Eyring formula (Eyring *et al.*, 1944)

$$E = Q \pm (\alpha^2 + \beta^2 + \gamma^2 - \alpha\beta - \alpha\gamma - \beta\gamma)^{1/2}. \tag{8.62}$$

B. *Configurations of Doubly Occupied Orbitals*

We denote such a configuration by K_0,

$$K_0\colon \quad \phi_1{}^2\phi_2{}^2 \cdots \phi_m{}^2. \tag{8.63}$$

Each doubly occupied orbital corresponds to a pairing of spins and is characterized by a Young diagram [2]. For a given ordering, the coupling of the orbitals in the m diagrams [2] corresponds uniquely to a total diagram $[\lambda] = [2^m]$. This is just a single state whose symmetry with respect to permutations of the orbitals is characterized by the Young tableau r_0:

$$\begin{array}{|c|c|} \hline 1 & 1 \\ \hline 2 & 2 \\ \hline \vdots & \vdots \\ \hline m & m \\ \hline \end{array} \tag{8.64}$$

The matrix elements of the operator F are independent of the symmetry of the wavefunction. The dependence of the matrix elements of G, (8.45), upon the symmetry of a state is contained only in the coefficients which precede the exchange integrals. Since all the orbitals of the configuration K_0 occur equivalently in the Young tableau (8.64), the coefficients $\tilde{N}_{ab}^{r_0r_0} \equiv \tilde{N}_{ab}^{[2^m]}$ must be identical for all pairs a, b, and cannot depend upon the particular orbitals. A simple calculation by formula (8.46) with the matrix elements $\langle r'\lambda'' | P_{ab}^{34} | r_0\rangle^{[2^2]}$ taken from Table 2 of Appendix 6 yields $\tilde{N}_{ab}^{[2^2]} = -\frac{1}{2}$. This remains true for any m.

On substituting $n_a = n_b = 2$, $\tilde{N}_{ab}^{[2^m]} = -\frac{1}{2}$ into formulae (8.35) and (8.45), we arrive at the following simple result:

$$\langle K_0 | \mathcal{H} | K_0 \rangle^{[2^m]} = \sum_a (2\epsilon_a + \alpha_{aa}) + 2 \sum_{a<b} (2\alpha_{ab} - \beta_{ab}). \tag{8.65}$$

The first term of this equation determines the energy of the system in the absence of any interactions between electrons in different orbitals. The energy of such interactions is given by the second term. As can be seen from (8.65), the energy of interaction between pairs of electrons in different orbitals is equal to the difference between twice the Coulomb and the exchange integrals, multiplied by two.

C. *Arbitrary Configurations*

Any configuration of nondegenerate orbitals can be represented as a combination of the two kinds of configuration just considered. We arrange the orbitals in a configuration K with the doubly occupied orbitals first, followed by the singly occupied orbitals. This configuration can then be written as

$$K = K_0 K_1: \quad \phi_1{}^2 \phi_2{}^2 \cdots \phi_m{}^2 \phi_{m+1} \cdots \phi_{N-m}. \tag{8.66}$$

The permutational symmetry of this configuration is characterized by Young tableaux r which consist of a constant tableau r_0 for the doubly occupied orbitals and of tableaux r_1 for the singly occupied orbitals. The number of Young tableaux r required to enumerate the linearly independent states of the configuration (8.66), which are characterized by a diagram $[\lambda]$, is equal to the number of different tableaux r_1, i.e., the number of tableaux required is equal to the dimension of the irreducible representation whose Young diagram is $[\lambda^{(2m)}]$ (see Fig. 6.2).

We employ the convention according to which orbitals of K_0 are denoted by the letters a or b, and orbitals of K_1 by c or d. Matrix elements of the Hamiltonian are derived with the aid of formulae (8.35) and (8.45), just as in the previous cases. These can be divided into terms which depend only upon the orbitals of K_0, only upon the orbitals of K_1, and upon "interference" terms which correspond to the interaction between the two configurations. We consider the terms determined by the orbitals of K_1 first.

In the general formula (8.45) with $K = K_0 K_1$ the coefficients in front of the exchange integrals for the configuration K_1 contain permutations which affect the electrons in orbitals $\phi_{m+1}, \ldots, \phi_{N-m}$. These coefficients are determined by Eq. (8.46) but with the matrices taken from a representation which is reduced with respect to the subgroup $\pi_{2m} \times \pi_{N-2m}$,

$$\begin{aligned} \tilde{N}_{cd}^{r_0 r_1, r_0 \bar{r}_1{}'} = \sum_{r_1{}', \lambda_1{}', \lambda''} & (-1)^{\kappa} \langle (r_0 r_1{}') \lambda' \lambda'' \,|\, P_{cd}^{N-1\,N} \,|\, r_0 r_1 \rangle^{[\lambda]} \\ & \times \langle (r_0 r_1{}') \lambda' \lambda'' \,|\, P_{cd}^{N-1\,N} \,|\, r_0 \bar{r}_1 \rangle^{[\lambda]}. \end{aligned} \tag{8.67}$$

The permutations $P_{cd}^{N-1\,N}$ act upon the last $N - 2m$ electrons only. Hence a transformation of the matrices in this equation to a form which is reduced with respect to permutations of the last $N - 2m$ numbers brings the equation to the form

$$\begin{aligned} \tilde{N}_{cd}^{r_0 r_1, r_0 \bar{r}_1} \equiv \tilde{N}_{cd}^{r_1 \bar{r}_1} = \sum_{\lambda_1{}', r_1{}', \lambda''} & (-1)^{\kappa} \langle r_1{}' \lambda'' \,|\, P_{cd}^{N-1\,N} \,|\, r_1 \rangle^{[\lambda^{(2m)}]} \\ & \times \langle r_1{}' \lambda'' \,|\, P_{cd}^{N-1\,N} \,|\, \bar{r}_1 \rangle^{[\lambda_1{}^{(2m)}]}, \end{aligned} \tag{8.68}$$

which is identical with coefficients that are determined by the representation $\Gamma^{[\lambda^{(2m)}]}$. Since the coefficients of the Coulomb integrals are always equal to $\delta_{r_1 \bar{r}_1}$, one concludes from (8.68) that in a state of the configuration $K = K_0 K_1$ the part of a matrix element of the Hamiltonian that depends upon the orbitals of K_1 has the same form as for the isolated configuration K_1.

In a similar way we conclude that Eq. (8.65), which was derived for an isolated configuration K_0, holds for the part of a matrix element of the

Hamiltonian that depends upon the orbitals of K_0 only. This last remains true even if we interchange K_0 and K_1 and write K in the form K_1K_0.

We now consider the terms that correspond to the interaction between the configurations K_0 and K_1. These are of the following form:

$$\langle Kr_0r_1|\mathcal{H}|Kr_0\bar{r}_1\rangle_{01}^{[\lambda]} = 2\sum_{a,c}(\delta_{r_1\bar{r}_1}\alpha_{ac} + \tilde{N}_{ac}^{r_0r_1,r_0\bar{r}_1}\beta_{ac}). \tag{8.69}$$

Now all the tableaux for the configuration K are constructed in the same way: Tableau r_1 is added to the bottom of tableau r_0. Hence in the total diagram $[\lambda]$ the orbitals of K_0 are all connected to those of K_1 in a unique fashion which is independent of the form of r_1 and $[\lambda]$. One expects, therefore, that the coefficients $\tilde{N}_{ac}^{r_0r_1,r_0\bar{r}_1}$ do not depend upon $[\lambda]$ or upon r_1 and that they are the same for all pairs a and c. This is confirmed by a calculation using Eq. (8.46) for different types of symmetry. One obtains an identical result in all cases:

$$\tilde{N}_{ac}^{r_0r_1,r_0\bar{r}_1} = -\tfrac{1}{2}\delta_{r_1\bar{r}_1}. \tag{8.70}$$

From this it follows that the part of a matrix element of the Hamiltonian that corresponds to the interaction between the configurations K_0 and K_1 does not depend upon the permutational symmetry of the state and is equal to

$$\langle Kr_0r_1|\mathcal{H}|Kr_0\bar{r}_1\rangle_{01}^{[\lambda]} = \delta_{r_1\bar{r}_1}\sum_{a,c}(2\alpha_{ac} - \beta_{ac}). \tag{8.71}$$

We emphasize that this expression is not the same as a matrix element for the interaction energy between two isolated electronic systems with configurations K_0 and K_1. Thus (8.71) does not contain any one-electron integrals corresponding to the interaction of the electrons of one system with the nuclei of the other. This is because the electrons of the configuration K cannot be divided into those that belong to K_0 and those that belong to K_1, since they all occur in a single antisymmetric function.

We thus obtain the following expression for the expectation value of the Hamiltonian for a configuration of electrons (8.66) in a state with permutational symmetry $[\lambda]$:

$$\begin{aligned}\langle Kr_0r_1|\mathcal{H}|Kr_0r_1\rangle^{[\lambda]} &= \sum_a(2\epsilon_a + \alpha_{aa}) + 2\sum_{a<b}(2\alpha_{ab} - \beta_{ab}) + \sum_c \epsilon_c \\ &\quad + \sum_{c<d}(\alpha_{cd} + \tilde{N}_{cd}^{r_1r_1}\beta_{cd}) + \sum_{a,c}(2\alpha_{ac} - \beta_{ac}), \\ &\quad 1 \le a, b \le m, \qquad m+1 \le c, d \le N-m.\end{aligned} \tag{8.72}$$

The Coulomb integral α_{aa} in this expression is defined for doubly occupied orbitals only.

Nondiagonal matrix elements of the Hamiltonian contain just the exchange integrals of configuration K_1, and are given by

$$\langle Kr_0r_1|\mathcal{H}|Kr_0\bar{r}_1\rangle^{[\lambda]} = \sum_{c<d}\tilde{N}_{cd}^{r_1\bar{r}_1}\beta_{cd}. \tag{8.73}$$

Consequently, the secular equation for an arbitrary configuration K reduces to a secular equation for the singly occupied orbitals in a configuration K_1, this configuration possessing the symmetry $[\lambda^{(2m)}]$. We denote the roots of the secular equation for K_1 by $E_i(K_1, S)$, where S denotes the total spin of the state and is uniquely determined by $[\lambda^{(2m)}]$. The total energy of a configuration K in a state with spin S consists of three parts: the energy of the configuration K_1, $E_i(K_1, S)$; the energy of the configuration K_0; and the energy of interaction between K_0 and K_1;

$$E_i(K, S) = E_i(K_1, S) + \sum_a (2\epsilon_a + \alpha_{aa}) + 2\sum_{a<b} (2\alpha_{ab} - \beta_{ab}) + \sum_{a,c} (2\alpha_{ac} - \beta_{ac}). \tag{8.74}$$

The dependence upon the spin of the system occurs only in the energy $E_i(K_1, S)$ characteristic of the configuration K_1. The matrices given in Appendix 6 allow one easily to write out the secular equation for a configuration K_1 containing up to six orbitals.

If the representation $\Gamma^{[\lambda^{(2m)}]}$ is one dimensional, there is no need to solve a secular equation. The energy of the configuration is given by formula (8.72). This arises in two cases: $[\lambda^{(2m)}] = [1^{N-2m}]$ and $[\lambda^{(2m)}] = [2]$. The coefficients $\tilde{N}_{cd}$ in front of the exchange integrals are then

$$\tilde{N}_{cd}^{[1^n]} = -1, \qquad \tilde{N}_{cd}^{[2]} = 1. \tag{8.75}$$

Matrix elements of the Hamiltonian that are nondiagonal in the configurations are found by formulae (8.37), (8.47), and (8.48). It is convenient in each configuration to separate off the doubly occupied orbitals that are common to both. The coefficients of the two-electron matrix elements, which are determined by the remaining orbitals, are then calculated with the aid of the matrices of the representation $\Gamma^{[\lambda^{(2m)}]}$, where m is now the number of doubly occupied orbitals common to both configurations. The part of the sum over c in (8.47) that corresponds to the common doubly occupied orbitals can be expressed similarly to Eq. (8.71):

$$\langle K_0 K_1 | \mathcal{H} | K_0 (K_1)_a{}^b \rangle_{01}^{[\lambda]} = [n_a(n_b + 1)]^{1/2} \sum_e (2g_{ea,eb} - g_{ea,be}), \tag{8.76}$$

where e runs over all doubly occupied orbitals of K_0.

EXAMPLE 1

We consider the configuration

$$K_I: \quad \phi_1{}^2\phi_2{}^2\phi_3{}^2\phi_4\phi_5.$$

Let us find the energy of this configuration in a state with spin $S = 1$. The coordinate Young diagram corresponding to this state is $[2^3\, 1^2]$. The Young diagram for the singly occupied orbitals is $[\lambda^{(6)}] = [1^2]$. The representation $\Gamma^{[1^2]}$ is one dimensional, and therefore the energy is found from formula

(8.72) with $\tilde{N}_{cd}^{[1^2]} = -1$,

$$E(K_I, S{=}1) = \sum_{a=1}^{3} (2\epsilon_a + \alpha_{aa}) + 2\sum_{a<b}^{3} (2\alpha_{ab} - \beta_{ab})$$
$$+ \sum_{c=4}^{5} \epsilon_c + (\alpha_{45} - \beta_{45}) + \sum_{a=1}^{3} \sum_{c=4}^{5} (2\alpha_{ac} - \beta_{ac}).$$

EXAMPLE 2

In the configuration

$$K_{\text{II}}:\quad \phi_1{}^2\phi_2{}^2\phi_3\phi_4\phi_5\phi_6$$

the Young diagram corresponding to the state with $S = 2$ is $[2^2\,1^4]$. The diagram for the singly occupied orbitals $[\lambda^{(4)}] = [1^4]$, and is therefore nondegenerate; its energy is given by formula (8.72) with $\tilde{N}_{cd}^{[1^4]} = -1$, just as in the previous example:

$$E(K_{\text{II}}, S{=}2) = \sum_{a=1}^{2} (2\epsilon_a + \alpha_{aa}) + 2(2\alpha_{12} - \beta_{12})$$
$$+ \sum_{c=3}^{6} \epsilon_c + \sum_{c<d}^{6} (\alpha_{cd} - \beta_{cd}) + \sum_{a=1}^{2} \sum_{c=3}^{6} (2\alpha_{ac} - \beta_{ac}).$$

In the state with spin $S = 1$, $[\lambda] = [2^3\,1^2]$, and $[\lambda^{(4)}] = [2\,1^2]$. The energy of the singly occupied orbitals is determined by solving a secular equation of order three. The matrix elements which occur in it are given by formula (8.57) with coefficients obtained from formula (8.46). With the aid of the matrices $\langle r'\lambda'' | P_{cd}^{34} | r\rangle^{[21^2]}$ taken from Table 2, Appendix 6, we obtain

$$\mathcal{H}_{11} = \sum_{c=3}^{6} \epsilon_c + \sum_{c<d} \alpha_{cd} + \beta_{34} - \beta_{56} - \tfrac{1}{2}(\beta_{46} + \beta_{36} + \beta_{45} + \beta_{35}),$$

$$\mathcal{H}_{22} = \sum_{c=3}^{6} \epsilon_c + \sum_{c<d} \alpha_{cd} - \beta_{34} - \tfrac{1}{6}(2\beta_{56} + 5\beta_{46} + 5\beta_{36} - 3\beta_{45} - 3\beta_{35}),$$

$$\mathcal{H}_{33} = \sum_{c=3}^{6} \epsilon_c + \sum_{c<d} \alpha_{cd} - (\beta_{34} + \beta_{35} + \beta_{45}) + \tfrac{1}{3}(\beta_{56} + \beta_{46} + \beta_{36}), \qquad (8.77)$$

$$\mathcal{H}_{12} = \mathcal{H}_{21}^* = \tfrac{1}{6}\sqrt{3}\,(\beta_{46} - \beta_{36} + 3\beta_{45} - 3\beta_{35}),$$
$$\mathcal{H}_{13} = \mathcal{H}_{31}^* = \tfrac{1}{3}\sqrt{6}\,(\beta_{36} - \beta_{46}),$$
$$\mathcal{H}_{23} = \mathcal{H}_{32}^* = \tfrac{1}{3}\sqrt{2}\,(2\beta_{56} - \beta_{46} - \beta_{36}).$$

We denote the solutions of the secular equation with these matrix elements by $E_i(K_1, S{=}1)$, the index i assuming three values. The energy levels of the configuration K_{II} are determined by adding a constant term to $E_i(K_1, S{=}1)$. According to formula (8.74),

$$E_i(K_{\text{II}}, S{=}1) = E_i(K_1, S{=}1) + \sum_{a=1}^{2} (2\epsilon_a + \alpha_{aa})$$
$$+ 2(2\alpha_{12} - \beta_{12}) + \sum_{a=1}^{2} \sum_{c=3}^{6} (2\alpha_{ac} - \beta_{ac}). \qquad (8.78)$$

In the state with spin $S = 0$, $[\lambda] = [2^4]$ and $[\lambda^{(4)}] = [2^2]$. There are two energy levels, these being determined by a formula similar to (8.78). The energy levels for the singly occupied orbitals $E_i(K_1, S{=}0)$ are given by formula (8.62), the matrix elements in this being calculated with the orbitals ϕ_3, ϕ_4, ϕ_5, and ϕ_6.

EXAMPLE 3

Let us obtain an expression for a matrix element of the Hamiltonian for the interaction of the configuration K_0, (8.63), with $(K_0)_a{}^b$. We place the new orbital ϕ_b next to ϕ_a, the permutation $P_a^{N-1\,N}$ coinciding in this case with $P_{ab}^{N-1\,N}$. In addition, from Eq. (8.49) we have that $N_{aa,ab}^{p\bar{p}} = 1$. Substituting the values $n_a = 2$ and $n_b = 0$ into (8.37) and (8.49) and taking into account Eq. (8.76), we obtain

$$\langle K_0 | \mathcal{H} | (K_0)_a{}^b \rangle^{[\lambda]} = \delta_{[\lambda],[2^m]} \sqrt{2} \{ f_{ab} + g_{aa,ab} + \sum_{e \neq a} (2g_{ea,eb} - g_{ea,be}) \}. \quad (8.79)$$

EXAMPLE 4

Consider a matrix element between the configurations

$$K\colon\ \phi_1{}^2 \cdots \phi_m{}^2 \phi_a \phi_b \qquad \text{and} \qquad K_b{}^c\colon\ \phi_1{}^2 \cdots \phi_m{}^2 \phi_a \phi_c. \quad (8.80)$$

The coefficients of the two-electron matrix elements over the singly occupied orbitals are found with the aid of the representations of the group π_2. There are two kinds of states: a singlet with $[\lambda] = [2^{m+1}]$ and a triplet with $[\lambda] = [2^m\, 1^2]$. Using Eqs. (8.37), (8.47), and (8.76), we obtain

$$\langle K | \mathcal{H} | K_b{}^c \rangle^{[\lambda]} = f_{bc} + g_{ab,ac} \pm g_{ab,ca} + \sum_{e=1}^{m} (2g_{eb,ec} - g_{eb,ce}). \quad (8.81)$$

The plus sign occurs for $[\lambda] = [2^{m+1}]$ and the minus sign for $[\lambda] = [2^m\, 1^2]$.

EXAMPLE 5

As a further example, we calculate the matrix element

$$\langle (K_0)_a{}^b | \mathcal{H} | (K_0)_c{}^d \rangle^{[\lambda]} \quad (8.82)$$

between configurations which differ by two orbitals. If we denote the configuration K_0 without the orbitals $\phi_a{}^2 \phi_c{}^2$ as K_0'', then the configurations occurring in (8.82) can be written as

$$K_0'' \phi_a \phi_b \phi_c{}^2 \qquad \text{and} \qquad K_0'' \phi_a{}^2 \phi_c \phi_d. \quad (8.83)$$

These configurations are arranged so that orbitals that are common to them both appear in the same order in the two configurations. We can now employ formulae (8.50a) and (8.50b) for calculating the coefficients. From (8.50a), (8.50b), and (8.48) it follows that (8.82) is equivalent to the matrix element

$$\langle \phi_a \phi_b \phi_c{}^2 | \mathcal{H} | \phi_a{}^2 \phi_c \phi_d \rangle^{[\lambda^{(2m-4)}]}. \quad (8.84)$$

If we denote the configuration on the left-hand side of (8.84) by K, then the configuration on the right is written K_{bc}^{ad}. The matrix element is given by Eqs. (8.48), (8.50a), and (8.50b), with the indices appropriately rearranged. We first consider the singlet state $[\lambda^{(2m-4)}] = [2^2]$. The symbols ρ and $\bar{\rho}$ in Eqs. (8.50a) and (8.50b) correspond to the Young tableau $[2^2]_1$ in this case. On taking the values of the matrix elements $\langle r'\lambda'' | P_{24}^{34} | r^{(1)}\rangle^{[2^2]}$ from Table 2 of Appendix 6, we find that

$$\langle \phi_a\phi_b\phi_c^2 | \mathcal{H} | \phi_a^2\phi_c\phi_d\rangle^{[2^2]} = 2g_{bc,ad} - g_{bc,da}. \tag{8.85}$$

The triplet state has $[\lambda^{(2m-4)}] = [21^2]$. The permutational symmetry of the configuration on the right-hand side is characterized by the tableau $[21^2]_1$, while the configuration on the left is symmetrized according to a nonstandard representation which is reduced with respect to $\pi_2 \times \pi_2$, and for this we have ρ: $[1^2][2]$. The matrix elements which are needed for formula (8.50) are taken from the first column of the matrix $\langle r'\lambda'' | P_{24}^{34} | r\rangle^{[21^2]}$ (see Table 2, Appendix 6), and from the second column of the matrix $\langle \lambda'\lambda'' | P_{24}^{34} | r_1 r_2\rangle^{[21^2]}$ (see Table 2, Appendix 7). As a result, we obtain

$$\langle \phi_a\phi_b\phi_c^2 | \mathcal{H} | \phi_a^2\phi_c\phi_d\rangle^{[21^2]} = -g_{bc,da}. \tag{8.86}$$

8.6. The Interaction of Two Subsystems in States with Definite Spins

We consider a system of electrons which is made up of two subsystems containing N_1 and N_2 electrons, respectively. Let each subsystem be in a state with a definite value for the resultant spin S_i. The total spin of the system S is determined by the rules for coupling the two spins S_1 and S_2 vectorially. This kind of coupling, in which the spins of the electrons within each subsystem are first coupled together and the resultants then coupled to each other, occurs, for example, when the two subsystems are spatially separated from each other (the average distance between the subsystems is substantially larger than the interelectronic separations within each subsystem).

When terms in the Hamiltonian that contain spin are neglected all properties of the system are adequately described by specifying a coordinate wavefunction which possesses the permutational symmetry of an appropriate Young diagram $[\lambda]$. Each subsystem is characterized by its own coordinate function

$$\Phi_{r_i p_i}^{[\lambda_i]}(K_i), \tag{8.87}$$

in which K_i stands for the configuration of the ith subsystem, r_i is the Young tableau characterizing the symmetry of the function with respect to the electronic coordinates, and p_i characterizes the symmetry of the function with respect to permutations of the orbitals.

The correct coordinate wavefunction for the system in the zeroth approximation is constructed from products of the functions (8.87) and must possess the permutational symmetry of a Young diagram $[\lambda]$. According to Eq. (2.85) with $(r)^A = r$, one can write such a function in the form

$$\Phi_r^{[\lambda]}(K_1\rho_1[\lambda_1], K_2\rho_2[\lambda_2]) = \left\{\frac{f_\lambda}{f_{\lambda_1}f_{\lambda_2}}\frac{N_1!\,N_2!}{N!}\right\}^{1/2} \sum_{r_1, r_2}\sum_{Q} \langle r|Q|r_1r_2\rangle^{[\lambda]} Q\Phi_{r_1\rho_1}^{[\lambda_1]}(K_1)\Phi_{r_2\rho_2}^{[\lambda_2]}(K_2). \tag{8.88}$$

The calculation of matrix elements of the Hamiltonian for a state with this coordinate wavefunction can be carried out by means of a fractional parentage decomposition of the function. Instead of writing this out, we shall make use of some results of previous sections. Thus in Section 8.5 it was shown for an arbitrary total configuration which consists of a configuration of doubly filled orbitals K_0 and of a configuration of singly occupied orbitals K_1 [see (8.66)] that the matrix elements of the Hamiltonian can be expressed as a sum of matrix elements calculated for the isolated configurations K_0 and K_1, plus terms which correspond to the interaction between the two configurations. This result must also hold for any two configurations K_1 and K_2.

We denote the orbitals of configuration K_1 by the letters a and b and the orbitals of K_2 by c and d. A matrix element of the Hamiltonian can be broken up into a sum of three terms:

$$\langle K_1\rho_1[\lambda_1], K_2\rho_2[\lambda_2]|\mathcal{H}|K_1\bar{\rho}_1[\lambda_1], K_2\bar{\rho}_2[\lambda_2]\rangle^{[\lambda]} = \delta_{\rho_2\bar{\rho}_2}\mathcal{H}_{\rho_1\bar{\rho}_1}^{[\lambda_1]} + \delta_{\rho_1\bar{\rho}_1}\mathcal{H}_{\rho_2\bar{\rho}_2}^{[\lambda_2]} + \mathcal{H}_{\rho_1\rho_2,\bar{\rho}_1\bar{\rho}_2}^{[\lambda]}, \tag{8.89}$$

where

$$\mathcal{H}_{\rho_1\bar{\rho}_1}^{[\lambda_1]} = \delta_{\rho_1\bar{\rho}_1}\sum_a (n_a\epsilon_a + \alpha_{aa}) + \sum_{a<b} n_an_b(\delta_{\rho_1\bar{\rho}_1}\alpha_{ab} + \tilde{N}_{ab}^{\rho_1\bar{\rho}_1}\beta_{ab}), \tag{8.90}$$

$$\mathcal{H}_{\rho_2\bar{\rho}_2}^{[\lambda_2]} = \delta_{\rho_2\bar{\rho}_2}\sum_c (n_c\epsilon_c + \alpha_{cc}) + \sum_{c<d} n_cn_d(\delta_{\rho_2\bar{\rho}_2}\alpha_{cd} + \tilde{N}_{cd}^{\rho_2\bar{\rho}_2}\beta_{cd}), \tag{8.91}$$

$$\mathcal{H}_{\rho_1\rho_2,\bar{\rho}_1\bar{\rho}_2}^{[\lambda]} = \sum_{a,c} n_an_c(\delta_{\rho_1\bar{\rho}_1}\delta_{\rho_2\bar{\rho}_2}\alpha_{ac} + \tilde{N}_{ac}^{\rho_1\rho_2,\bar{\rho}_1\bar{\rho}_2}\beta_{ac}). \tag{8.92}$$

The index a runs over all orbitals of configuration K_1 and index c over all orbitals of K_2. It should be recalled that α_{aa} and α_{cc} are only defined for doubly occupied orbitals. $\tilde{N}_{ab}^{\rho_1\bar{\rho}_1}$ and $\tilde{N}_{cd}^{\rho_2\bar{\rho}_2}$ in (8.90) and (8.91) are given by formula (8.46) in which N must be replaced by N_1 or N_2, respectively. The coefficient $\tilde{N}_{ac}^{\rho_1\rho_2,\bar{\rho}_1\bar{\rho}_2}$ is given by the following expression, which results directly from formula (8.46):

$$\tilde{N}_{ac}^{\rho_1\rho_2,\bar{\rho}_1\bar{\rho}_2} = \sum_{\rho_1', \rho_2'}\sum_{\lambda', \lambda''} (-1)^\kappa \langle(\rho_1'\rho_2')\lambda'\lambda''|P_{ac}^{N-1\,N}|\rho_1\rho_2\rangle^{[\lambda]} \times \langle(\rho_1'\rho_2')\lambda'\lambda''|P_{ac}^{N-1\,N}|\bar{\rho}_1\bar{\rho}_2\rangle^{[\lambda]}. \tag{8.93}$$

The summation in this equation corresponds to the multiplication of columns $\rho_1\rho_2$ and $\bar{\rho}_1\bar{\rho}_2$ of the matrix $\langle|P_{ac}^{N-1\,N}|\rangle$, together with an additional multiplication of each product by ± 1, depending upon λ''. $P_{ac}^{N-1\,N}$ denotes a permutation which puts electron $N-1$ into orbital ϕ_a of configuration K_1 and electron N into orbital ϕ_c of K_2. The matrices $\langle|P_{ac}^{N-1\,N}|\rangle$ are defined in terms of two sets of basis functions, each of which belongs to a representation with a different mode of reduction with respect to subgroups. Thus the symbols ρ_1 and ρ_2 characterize the symmetry of the configurations K_1 and K_2, and ρ_1' and ρ_2' characterize the symmetry of the same configurations but with an orbital missing from each.

When calculating the matrix elements (8.90) and (8.91) for the configurations K_1 and K_2 one should bear in mind that it is convenient to place the doubly occupied orbitals together. This is because for each configuration the matrix element that corresponds to the interaction between the electrons in the doubly filled orbitals and those in singly occupied orbitals does not depend upon the permutational symmetry of the state, and is given by formula (8.71). Clearly, the interaction between the doubly filled orbitals of one configuration and the singly occupied orbitals of the other can be expressed by the same formula. Thus the calculation of expression (8.92) reduces to the calculation of the matrix element between the singly occupied orbitals of K_1 and K_2.

Let the index a enumerate the doubly occupied orbitals of K_1 and b the singly occupied orbitals. Similarly, let c enumerate the doubly occupied orbitals of K_2 and d the singly occupied ones. Then for diagonal matrix elements expression (8.92) can be put into the following form:

$$\mathcal{H}^{[\lambda]}_{\rho_1\rho_2,\rho_1\rho_2} = 2\sum_{a,c}(2\alpha_{ac} - \beta_{ac}) + \sum_{a,d}(2\alpha_{ad} - \beta_{ad}) + \sum_{b,c}(2\alpha_{bc} - \beta_{bc}) + \sum_{b,d}(\alpha_{bd} + \tilde{N}_{bd}^{r_1r_2,r_1r_2}\beta_{bd}), \tag{8.94}$$

where r_1 and r_2 are the Young tableaux for the singly occupied orbitals of K_1 and K_2, respectively.

For the nondiagonal elements we have

$$\mathcal{H}^{[\lambda]}_{\rho_1\rho_2,\bar{\rho}_1\bar{\rho}_2} = -2\sum_{a,c}\beta_{ac} - \sum_{a,d}\beta_{ad} - \sum_{c,c}\beta_{bc} + \sum_{b,d}\tilde{N}_{bd}^{r_1r_2,\bar{r}_1\bar{r}_2}\beta_{bd}. \tag{8.95}$$

The coefficients in front of the exchange integrals between the singly occupied orbitals are given by

$$\tilde{N}_{bd}^{r_1r_2,\bar{r}_1\bar{r}_2} = \sum_{r_1',r_2'}\sum_{\lambda',\lambda''}(-1)^{\kappa}\langle(r_1'r_2')\lambda'\lambda''|P_{bd}^{N-1\,N}|r_1r_2\rangle^{[\lambda^{(2m)}]} \times \langle(r_1'r_2')\lambda'\lambda''|P_{bd}^{N-1\,N}|\bar{r}_1\bar{r}_2\rangle^{[\lambda^{(2m)}]}, \tag{8.96}$$

where the permutations $P_{bd}^{N-1\,N}$ are defined only for the singly occupied orbitals, N here denotes the total number of singly occupied orbitals in the two configurations, $[\lambda^{(2m)}]$ denotes the Young diagram which characterizes them, and m denotes the total number of doubly filled orbitals in K_1 and K_2. The matrices $\langle|P_{bd}|\rangle$ for systems with $N = 3, 4, 5$, and 6 are given in Appendix 7.

The total energy of a system in a state with the coordinate wavefunction (8.88) is determined by solving a secular equation whose matrix elements are found from formulae (8.89)–(8.96). The order of this secular equation is equal to the product of the degrees of degeneracy of the subsystems. We define the energy of interaction between the subsystems as the difference between the total energy of the system and the sum of the energies of the isolated systems. When the states of the subsystems are nondegenerate the interaction energy between the electrons of the subsystems is given directly by Eq. (8.94). This expression is considerably simplified if all the orbitals of one of the subsystems are doubly occupied. Let the first subsystem possess doubly filled orbitals only, i.e., let it be characterized by the configuration K_0 [see (8.63)]. This means that in (8.94) one must discount all terms that involve a summation over b. As a result we arrive at the important conclusion that the interaction energy between the electrons of a subsystem with a configuration K_0 and the electrons of an arbitrary subsystem does not depend upon the permutation symmetry (or upon the value of the spin) of the latter, and is given by‡

$$E_{\text{int}}(K_0 S_1=0, K_2 S_2; S_2) = 2 \sum_{a,c} (2\alpha_{ac} - \beta_{ac}) + \sum_{a,d} (2\alpha_{ad} - \beta_{ad}), \tag{8.97}$$

in which a enumerates the orbitals of K_0, c the doubly occupied orbitals of K_2, and d the singly occupied orbitals of K_2.

All the prcceding formulae were derived under the assumption that the one-electron orbitals of the subsystems are orthogonal. In the case of two interacting molecules this assumption no longer holds: Orbitals belonging to different molecules overlap with each other.

If one neglects exchange of electrons between subsystems and takes for the wavefunction of the system a simple product of wavefunctions for the subsystems, then the expression for the interaction energy must be independent of the spins of the subsystems. In fact, the interaction energy for the electrons is determined by just the Coulomb integrals, and is equal to

$$E_{\text{int}}(K_1 S_1, K_2 S_2) = \sum_{a,c} n_a n_c \alpha_{ac}, \tag{8.98}$$

where K_1 and K_2 are two arbitrary configurations of electronic orbitals, a runs over all the orbitals of K_1, and c runs over all the orbitals of K_2.

This result becomes more meaningful physically if one notes that the potential of an electrostatic field at some point in space $\mathbf{R}$ created by a quantal system of charges does not depend upon the spin of the system. This is because the operator for the potential,

$$V(\mathbf{R}) = \sum_i (e/|\mathbf{R} - \mathbf{r}_i|),$$

‡The total interaction energy between the subsystems includes a term which also does not depend upon the permutational symmetry of the states. This term involves integrals for the interaction of the electrons of one subsystem with the nuclei of the other; it is included in ϵ_a [see (8.55)].

belongs to the class of one-electron operators of type F. According to formula (8.35),

$$\bar{V}(\mathbf{R}) = \langle K\rho \,|\, V(\mathbf{R}) \,|\, K\rho \rangle^{[\lambda]} = \sum_a n_a \langle \phi_a \,|\, (e/|\mathbf{R} - \mathbf{r}|) \,|\, \phi_a \rangle, \qquad (8.99)$$

i.e., the expectation value of the potential at a point $\mathbf{R}$ does not depend upon the permutation symmetry of the coordinate wavefunction of the system of charges, and, consequently, does not depend upon the spin of the system. The interaction energy between two systems of charges in the absence of exchange is equal to the expectation value of the potential created by one system, averaged in turn over the wavefunction of the other system, i.e., it is equal to the expectation value of the operator $\sum_j e_j \bar{V}(R_j)$, where j enumerates the electrons in the second system. This operator is also of type F and so causes the interaction energy to be independent of the permuational symmetry of the states.

EXAMPLE 1

Consider the interaction between two subsystems, one of which consists only of doubly filled orbitals:

$$K_1:\ \phi_1{}^2\phi_2{}^2, \qquad [\lambda_1] = [2^2], \quad S_1 = 0,$$
$$K_2;\ \phi_3\phi_4, \qquad [\lambda_2] = [1^2], \quad S_2 = 1.$$

The interaction energy is given by the second term of formula (8.97),

$$E(K_1, S_1{=}0; K_2, S_2{=}1; S{=}1) = \sum_{a=1}^{2} \sum_{d=3}^{4} (2\alpha_{ad} - \beta_{ad}).$$

EXAMPLE 2

Consider the interaction of two subsystems with singly occupied orbitals:

$$K_1:\ \phi_1\phi_2\phi_3, \qquad [\lambda_1] = [1^3], \quad S_1 = \tfrac{3}{2},$$
$$K_2:\ \phi_4\phi_5\phi_6, \qquad [\lambda_2] = [1^3], \quad S_2 = \tfrac{3}{2},$$

in a state with $[\lambda] = [2^3]$, $S = 0$. The interaction energy is given by the last term in formula (8.94). The coefficients $\tilde{N}_{bd}^{[1^3][1^3],[1^3][1^3]}$ are given by formula (8.96) and are determined by the last column of the matrix $\langle | P_{bd}^{56} | \rangle^{[2^3]}$ in Table 4 of Appendix 7. As a result, we obtain

$$E(K_1, S_1{=}\tfrac{3}{2}; K_2, S_2{=}\tfrac{3}{2}; S{=}0) = \sum_{b=1}^{3} \sum_{d=4}^{6} (\alpha_{bd} + \tfrac{1}{3}\beta_{bd}).$$

EXAMPLE 3

Consider the interaction of the subsystems

$$K_1:\ \phi_1{}^2\phi_2{}^2\phi_3\phi_4, \qquad [\lambda_1] = [2^2 1^2], \qquad S_1 = 1,$$
$$K_2:\ \phi_5{}^2\phi_6\phi_7, \qquad [\lambda_2] = [21^2], \qquad S_2 = 1,$$

in a state with $[\lambda] = [2^3 1^4]$, $S = 2$. The interaction energy is determined by formula (8.94). The coefficients in front of the exchange integrals, which are determined by the singly occupied orbitals, are given by formula (8.75) and are equal to $N_{bd}^{[1^4]} = -1$. Hence

$$E(K_1, S_1{=}1; K_2, S_2{=}1; S{=}2) = 2\sum_{a=1}^{2} (2\alpha_{a5} - \beta_{a5}) + \sum_{a=1}^{2}\sum_{d=6}^{7} (2\alpha_{ad} - \beta_{ad}) + \sum_{b=3}^{4} (2\alpha_{b5} - \beta_{b5}) + \sum_{b=3}^{4}\sum_{d=6}^{7} (\alpha_{bd} - \beta_{bd}).$$

EXAMPLE 4

Consider a system consisting of two subsystems, one of which is in a degenerate state,

$$K_1{:}\quad \phi_1\phi_2\phi_3, \qquad [\lambda_1] = [21], \qquad S_1 = \tfrac{1}{2},$$
$$K_2{:}\quad \phi_4\phi_5, \qquad [\lambda_2] = [1^2], \qquad S_2 = 1.$$

The state of the first system is doubly degenerate in accordance with the dimension of the representation $\Gamma^{[21]}$. The energy of a state with $[\lambda] = [2^2 1]$, $S = \frac{1}{2}$, is determined by solving a secular equation of order two. According to formula (8.89), the form of the energy matrix is as follows:

$$\begin{bmatrix} \mathcal{H}_{11}^{[21]} + \mathcal{H}^{[1^2]} + \mathcal{H}_{[21]_1[1^2],[21]_1[1^2]}^{[2^2 1]} & \mathcal{H}_{12}^{[21]} + \mathcal{H}_{[21]_1[1^2],[21]_2[1^2]}^{[2^2 1]} \\ \mathcal{H}_{21}^{[21]} + \mathcal{H}_{[21]_2[1^2],[21]_1[1^2]}^{[2^2 1]} & \mathcal{H}_{22}^{[21]} + \mathcal{H}^{[1^2]} + \mathcal{H}_{[21]_2[1^2],[21]_2[1^2]}^{[2^2 1]} \end{bmatrix}.$$

The matrix elements for the configurations K_1 and K_2 are given by formula (8.57), the coefficients which precede the exchange integrals in them being found by formula (8.46). With the aid of the matrices $\langle r'\lambda'' | P_{ab}^{23} | r\rangle^{[21]}$ from Appendix 6 we obtain

$$\begin{aligned} \mathcal{H}_{11}^{[21]} &= \sum_{a<b}^{3} \alpha_{ab} + \beta_{12} - \tfrac{1}{2}(\beta_{13} + \beta_{23}), \\ \mathcal{H}_{22}^{[21]} &= \sum_{a<b}^{3} \alpha_{ab} - \beta_{12} + \tfrac{1}{2}(\beta_{13} + \beta_{23}), \\ \mathcal{H}_{12}^{[21]} &= \mathcal{H}_{21}^{[21]} = \tfrac{1}{2}\sqrt{3}(-\beta_{13} + \beta_{23}). \end{aligned} \tag{8.100}$$

For the configuration K_2 we have

$$\mathcal{H}^{[1^2]} = \alpha_{45} - \beta_{45}.$$

Since the configurations K_1 and K_2 both consist of just singly occupied orbitals, the diagonal matrix elements for the interaction between them is given by the last term of formula (8.94), and the nondiagonal elements by the last term of (8.95). The coefficients in front of the exchange integrals are found from Eq. (8.96) with the aid of the matrices $\langle | P_{bd}^{45} | \rangle^{[2^2 1]}$ taken from Table 3 of

Appendix 7:

$$\mathcal{H}^{[2^2 1]}_{[21]_1[1^2],[21]_1[1^2]} = \sum_{b=1}^{3}\sum_{d=4}^{5} \alpha_{ad} - \tfrac{1}{2}(\beta_{14} + \beta_{15} + \beta_{24} + \beta_{25} - \beta_{34} - \beta_{35}),$$

$$\mathcal{H}^{[2^2 1]}_{[21]_2[1^2],[21]_2[1^2]} = \sum_{b=1}^{3}\sum_{d=4}^{5} \alpha_{bd} + \tfrac{1}{6}(\beta_{14} + \beta_{15} + \beta_{24} - \beta_{25} - 5\beta_{34} - 5\beta_{35}),$$

$$\mathcal{H}^{[2^2 1]}_{[21]_1[1^2],[21]_2[1^2]} = \mathcal{H}^{[2^2 1]}_{[21]_2[1^2],[21]_1[1^2]} = \tfrac{1}{3}\sqrt{3}\,(\beta_{14} + \beta_{15} - \beta_{24} - \beta_{25}).$$

PART 3. Symmetric Systems

8.7. Construction of Basis Functions in the Molecular Orbital Method for the Molecular Symmetry Group

In symmetric molecules the Hamiltonian of the system is invariant with respect to the spatial transformations of the symmetry point group of the molecule. By selecting variational functions in the form of basis functions for irreducible representations of the point group, the order of the secular equation is considerably lowered. According to the results of Section 5.4, the original secular equation then factorizes into separate secular equations, one for each set of equivalent irreducible representations of the point group. The order of a secular equation corresponding to an irreducible representation $\Gamma^{(\alpha)}$ is equal to the number of times $\Gamma^{(\alpha)}$ occurs in the decomposition of the reducible representation generated by the initial set of variational functions. If each irreducible representation occurs just once, then the secular equation is completely diagonal.

When the one-electron molecular orbitals are determined in the form of linear combinations of atomic orbitals (LCAO–MO method) one has to solve a secular equation whose order is equal to the number of atomic orbitals in the basis set chosen. Basis functions for irreducible representations of the molecular point symmetry group are easily constructed from a set of orbitals which transform among themselves by using the projection operators $\epsilon_{ik}^{(\alpha)}$ (see Section 1.19).‡ For this purpose the operator $\epsilon_{ik}^{(\alpha)}$ is applied to an orbital which has been selected arbitrarily from the given set. If the action of $\epsilon_{ik}^{(\alpha)}$ gives zero, another orbital is chosen, etc., continuing in this way until a non-zero result is obtained. In cases when the initial atomic orbitals are orthonormal and form a regular representation one obtains normalized basis functions by applying the normalized operator

$$\epsilon_{ik}^{(\alpha)} = (f_\alpha/g)^{1/2} \sum_R \Gamma_{ik}^{(\alpha)}(R)^* R \tag{8.101}$$

‡For an alternative method of constructing symmetrized molecular orbitals with the aid of a reducing matrix see Bolotin and Levinson (1960).

to any orbital in the set. In other cases it is necessary to normalize the basis functions which are obtained.

As an example, we consider a symmetric system of four identical atoms, with an orbital of the same type on each center (Fig. 8.5). This atomic configuration is invariant under the operations of the point group $\mathbf{D}_4$. The set

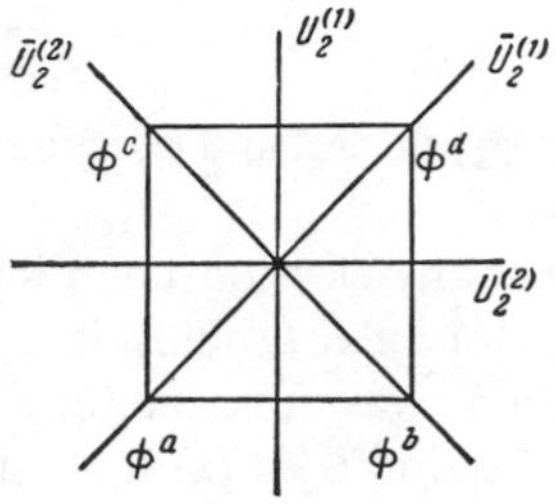

Figure 8.5.

of four orbitals ϕ_a, ϕ_b, ϕ_c, and ϕ_d constitutes a basis for a reducible representation of this group. Let us write out the characters of this representation, remembering that each character is equal to the number of orbitals that remain invariant under the particular operation of the group:

$$\begin{array}{c|ccccc} & E & C_2 & 2C_4 & 2U_2 & 2\bar{U}_2 \\ \hline \chi^\Gamma & 4 & 0 & 0 & 0 & 2 \end{array} \tag{8.102}$$

The representation is now decomposed into irreducible components by the standard procedure (1.63) with the aid of a table of characters for the group $\mathbf{D}_4$. The result is

$$\Gamma \doteq A_1 + B_2 + E. \tag{8.103}$$

Since each irreducible representation occurs in this decomposition once only, the original secular equation factorizes into three equations, each of order one, when appropriately symmetrized variational functions are used. It is immaterial from which of the two basis functions for the doubly degenerate representation one calculates the repeated root of the secular equation.

Basis functions for the irreducible representations which occur in the decomposition (8.103) are constructed by applying the appropriate operators (8.101) to the orbital ϕ_a. For convenience, we write out the orbitals into which ϕ_a is sent by the eight operations of $\mathbf{D}_4$:

$$\begin{array}{c|cccccccc} & E & C_4 & C_2 & C_4{}^3 & U_2^{(1)} & U_2^{(2)} & \bar{U}_2^{(1)} & \bar{U}_2^{(2)} \\ \hline R\phi_a & \phi_a & \phi_b & \phi_d & \phi_c & \phi_b & \phi_c & \phi_a & \phi_d \end{array}$$

A character table alone is sufficient for constructing basis functions for the one-dimensional representations. We find basis functions for the two-dimensional representation with the aid of matrices for the E representation of $\mathbf{D}_4$ (Appendix 2), having applied the two operators ϵ^E_{11} and ϵ^E_{21}. As a result, we obtain the following functions:

$$\sigma^{A_1} = \tfrac{1}{2}(\phi_a + \phi_b + \phi_c + \phi_d), \qquad \sigma_1{}^E = (1/\sqrt{2})(\phi_a - \phi_d),$$
$$\sigma^{B_2} = \tfrac{1}{2}(\phi_a - \phi_b - \phi_c + \phi_d), \qquad \sigma_2{}^E = (1/\sqrt{2})(-\phi_b + \phi_c). \tag{8.104}$$

If the set of atomic orbitals ϕ_a, ϕ_b, ϕ_c, and ϕ_d is taken to be orthonormal, the set (8.104) is also normalized.

In the MO method the many-electron coordinate wavefunction is constructed from a product of molecular orbitals. In a state with a definite value of the total spin the coordinate wavefunction for a symmetric molecule must satisfy two conditions (see Section 6.9): (a) It must belong to a basis for an irreducible representation $\Gamma^{[\lambda]}$ of the permutation group for the electrons, π_N; and (b) it must belong to a basis for an irreducible representation $\Gamma^{(\alpha)}$ of the point symmetry group of the molecule.

In Section 8.5 we derived formulae for matrix elements of the Hamiltonian for an arbitrary electronic system with coordinate wavefunctions which possess a specified permutational symmetry, $\Phi^{[\lambda]}_{r\rho}(K)$. The index ρ in this function characterizes its symmetry with respect to permutations of the orbitals, and r its symmetry with respect to permutation of the electrons. The possible Young diagrams $[\lambda]$ are determined by the form of the electronic configuration K. Hence, when one constructs a function which simultaneously is to satisfy the requirements of permutational and point symmetry, it is desirable to take as the starting point functions $\Phi^{[\lambda]}_{r\rho}(K)$ which already possess the appropriate permutational symmetry. The required symmetry with respect to the operations of the point group can then be attained by applying to the $\Phi^{[\lambda]}_{r\rho}(K)$ the operators $\epsilon^{(\alpha)}_{ik}$. Since the spatial transformation operations of the point group and the operations of permuting the electronic coordinates commute with one another, instead of applying the $\epsilon^{(\alpha)}_{ik}$ to the $\Phi^{[\lambda]}_{r\rho}(K)$, one can apply the $\epsilon^{(\alpha)}_{ik}$ to a product of molecular orbitals which is unsymmetrized as regards permutations. This is then followed by symmetrization with respect to permutations. We note that the application of a point symmetry operation to a product of functions is defined as applying the operation successively to each function in the product, i.e.,

$$R(\phi_1\phi_2) = (R\phi_1)(R\phi_2).$$

If some of the molecular orbitals belong to a single basis for an irreducible representation, i.e., if the orbitals transform among themselves under the operations of the point group, then the basis functions for a representation $\Gamma^{(\alpha)}$ obtained by applying the $\epsilon^{(\alpha)}_{ik}$ operators can also possess a definite permutation symmetry which must be taken into account when the functions are subsequently symmetrized with respect to permutations.‡ If all the molecular orbitals belong to different irreducible representations, the basis functions for $\Gamma^{(\alpha)}$ which are obtained will consist of products of molecular orbitals which do not transform into one another under permutations of the electron coordinates. Every term in such linear combinations must then be further symmetrized with respect to permutations. This is carried out by formally replacing the molecular orbital products by the symmetrized functions $\Phi_{rp}^{[\lambda]}$ that are allowed for the given configuration. Matrix elements of the Hamiltonian can then be calculated according to the equations derived in Section 8.5.

We illustrate this procedure by an example of a system of four electrons which occupy the molecular orbitals σ^{A_1} and σ^E of formulae (8.104). First we put

$$\sigma^{A_1} = \sigma_0, \qquad \sigma_1{}^E = \sigma_1, \qquad \sigma_2{}^E = \sigma_2. \tag{8.105}$$

for short. We construct the following four-electron configurations for these orbitals:

$$\sigma_0{}^2\sigma_1{}^2, \qquad \sigma_0{}^2\sigma_2{}^2, \qquad \sigma_0{}^2\sigma_1\sigma_2. \tag{8.106}$$

The product $\sigma_0(1)\sigma_0(2)$ of orbitals is totally symmetric, and therefore the coordinate functions corresponding to (8.106) form a basis for the direct product $E \times E$. Decomposing this into irreducible components, we have

$$E \times E \doteq A_1 + A_2 + B_1 + B_2. \tag{8.107}$$

Consequently, basis functions belonging to four one-dimensional representations can be constructed from the orbital products (8.106). These basis functions are obtained with the aid of the projection operators for the one-dimensional representations, $\epsilon^{(\alpha)}$. A knowledge of the characters is sufficient for this purpose. We write out first the result of applying the operations of $\mathbf{D}_4$ to the orbitals σ_1 and σ_2, making use of the definition (1.24) of a basis function of an irreducible representation, and of the matrices for the E representation of $\mathbf{D}_4$ in Appendix 2:

	E	C_2	C_4	$C_4{}^3$	$U_2^{(1)}$	$U_2^{(2)}$	$\bar{U}_2^{(1)}$	$\bar{U}_2^{(2)}$
$R\sigma_1$	σ_1	$-\sigma_1$	$-\sigma_2$	σ_2	$-\sigma_2$	σ_2	σ_1	$-\sigma_1$
$R\sigma_2$	σ_2	$-\sigma_2$	σ_1	$-\sigma_1$	$-\sigma_1$	σ_1	$-\sigma_2$	σ_2

As a result, we obtain the following four orthonormal functions, each of

‡ One can also construct from binary products of functions which transform according to a direct product of the point group basis functions for irreducible representations of the group by means of Clebsch–Gordan coefficients (see Section 1.17). Griffith (1961) gives values of coefficients called "*V* coefficients," which are proportional to the Clebsch–Gordan coefficients, for all point groups. *Translator's note:* Tables of Clebsch–Gordan coefficients for all point groups are also given by Koster *et al.* (1963).

which belongs to one of the one-dimensional irreducible representations of $\mathbf{D}_4$:

$$\begin{aligned}
\Phi^{A_1} &= (1/\sqrt{2})\sigma_0(1)\sigma_0(2)[\sigma_1(3)\sigma_1(4) + \sigma_2(3)\sigma_2(4)],\\
\Phi^{A_2} &= (1/\sqrt{2})\sigma_0(1)\sigma_0(2)[\sigma_1(3)\sigma_2(4) - \sigma_2(3)\sigma_1(4)],\\
\Phi^{B_1} &= (1/\sqrt{2})\sigma_0(1)\sigma_0(2)[\sigma_1(3)\sigma_2(4) + \sigma_2(3)\sigma_1(4)],\\
\Phi^{B_2} &= (1/\sqrt{2})\sigma_0(1)\sigma_0(2)[\sigma_1(3)\sigma_1(4) - \sigma_2(3)\sigma_2(4)].
\end{aligned} \tag{8.108}$$

These functions possess a definite symmetry with respect to permutations of both the first two ($[\lambda_1] = [2]$) and the last two electrons. In this particular case this fixes the total Young diagram $[\lambda]$ and hence the value of the resultant spin S. From the form of the functions (8.108) it follows that there will be three singlet states $^1A_1$, $^1B_1$, and $^1B_2$ and one triplet state $^3A_2$, the coordinate wavefunctions of which are

$$\begin{aligned}
\Phi(^1A_1) &= (1/\sqrt{2})\{\Phi^{[2^2]}(\sigma_0{}^2\sigma_1{}^2) + \Phi^{[2^2]}(\sigma_0{}^2\sigma_2{}^2)\},\\
\Phi(^1B_2) &= (1/\sqrt{2})\{\Phi^{[2^2]}(\sigma_0{}^2\sigma_1{}^2) - \Phi^{[2^2]}(\sigma_0{}^2\sigma_2{}^2)\},\\
\Phi(^1B_1) &= \Phi^{[2^2]}(\sigma_0{}^2, \sigma_1\sigma_2[2]),\\
\Phi(^3A_2) &= \Phi^{[21^2]}(\sigma_0{}^2, \sigma_1\sigma_2[1^2]).
\end{aligned} \tag{8.109}$$

Since the matrix elements of the Hamiltonian do not depend upon the tableau r which characterizes the symmetry of a function with respect to permutations of the electron coordinates (in the formulae derived in Section 8.4 one always sums over all r), the value of this tableau is dropped from the functions in Eq. (8.109). Only the symmetry with respect to permutations of the singly occupied orbitals is shown in the arguments of these functions, since $[\lambda] = [2]$ is the sole type of symmetry possible for doubly occupied orbitals.

The energy of the system in states with coordinate wavefunctions (8.109) is given by the expectation value of the Hamiltonian in these states. Matrix elements diagonal in the configurations are found by formulae (8.72) and (8.75), and nondiagonal elements by formula (8.48b) with $N^{\bar{p}p}_{aa,bb} = 1$, since the permutations $P_a^{34} = P_b^{34} = 1$. Making use of the fact that $g_{11,22} = \beta_{12}$, we obtain finally

$$\begin{aligned}
E(^1A_1) &= 2\epsilon_0 + \epsilon_1 + \epsilon_2 + \alpha_{00} + \tfrac{1}{2}(\alpha_{11} + \alpha_{22})\\
&\quad + 2(\alpha_{01} + \alpha_{02}) - (\beta_{01} + \beta_{02}) + \beta_{12}\\
E(^1B_2) &= 2\epsilon_0 + \epsilon_1 + \epsilon_2 + \alpha_{00} + \tfrac{1}{2}(\alpha_{11} + \alpha_{22})\\
&\quad + 2(\alpha_{01} + \alpha_{02}) - (\beta_{01} + \beta_{02}) - \beta_{12},\\
E(^1B_1) &= 2\epsilon_0 + \epsilon_1 + \epsilon_2 + \alpha_{00} + 2(\alpha_{01} + \alpha_{02})\\
&\quad + \alpha_{12} - (\beta_{01} + \beta_{02}) + \beta_{12},\\
E(^3A_2) &= 2\epsilon_0 + \epsilon_1 + \epsilon_2 + \alpha_{00} + 2(\alpha_{01} + \alpha_{02})\\
&\quad + \alpha_{12} - (\beta_{01} + \beta_{02}) - \beta_{12},
\end{aligned} \tag{8.110}$$

in which all integrals are defined in terms of the molecular orbitals (8.105).

8.8. Method of Calculation in the Valence Bond Approximation

In the valence bond method the one-electron orbitals are localized on particular atoms. In the simplest variant one uses atomic orbitals as the one-electron orbitals and neglects overlap integrals. In more accurate calculations orthogonal sets of localized orbitals such as Wannier [(1937), also Slater (1963)] or Löwdin (1950) functions are used.

The many-electron coordinate wavefunction is constructed in the form of a product of the one-electron orbitals. When spin–orbit interactions are neglected the wavefunction for symmetric molecules must satisfy two requirements, just as in the molecular orbital method: (a) The wavefunction must belong to a basis for an irreducible representation $\Gamma^{[\lambda]}$ for the permutation group of the electrons, and (b) it must belong to a basis for an irreducible representation $\Gamma^{(\alpha)}$ of the symmetry point group of the molecule.

Under the operations of the point group, orbitals which were originally localized on certain nuclei become localized on others. As a result, we can associate with each operation R of the point group a permutation $\bar{P}$ of the orbital labels.

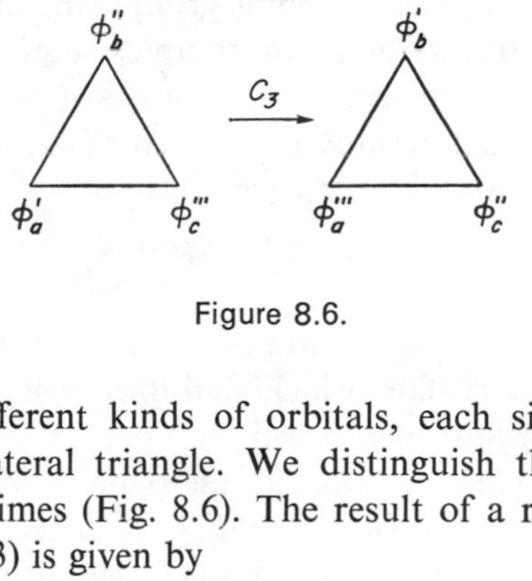

Figure 8.6.

Consider three different kinds of orbitals, each situated at one of the vertices of an equilateral triangle. We distinguish the orbitals by single, double, and triple primes (Fig. 8.6). The result of a rotation C_3 (clockwise rotation through $2\pi/3$) is given by

$$C_3\phi_a'(1)\phi_b''(2)\phi_c'''(3) = \phi_b'(1)\phi_c''(2)\phi_a'''(3) \equiv \bar{P}_{abc}\phi_a'(1)\phi_b''(2)\phi_c'''(3).$$

The operation C_3 applied to the orbitals is therefore equivalent to a permutation $\bar{P}_{abc}$ of the nuclei. The same operation applied to the nuclei is equivalent to the inverse permutation $\bar{P}_{acb} = \bar{P}_{abc}^{-1}$. In the case when a nondegenerate orbital of the same type is situated on each atom symmetrization of the coordinate wavefunction with respect to permutations of the electronic coordinates automatically gives rise to a specific symmetry with respect to permutations of the orbital labels (see Section 2.9) and, consequently, with respect to the operations of the point symmetry.‡

‡This situation only arises when identical nondegenerate orbitals are situated on each atom. When different kinds of orbitals are centered on the atoms permutations of the

This circumstance must be taken into account when constructing functions which simultaneously belong to bases for representations $\Gamma^{[\lambda]}$ and $\Gamma^{(\alpha)}$. We now consider the process of constructing such functions for covalent and ionic structures in the case when the number of interacting electrons is equal to the number of one-electron orbitals, so that there is a single type of orbital on each atom.

A. *Covalent Structures*

In the present case the covalent structures all correspond to a single electronic configuration with singly occupied orbitals of the form (8.56). The coordinate wavefunction corresponding to a particular value of the total spin S is obtained by applying the Young operator $\omega_{rt}^{[\lambda]}$ to an unsymmetrized product of orbitals (see Section 2.9). In the functions $\Phi_{rt}^{[\lambda]}$ which result, the Young tableau r characterizes the symmetry with respect to permutations of the electron coordinates and the tableau t the symmetry with respect to permutations of the orbitals. Consequently, the functions $\Phi_{rt}^{[\lambda]}$ form a basis for irreducible representations of two commuting groups: the permutation group of the electron coordinates $\boldsymbol{\pi}_N$ and the permutation group of the orbitals $\bar{\boldsymbol{\pi}}_N$. Since the operations of a discrete point group generally do not include all permutations of the orbitals, the point group is isomorphic with a subgroup of $\bar{\boldsymbol{\pi}}_N$. The irreducible representations $\Gamma^{(\alpha)}$ into which the representation $\Gamma^{[\lambda]}$ decomposes when the total permutation group of the orbitals is restricted to its point subgroup are given by the decomposition

$$\Gamma^{[\lambda]} \doteq \sum_{\alpha} a_{\lambda}^{(\alpha)} \Gamma^{(\alpha)}, \tag{8.111}$$

which may be carried out with the aid of a character table. This decomposition enables one to determine which multiplets can be constructed from covalent structure functions [see (6.66)].

The construction of eigenfunctions of the point symmetry proceeds as follows: We begin with a set of f_λ functions $\Phi_{rot}^{[\lambda]}$ which belong to a single basis for a representation $\Gamma^{[\lambda]}$ of the group $\bar{\boldsymbol{\pi}}_N$. Basis functions which belong simultaneously to the representations $\Gamma^{[\lambda]}$ and $\Gamma^{(\alpha)}$ are obtained by applying the operators $\epsilon_{ik}^{(\alpha)}$ [see (8.101)] to an arbitrary trial function $\Phi_{r_0 t_0}^{[\lambda]}$ from the given set. If the result of this operation is zero, another function is selected from the set, etc. The operators $\epsilon_{ik}^{(\alpha)}$ are picked from just those representations $\Gamma^{(\alpha)}$ that occur in the decomposition (8.111).

---

electron coordinates are not connected with any permutations of the orbitals. Thus for two orbitals of the same kind we have

$$P_{12}\phi_a(1)\phi_b(2) = \phi_a(2)\phi_b(1) \equiv \bar{P}_{ab}\phi_a(1)\phi_b(2),$$

but for two different kinds of orbital ϕ' and ϕ'' we have

$$P_{12}\phi_a'(1)\phi_b''(2) = \phi_a'(2)\phi_b''(1) \neq \bar{P}_{ab}\phi_a'(1)\phi_b''(2).$$

The operations R of the point group in the $\epsilon_{ik}^{(\alpha)}$ are replaced by the corresponding permutations of the orbitals $\bar{P}$. The desired linear combinations are then found with the aid of Eq. (2.48):

$$\Phi_{ik}^{(\alpha)}([\lambda]r_0) = c\sum_R \Gamma_{ik}^{(\alpha)}(R)^*\bar{P}\Phi_{r_0t_0}^{[\lambda]} = c\sum_R\sum_t \Gamma_{ik}^{(\alpha)}(R)^*\Gamma_{tt_0}^{[\lambda]}(\bar{P})\Phi_{r_0t}^{[\lambda]}$$

or

$$\Phi_{ik}^{(\alpha)}([\lambda]r_0) = \sum_t a_{ik,t}^{(\alpha)}\Phi_{r_0t}^{[\lambda]}, \tag{8.112}$$

$$a_{ik,t}^{(\alpha)} = c\sum_R \Gamma_{ik}^{(\alpha)}(R)^*\Gamma_{tt_0}^{[\lambda]}(\bar{P}), \tag{8.113}$$

where the coefficient c is determined by the requirement that function (8.112) be normalized. Matrices of the irreducible representations of the point groups are given in Appendix 2. Matrices of the permutations $\bar{P}$ which correspond to operations of point groups can be calculated from the transposition matrices $\Gamma^{[\lambda]}(P_{i-1\,i})$, these being obtained by means of the rules described in Section 2.5. Appendix 5 gives all the matrices for the irreducible representations of the groups $\boldsymbol{\pi}_3$ and $\boldsymbol{\pi}_4$ and the transposition matrices for the irreducible representations of the groups $\boldsymbol{\pi}_5$ and $\boldsymbol{\pi}_6$.

Once the coefficients $a_{ik,t}^{(\alpha)}$ have been evaluated the calculation of the matrix elements of the Hamiltonian in states with functions (8.112) is carried out with the aid of the formulae of Section 8.5. As an example, we consider the covalent configuration formed by the six π electrons of the C_6H_6 molecule in a state with spin $S = 0$. The number of linearly independent covalent structures with this value of S is determined by the dimension of the irreducible representation $\Gamma^{[2^3]}$ (see Section 6.10) and is given by $f_{[2^3]} = 5$. The point symmetry group of the benzene molecule is $\mathbf{D}_{6h}$. Since all the one-electron orbitals behave similarly under a reflection in the plane of the ring, it is sufficient to consider just the group $\mathbf{D}_6$ instead of $\mathbf{D}_{6h}$. According to (6.76), the decomposition (8.111) is of the form

$$\Gamma^{[2^3]} \doteq 2A_1 + B_2 + E_2. \tag{8.114}$$

We obtain linear combinations of the functions $\Phi_{r_0t}^{[2^3]}$ which transform according to these irreducible representations by making use of formulae (8.112) and (8.113). The matrices $\Gamma_{tt_0}^{[2^3]}(\bar{P})$ are given in Table 9 of Appendix 5 and the matrices $\Gamma_{ik}^{(\alpha)}(R)$ for the one-dimensional representations are taken from the character tables in Appendix 1, those for the two-dimensional representation E_2 being found from Appendix 2. Two linearly independent and orthogonal functions $\Phi^{A_1(1)}$ and $\Phi^{A_1(2)}$ are obtained by choosing as trial functions $\Phi_{r_0t_1}^{[2^3]}$ and $\Phi_{r_0t_5}^{[2^3]}$ (the value of r_0 is unimportant). A function Φ^{B_2} is obtained by choosing $\Phi_{r_0t_1}^{[2^3]}$ as a trial function also. Basis functions for the E_2 representation are obtained with the aid of the operators $\epsilon_{11}^{E_2}$ and $\epsilon_{21}^{E_2}$ applied to the trial function $\Phi_{r_0t_3}^{[2^3]}$ (the application of the $\epsilon_{ik}^{E_2}$ to $\Phi_{r_0t_1}^{[2^3]}$ and $\Phi_{r_0t_2}^{[2^3]}$ gives zero). As a result, we obtain the following linear combinations:

$$
\begin{aligned}
\Phi_1(^1A_1) &= (1/\sqrt{40})(5\Phi_1 + \sqrt{3}\,\Phi_2 + \sqrt{3}\,\Phi_3 + 3\Phi_4), \\
\Phi_2(^1A_1) &= (1/\sqrt{45})(\sqrt{6}\,\Phi_2 + \sqrt{6}\,\Phi_3 - 2\sqrt{2}\,\Phi_4 + 5\Phi_5), \\
\Phi(^1B_2) &= (1/\sqrt{24})(3\Phi_1 - \sqrt{3}\,\Phi_2 - \sqrt{3}\,\Phi_3 - 3\Phi_4), \\
\Phi_1(^1E_2) &= (1/\sqrt{18})(-3\Phi_2 + \sqrt{3}\,\Phi_4 + \sqrt{6}\,\Phi_5), \\
\Phi_2(^1E_2) &= (1/\sqrt{18})(2\sqrt{3}\,\Phi_2 - \sqrt{3}\,\Phi_3 - \Phi_4 - \sqrt{2}\,\Phi_5),
\end{aligned}
\tag{8.115}
$$

in which Φ_i stands for $\Phi^{[2^3]}_{rot_i}$.

Matrix elements of the Hamiltonian with the functions $\Phi^{[2^3]}_{rot}$ are given by formula (8.57). The coefficients which precede the exchange integrals in this formula are calculated according to Eq. (8.46) with the aid of the matrices $\langle r'\lambda'' | P^{56}_{ab} | r\rangle^{[2^3]}$ taken from Table 4 of Appendix 6. If we number the atoms consecutively around the benzene ring then, because of symmetry, the following equalities hold among the exchange (and Coulomb) integrals:

$$
\begin{aligned}
&\beta_{12} = \beta_{23} = \beta_{34} = \beta_{45} = \beta_{56} = \beta_{16}, \\
&\beta_{13} = \beta_{24} = \beta_{35} = \beta_{46} = \beta_{15} = \beta_{26}, \\
&\beta_{14} = \beta_{25} = \beta_{36}.
\end{aligned}
\tag{8.116}
$$

The matrix elements of the Hamiltonian, taking these equalities into account, are given in Table 8.1, in which Q stands for $\sum_a \epsilon_a + \sum_{a<b} \alpha_{ab}$.

Table 8.1

Table of Matrix Elements of the Hamiltonian for the π Electrons of the Benzene Molecule for a State with Spin $S = 0$

$$
\begin{aligned}
&\mathcal{H}_{11} = Q + \tfrac{3}{2}\beta_{12} - 3\beta_{13} - \tfrac{3}{2}\beta_{14} \\
&\mathcal{H}_{22} = Q - \tfrac{3}{2}\beta_{12} - \beta_{13} - \tfrac{1}{2}\beta_{14} \\
&\mathcal{H}_{33} = Q - \tfrac{3}{2}\beta_{12} - \beta_{13} - \tfrac{1}{2}\beta_{14} \\
&\mathcal{H}_{44} = Q - \tfrac{7}{6}\beta_{12} - \tfrac{1}{3}\beta_{13} - \tfrac{3}{2}\beta_{14} \\
&\mathcal{H}_{55} = Q - \tfrac{10}{3}\beta_{12} - \tfrac{2}{3}\beta_{13} + \beta_{14} \\
&\mathcal{H}_{12} = \mathcal{H}_{21} = \sqrt{3}(\tfrac{1}{2}\beta_{12} - \beta_{13} + \tfrac{1}{2}\beta_{14}) \\
&\mathcal{H}_{13} = \mathcal{H}_{31} = \sqrt{3}(\tfrac{1}{2}\beta_{12} - \beta_{13} + \tfrac{1}{2}\beta_{14}) \\
&\mathcal{H}_{14} = \mathcal{H}_{41} = \tfrac{1}{2}\beta_{12} - \beta_{13} + \tfrac{1}{2}\beta_{14} \\
&\mathcal{H}_{15} = \mathcal{H}_{51} = \sqrt{2}(\tfrac{1}{2}\beta_{12} - \beta_{13} + \tfrac{1}{2}\beta_{14}) \\
&\mathcal{H}_{23} = \mathcal{H}_{32} = \tfrac{1}{2}\beta_{12} - \beta_{13} + \tfrac{1}{2}\beta_{14} \\
&\mathcal{H}_{24} = \mathcal{H}_{42} = \sqrt{3}(\tfrac{5}{6}\beta_{12} - \tfrac{1}{3}\beta_{13} - \tfrac{1}{2}\beta_{14}) \\
&\mathcal{H}_{25} = \mathcal{H}_{52} = \sqrt{6}(-\tfrac{1}{6}\beta_{12} - \tfrac{1}{3}\beta_{13} + \tfrac{1}{2}\beta_{14}) \\
&\mathcal{H}_{34} = \mathcal{H}_{43} = \sqrt{3}(\tfrac{5}{6}\beta_{12} - \tfrac{1}{3}\beta_{13} - \tfrac{1}{2}\beta_{14}) \\
&\mathcal{H}_{35} = \mathcal{H}_{53} = \sqrt{6}(-\tfrac{1}{6}\beta_{12} - \tfrac{1}{3}\beta_{13} + \tfrac{1}{2}\beta_{14}) \\
&\mathcal{H}_{45} = \mathcal{H}_{54} = \sqrt{2}(\tfrac{5}{6}\beta_{12} - \tfrac{1}{3}\beta_{13} - \tfrac{1}{2}\beta_{14})
\end{aligned}
$$

When the functions (8.115) are used as variational functions the initial secular equation breaks up into two equations of order one and one equation of order two. From the one-dimensional equations we obtain the energies of the B_2 and E_2 states (the $^1E_2$ level is doubly degenerate):

$$E(^1B_2) = Q - 3\beta_{14}, \qquad E(^1E_2) = Q - 2\beta_{12} - \beta_{14}. \tag{8.117}$$

The energies of the two states of A_1 symmetry are found by solving a secular equation of order two with matrix elements

$$\begin{aligned} \mathcal{H}_{11}^{A_1} &= Q + \tfrac{12}{5}\beta_{12} - \tfrac{24}{5}\beta_{13} - \tfrac{3}{5}\beta_{14}, \\ \mathcal{H}_{22}^{A_1} &= Q - \tfrac{22}{5}\beta_{12} - \tfrac{6}{5}\beta_{13} + \tfrac{13}{5}\beta_{14}, \\ \mathcal{H}_{12}^{A_1} &= \mathcal{H}_{21}^{A_1} = \tfrac{6}{5}\beta_{12} - \tfrac{12}{5}\beta_{13} + \tfrac{6}{5}\beta_{14}. \end{aligned} \tag{8.118}$$

If all exchange integrals except those between adjacent atoms are neglected in these equations, the values of the energy which are obtained are the same as those given for the electronic states of the C_6H_6 molecule in the book by Eyring *et al.* (1944, p. 251).

If the point symmetry group is isomorphic with the permutation group, the secular equation corresponding to a particular value of the spin S no longer factorizes and the term is degenerate. As an example, consider a system of four electrons in a tetrahedral field. The point group $\mathbf{T}_d$ consists of 24 elements and is isomorphic with the permutation group π_4.

There is a one-to-one correspondence between the irreducible representations of these groups [see (6.25)]. This makes is possible to write down the allowed multiplets immediately, since an irreducible representation of the point group corresponds uniquely to each value of the spin S. The allowed terms are

$$^5A_2, \qquad ^3F_1, \qquad \text{and} \qquad ^1E. \tag{8.119}$$

Only the quintet term is nondegenerate, the triplet terms is triply degenerate, and the singlet is doubly degenerate—in accordance with the dimensions of the irreducible representations.

Electrons which are localized at the vertices of a tetrahedron are equidistant from one another; hence all Coulomb and, correspondingly, all exchange integrals are equal:

$$\alpha_{ab} = \alpha, \qquad \beta_{ab} = \beta \qquad \text{for} \quad a, b = 1, 2, 3, 4. \tag{8.120}$$

The coordinate wavefunction corresponding to the $^5A_2$ term is totally antisymmetric and belongs to the $\Gamma^{[1^4]}$ representation. Its energy is found by formula (8.57) in which $\tilde{N}_{ab}^{[1^4]} = -1$ for all a and b. Taking Eq. (8.120) into account, we obtain

$$E(^5A_2) = 4\epsilon + 6\alpha - 6\beta. \tag{8.121}$$

The coordinate wave functions corresponding to the $^3F_1$ and ^1E terms are of symmetry $[\lambda] = [21^2]$ and $[\lambda] = [2^2]$, respectively. Matrix elements for states with this kind of permutational symmetry were given in Section 8.5 [formulae (8.60) and (8.77); in formulae (8.77) it is necessary to lower the numbering of the orbitals by two]. According to Eq. (5.15), all diagonal matrix elements of the Hamiltonian between representations $\Gamma^{(\alpha)}$ must be the same and all off-diagonal elements must be zero. Substitution of Eqs. (8.120) into (8.60) and (8.77) shows that this is indeed the case. We obtain for the energies of the terms

$$E(^3F_1) = 4\epsilon + 6\alpha - 2\beta, \qquad E(^1E) = 4\epsilon + 6\alpha. \qquad (8.122)$$

B. Ionic Structures

If the covalent structures in a particular case can all be associated with just a single configuration, then usually a fairly large number of different electronic configurations correspond to the ionic structures. The number of ionic structures is given by formula (6.68). In order to construct coordinate functions which are simultaneously basis functions for the representations $\Gamma^{[\lambda]}$ and $\Gamma^{(\alpha)}$, we must divide the ionic configurations into separate sets, the functions within a given set transforming into each other under the operations of the point symmetry group of the molecule. In addition to this, we form for each configuration the coordinate wavefunctions $\Phi_{r\rho}^{[\lambda]}(K)$, in which the suffix ρ characterizes the symmetry of the function with respect to permutations of the orbitals, this being determined to a considerably extent by the electronic configuration K. The subscript r denotes the Young tableau which characterizes the symmetry of the function with respect to permutations of the electronic coordinates; its explicit form is unimportant since the matrix elements of the Hamiltonian are independent of it. It is convenient to write the electronic configuration as a product of two configurations K_0 and K_1 [see (8.66)], so that when symmetrized with respect to permutations a coordinate wavefunction can be written in the form $\Phi^{[\lambda]}(Kr_0r_1)$, in which the tableau r_0 characterizes its symmetry with respect to permutations of the orbitals of configuration K_0, and tableau r_1 the symmetry with respect to permutations of the orbitals of K_1. The spin is determined by the permutational symmetry of K_1, whose Young tableau is denoted, as in previous sections, by $[\lambda^{(2m)}]$ (m specifies the number of doubly occupied orbitals and is equal to the number of pairs of ions in the particular molecular configuration). Linear combinations of the $\Phi^{[\lambda]}(Kr_0r_1)$ functions which transform according to irreducible representations of the point group are obtained by applying the operator $\epsilon_{ik}^{(\alpha)}$ to an arbitrary function in the set. A nonzero result is obtained only if $\epsilon_{ik}^{(\alpha)}$ corresponds to one of the representations $\Gamma^{(\alpha)}$ which occurs in the decomposition of the reducible representation of the point group generated by the particular set of ionic structures. The method of finding the characters of representations of this kind was described in Section 6.10B.

Let the ionic configurations under consideration form a regular set. A set such as this consists of g configurations, where g is the order of the point group. We denote an arbitrary configuration in the set by K, to which we shall refer as the generating configuration. All the other configurations K_R are derived from it by the operations R of the group. Symmetry-adapted functions for the molecular multiplet states are easily obtained by applying the operators (8.101) to the function $\Phi^{[\lambda]}(Kr_0 r_1)$ and are given by

$$\Phi_{ik}^{(\alpha)}([\lambda]r_0 r_1) = (f_\alpha/g)^{1/2} \sum_R \Gamma_{ik}^{(\alpha)}(R)^* \Phi^{[\lambda]}(K_R r_0 r_1). \tag{8.123}$$

An example of the construction of such functions is described in the following section.

The construction of symmetry-adapted functions for molecular multiplet states is somewhat more complicated if the ionic configurations possess some elements of symmetry. Basis functions for irreducible representations are then obtained as follows.

First we write out the elements of the subgroup **H** under which the chosen generating configuration is invariant. All the elements of the point group **G** are divided into left cosets of **H**. As a result, every element of **G** can be written as a product $R_a Q$, where Q denotes an operation of the point group which leaves the generating configuration invariant, i.e., $Q \in \mathbf{H}$. The number of distinct configurations in the set is equal to the number of operations R_a, i.e., is equal to g/h. In addition, we make use of the fact that the application of an operation Q leads to a permutation of the singly occupied orbitals. Denoting such a permutation by $\bar{Q}$, we have, by Eq. (2.48),

$$\bar{Q}\Phi^{[\lambda]}(Kr_0 \bar{r}_1) = \sum_{r_1} \Gamma_{r_1 \bar{r}_1}^{[\lambda^{(2m)}]}(\bar{Q}) \Phi^{[\lambda]}(Kr_0 r_1). \tag{8.124}$$

As a result, we obtain

$$\epsilon_{ik}^{(\alpha)} \Phi^{[\lambda]}(Kr_0 \bar{r}_1) = (f_\alpha/g)^{1/2} \sum_{R_a Q} \sum_{r_1} \Gamma_{ik}^{(\alpha)}(R_a Q)^* \Gamma_{r_1 \bar{r}_1}^{[\lambda^{(2m)}]}(\bar{Q}) \Phi^{[\lambda]}(K_{R_a} r_0 r_1) \tag{8.125}$$

or

$$\Phi_{ik}^{(\alpha)}([\lambda]) = \sum_{R_a} \sum_{r_1} a_{ik,r_1}^{(\alpha)}(R_a) \Phi^{[\lambda]}(K_{R_a} r_0 r_1), \tag{8.126}$$

in which the coefficients in this linear combination are given by

$$a_{ik,r_1}^{(\alpha)}(R_a) = c \sum_Q \Gamma_{ik}^{(\alpha)}(R_a Q)^* \Gamma_{r_1 \bar{r}_1}^{[\lambda^{(2m)}]}(\bar{Q}), \tag{8.127}$$

the constant c being determined from the condition that function (8.126) be normalized.

As an example, we consider the construction of symmetry-adapted functions for the molecular multiplet states which arise from the four configurations with a single ion pair of the type shown in Fig. 8.7. The symmetry point group of the equilibrium nuclear configuration is $\mathbf{D}_4$. The electronic configuration shown in this figure is invariant under the operations of a subgroup

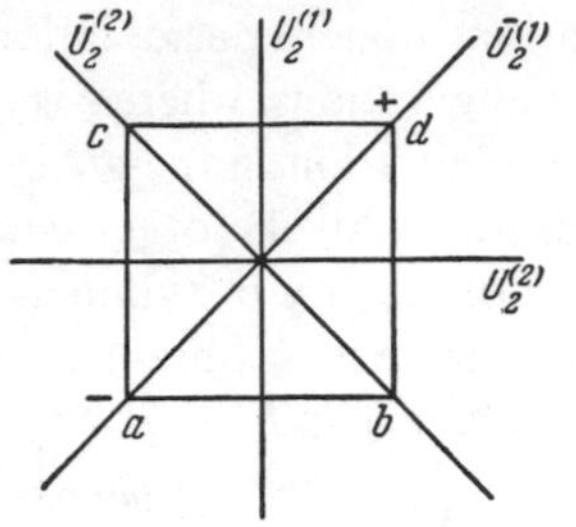

Figure 8.7.

of $\mathbf{D}_4$ whose two elements Q are E and $\bar{U}_2^{(1)}$. We write out all the elements of $\mathbf{D}_4$ in the form of a product R_aQ‡:

$$\begin{array}{lcccccccc} R_aQ: & E & \bar{U}_2^{(1)} & C_4 & C_4\bar{U}_2^{(1)} & C_2 & C_2\bar{U}_2^{(1)} & C_4{}^3 & C_4{}^3\bar{U}_2^{(1)} \\ R: & E & \bar{U}_2^{(1)} & C_4 & U_2^{(1)} & C_2 & \bar{U}_2^{(2)} & C_4{}^3 & U_2^{(2)} \end{array}$$

There are four linearly independent ionic configurations:

$$\begin{array}{ll} K\colon\ \phi_a{}^2\phi_b\phi_c, & K_{C_4}\colon\ \phi_c{}^2\phi_a\phi_d, \\ K_{C_2}\colon\ \phi_d{}^2\phi_c\phi_b, & K_{C_4{}^3}\colon\ \phi_b{}^2\phi_d\phi_a. \end{array} \tag{8.128}$$

Two spin states are possible, $S = 0$ and $S = 1$. We consider the state with spin $S = 1$, the coordinate wavefunction for which is symmetrized according to the Young diagram $[\lambda] = [21^2]$. The diagram for the singly occupied orbitals is $[\lambda^{(2)}] = [1^2]$. According to the theorem of Section 6.10B, the characters of the representation generated by the coordinate wavefunctions of the configurations (8.128) in a state with $[\lambda] = [21^2]$ are given by

	E	C_2	$2C_4$	$2U_2$	$2\bar{U}_2$
χ^Γ	4	0	0	0	−2

(8.129)

This is decomposed into the following irreducible components:

$$\Gamma \doteq A_2 + B_1 + E. \tag{8.130}$$

Basis functions for these irreducible representations are found by means of formulae (8.126) and (8.127), noting that

$$\Gamma^{[1^2]}(E) = 1, \qquad \Gamma^{[1^2]}(\bar{U}_2^{(1)}) = -1.$$

The matrix elements $\Gamma_{ik}^{(\alpha)}(R_aQ)$ are taken from Appendices 1 and 2. Basis functions for the two-dimensional representation E are constructed with the

‡This process is equivalent to the division of $\mathbf{D}_4$ into cosets with respect to a subgroup consisting of the two elements E and $\bar{U}_2^{(1)}$. We regard the operations of the point group as being applied to the electronic coordinates with the nuclei fixed. All rotations are carried out in a clockwise sense.

aid of the two operators ϵ_{12}^{E} and ϵ_{22}^{E} {the application of ϵ_{11}^{E} and ϵ_{21}^{E} to $\Phi^{[21^2]}(K[2][1^2])$ gives zero}. As a result, we obtain the following normalized combinations:

$$\begin{aligned}
\Phi(^3A_2) &= \tfrac{1}{2}(\Phi_1 + \Phi_2 + \Phi_3 + \Phi_4),\\
\Phi(^3B_1) &= \tfrac{1}{2}(\Phi_1 - \Phi_2 + \Phi_3 - \Phi_4),\\
\Phi_1(^3E) &= (1/\sqrt{2})(\Phi_2 - \Phi_4),\\
\Phi_2(^3E) &= (1/\sqrt{2})(\Phi_1 - \Phi_3).
\end{aligned} \tag{8.131}$$

in which, for compactness, the following notation has been employed:

$$\begin{aligned}
\Phi_1 &= \Phi^{[21^2]}(K[2][1^2]), &\quad \Phi_2 &= \Phi^{[21^2]}(K_{C_4}[2][1^2]),\\
\Phi_3 &= \Phi^{[21^2]}(K_{C_2}[2][1^2]), &\quad \Phi_4 &= \Phi^{[21^2]}(K_{C_4^3}[2][1^2]).
\end{aligned} \tag{8.132}$$

By means of a linear transformation the original secular equation of order four has been factorized into four one-dimensional equations, two of which refer to the E representation and are therefore identical. Matrix elements of the Hamiltonian which are diagonal is the configurations (8.128) are easily found from formula (8.72). Matrix elements off-diagonal in the configurations are determined by formulae (8.48) and (8.48a)–(8.48c) since the configurations (8.128) differ from one another in two orbitals. Thus to within the order of the orbitals

$$K_{C_4^3} = K_{ac}^{bd}, \qquad K_{C_2} = K_{aa}^{dd}, \qquad K_{C_4} = K_{ab}^{cd}.$$

In this particular case the Young diagrams for the configurations which remain when two orbitals are removed are $[\lambda'] = [2]$ or $[\lambda'] = [1^2]$, and therefore in formulae (8.52)

$$\sum_{\rho',\bar{\rho}'} \longrightarrow \delta_{\rho'\bar{\rho}'} \sum_{\rho'}.$$

The matrix elements which occur in (8.52) are taken from Table 2 of Appendix 6. For example, in the matrix element between the configurations K and K_{C_4} the coefficients $N_{ac,bd}^{[2][1^2],[2][1^2]}$ are found by multiplying together the first columns of the matrices for the permutations P_{24}^{34} and P_{23}^{34}, with an additional multiplication by $\Gamma^{[\lambda']}(P_{12}) = \pm 1$, the sign depending upon $[\lambda']$ ($[\lambda']$ is determined by the Young tableaux for the representation $\Gamma^{[21^2]}$). This leads to

$$\mathcal{H}_{12} \equiv \langle K[2][1^2] \,|\, \mathcal{H} \,|\, K_{C_4}[2][1^2] \rangle^{[21^2]} = g_{ab,cd}. \tag{8.133}$$

The coefficients $N_{aa,dd}^{[2][1^2],[2][1^2]}$ in the matrix element between the configurations K and K_{C_2} are determined by squaring the first column of the matrix of the permutation P_{12}^{34} and multiplying this by $\Gamma^{[\lambda']}(P_{12})$. The matrix element is then given by formula (8.48c):

$$\mathcal{H}_{13} \equiv \langle K[2][1^2] \,|\, \mathcal{H} \,||\, K_{C_2}[2][1^2] \rangle^{[21^2]} = -g_{aa,dd}. \tag{8.134}$$

Because of the symmetry of the problem, we have the following equalities among the one-electron, Coulomb, and exchange integrals:

$$\alpha_{aa} = \alpha_{bb} = \alpha_{cc} = \alpha_{dd} = \alpha_0, \qquad \alpha_{ab} = \alpha_{ac} = \alpha_{cd} = \alpha_{bd} = \alpha_1,$$
$$\epsilon_a = \epsilon_b = \epsilon_c = \epsilon_d = \epsilon, \qquad \beta_{ab} = \beta_{ac} = \beta_{cd} = \beta_{bd} = \beta_1, \tag{8.135}$$
$$\alpha_{ad} = \alpha_{bc} = \alpha_2, \qquad \beta_{ad} = \beta_{bc} = \beta_2.$$

It is easily seen that

$$g_{aa,\,dd} = g_{cc,\,bb} = \beta_2. \tag{8.136}$$

In addition, we put

$$g_{ab,\,cd} = \gamma. \tag{8.137}$$

By taking Eqs. (8.135)–(8.137) into account, we arrive at the following expressions for the matrix elements of the Hamiltonian between the functions (8.132):

$$\mathcal{H}_{11} = \mathcal{H}_{22} = \mathcal{H}_{33} = \mathcal{H}_{44} = 4\epsilon + \alpha_0 + 4\alpha_1 + \alpha_2 - 2\beta_1 - \beta_2,$$
$$\mathcal{H}_{12} = \mathcal{H}_{23} = \mathcal{H}_{34} = \mathcal{H}_{14} = \gamma, \tag{8.138}$$
$$\mathcal{H}_{13} = \mathcal{H}_{24} = -\beta_2.$$

As a result, we obtain the following expressions for the energies of the ionic triplet states characterized by the coordinate wavefunctions (8.131):

$$E(^3\mathrm{A}_2) = 4\epsilon + \alpha_0 + 4\alpha_1 + \alpha_2 - 2\beta_1 - 2\beta_2 + 2\gamma,$$
$$E(^3\mathrm{B}_1) = 4\epsilon + \alpha_0 + 4\alpha_1 + \alpha_2 - 2\beta_1 - 2\beta_2 - 2\gamma, \tag{8.139}$$
$$E(^3\mathrm{E}) = 4\epsilon + \alpha_0 + 4\alpha_1 + \alpha_2 - 2\beta_1.$$

8.9. Calculation on the H_3 Molecule with Full Interaction of All Configurations of 1s Orbitals

We now apply the method described in the previous section to the determination of the energy of a system of three H atoms assuming that a 1s orbital is centered on each nucleus, the nuclei being placed at the vertices of an equilateral triangle. We choose as the one-electron orbitals an orthonormal set of three localized orbitals ϕ_a, ϕ_b, and ϕ_c constructed from the atomic 1s orbitals by the Wannier method [see Slater (1963, Chapter 9, §5)].

There is one covalent configuration

$$K_0\!: \quad \phi_a\phi_b\phi_c. \tag{8.140}$$

When the spin $S = \frac{3}{2}$ there is a single state whose coordinate wavefunction is $\Phi(K_0[1^3])$. Two states correspond to a spin $S = \frac{1}{2}$, the coordinate wavefunctions of which are $\Phi(K_0[21]_1)$ and $\Phi(K_0[21]_2)$. The point group of the equilibrium nuclear configuration is $\mathbf{D}_3$. This group is isomorphic with the permutation group π_3. On comparing character tables, we observe the following one-to-one correspondence between the irreducible representations of the two groups:

$$\Gamma^{[1^3]} \longleftrightarrow A_2, \qquad \Gamma^{[21]} \longleftrightarrow E, \qquad \text{and} \qquad \Gamma^{[3]} \longleftrightarrow A_1 \tag{8.141}$$

Hence it follows that if one takes into account the covalent configuration only, two terms arise: ${}^4A_2$ and ${}^2E$. The coordinate wavefunction (8.142) corresponds to the ${}^4A_2$ term:

$$\Phi({}^4A_2) \equiv \Phi(K_0[1^3]). \tag{8.142}$$

The expectation value of the Hamiltonian in a state with this wavefunction is found from formula (8.57), in which all the coefficients $\tilde{N}_{ab} = -1$:

$$E({}^4A_2) = \langle K_0 | \mathcal{H} | K_0 \rangle^{[1^3]} = 3(\epsilon + \alpha - \beta). \tag{8.143}$$

In writing this expression, we have made use of the fact that since all the orbitals are equivalent, it follows from the symmetry of the nuclear configuration that

$$\begin{gathered} \epsilon_a = \epsilon_b = \epsilon_c = \epsilon, \qquad \alpha_{ab} = \alpha_{ac} = \alpha_{bc} = \alpha, \\ \beta_{ab} = \beta_{ac} = \beta_{bc} = \beta. \end{gathered} \tag{8.144}$$

The ${}^2E$ term is doubly degenerate, and we denote the corresponding coordinate wavefunctions by

$$\Phi_{10}({}^2E) \equiv \Phi(K_0[21]_1), \qquad \Phi_{20}({}^2E) \equiv \Phi(K_0[21]_2). \tag{8.145}$$

Matrix elements of the Hamiltonian in states with these wavefunctions are given by formula (8.57), in which the coefficients preceding the exchange integrals are calculated by formula (8.46). Making use of the equalities (8.144), we find, with the aid of the matrices in Table 1 of Appendix 6,

$$\begin{aligned} E({}^2E) = \langle K_0 | 21]_1 | \mathcal{H} | K_0[21]_1 \rangle &= \langle K_0[21]_2 | \mathcal{H} | K_0[21]_2 \rangle \\ &= 3(\epsilon + \alpha). \end{aligned} \tag{8.146}$$

We now consider the ionic configurations. In this particular case these do not possess any elements of point symmetry and they therefore form regular sets. All the configurations within a set can be obtained from one of the members by applying the elements of the group to it (see Fig. 8.8).

Since the spins of the electrons in a doubly filled orbital are paired, the only allowed terms are doublets. We write

$$\Phi_i = \Phi(K_i[21]_1) \tag{8.147}$$

for compactness. The decomposition of a regular representation contains all the irreducible representations of the group. Hence the required coordinate wavefunctions for the multiplets which can be constructed from the ionic configurations are easily found by formula (8.123). Taking the matrix ele-

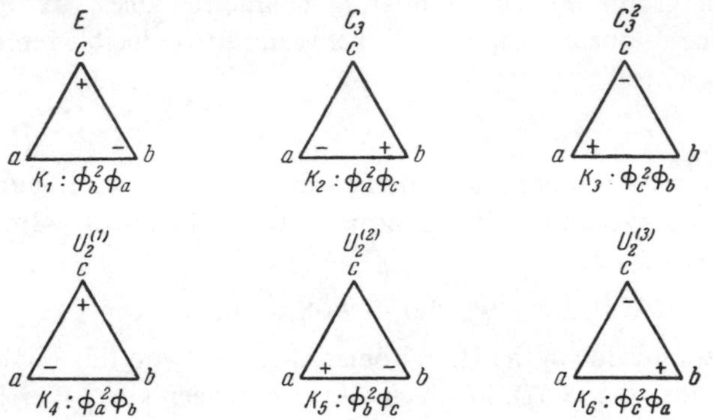

Figure 8.8.

ments $\Gamma_{ik}^{(\alpha)}(R)$ from Appendices 1 and 2, we obtain

$$\begin{aligned}\Phi({}^2A_1) &= (1/\sqrt{6})(\Phi_1 + \Phi_2 + \Phi_3 + \Phi_4 + \Phi_5 + \Phi_6),\\ \Phi({}^2A_2) &= (1/\sqrt{6})(\Phi_1 + \Phi_2 + \Phi_3 - \Phi_4 - \Phi_5 - \Phi_6),\end{aligned} \tag{8.148}$$

$$\begin{aligned}\Phi_{11}({}^2E) &= (1/\sqrt{3})(\Phi_1 - \tfrac{1}{2}\Phi_2 - \tfrac{1}{2}\Phi_3 + \Phi_4 - \tfrac{1}{2}\Phi_5 - \tfrac{1}{2}\Phi_6),\\ \Phi_{21}({}^2E) &= \tfrac{1}{2}(\Phi_2 - \Phi_3 - \Phi_5 + \Phi_6),\end{aligned} \tag{8.149}$$

$$\begin{aligned}\Phi_{12}({}^2E) &= \tfrac{1}{2}(-\Phi_2 + \Phi_3 - \Phi_5 + \Phi_6),\\ \Phi_{22}({}^2E) &= (1/\sqrt{3})(\Phi_1 - \tfrac{1}{2}\Phi_2 - \tfrac{1}{2}\Phi_3 - \Phi_4 + \tfrac{1}{2}\Phi_5 + \tfrac{1}{2}\Phi_6).\end{aligned} \tag{8.150}$$

The pairs of equations (8.149) and (8.150) are coordinate wavefunctions which describe degenerate states. Thus, when ionic configurations are included we obtain two extra ${}^2E$ terms, a ${}^2A_1$ term and a ${}^2A_2$ term. In order to find their energies, it is first necessary to determine the matrix elements of the Hamiltonian defined by the functions (8.147).

We calculate matrix elements that are diagonal in the configurations by means of formula (8.72). Denoting, in addition to (8.144),

$$\alpha_{aa} = \alpha_{bb} = \alpha_{cc} = \alpha_0, \tag{8.151}$$

we find that

$$\mathcal{H}_{ii} = 3\epsilon + \alpha_0 + 2\alpha - \beta, \qquad i = 1, 2, \ldots, 6. \tag{8.152}$$

Matrix elements that are off-diagonal in the configurations are found from (8.37), (8.47), and (8.48). The matrix elements of the permutations that are required in calculating the coefficients (8.49) and (8.50) are taken from the first column of the appropriate matrices in Table 1 of Appendix 6. Thus the elements $\mathcal{H}_{12}$ and $\mathcal{H}_{16}$ are found from formulae (8.48a) and (8.48c), respectively:

$$\mathcal{H}_{12} = \langle \phi_b{}^2\phi_a | \mathcal{H} | \phi_a{}^2\phi_c \rangle^{[21]} = -g_{bb,ac} \equiv -\gamma_1, \tag{8.153}$$

$$\mathcal{H}_{16} = \langle \phi_b{}^2\phi_a | \mathcal{H} | \phi_c{}^2\phi_a \rangle^{[21]} = g_{bb,cc} \equiv \beta; \tag{8.154}$$

the element $\mathcal{H}_{14}$ is found from (8.37b) and (8.47), and element $\mathcal{H}_{15}$ from (8.37) and (8.47):

$$\begin{aligned} \mathcal{H}_{14} &= \langle \phi_b{}^2\phi_a | \mathcal{H} | \phi_a{}^2\phi_b \rangle^{[21]} \\ &= -f_{ba} - (g_{ba,aa} + g_{bb,ab}) \equiv -(f + 2\gamma_2), \end{aligned} \tag{8.155}$$

$$\begin{aligned} \mathcal{H}_{15} &= \langle \phi_b{}^2\phi_a | \mathcal{H} | \phi_b{}^2\phi_c \rangle \\ &= f_{ac} + 2g_{ba,bc} - g_{ba,cb} \equiv f + 2\gamma_3 - \gamma_1. \end{aligned} \tag{8.156}$$

It is not difficult to convince oneself that because of the symmetry of the problem, all the other matrix elements are equal to one or other of the elements (8.153)–(8.156). Thus

$$\begin{aligned} &\mathcal{H}_{12} = \mathcal{H}_{13} = \mathcal{H}_{23} = \mathcal{H}_{45} = \mathcal{H}_{46} = \mathcal{H}_{56} = -\gamma_1, \\ &\mathcal{H}_{14} = \mathcal{H}_{26} = \mathcal{H}_{35} = -(f + 2\gamma_2), \\ &\mathcal{H}_{15} = \mathcal{H}_{24} = \mathcal{H}_{36} = f + 2\gamma_3 - \gamma_1, \\ &\mathcal{H}_{16} = \mathcal{H}_{25} = \mathcal{H}_{34} = \beta. \end{aligned} \tag{8.157}$$

The ${}^2A_1$ and ${}^2A_2$ terms each occur once. Their energies are given by the expectation values of the Hamiltonian in states with the coordinate wavefunctions (8.148). Making use of (8.157), the expressions for these terms are

$$E({}^2A_1) = 3\epsilon + \alpha_0 + 2\alpha - 3\gamma_1 - 2\gamma_2 + 2\gamma_3, \tag{8.158}$$

$$E({}^2A_2) = 3\epsilon + \alpha_0 + 2\alpha - 2\beta - \gamma_1 + 2\gamma_2 - 2\gamma_3. \tag{8.159}$$

The interaction between the covalent and ionic configurations gives rise to three ${}^2E$ terms. Their energies are given by the roots of a secular equation of order three. In order to write down this equation, it is necessary to choose one function from each of the sets (8.145), (8.149), and (8.150). We pick the following three basis functions:

$$\Phi_{10}({}^2E), \qquad \Phi_{11}({}^2E), \qquad \text{and} \qquad \Phi_{12}({}^2E). \tag{8.160}$$

In order to determine the off-diagonal matrix elements between covalent and ionic structure functions, it is necessary to know, apart from the matrix elements (8.157), the matrix elements between the functions (8.147) and (8.145). We obtain, with the aid of formulae (8.37a) and (8.47),

$$\begin{aligned} &\mathcal{H}_{01} = \mathcal{H}_{04} = \sqrt{2}(-\tfrac{1}{2}f + \gamma_1 - \tfrac{1}{2}\gamma_2 - \tfrac{1}{2}\gamma_3), \\ &\mathcal{H}_{02} = \mathcal{H}_{05} = \sqrt{2}(f - \tfrac{1}{2}\gamma_1 + \gamma_2 + \gamma_3), \\ &\mathcal{H}_{03} = \mathcal{H}_{06} = -\tfrac{1}{2}\sqrt{2}(f + \gamma_1 + \gamma_2 + \gamma_3). \end{aligned} \tag{8.161}$$

As a result, we obtain the following expressions for the matrix elements of the secular equation:

$$\begin{aligned}
\langle\Phi_{10}|\mathcal{H}|\Phi_{10}\rangle &= 3(\epsilon+\alpha),\\
\langle\Phi_{11}|\mathcal{H}|\Phi_{11}\rangle &= 3\epsilon-\tfrac{3}{2}f+\alpha_0+2\alpha-\tfrac{3}{2}\beta+\tfrac{3}{2}\gamma_1-2\gamma_2-\gamma_3,\\
\langle\Phi_{12}|\mathcal{H}|\Phi_{12}\rangle &= 3\epsilon+\tfrac{3}{2}f+\alpha_0+2\alpha-\tfrac{1}{2}\beta+\tfrac{1}{2}\gamma_1+2\gamma_2+\gamma_3,\\
\langle\Phi_{10}|\mathcal{H}|\Phi_{11}\rangle &= \langle\Phi_{11}|\mathcal{H}|\Phi_{10}\rangle=\sqrt{6}\,(-\tfrac{1}{2}f+\gamma_1-\tfrac{1}{2}\gamma_2-\tfrac{1}{2}\gamma_3),\\
\langle\Phi_{10}|\mathcal{H}|\Phi_{12}\rangle &= \langle\Phi_{12}|\mathcal{H}|\Phi_{10}\rangle=-\tfrac{3}{2}\sqrt{2}\,(f+\gamma_2+\gamma_3),\\
\langle\Phi_{11}|\mathcal{H}|\Phi_{12}\rangle &= \langle\Phi_{12}|\mathcal{H}|\Phi_{11}\rangle=\tfrac{1}{2}\sqrt{3}\,(-f+\beta+\gamma_1-2\gamma_3).
\end{aligned}\qquad(8.162)$$

In this calculation we made maximum use of the symmetry of the particular system. In order to obtain numerical results, we must further calculate the molecular integrals which occur in expressions (8.143), (8.158), and (8.162).‡

PART 4. The Self-Consistent Field Method

8.10. The Hartree–Fock Equations

In the valence bond (VB) and simple LCAO-MO methods the form of the one-electron functions is already specified before the many-electron problem is attempted. Thus in the VB method one takes atomic orbitals as the starting point. In the simple LCAO-MO method one seeks the molecular orbitals in the form of linear combinations of atomic orbitals, the coefficients in which are determined by the requirement that the average energy of a single electron in the field of the molecular framework be a minimum. However, it is possible to formulate the many-electron problem without restricting the form of the one-electron functions. This approach, which was first proposed and developed for atoms by Hartree (1928) and Fock (1930a, b), is usually referred to as the Hartree–Fock or self-consistent-field (SCF) method. Each electron is regarded as moving in some kind of "self-consistent field" due to the nuclei and all the other electrons.

In the Hartree–Fock method the equations which the one-electron functions must satisfy are derived from the variational principle

$$\bar{E}=\int\psi^*\mathcal{H}\psi\,dV=\text{a minimum},\qquad(8.163)$$

with the additional condition that

$$\int\psi^*\psi\,dV=1.\qquad(8.164)$$

‡Kaplan and Rodimova (1970, 1971) have applied a similar method in a calculation of the electronic energies of the H_4^+ and H_5^+ ions, including complete interation between all configurations of Slater 1s orbitals.

The trial function ψ is taken to be a single determinant of one-electron functions. If ψ were completely arbitrary, the variation principle applied to Eqs. (8.163) and (8.164) would lead to the Schrödinger equation [see Landau and Lifshitz (1965)]. Hence by solving the variational problem with a determinantal wavefunction, we obtain the best possible function of this particular form.

A one-electron approximation to the Hamiltonian for an N-electron system, (8.53),‡

$$\hat{\mathcal{H}} = \sum_{i=1}^{N} \hat{f}(i) + \sum_{i<j} \hat{g}(i,j), \tag{8.165}$$

is vitiated by the presence of terms in $\mathcal{H}$ which represent the interelectronic interactions. The objective of the Hartree–Fock method is to find the best approximation by one-electron operators to the two-electron operators, i.e., one seeks an approximate equation

$$\sum_{i<j} \hat{g}(i,j) \equiv \sum_{i=1}^{N} (\tfrac{1}{2} \sum_{j \neq i} \hat{g}(i,j)) \simeq \sum_{i=1}^{N} \hat{g}(i). \tag{8.166}$$

In calculations on molecules the Hartree–Fock equations, which are obtained as a result of describing the system by a single determinant, are usually referred to as the SCF equations. A single determinant is a useful representation for configurations which consist of completely filled orbitals (closed-shell configurations), and hence such configurations can only occur in nondegenerate singlet states. In configurations which contain open shells a single-determinant wavefunction is generally not an eigenfunction of $\hat{\mathbf{S}}^2$ but contains contributions from all the allowable multiplet states.

It is possible to derive the Hartree–Fock equations without using a determinantal representation for the total wavefunction. It has been remarked a number of times (see, for example, Section 6.9) that all spin-independent properties of a system of identical particles are completely described by specifying just the coordinate wavefunction, the permutational symmetry of which carries all the information about the spin. The Hartree–Fock equations possess a comparatively simple form for closed-shell configurations (K_0). The permutational symmetry of the coordinate wavefunction for such a configuration is characterized by the Young diagram $[\lambda] = [2^m]$ $(m = N/2)$. The expectation value of the energy for the configuration K_0 is given by Eq. (8.65). We rewrite this in a form which is symmetric in the subscripts a and b,

$$\bar{E} = \langle K_0 | \mathcal{H} | K_0 \rangle^{[2^m]} = 2 \sum_{a} h_a + \sum_{a,b} (2\alpha_{ab} - \beta_{ab}), \tag{8.167}$$

where the one-electron integral (8.55) is now denoted by h_a. Following Roothaan (1951), we introduce the Coulomb and exchange operators $\hat{\alpha}_b$ and

‡In this section we distinguish operators by carets.

$\hat{\beta}_b$, which are defined by the relations

$$\begin{aligned} \hat{\alpha}_b(i)\phi_a(i) &= \left[\int \phi_b^*(j)\phi_b(j)\hat{g}(i,j)\, dV_j\right]\phi_a(i), \\ \hat{\beta}_b(i)\phi_a(i) &= \left[\int \phi_b^*(j)\phi_a(j)\hat{g}(i,j)\, dV_j\right]\phi_b(i). \end{aligned} \tag{8.168}$$

It is not difficult to verify that these operators are Hermitian‡

$$\langle\psi|\hat{\alpha}_b|\phi_a\rangle = \langle\phi_a^*|\hat{\alpha}_b^*|\psi^*\rangle, \qquad \langle\psi|\hat{\beta}_b|\phi_a\rangle = \langle\phi_a^*|\hat{\beta}_b^*|\psi^*\rangle \tag{8.169}$$

for any function ψ. With the aid of the $\hat{\alpha}_b$ and $\hat{\beta}_b$ operators we can rewrite the two-electron Coulomb and exchange integrals in the form of one-electron integrals. Thus

$$\begin{aligned} \alpha_{ab} &= \langle\phi_a|\hat{\alpha}_b|\phi_a\rangle = \langle\phi_b|\hat{\alpha}_a|\phi_b\rangle, \\ \beta_{ab} &= \langle\phi_a|\hat{\beta}_b|\phi_a\rangle = \langle\phi_b|\hat{\beta}_a|\phi_b\rangle. \end{aligned} \tag{8.170}$$

Taking these equations into account, and also Eqs. (8.54) and (8.55), we can rewrite expression (8.167) for the energy in the form

$$\bar{E} = 2\sum_a \langle\phi_a|\hat{f}|\phi_a\rangle + \sum_{a,b}\langle\phi_a|2\hat{\alpha}_b - \hat{\beta}_b|\phi_a\rangle. \tag{8.171}$$

The problem is now to find a set of orbitals that minimizes the energy (8.171) and satisfies the orthonormality conditions

$$\langle\phi_a|\phi_b\rangle = \delta_{ab}. \tag{8.172}$$

This reduces to a search for an unconditional extremum of the functional

$$J = \sum_a \langle\phi_a|2\hat{f} + \sum_b (2\hat{\alpha}_b - \hat{\beta}_b)|\phi_a\rangle - \sum_{a,b}\epsilon_{ba}\langle\phi_a|\phi_b\rangle, \tag{8.173}$$

in which the ϵ_{ba} are Lagrange multipliers. We determine the first-order variation of this functional and equate it to zero [for details see Roothaan (1951)]. From this we find that the orbitals that minimize the energy of a closed-shell configuration must be solutions of the equation

$$\hat{\mathcal{H}}^{\text{HF}}(i)\phi_a(i) = \epsilon_a\phi_a(i), \tag{8.174}$$

where the one-electron operator $\hat{\mathcal{H}}^{\text{HF}}$ is called the *Hartree–Fock Hamiltonian* and is given by

$$\hat{\mathcal{H}}^{\text{HF}} = \hat{f} + \sum_b (2\hat{\alpha}_b - \hat{\beta}_b). \tag{8.175}$$

As can be seen from (8.168), the operator $\hat{\mathcal{H}}^{\text{HF}}$ itself depends upon the orbitals being sought. This circumstance greatly hampers the solution of Eq. (8.174). The Hartree–Fock equations belong to a class of nonlinear integro-differential equations for which there are only numerical solutions. Equations of this kind must be solved by a method of successive approximations. One

‡Recall that $\langle\psi|\hat{\alpha}|\phi\rangle \equiv \int \psi^*\hat{\alpha}\phi\, dV$.

chooses a starting set of orbitals which reproduces as closely as possible the particular features of the problem and substitutes into the expression for the operator $\hat{\mathcal{H}}^{HF}$. Upon solving the subsequent eigenvalue problem, a new set of orbitals is found which is once more substituted into $\hat{\mathcal{H}}^{HF}$, this process being continued until self-consistency is reached.

In the case of atoms the presence of central symmetry causes the variables in the Hartree–Fock equation to factorize. Integration over the angular variables reduces the problem to Hartree–Fock equations for the radial functions only (Hartree, 1957). Factorization of the variables is not possible for molecules, and the process of solving the equations becomes extremely complicated. Roothaan (1951) suggested that one constrict the class of functions to be varied to those of the LCAO type, and that one vary only the coefficients in the linear combinations. As a result, the integro-differential Hartree–Fock equations reduce to a system of nonlinear algebraic equations for the coefficients (the Roothaan equations). These are also solved by a process of self-consistency, and the solutions represent the best molecular orbitals that can be constructed from the given basis. The larger the set of basis functions, the more accurate is the method. In a number of cases the self-consistent orbitals for atoms (Clementi, 1965) determined by the Roothaan method are practically indistinguishable from those that are obtained by solving the Hartree–Fock equations numerically. It is difficult to achieve such an accuracy in molecules, since the process of self-consistency involves the extremely laborious process of calculating multicenter molecular integrals. The number of two-electron repulsion integrals in the Roothaan equations increases as the fourth power of the number of functions in the basis set (Nesbet, 1967).

The orbitals that satisfy the Hartree–Fock equations (8.174) are all eigenfunctions of a single Hamiltonian. From this it follows immediately that: (1) Self-consistent orbitals that belong to different eigenvalues of Eq. (8.174) are orthogonal to one another, and (2) orbitals that correspond to the same eigenvalue transform according to irreducible representations of the symmetry group of the Hamiltonian $\hat{\mathcal{H}}^{HF}$ (as long as there is no accidental degeneracy). The operator $\hat{\mathcal{H}}^{HF}$ consists of the one-electron operator $\hat{f}$, the symmetry of which is determined by the equilibrium configuration of the nuclei—i.e., the symmetry of $\hat{f}$ coincides with that of the symmetry group of the molecule—and depends upon the orbitals in the sum over the Coulomb and exchange operators. It is not difficult to show [see Roothaan (1951)] that this sum is invariant under any unitary transformation of the orbitals, i.e., the sum is a scalar with respect to operations of the point group. The symmetry of the Hamiltonian $\hat{\mathcal{H}}^{HF}$ therefore coincides with that of the operator $\hat{f}$, and solutions of the Hartree–Fock equations must transform according to irreducible representations of the point symmetry group of the molecule.

The Hartree–Fock equations are obtained by equating to zero the first variation of the energy. As is well known, this is a necessary but not a sufficient condition for the existence of an absolute minimum in the energy. This became a matter for serious concern after the appearance of a paper by Overhauser (1960, 1962) in which he showed that for an infinite one-dimensional chain of fermions interacting through a potential of the form of a δ function there exists a previously unknown type of solution of the Hartree–Fock equations which gives a lower energy than the generally accepted plane wave solution. Starting from the condition that the second variation of the energy be positive, Thouless (1961) derived a condition by which one can determine whether a particular solution does, in fact, correspond to an absolute minimum. By applying this criterion to the theory of collective motions in a many-particle system, he showed that if a Hartree–Fock state does not correspond to an absolute minimum, the state is unstable with respect to the collective modes. Following Thouless, the condition for the existence of an absolute minimum in the energy came to be called the *stability condition* for a Hartree–Fock state. The general form of stability conditions was examined by Adams (1962) [see also Löwdin (1963) and Čižek and Paldus (1967)].

The problem of finding the set of self-consistent orbitals that yields the lowest value of the energy of a system always occurs in calculations on finite systems. The Hartree–Fock equations in fact possess an infinite number of solutions, each of which corresponds to a particular value of the one-electron energy ϵ_a. One has to pick from these a set of $N/2$ orbitals which corresponds to the ground state of the system. The condition that the lowest one-electron levels are successively filled does not always lead to the lowest total energy. The derangement in the order in which the shells are filled in the iron-group elements provides an example of this: Filling of the 4s shell begins before the 3d shell is completed. The widely held opinion (Schiff, 1955; Pauling, 1960) that in the heavy elements the energy of the 4s orbitals is always lower than that of the 3d orbitals is incorrect. Experimental results (Slater, 1955) and Hartree–Fock calculations (Watson, 1960a, b) both show that in the iron-group elements the energy of the 4s electrons is higher than that of the 3d electrons, but that the energy of the configuration $3d^{n-2}4s^2$ is nevertheless lower than that of the configuration $3d^n$ (Watson, 1960a, b). Consequently, the state with configuration $3d^{n-2}4s^2$ is stable in this case. The conditions given by Adams (1962) and Čižek and Paldus (1967) may turn out to be useful in other similar cases, since if a solution which has been obtained is stable, there is no need to consider other possible configurations. However, as shown by Adams (1962), the nonlinear character of the Hartree–Fock equations means that there may be several states which give an absolute minimum in the energy. Hence one can never be sure, having found one of them, that there is not a lower state.

Let us assume that by solving the Hartree–Fock equations we have found a set of m orbitals for the ground state of the system (the configuration K_0), and that we wish further to calculate the contribution to the ground state of singly excited configurations, $(K_0)_a^e$, which are obtained by transferring an electron from an orbital of the ground state ϕ_a to one of the empty self-consistent orbitals ϕ_e. The extent to which a configuration is mixed is proportional to the off-diagonal matrix element of the Hamiltonian of the system between the two configurations K_0 and $(K_0)_a^e$. An expression for this matrix element in terms of electron interaction integrals is given by formula (8.79). We take the integral $g_{aa,ab}$ in this formula into the summation, having rearranged indices beforehand, and make use of the definition of Coulomb and exchange operators, (8.168):

$$\begin{aligned}\langle K_0 | \hat{\mathcal{H}} | (K_0)_a^e \rangle^{[2m]} &= \sqrt{2}\{f_{ae} + \sum_b (2g_{ba,be} - g_{ba,eb})\} \\ &= \sqrt{2}\langle \phi_a | \hat{f} + \sum_b (2\hat{\alpha}_b - \hat{\beta}_b) | \phi_e \rangle. \qquad (8.176)\end{aligned}$$

The operator on the right-hand side of this equation is just the Hartree–Fock operator (8.175). Since the orbital ϕ_e is a solution of the Hartree–Fock equations and is orthogonal to all the orbitals of the ground state, we find that

$$\langle K_0 | \hat{\mathcal{H}} | (K_0)_a^e \rangle^{[2m]} = \sqrt{2}\,\epsilon_e \langle \phi_a | \phi_e \rangle = 0, \qquad (8.177)$$

i.e., singly excited configurations do not mix directly with a Hartree–Fock state derived from a configuration of closed shells. This result is known as *Brillouin's theorem* (Brillouin, 1934), and plays an important role in configuration interaction calculations with Hartree–Fock orbitals. We emphasize that Brillouin's theorem only holds for orbitals which are *exact* solutions of the Hartree–Fock equations. It is not difficult to show that Brillouin's theorem also holds when the orbitals are exact solutions of the Roothaan equations. We note that singly excited configurations are, in fact, mixed into the ground-state configuration due to their interaction with more highly excited configurations. This may cause the contribution from singly excited orbitals to the ground state to remain equal in magnitude to the contribution from doubly excited configurations [see, for example, Karo (1959)].

From a mathematical point of view the determination of the self-consistent orbitals for a closed-shell configuration is a pseudoeigenvalue problem for the Hamiltonian $\hat{\mathcal{H}}^{HF}$ (it is a pseudoeigenvalue problem because $\hat{\mathcal{H}}^{HF}$ itself depends upon the orbitals being sought). The self-consistent orbitals ϕ_a are eigenfunctions of a single Hamiltonian. This is no longer so for configurations which contain open shells. Thus for a configuration which consists of singly occupied orbitals we have a system of equations [see (8.213)]

$$\hat{\mathcal{H}}_a(i)\phi_a(i) = \sum_b \epsilon_{ba}\phi_b(i), \qquad a = 1, 2, \ldots, N. \qquad (8.178)$$

Each orbital ϕ_a possesses its "own" Hamiltonian $\hat{\mathcal{H}}_a$. Therefore orbitals which belong to a configuration with open shells cannot be grouped into bases for irreducible representations of the molecular symmetry group, and in general the orbitals are not even orthogonal to each other.

Since the matrix of Lagrange multipliers ϵ_{ba} is not diagonal, Eqs. (8.178) no longer reduce to pseudoeigenvalue equations and this circumstance makes the process of solving the equations considerably more difficult. In some special cases where the configuration contains a single open shell, Roothaan (1960) managed to derive SCF equations in which the off-diagonal Lagrange multipliers are included in the Hamiltonian by means of specially defined Coulomb and exchange *coupling* operators. This method has been further developed and generalized by Huzinaga (1960, 1961), Birss and Fraga (1963), and Dyadusha and Kuprievich (1965a, b). Actual calculations based upon it, however, have only been carried out for simple configurations with one or two open shells. In more complicated configurations the self-consistency procedure of solving for the pseudoeigenvalues is usually either very slow or even diverges (Hinze and Roothaan, 1967; Wessel, 1967). It seems to be preferable to deal directly with SCF equations which contain the off-diagonal Lagrange multipliers. This approach was already used in earlier calculations by Lefebvre (1957) on open-shell configurations. Hinze and Roothaan (1967), and Wessel (1967) developed a quadratically convergent iterative procedure for solving the systems of algebraic equations which one obtains on applying Roothaan's method to the general SCF equations.

The exact many-electron wavefunction can always be written without any loss in generality in the form

$$\psi = c\psi_0 + \alpha\chi, \tag{8.179}$$

where ψ_0 is the Hartree–Fock function, χ is a correction function, orthogonal to ψ_0, and α is some small parameter. If ψ_0 is a good approximation to ψ, then $\alpha \ll 1$. The factor c ensures that ψ is normalized and is equal to $(1 - \alpha^2)^{1/2}$. Upon normalizing to an accuracy of α^4, we have

$$\psi = (1 - \tfrac{1}{2}\alpha^2)\psi_0 + \alpha\chi, \tag{8.180}$$

or

$$\psi_0 = (1 - \tfrac{1}{2}\alpha^2)^{-1}(\psi - \alpha\chi) \simeq (1 + \tfrac{1}{2}\alpha^2)(\psi - \alpha\chi). \tag{8.181}$$

Let us write out an expression for the energy of a state with wavefunction (8.181), taking into account the Hermitian character of the Hamiltonian and the fact that ψ is an eigenfunction of it, i.e., $\hat{\mathcal{H}}\psi = E\psi$,

$$\begin{aligned} E^{\text{HF}} &= \langle\psi_0|\hat{\mathcal{H}}|\psi_0\rangle = (1 + \alpha^2)[\langle\psi|\hat{\mathcal{H}}|\psi\rangle - 2\alpha\langle\chi|\hat{\mathcal{H}}|\psi\rangle + \alpha^2\langle\chi|\hat{\mathcal{H}}|\chi\rangle] \\ &= E - \alpha^2(E - \langle\chi|\hat{\mathcal{H}}|\chi\rangle). \end{aligned} \tag{8.182}$$

Thus for a function which is defined to first order in α the energy is correct to second order in α. It should be noted that nowhere in this derivation was any use made of the specific characteristics of the Hartree–Fock function. This result is a particular case of a more general situation in perturbation theory according to which the definition of a function to order n in a parameter of smallness yields the energy to order $2n + 1$.

In general, one does not expect the same accuracy for the expectation value of a one-electron operator $\hat{F}$. Substitution of function (8.181) into the expression for a diagonal matrix element of F gives a correction to the exact value which is first order in α. However, as shown by Goodisman and Klemperer (1963), because the Hartree–Fock function satisfies Brillouin's theorem, terms $\langle\psi_0|\hat{F}|\chi\rangle$ which occur to first order in α are in fact themselves of order α, and the expectation value of $\hat{F}$ is given by

$$\langle\psi_0|\hat{F}|\psi_0\rangle = \langle\psi|\hat{F}|\psi\rangle + \alpha^2[\langle\psi|\hat{F}|\psi\rangle - 2\langle\psi_0|\hat{F}|\chi\rangle - \langle\chi|\hat{F}|\chi\rangle]$$

This circumstance leads one to expect that the use of Hartree–Fock functions to calculate quantities which are described by one-electron operators should yield fairly accurate results. In fact, for small molecules, for which the Roothaan orbitals are close to the true Hartree–Fock orbitals, the accuracy of the calculated values of dipole moments, quadrupole moments, and other one-electron properties is about 1% (Nesbet, 1967; Gerratt and Mills, 1968; Gerratt, 1968).

The calculated values of the total energies of molecules are extremely accurate. Thus a calculation by the Hartree–Fock–Roothaan method of the energy of the ground state of N_2 gives 99.5% of the experimental value (Cade *et al.*, 1966), 99% in the case of LiH (Kahalas and Nesbet, 1963), and 96% for H_2 (Kołos and Roothaan, 1960). However, in spite of such accuracy the error is still comparable to the characteristic molecular parameters which are of interest to us, i.e., the dissociation energy and electronic transition frequencies. Energy level differences thus are not always adiabatic in character; the errors in calculating different energy levels are often different. The magnitude of the error also increases with increasing internuclear distance, due to the deterioration in the LCAO-MO approximation as the nuclei separate.

As a result, Hartree–Fock calculations for some molecules give the order of the energy levels incorrectly. The ground state of the stable F_2 molecule is even found to be repulsive.‡ For most molecules incorrect dissociation products are predicted, as, for example, in LiH and N_2 (Nesbet, 1961). The error in the calculated values of dissociation energies is particularly severe. Thus the error in the value of the dissociation energy of H_2 is 7.74 eV (Löwdin, 1959), which is 160% of the experimental value.

These errors are all a consequence of the one-particle approximation upon which the Hartree–Fock method is based. The increase in the error as the nuclei are separated is also due to the incorrect asymptotic behavior of the Hartree–Fock orbitals. In order to improve matters, it is necessary either to

‡See Wahl (1964).

abandon the one-particle approximation altogether, or, within this approximation, to take into account more fully the influence which the electrons exert upon each other. In other words, it is necessary to take electron correlation more completely into account.

8.11. Survey of Methods of Treating Electron Correlation

In a real molecule the motion of each electron is correlated with the motion of all the remaining electrons. Indeed, this can be seen directly from the interaction law for electrons. As a consequence of the Coulomb repulsion, the approach of two electrons is energetically unfavorable, since their energy of interaction e^2/r_{12} tends to infinity as r_{12} tends to zero. One might say that each electron is surrounded by a "Coulomb hole" with respect to the other electrons.

However, the Hartree–Fock method does take into account electron correlation to some extent, since the method allows for electron exchange. Thus antisymmetrization of the total wavefunction imposes a certain amount of correlation between the motions of those electrons whose spins are oriented in the same direction. This correlation is present independently of whether or not the wavefunction is constructed in a single-particle or many-particle approximation.

Consider some antisymmetric function of N electrons, $\psi(q_1, q_2, \ldots, q_N)$, where q_i stands for the set of three spatial and one spin coordinates. We write the two-electron density matrix as

$$\rho(q_1'q_2' \mid q_1q_2) = \int \psi^*(q_1'q_2'q_3, \ldots, q_N)\psi(q_1, q_2, q_3, \ldots, q_N)\, dq_3 \cdots dq_N. \tag{8.183}$$

From its form it can be seen that this density matrix is antisymmetric with respect to the permutations $(q_1'q_2')$ and (q_1q_2). Hence the "diagonal" element of the density matrix $\rho(q_1q_2 \mid q_1q_2)$, which defines the probability of locating electrons 1 and 2, reduces to zero when $q_1 = q_2$. Consequently, the probability of finding two electrons with the identical spin orientation at the same point in space is equal to zero (the "Fermi hole"). The antisymmetry of the total wavefunction does not restrict the motion of electrons with oppositely oriented spins.

Thus within the one-particle approximation, Hartree–Fock calculations allow for correlation between electrons with identically oriented spins. It is customary to call the *correlation energy* E^{corr} the correction to the energy obtained in the Hartree–Fock method. Löwdin (1959) gives the following definition of the correlation energy:

The correlation energy of a state of a particular Hamiltonian is defined as the difference between the exact eigenvalue of the Hamiltonian for that state and the value which is obtained in the Hartree–Fock approximation.

From this definition it follows that the correlation energy depends upon the form of the Hamiltonian. Usually in deriving the Hartree–Fock equations, one takes as the starting point the nonrelativistic Hamiltonian in the Born–Oppenheimer approximation, so that

$$E^{\text{corr}} = E^{\text{exact}} - E^{\text{rel}} - E^{\text{na}} - E^{\text{HF}}. \tag{8.184}$$

In the case of open-shell configurations there are several modifications of the SCF equations, each of which gives a different value for E^{HF}. This leads to a nonuniqueness in the definition of E^{corr} [see Sinanoğlu (1964)]. One should also note that for molecules E^{corr} is a function of the internuclear separations.

The precise determination of the correlation energy of the electrons in molecules is very difficult due to the absence of reliable values for E^{exact} and to the fact that the other corrections appearing in (8.184) can only be determined approximately. For H_2 in its ground state at the equilibrium internuclear separation $E^{\text{corr}} = -1.06$ eV (Löwdin, 1959). This is close to the correlation energy for the $1s^2$ electrons in light atoms 1.19 eV (Allen *et al.*, 1963), this value being practically independent of Z. The correlation energy for a $2p^2$ pair of electrons is estimated to be about 1.7 eV (Allen *et al.*, 1963). The correlation energy increases with increasing atomic number. Thus the average correlation energy per pair of electrons in the Cl atom is equal to 2.29 eV (Clementi, 1963a; Slater, 1967). In molecules which do not contain any heavy elements the correlation energy per pair of electrons is probably about 2 eV. Clementi (1963b) estimates that the correlation energy per pair of valence electrons in the CH_4 molecule is 2.02 eV. Sinanoğlu (1969) constructed a model of the four bonds in CH_4 by forming sp^3 hybrids from the atomic orbitals of the Ne atom, this atom containing the same number of paired electrons (ten) as CH_4. The correlation energy for the eight electrons which occupy the four hybridized orbitals is found to be equal to -8.64 or -2.16 eV per pair. In addition, the magnitude of the correlation energy for a pair of electrons with antiparallel spins is of the same order as the correlation energy between pairs.

The first calculations in which electron correlation was included were carried out by Hylleraas (1928, 1929) in his classic work on the He atom. These papers in fact deal with all the three main methods of treating correlation: (1) The introduction of interelectronic separations into the wave function; (2) the use of configuration interaction; and (3) the use of different orbitals for electrons with different spin orientations. We briefly consider all these three methods.

(1) In a variational calculation on the ground state of the helium atom Hylleraas wrote the wavefunction in the form of an infinite series

$$\Phi(\mathbf{r}_1, \mathbf{r}_2) = e^{-(r_1+r_2)/2} \sum_{l,m,n=0}^{\infty} c_{lmn}(r_1 + r_2)^l (r_1 - r_2)^m r_{12}^n. \tag{8.185}$$

Eight terms of this series prove to be sufficient to obtain very good agreement with experiment. Following Hylleraas' successful calculation, James and

Coolidge (1933) carried out a similar variational calculation on the ground state of the H_2 molecule. The James–Coolidge function included the internuclear distance R_{ab} as a parameter, the function going over into the Hylleraas function (8.185) when $R_{ab} = 0$. A calculation with a 13-term function gave a value for the dissociation energy of $E_D = 4.7200$ eV which agrees very well with the experimental value $E_D^{\exp} = 4.7466 \pm 0.0007$ eV (Herzberg, 1950). In order to obtain a more accurate result, Kołos and Roothaan (1960) carried out a calculation with a James–Coolidge-type function with up to 50 terms. As a result, they obtained a value of $E_D = 4.7467$ eV for the dissociation energy, in complete agreement with all the proven figures in the experimental quantity.‡ These calculations demonstrate conclusively that both atoms and molecules are completely described, not only qualitatively, but also quantitatively, by the laws of quantum mechanics.

Although it is simple to include the interelectronic separation in the wavefunction for two electrons, the situation becomes complicated in many-electron systems. In an N-electron system there are $\frac{1}{2}N(N-1)$ interelectronic separations, which, when N is greater than seven, exceed the number of independent spatial coordinates $3N$.§ As was first shown by Wigner (1938), a large part of the correlation energy arises from electron pair correlations. One would naturally construct a wavefunction which allows for this kind of correlation from two-electron functions $\psi(q_i, q_j)$. Functions such as these are known as *geminals*, and in a nonrelativistic approximation they can always be written as a product of a coordinate and a spin function:

$$\psi(q_i, q_j) = \phi(\mathbf{r}_i, \mathbf{r}_j)\chi(\sigma_i, \sigma_j). \tag{8.186}$$

The coordinate functions $\phi(\mathbf{r}_i, \mathbf{r}_j)$ are called *biorbitals*,¶ in analogy with orbitals. The biorbitals are symmetric with respect to permutations of the spatial electronic coordinates when the spins are paired ($S_{ij} = 0$) and are antisymmetric when the spins are parallel ($S_{ij} = 1$).

The most rigorous method of treating pair correlation is to determine the biorbitals as solutions to equations which describe the motion of an electron pair in a self-consistent field created by all the other electrons. An SCF equation for symmetric biorbitals was first derived by Fock *et al.* (1940), who described the remaining $N - 2$ electrons which constitute the self-consistent field by Hartree–Fock orbitals. This equation can be written in the form

‡For a survey of more recent, very accurate calculations on the H_2 molecule see Kołos (1968).

§Conroy (1964, 1967) successfully introduced interelectronic coordinates in a variational function in order to improve the rate of convergence of configuration interaction expansions of molecular wavefunctions for two- and three-electron systems.

¶There is so far no clearly agreed upon terminology; there is often confusion in the literature between the concepts of biorbitals and geminals (Daudel and Durand, 1965; Bratož and Durand, 1965).

(Labzovskii, 1963)

$$[\hat{\mathcal{H}}^{\mathrm{HF}}(i) + \hat{\mathcal{H}}^{\mathrm{HF}}(j) + \hat{g}(i,j) + \hat{R}_{ab}(i,j)]\phi_{ab}(i,j) = \epsilon_{ab}\phi_{ab}(i,j), \tag{8.187}$$

in which $\hat{\mathcal{H}}^{\mathrm{HF}}(i)$ is a one-electron Hartree–Fock operator, $\hat{g}(i,j)$ is the operator for the two-electron repulsion, and $\hat{R}_{ab}(i,j)$ is a two-electron integral operator which depends upon the biorbitals ϕ_{ab} being sought. The exact solution of these equations is more difficult than the solution of the Hartree–Fock equations (8.174). Equations (8.187) were written in a density matrix formalism by Labzovskii (1963), who devised an approximate method of solution using perturbation theory. A calculation on the lithium atom showed that the correlation energy obtained in the second order of perturbation theory accounts for 85% of the total correlation energy. Kutzelnigg (1964) has also discussed the procedure for solving the SCF equations for orthogonal geminals. Expressions for the one- and two-electron density matrices for an antisymmetric product of geminals were derived by Mestechkin (1967). He showed that it is unnecessary to expand the geminals in a one-electron basis in order to calculate either these matrices or the energy. This means that one could construct models in which the correlation is described by a certain type of r_{12} dependence in the geminal.

We note that with nuclei one of the methods of treating two-nucleon correlations is to solve the Brueckner equation (Brueckner, 1954, 1955; Bethe, 1956). This equation was derived for an infinite medium with a short-range potential. Its differential variant is called the Bethe–Goldstone equation (Bethe and Goldstone, 1957). If the operator $\hat{R}_{ab}(i,j)$ in (8.187) is taken to be the same for all ϕ_{ab} and the background potential is taken to be of Brueckner type instead of Hartree–Fock, then, as shown by Szasz (1959), the Fock–Veselov–Petrashen equation becomes the Bethe–Goldstone equation. This kind of averaging is natural for a uniform nuclear medium. In general, its application to atoms and molecules requires special study (Nesbet, 1965).

Sinanoğlu (1964), Kelly (1963, 1964, 1966), Silverstone and Sinanoğlu (1965, 1966) and Freed (1968) have developed a more general approach to the many-electron problem in which one can treat contributions to the correlation in different orders. The variational function consists of the Hartree–Fock function plus a correlation function which is orthogonal to it:

$$\psi = \psi^{\mathrm{HF}} + \chi. \tag{8.188}$$

The equations for χ which are obtained from the variation principle are solved with the aid of many-particle perturbation theory. The function χ separates into one-, two- and many-electron terms, these giving the corresponding orders in the correlation. Calculations of the correlation energy for the ground states of closed-shell systems have shown that this energy is almost completely determined by the two-electron correlations (Sinanoğlu, 1964). Thus the energy due to pair correlations in the ground state of the Be atom calculated by Sinanoğlu's method amounts to 97% of the experimental value for the total correlation energy (Geller *et al.*, 1965). However, many-electron correlations may be important in some open-shell systems (Silverstone and Sinanoğlu, 1965, 1966). Accurate calculations of the pair correlation energy in molecules encounter a number of numerical difficulties due to the need for calculating multicenter integrals over two-electron functions. Several semi-empirical variants of the many-electron theory have been suggested in connection with this problem (Hollister and Sinanoğlu, 1965; Veselov and Labzovskii, 1968; Bessis *et al.*, 1969).

(2) The simple configuration interaction method (without self-consistency) has been widely applied in molecular calculations for treating the correlation energy (see Section 6.9). The self-consistent variant of this method (which we denote by MC SCF for short) was first discussed in the book by Frenkel (1934). Frenkel noted the importance of the linear independence of the variational conditions which one imposes upon the coefficients of the configurations and upon the one-electron functions. Yutsis (1952) [see also Hartree, Hartree, and Swirles (1939)] has given a rigorous derivation of the MC SCF equations for atoms, and has obtained solutions of the two-configuration equations for the Be atom (Kibertas and Yutsis, 1953). This work shows that it is an extremely arduous task to solve the MC SCF equations for the radial functions in a central field, even in simple cases. However, an approximate method in which simplified equations are used has been developed [see Yutsis *et al.* (1962b)].

Löwdin (1955a) and McWeeny (1955) have discussed the general foundations of the MC SCF method and the possibility of applying it to molecular problems. However, for a long time self-consistency was either completely neglected in configuration interaction calculations (if one does not count the construction of molecular orbitals from atomic orbitals purely by symmetry), or was treated by the "average configuration" method (Slater, 1963). In this method one picks out from all the configurations under consideration a single configuration corresponding to a state which can be described by a single determinant. Such a configuration may either be the closed-shell ground-state configuration or a configuration of singly occupied orbitals in a state with the maximum spin. [Slater (1963) points out that this last configuration gives the best average description of the whole set of configurations.] The SCF equations for the chosen configuration are then solved, and the self-consistent orbitals which are obtained are used to construct all the possible configurations. The interaction of these configurations leads to the solution of a secular equation. However, the series of configurations constructed in this manner usually converges very slowly. The rate of convergence would be considerably increased if the orbitals and the coefficients of the configurations were optimized simultaneously.

The first molecular configuration interaction calculations in which self-consistency was included were carried out by Das and Wahl (1966) and Das (1967) for the ground states of diatomic molecules with two valence electrons (H_2, Li_2, and F_2 molecules). The calculations were performed in the LCAO-MO approximation, and only doubly excited configurations were considered. Veillard and Clementi (1967) and Clementi (1967) extended the method to polyatomic molecules. Let us consider a molecule in its ground state with a configuration K_0 of m doubly filled orbitals. In addition, there are $k - m$ excited orbitals. We consider the doubly excited configurations $(K_0)_{aa}^{ee}$ formed by transferring two electrons from an orbital ϕ_a to an excited orbital ϕ_e. All configurations consist of doubly occupied orbitals only, and each one can be described by a single Slater determinant. The variational wavefunction is constructed in the form of a superposition of wavefunctions for the configurations,

$$\psi = A_0\psi(K_0) + \sum_{a=1}^{m} \sum_{e=m+1}^{k} A_{ae}\psi[(K_0)_{aa}^{ee}]. \tag{8.189}$$

The molecular orbitals are expanded in a set of Gaussian atomic orbitals $\phi_{q\mu}$,

$$\phi_a = \sum_{q,\mu} c_{aq\mu}\phi_{q\mu}, \tag{8.190}$$

in which the index q enumerates the atoms and the index μ the orbitals on a given atom. The solution of the equations which are obtained requires the simultaneous optimization of the sets of coefficients A_{ae} and $c_{aq\mu}$. Calculations on polyatomic molecules by the method of Veillard and Clementi (1967) and Clementi (1967) have, apparently, so far not been carried out.

The methods of Das and Wahl (1966), Das (1967), Veillard and Clementi (1967), and Clementi (1967) are only applicable to the ground states of molecules. Furthermore, the extent of the configuration interaction is severely limited, since one only considers doubly excited configurations with closed shells. General formulations of the MC SCF problem have been given by Adams (1967), Hinze and Roothaan (1967), and Huzinaga (1969). Hinze and Roothaan (1967) have proposed a quadratically convergent iterative method for solving the system of nonlinear algebraic equations which are obtained when the orbitals in the MC SCF method are expanded in a given basis. The method consists in successively refining the first guess for the matrix of orbital expansion coefficients. The correction matrix is found from a system of linear algebraic equations.

It should be noted that Brillouin's theorem in the form (8.177) does not hold for MC SCF wavefunctions (nor for configurations with open shells). The theorem was generalized to the multiconfiguration approximation by Levy and Berthier (1968). In the MC SCF method Eq. (8.177) is replaced by

$$\langle\psi_0|\hat{\mathcal{H}}|\psi_0(a \to b)\rangle = 0, \tag{8.191}$$

where

$$\psi_0 = \sum_K A_K \psi(K) \tag{8.192}$$

is a solution of the MC SCF equations corresponding to the ground state of the system and

$$\psi_0(a \to b) = \sum_K A_K[\psi(K_a^{\ b}) - \psi(K_b^{\ a})] \tag{8.193}$$

is a linear combination of singly excited configurations. The function $\psi(K_a^{\ b})$ in (8.193) is taken to be equal to zero for configurations K that do not contain the orbital ϕ_a or in which the orbital ϕ_b is doubly occupied. We note that some of the functions $\psi(K_a^{\ b})$ in (8.193) may coincide with the functions $\psi(K)$ in (8.192). The generalized Brillouin theorem does not impose any restrictions upon the configurations which occur in the ground-state wavefunction (8.192). This expansion may therefore also include singly excited configurations.

(3) The third group of methods for treating electron correlation is known as "different orbitals for different spins." As pointed out before, the antisymmetry of the total wavefunction in the Hartree–Fock method gives rise, within the one-particle approximation, to correlation between electrons which have the same spin orientation: Electrons with the same spin may not be placed in the same orbital. In a similar way, in order to introduce correlation between electrons with oppositely oriented spins, one would naturally place them in different orbitals. This kind of approach was used by Hylleraas (1928, 1929) and Eckart (1930b) in calculations on the ground state of the helium atom. The SCF equations for molecular states described by a single determinant but with different orbitals for electrons with different spin orientations were obtained by Pople and Nesbet (1954) and Berthier (1954). This method is known as the *unrestricted Hartree–Fock method* (UHF method). The wavefunction in the UHF method corresponds to a definite value for the projection of the total electron spin, i.e., the function is an eigenfunction of the operator $\hat{S}_z$ but not of the operator $\hat{\mathbf{S}}^2$. A function such as this contains contributions from all the multiplet states which arise from the given system of spins.

Löwdin (1955b) proposed that one should take as the initial variation function a linear combination of determinants which does form an eigenfunction of $\hat{\mathbf{S}}^2$, the combination being obtained by applying a specially defined projection operator $\mathfrak{O}^{\mathrm{S}}$ to the determinant, $\mathfrak{D}_M$,

$$\psi_M^{\mathrm{S}} = \mathfrak{O}^{\mathrm{S}}\mathfrak{D}_M. \tag{8.194}$$

This method is sometimes known as the *projected unrestricted* Hartree–Fock method (PUHF method). However, Löwdin did not derive the explicit form of the SCF equations which result from the functions (8.194). Subsequent calculations on molecular systems by the PUHF method have been carried out using a restricted atomic basis which does not require optimization.

The method which has achieved the widest currency is the *alternant molecular orbital method* (AMO method), having been successfully applied to so-called *alternant systems* (Coulson and Longuet-Higgins, 1948).‡ The atoms of an alternant system can always be divided into two subsets, the atoms of which alternate in such a way so that if we move by the same step from one atom to another of subset I, we never land upon an atom of subset II (examples are conjugated chains of carbon atoms, aromatic hydrocarbons, and crystals with two cubic sublattices). For every MO of the ground state of an alternant system ϕ_a there is a corresponding excited MO $\phi_{\bar{a}}$ which is obtained from ϕ_a by reversing the sign of the coefficients of the atoms belonging to one of the subsets, for example, by reversing the sign of the atoms of subset II:

$$\phi_a = \sum_q^{\mathrm{I}} c_{qa}\chi_q + \sum_p^{\mathrm{II}} c_{pa}\chi_p, \qquad \phi_{\bar{a}} = \sum_q^{\mathrm{I}} c_{qa}\chi_q - \sum_p^{\mathrm{II}} c_{pa}\chi_p. \tag{8.195}$$

The orbitals in the AMO method are constructed in the form of nonorthogonal (but normalized) linear combinations

$$\psi_a = (\cos\theta)\phi_a + (\sin\theta)\phi_{\bar{a}}, \qquad \psi_{\bar{a}} = (\cos\theta)\phi_a - (\sin\theta)\phi_{\bar{a}}, \tag{8.196}$$

in which the form of the coefficients merely reflects the fact that the sum of their squares must be equal to unity, and the angle θ is a variational parameter. On substituting (8.195) into (8.196), it is easily seen that the orbitals of (8.196) are characterized by alternating maximima and minima in the amplitudes between adjacent atoms (alternation in amplitude). In addition, the phases of the functions ϕ_a and $\phi_{\bar{a}}$ alternate oppositely. Electrons with α spin projection are assigned to the orbitals ϕ_a and electrons with β spin to the orbitals $\phi_{\bar{a}}$. The determinant of alternant spin-orbitals is then projected with the aid of Löwdin's operator $\mathcal{O}^S$ onto a state with a definite value for the total spin S. When $\theta = 0$ the orbitals (8.196) become identical and the singlet AMO wavefunction goes over into the usual single-determinant MO wavefunction.

A calculation on the benzene molecule by the AMO method in which just a single parameter was varied (Itoh and Yoshizumi, 1955) gave an energy lowering of 2.35 eV relative to the simple MO method [for comparison, a configuration interaction calculation with nine of the most important configurations gave an energy lowering of 2.70 eV (Parr *et al.*, 1950). A similar AMO calculation on the lowest state of a hexagonal ring of six H atoms (Moskowitz, 1963) at $R = 2$ a.u. gave an energy lowering of 1.29 eV compared to the MO method [a complete configuration interaction calculation with 1s orbitals gave an energy lowering of 2.08 eV (Moskowitz, 1963). The AMO method is therefore equivalent to a restricted configuration interaction treatment. Introduction of several variational parameters into the AMO function leads to a substantial improvement in the results (Pauncz, 1965, 1967).

‡The AMO method can also be applied to nonalternant systems (Pauncz, 1967).

Another variant of the "different orbitals for different spins" method was proposed by Hirst and Linnett (1961) and Linnett (1964) and is known as the *nonpaired spatial orbital method* (NPSO method). The orbitals in the NPSO method are taken to be two-center semilocalized orbitals of the Coulson–Fischer type as in H_2 (Coulson and Fischer, 1949). An example of Coulson–Fischer orbitals is provided by the orthogonal localized orbitals in (8.25). Accommodation of electrons with different spin orientations in semilocalized orbitals is equivalent to separating the electrons spatially. Calculations by the NPSO method on a series of simple aromatic and conjugated molecules (Empedocles and Linnett, 1964) have yielded results which are substantially closer to complete configuration interaction calculations than calculations by the AMO method. An NPSO calculation on the $^3A_2$ ground state of the trimethylene methyl radical provides an example where the result is particularly close to those given by configuration interaction calculations. With the particular atomic basis the energy given by the NPSO method differs from the configuration interaction result by 0.04 eV in all. At the same time the corresponding difference in the energy for the AMO method amounts to 1.43 eV (Chong, 1965). In a comparative analysis of the AMO and NPSO methods Pauncz (1965) concluded that the NPSO method gives better results for small molecules, but that it is possible to apply the AMO method to more complex systems.

The SCF equations for the PUHF method were derived by Goddard (1967b). In constructing the function to be varied, he started from a form in which the function is factored into products of coordinate and spin functions. By means of the so-called *orthogonal units* (Rutherford, 1947) [which are identical to the Young operators (2.30) to within a multiplicative factor], the functions in the products were symmetrized according to mutually dual irreducible representations of the permutation group. The wavefunction which is obtained in this way is an eigenfunction of $\hat{\mathbf{S}}^2$ and simultaneously satisfies the exclusion principle (see Section 6.9). On equating the first variation of the energy to zero and assuming the orbitals to be nonorthogonal, Goddard arrived at the set of N SCF equations

$$\hat{\mathcal{H}}_a \phi_a = \epsilon_a \phi_a, \qquad a = 1, 2, \ldots, N. \tag{8.197}$$

The form of the operators $\hat{\mathcal{H}}_a$ is given by Goddard (1967b). Since the orbitals are nonorthogonal, the matrix of Lagrange multipliers is diagonal. However, the determination of the $\hat{\mathcal{H}}_a$ operators now becomes very complicated, due to the need for calculating cumbersome linear combinations of overlap integrals.

For configurations of singly occupied orbitals, i.e., for the configurations occurring in the "different orbitals for different spins" method, there are always several linearly independent ways of vector-coupling the spins of the individual electrons to form a given resultant spin S (the case when S takes on its maximum value forms an exception). Several linearly independent variational functions, each of which is an eigenfunction of $\hat{\mathbf{S}}^2$, can therefore be constructed. The number of such functions is determined by the dimension f_λ, of the irreducible representation of the permutation group $\Gamma^{[\lambda]}$ and is given by formula (8.61). There are therefore f_λ systems of SCF equations (Goddard, 1967b).

Equations (8.197) become simplified for a coordinate wavefunction which is antisymmetric with respect to the permutations of the first n electron coordinates and with respect to the last $N - n$ coordinates. In this case the orbitals can be divided into two sets, the orbitals of each set forming eigenfunctions of their own effective Hamiltonian [the Goddard GF equations (Goddard, 1968)]:

$$\begin{aligned} \hat{\mathcal{H}}^{\alpha}\phi_a &= \epsilon_a\phi_a, \qquad a = 1, 2, \ldots, n, \\ \hat{\mathcal{H}}^{\beta}\phi_b &= \epsilon_b\phi_b, \qquad b = n + 1, \ldots N. \end{aligned} \tag{8.198}$$

The antisymmetry of the coordinate wavefunction with respect to permutations of the electrons in the two subsets means that the spins of the electrons in each subset are coupled to form the maximum value. The total spin of the state is obtained as the difference between the spins of the subsets

$$S = \tfrac{1}{2}n - \tfrac{1}{2}(N - n). \tag{8.199}$$

This mode of coupling is the same as that in the AMO method; for the ground state $n = N/2$ and $S = 0$. The orbitals in the AMO method should in fact be solutions of Eqs. (8.198), but because of the choice of atomic basis it is unnecessary to solve these equations, the alternant orbitals being found by symmetry arguments. Improvement of the AMO method by extending the basis set would then require the solution of these equations.

The treatment of the linearly independent eigenfunctions of $\hat{\mathbf{S}}^2$ in the AMO method is discussed by Lunnell and Lindner (1968) and Tamir and Pauncz (1968). In both of these works it was found that when all the linearly independent eigenfunctions are taken into account in the benzene molecule the energy lowering which results for the ground state and for a number of excited states is insignificant. Gerratt (1968) considers that this is due to the degeneracy of the benzene orbitals. However, it should be borne in mind that even though the energy lowering is small, inclusion of all the independent eigenfunctions of $\hat{\mathbf{S}}^2$ can lead to a significant change in the spin density distribution. Thus in the case of the ^{2}S ground state of the Li atom inclusion of the second linearly independent function substantially improves the calculated value of the hyperfine interaction constant although the change in energy is insignificant (Lunnell, 1968; Larsson, 1968). It is therefore desirable in molecular calculations to take into account all the independent functions which correspond to the state with the given S. The function to be varied is then of course constructed as a linear combination of eigenfunctions of $\hat{S}^2$. The coefficients in this linear combination are solutions of a secular equation, and the orbitals are solutions of SCF equations, as in the MC SCF method.

This approach was proposed by Gerratt (1968, 1971), Hameed *et al.* (1969), Poshusta and Kramling (1968), Kaplan and Maximov (1969, 1970). The SCF equations derived by Gerratt (1968, 1971) reduce to Goddard's equations upon restriction to a single independent eigenfunction of $\hat{S}^2$. The same problem is considered for atoms by Hameed *et al.* (1969). The orbitals obtained there for an N-electron atom are referred to by Hameed *et al.* (1969) as *the best radial N orbitals* (BRNO) since they represent the best orbitals which correspond to an independent-particle model (i.e., to a single-configuration approximation). In the work of Gerratt (1968, 1971) and Hameed *et al.* (1969) the orbitals are assumed to be nonorthogonal. SCF

Table 8.2

Calculations on the Ground State of H_2 in Various Approximations

Method and details of calculation	R_0 (a.u.)	$-E$ (a.u.)	D (eV)	Reference
SCF-LCAO-MO; single configuration; 1s, 2s, 2pσ atomic orbital basis	1.402	1.133494	3.632	Fraga and Ransil (1961)
SCF MO; single-configuration analytic nine-term function	1.400	1.133630	3.636	Kołos and Roothaan (1960)
SCF-LCAO-MO; single configuration, "different orbitals for different spins"; 1s, 2s, 2pσ atomic orbital basis	1.400	1.151526	4.130	Goddard *et al.* (1967b)
Configuration interaction; 12 configurations included from SCF-LCAO-MO with 1s, 2s, 2pσ atomic orbitals	1.402	1.159187	4.331	Fraga and Ransil (1961)
Configuration interaction; five most important configurations from symmetrized LCAO-MO with 1s, 2s, 2pσ, 2pπ atomic orbitals	1.400	1.167236	4.550	Mclean *et al.* (1960)
MC SCF-LCAO-MO with atomic basis as in the work of Mclean *et al.* (1960)	1.400	1.1698	4.63	Das and Wahl (1966), Das (1967)
Introduction of interelectronic separation in variational function; 50-term James–Coolidge function	1.4009	1.174448	4.7467	Kołos and Roothaan (1960)
Experiment	1.4006	1.174443	4.7466 $\pm$0.0007	Herzberg (1950)

equations which take into account all the independent eigenfunctions of $\hat{\mathbf{S}}^2$ for configurations of orthogonal orbitals are given by Poshusta and Kramling (1968) and Kaplan and Maximov (1969, 1970) (see the next section).

A somewhat different approach to the problem of including all the linearly independent eigenfunctions of $\hat{\mathbf{S}}^2$ has been developed by Ladner and Goddard (1969). They subject the spin functions to a certain orthogonal transformation **L** whose matrix is found by the variation principle at the same time as the orbitals. The orbitals satisfy a system of N SCF equations of the kind derived previously by Goddard (1967b), and the matrix **L** is found by a process of self-consistency from a system of f_λ nonlinear third-order algebraic equations. This set of $N+f_\lambda$ equations has been solved for the LiH, H_3, and H_4 molecules. In Table 8.2 are shown the results of some calculations on the ground state of the H_2 molecule by the various methods discussed in this section.

8.12. Self-Consistent Field Equations for Spin-Degenerate States

We now derive SCF equations for an arbitrary configuration of nondegenerate orbitals in a state with a definite value for the spin **S** (Kaplan and Maximov, 1969, 1970). Since there are usually several linearly independent wavefunctions for a particular **S**, these must all be taken into account when forming the function to be varied. We give the derivation first for the case of a configuration of singly occupied orbitals K_1 [see (8.56)].

A. *Configuration of Singly Occupied Orbitals*

The linearly independent states which arise from a configuration K_1 and possess the permutational symmetry of a Young diagram $[\lambda]$ (and, correspondingly, with a spin **S**), are enumerated by the Young tableaux r. These tableaux characterize the symmetry of the coordinate wavefunctions $\Phi_r^{[\lambda]}$ with respect to permutations of the orbitals.‡ The number of such states is equal to the dimension f_λ of the irreducible representation $\Gamma^{[\lambda]}$ [see (6.61)].

We seek the coordinate wavefunction of a state with permutational symmetry $[\lambda]$ in the form of a superposition of the functions $\Phi_r^{[\lambda]}$

$$\Phi^{[\lambda]}(K_1) = \sum_r A_r \Phi_r^{[\lambda]}(K_1). \qquad (8.200)$$

The coefficients A_r and the orbitals ϕ_a are determined by the requirement that the expectation value of the energy for a state with this coordinate wavefunction be a minimum:

$$\bar{E} = \langle K_1 | \hat{\mathcal{H}} | K_1 \rangle^{[\lambda]} = \sum_{r,\tilde{r}} A_r A_{\tilde{r}} \mathcal{H}_{r\tilde{r}}^{[\lambda]}(K_1). \qquad (8.201)$$

‡The index of the Young tableau that characterizes the symmetry with respect to permutations of the electron coordinates is dropped since it is irrelevant for the determination of the energy (see footnote on p. 214).

An expression for $\mathcal{H}_{r\bar{r}}^{[\lambda]}(K_1)$ in terms of molecular integrals was given in formula (8.57). In order to facilitate the subsequent variation, we rewrite this equation as a sum which is symmetric in the indices a and b, formally put $\tilde{N}_{aa}^{r\bar{r}} = -\delta_{r\bar{r}}$ and replace ϵ_a by h_a. Expression (8.201) then becomes

$$\bar{E} = \sum_{r,\bar{r}} A_r A_{\bar{r}} \{\delta_{r\bar{r}} \sum_a h_a + \tfrac{1}{2} \sum_{a,b} (\delta_{r\bar{r}} \alpha_{ab} + \tilde{N}_{ab}^{r\bar{r}} \beta_{ab})\}. \tag{8.202}$$

There are two kinds of quantity to be varied: the f_λ coefficients A_r, and the N orbitals ϕ_a. These are subjected to the following constraints:

$$\langle \phi_a | \phi_b \rangle = \delta_{ab}, \tag{8.203}$$

$$\sum_{r\bar{r}} A_r A_{\bar{r}} \delta_{r\bar{r}} = \sum_r |A_r|^2 = 1. \tag{8.204}$$

The orthogonality conditions for the orbitals (8.203) are necessary for Eq. (8.57) to be applicable. Conditions (8.204) are a consequence of the requirement that function (8.200) be normalized and the $\Phi_r^{[\lambda]}$ be orthonormal.

Variation of the coefficients A_r can be carried out independently of the variation of the orbitals, although the systems of equations which are obtained must be solved simultaneously. The A_r coefficients are obtained from the usual Ritz equations for orthogonal functions

$$\sum_{\bar{r}} A_{\bar{r}} (\mathcal{H}_{r\bar{r}}^{[\lambda]} - \Lambda \delta_{r\bar{r}}) = 0, \qquad r = 1, \ldots, f_\lambda, \tag{8.205}$$

in which the parameter Λ is determined by solving the secular equation

$$|\mathcal{H}_{r\bar{r}}^{[\lambda]} - \Lambda \delta_{r\bar{r}}| = 0. \tag{8.206}$$

We now derive the equations satisfied by the orbitals that minimize the energy of the system (8.202) under the constraints (8.203). We put

$$\tfrac{1}{2} \sum_{r,\bar{r}} A_r A_{\bar{r}} \tilde{N}_{ab}^{r\bar{r}} = M_{ab} \tag{8.207}$$

and express the Coulomb and exchange integrals in terms of the corresponding operators as in (8.170). The problem reduces to one of finding an unconditional extremum in the functional

$$J = \sum_a \langle \phi_a | \hat{f} + \sum_b (\tfrac{1}{2}\hat{\alpha}_b + M_{ab}\hat{\beta}_b) | \phi_a \rangle - \sum_{a,b} \epsilon_{ba} \langle \phi_a | \phi_b \rangle, \tag{8.208}$$

where ϵ_{ba} is a Lagrange multiplier. For this purpose it is necessary to find the first variation of the functional. When varying ϕ_b and $\phi_b{}^*$ it is useful to exploit the equivalence (8.170) of the two forms of writing the Coulomb and exchange integrals:

$$\begin{aligned} \delta J = \sum_a & [\langle \delta\phi_a | \hat{f} + \sum_b (\tfrac{1}{2}\hat{\alpha}_b + M_{ab}\hat{\beta}_b) | \phi_a \rangle \\ & + \langle \phi_a | \hat{f} + \sum_b (\tfrac{1}{2}\hat{\alpha}_\beta + M_{ab}\hat{\beta}_\beta) | \delta\phi_a \rangle] \\ & + \sum_{a,b} [\langle \delta\phi_b | \tfrac{1}{2}\hat{\alpha} + M_{ab}\hat{\beta}_a | \phi_b \rangle + \langle \phi_\beta | \tfrac{1}{2}\hat{\alpha}_a + M_{ab}\hat{\beta}_a | \delta\phi_\beta \rangle] \\ & - \sum_{a,b} \epsilon_{ba} [\langle \delta\phi_a | \phi_b \rangle + \langle \phi_a | \delta\phi_b \rangle]. \end{aligned} \tag{8.209}$$

Noting that the operators $\hat{f}$, $\hat{\alpha}_b$, and $\hat{\beta}_b$ are Hermitian [see Eq. (8.169)] and that $M_{ab} = M_{ba}$, we can rewrite (8.209) in the form

$$\begin{aligned}\delta J = &\sum_a \langle \delta\phi_a | \hat{f} + \sum_b (\hat{\alpha}_b + 2M_{ab}\hat{\beta}_b) | \phi_a \rangle - \sum_{a,b} \langle \delta\phi_a | \epsilon_{ba}\phi_b \rangle \\ &+ \sum_a \langle \delta\phi_a^* | \hat{f}^* + \sum_b (\hat{\alpha}_b + 2M_{ab}\hat{\beta}_b^*) | \phi_a^* \rangle \\ &- \sum_{a,b} \langle \delta\phi_a^* | \epsilon_{ab}\phi_b^* \rangle. \end{aligned} \tag{8.210}$$

Since the variations $\delta\phi_a$ and $\delta\phi_a^*$ are linearly independent, on equating δJ to zero, we arrive at two systems of equations

$$[\hat{f} + \sum_b (\hat{\alpha}_b + 2M_{ab}\hat{\beta}_b)]\phi_a = \sum_b \epsilon_{ba}\phi_b, \tag{8.210a}$$

$$[f^* + \sum_b (\hat{\alpha}_b^* + 2M_{ab}\hat{\beta}_b^*)]\phi_a^* = \sum_b \epsilon_{ab}\phi_b^*, \tag{8.210b}$$

$$a = 1, 2, \ldots, N.$$

Taking the complex conjugate of (8.210b) and subtracting it from (8.210a), we obtain

$$\sum_b (\epsilon_{ba} - \epsilon_{ab}^*)\phi_b = 0. \tag{8.211}$$

Since the orbitals are linearly independent, it follows that

$$\epsilon_{ba} = \epsilon_{ab}^*, \tag{8.212}$$

i.e., the matrix of Lagrange coefficients is Hermitian. Equation (8.212) is the condition that Eqs. (8.210a) and (8.210b) arc equivalent.

The set of orbitals that minimizes the energy must thus satisfy the system of N integro-differential equations

$$\hat{\mathcal{H}}_a\phi_a = \sum_b \epsilon_{ba}\phi_b, \qquad a = 1, 2, \ldots, N, \tag{8.213}$$

with the Hartree–Fock Hamiltonian

$$\hat{\mathcal{H}}_a = \hat{f} + \sum_b (\hat{\alpha}_b + 2M_{ab}\hat{\beta}_b). \tag{8.214}$$

There is a different Hamiltonian for each orbital. The operators $\hat{\mathcal{H}}_a$ are not invariant with respect to a unitary transformation of the orbitals, due to the presence of the factors M_{ab} in front of the exchange operators. Hence a unitary transformation of the orbitals which diagonalizes the matrix of Lagrange multipliers [as is carried out in the closed-shell case (Roothaan, 1951)] does not result in any simplification of Eqs. (8.213).

The system of N equations (8.213) contains N independent functions ϕ_a and N^2 independent parameters ϵ_{ba}. Another set of N^2 equations is therefore necessary for completeness. These are provided by the N^2 conditions (8.203). However, instead of these equations it is more convenient to use expressions for the Lagrange multipliers ϵ_{ba} in terms of the orbitals being sought. For this purpose we multiply Eq. (8.213) through from the left by ϕ_b^* and integrate over all space. On taking (8.203) into account, we obtain N^2 relations which supplement the system of equations (8.213),

$$\epsilon_{ba} = \langle \phi_b | \hat{\mathcal{H}}_a | \phi_a \rangle. \tag{8.215}$$

As a consequence of the Hermitian character of the operator $\hat{\mathcal{H}}_a$, we can obtain another expression for ϵ_{ba} from (8.212):

$$\epsilon_{ba} = \epsilon_{ab}^* = \langle \phi_a^* | \hat{\mathcal{H}}_b^* | \phi_b^* \rangle = \langle \phi_b | \hat{\mathcal{H}}_b | \phi_a \rangle. \tag{8.216}$$

On averaging formulae (8.215) and (8.216), we obtain an expression for ϵ_{ba} which is symmetric in the indices a and b,

$$\epsilon_{ba} = \tfrac{1}{2}\langle \phi_b | \hat{\mathcal{H}}_a + \hat{\mathcal{H}}_b | \phi_a \rangle. \tag{8.217}$$

B. *An Arbitrary Configuration of Molecular Orbitals*

We consider an arbitrary configuration of N electrons occupying $N - m$ nondegenerate orbitals. A configuration of this kind can always be written as a product of two configurations K_0 and K_1 [see (8.66)]. The number of independent coordinate functions $\Phi_{r_0 r_1}^{[\lambda]}(K_0 K_1)$ corresponding to a state with spin S is equal to the dimension of the irreducible representation $\Gamma^{[\lambda^{(2m)}]}$ of the permutation group for the unpaired electrons and is given by formula (6.70). The desired wavefunction is in the form of a superposition of functions $\Phi_{r_0 r_1}^{[\lambda]}$,

$$\Phi^{[\lambda]}(K) = \sum_{r_1} A_{r_1} \Phi_{r_0 r_1}^{[\lambda]}(K_0 K_1). \tag{8.218}$$

The coefficients A_{r_1} are found from homogeneous algebraic equations similar to (8.205). However, the form of the orbital equations will not be the same as that of (8.213).

The form of the energy matrix for a configuration $K = K_0 K_1$ has been given by formulae (8.72) and (8.73). We rewrite all the summations in a form which is symmetric in the indices, put the coefficients $\tilde{N}_{cc}^{r_1 \bar{r}_1} = -\delta_{r\bar{r}}$, and combine the two formulae into one,

$$\begin{aligned} \mathcal{H}_{r_0 r_1, r_0 \bar{r}_1}^{[\lambda]}(K) = {} & 2\delta_{r_1 \bar{r}_1} \sum_a h_a + \delta_{r_1 \bar{r}_1} \sum_{a,b} (2\alpha_{ab} - \beta_{ab}) \\ & + \delta_{r_1 \bar{r}_1} \sum_c h_c + \tfrac{1}{2} \sum_{c,d} (\delta_{r_1 \bar{r}_1} \alpha_{cd} + \tilde{N}_{cd}^{r_1 \bar{r}_1} \beta_{cd}) \\ & + \delta_{r_1 \bar{r}_1} \sum_{a,c} (2\alpha_{ac} - \beta_{ac}). \end{aligned} \tag{2.219}$$

In this and subsequent formulae in this subsection the indices a and b refer to doubly occupied orbitals only, and indices c and d to singly occupied orbitals. The letters n and l are used for indices which run over all the $N - m$ orbitals.

We put

$$\tfrac{1}{2} \sum_{r_1, \bar{r}_1} A_{r_1} A_{\bar{r}_1} N_{cd}^{r_1 \bar{r}_1} = M_{cd} \tag{8.220}$$

as in (8.207). In order to arrive at the SCF equations which must be satisfied by the orbitals, it is necessary to find an extremum of the functional

$$\begin{aligned} J = {} & \sum_a \langle \phi_a | 2\hat{f} + \sum_b (2\hat{\alpha}_b - \hat{\beta}_b) + \sum_c (2\hat{\alpha}_c - \hat{\beta}_c) | \phi_a \rangle \\ & + \sum_c \langle \phi_c | \hat{f} + \sum_d (\tfrac{1}{2}\hat{\alpha}_d + M_{cd}\hat{\beta}_d) | \phi_c \rangle - \sum_{n,l} \epsilon_{nl} \langle \phi_l | \phi_n \rangle. \end{aligned} \tag{8.221}$$

By making use of the Hermitian character of the operators and of the symmetry of the M_{cd}, we arrive at the system of equations

$$\begin{aligned}\hat{\mathcal{H}}_0\phi_a &= \sum_n \epsilon_{na}\phi_n, \qquad a = 1, 2, \ldots, m,\\ \hat{\mathcal{H}}_c\phi_c &= \sum_n \epsilon_{nc}\phi_n, \qquad c = m+1, \ldots, N-m,\end{aligned} \tag{8.222}$$

with the Hartree–Fock operators

$$\begin{aligned}\hat{\mathcal{H}}_0 &= 2\hat{f} + 2\sum_b (2\hat{\alpha}_b - \hat{\beta}_b) + \sum_d (2\hat{\alpha}_d - \hat{\beta}_d),\\ \hat{\mathcal{H}}_c &= \hat{f} + \sum_b (2\hat{\alpha}_b - \hat{\beta}_b) + \sum_d (\hat{\alpha}_d + 2M_{cd}\hat{\beta}_d).\end{aligned} \tag{8.223}$$

The operator $\mathcal{H}_0$ is invariant under a unitary transformation of the orbitals—in contrast to the operators $\hat{\mathcal{H}}_c$. All the doubly occupied orbitals ϕ_a satisfy a single equation. At the same time, however, it should be pointed out that the operator $\hat{\mathcal{H}}_0$, just as $\hat{\mathcal{H}}_c$, depends upon the whole set of doubly and singly occupied orbitals.

Equations (8.222) are supplemented by $(N-m)^2$ relations for the Lagrange multipliers

$$\epsilon_{nl} = \tfrac{1}{2}\langle\phi_n|\hat{\mathcal{H}}_n + \hat{\mathcal{H}}_l|\phi_l\rangle, \tag{8.224}$$

the operators in which are determined by Eqs. (8.223).

C. *Algebraic Approximation*

The difficulties in solving the SCF equations derived earlier are compounded by the impossibility of factorizing the variables in a molecular field. A practicable method for solving these equations is to expand the molecular orbitals being sought in a certain set of basis functions. The integro-differential equations for the orbitals then become algebraic equations for the coefficients in the expansion. One method of deriving equations of this kind is to determine a conditional extremum in the energy as a function of the coefficients (Roothaan, 1951). However, we prefer to obtain the equations for the coefficients from the orbital equations by a method which is simpler than the direct approach.

We now derive the algebraic equivalent of Eqs. (8.213). These last are written in explicit form

$$\begin{aligned}\hat{f}\phi_a(i) &+ \sum_b \phi_a(i)\int \phi_b^*(j)\phi_b(j)\hat{g}(i,j)\,dV_j\\ &+ 2\sum_b M_{ab}\phi_b(i)\int \phi_b^*(j)\phi_a(j)\hat{g}(i,j)\,dV_j = \sum \epsilon_{ba}\phi_b(i).\end{aligned} \tag{8.225}$$

We expand the orbitals ϕ_a in a finite set of ν basis functions χ_q. Since the set of N orbitals ϕ_a are linearly independent, it follows that the set of ν functions χ_q must not be smaller than the set of ϕ_a, i.e., $\nu \geq N$:

$$\phi_a = \sum_q c_{qa}\chi_q \qquad \phi_a^* = \sum_p c_{pa}^*\chi_p^*. \tag{8.226}$$

We substitute these expansions into (8.225), multiply through by $\chi_p{}^*(i)$, and integrate over dV_i. Denoting

$$s_{pq} = \langle \chi_p | \chi_q \rangle, \qquad f_{pq} = \langle \chi_p | \hat{f} | \chi_q \rangle, \qquad g_{pp',qq'} = \langle \chi_p \chi_{p'} | \hat{g} | \chi_q \chi_{q'} \rangle, \tag{8.227}$$

we obtain instead of the N equations (8.225) νN equations for the coefficients c_{qa}:

$$\sum_q c_{qa} f_{pq} + \sum_b \sum_{q,p',q'} c_{qa} c^*_{p'b} c_{q'b} g_{pp',qq'} + 2 \sum_b \sum_{q,p',q'} M_{ab} c_{qa} c_{p'b} c_{q'b} g_{pp',q'q} = \sum_b \sum_{q'} \epsilon_{ba} c_{q'b} s_{pq'}. \tag{8.228}$$

We now introduce the matrices $\boldsymbol{\alpha}_b$ and $\boldsymbol{\beta}_b$, which are defined by their matrix elements as

$$(\boldsymbol{\alpha}_b)_{pq} = \sum_{p',q'} c^*_{p'b} g_{pp',qq'} c_{q'b}, \qquad (\boldsymbol{\beta}_b)_{pq} = \sum_{p',q'} c^*_{p'b} g_{pp',q'q} c_{q'b}. \tag{8.229}$$

These matrices are the matrix analogs of the operators $\hat{\alpha}_b$ and $\hat{\beta}_b$, since with their aid we can represent the Coulomb and exchange integrals in a form similar to that of (8.170):

$$\alpha_{ab} = \sum_{p,q} c^*_{pa} (\boldsymbol{\alpha}_b)_{pq} c_{qa}, \qquad \beta_{ab} = \sum_{p,q} c^*_{pa} (\boldsymbol{\beta}_b)_{pq} c_{qa}. \tag{8.230}$$

Taking (8.229) into account, Eqs. (8.228) assume the form

$$\sum_q f_{pq} c_{qa} + \sum_b \sum_q (\boldsymbol{\alpha}_b)_{pq} c_{qa} + 2 \sum_b \sum_q M_{ab} (\boldsymbol{\beta}_b)_{pq} c_{qa} = \sum_b \sum_{q'} s_{pq'} c_{q'b} \epsilon_{ba}. \tag{8.231}$$

The coefficients c_{qa} form a rectangular matrix $\mathbf{C}$ of order $\nu \times N$, the overlap integrals a square matrix $\mathbf{S}$ of order ν, and the Lagrange multipliers a square matrix $\boldsymbol{\epsilon}$ of order N. The matrices $\mathbf{f}$, $\boldsymbol{\alpha}_b$, and $\boldsymbol{\beta}_b$ are square and of order ν. Equations (8.231) can clearly be expressed in terms of elements of products of these matrices:

$$(\mathbf{fC})_{pa} + \sum_b (\boldsymbol{\alpha}_b \mathbf{C})_{pa} + 2 \sum_b M_{ab} (\boldsymbol{\beta}_b \mathbf{C})_{pa} = (\mathbf{SC}\boldsymbol{\epsilon})_{pa} \tag{8.232}$$

$$p = 1, 2, \ldots, \nu; \qquad a = 1, 2, \ldots, N.$$

There are thus νN equations for the νN unknown coefficients, these constituting a matrix $\mathbf{C}$ which also occurs in the matrices $\boldsymbol{\alpha}_b$ and $\boldsymbol{\beta}_b$. Equations (8.232) form a system of algebraic equations of order three.

Each column of the matrix $\mathbf{C}$ can be regarded as a column matrix, which we denote as $\mathbf{c}_a$. We also define the following square matrix of order ν:

$$\mathcal{H}_a = \mathbf{f} + \sum_b (\boldsymbol{\alpha}_b + 2M_{ab} \boldsymbol{\beta}_b). \tag{8.233}$$

The νN equations (8.232) for the matrix elements c_{qa} now reduce to N equations for the columns $\mathbf{c}_a$

$$\mathcal{H}_a \mathbf{c}_a = \mathbf{S} \sum_b \epsilon_{ba} \mathbf{c}_b . \tag{8.234}$$

Comparison of these equations with (8.213) shows that in order to transform the system of integro-differential equations (8.213) to a system of algebraic equations (8.234), the operator $\hat{\mathcal{H}}_a$ in the former is replaced by the matrix $\mathcal{H}_a$, the orbitals ϕ_a by the column matrices $\mathbf{c}_a$, and the right-hand side multiplied by the matrix of overlap integrals. The algebraic equivalents of Eqs. (8.222) are found similarly.

Substitution of expansions (8.226) into the coupling conditions (8.203) which are imposed upon the orbitals leads to the following coupling conditions for the coefficients:

$$\sum_{p,q} c_{pa}^* S_{pq} c_{qb} = \delta_{ab} . \tag{8.235}$$

The Hermitian conjugate $\mathbf{C}^\dagger$ of the matrix $\mathbf{C}$ is defined in the usual fashion,

$$c_{ap}^\dagger = c_{pa}^* , \tag{8.236}$$

but since $\mathbf{C}$ is a rectangular matrix, the dimensions of $\mathbf{C}^\dagger$ are not $\nu \times N$ as for $\mathbf{C}$, but $N \times \nu$. If these matrices are represented schematically by rectangles, then

$\mathbf{C} =$  , $\mathbf{C}^\dagger =$ 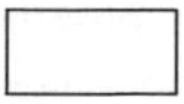 .

The equation equivalent to (8.235) in terms of rows and columns is

$$\mathbf{c}_a^\dagger \mathbf{S} \mathbf{c}_b = \delta_{ab} , \tag{8.237}$$

and for the complete matrices,

$$\mathbf{C}^\dagger \mathbf{S} \mathbf{C} = \mathbf{E}_N , \tag{8.238}$$

where $\mathbf{E}_N$ is the unit matrix of order N.

We now derive the analog of relation (8.215). By multiplying (8.234) by $\mathbf{c}_b^\dagger$ from the left and making use of Eq. (8.237), we obtain

$$\epsilon_{ba} = \mathbf{c}_b^\dagger \mathcal{H}_a \mathbf{c}_a . \tag{8.239}$$

The symmetric expression

$$\epsilon_{ba} = \tfrac{1}{2} \mathbf{c}_b^\dagger (\mathcal{H}_a + \mathcal{H}_b) \mathbf{c}_a \tag{8.240}$$

is more useful in practical applications.

The system of nonlinear algebraic equations (8.234) can be solved by a quadratically convergent iterative procedure (Hinze and Roothaan, 1967; Wessel, 1967), in which an initial matrix of coefficients $\mathbf{C}_0$ is successively refined by means of a correction matrix $\delta\mathbf{C}$. The correction matrix is determined from a system of linear inhomogeneous equations in which the coefficients and independent terms depend upon the matrix $\mathbf{C}_0$. The convergence of the solution depends heavily upon a felicitous choice for $\mathbf{C}_0$. We note that during the process of solving the equations it is necessary to adjust both the sets of coefficients A_r and $\mathbf{C}$.

8.13. Configuration of Singly Occupied Nonorthogonal Orbitals

As was pointed out in Section 8.11, the spatial separation between electrons which is achieved by assigning electrons with different spin orientations to different orbitals (DODS method) allows for fuller correlation between electrons. The configurations which occur in this method consist solely of singly occupied orbitals [configurations of the type (8.56)]. We now consider such configurations, but remove the requirement that the orbitals be orthogonal. The assumption of nonorthogonality enlarges the class of functions to be varied in an SCF method, and hence must lead to a lowering of the energy.

When the orbitals of a configuration are orthogonal the matrix elements of the Hamiltonian can be written in very compact form for states with any value of the spin S (see Section 8.5). However, when the orbitals are nonorthogonal the expressions become somewhat more complicated. Expressions for the matrix elements of operators which are symmetric in the electron coordinates were derived by Löwdin (1955a) for the case of wavefunctions in the form of single determinants of nonorthogonal orbitals. States with a particular value of the spin S as a rule are described by a linear combination of such determinants. As in the case when the orbitals are orthogonal, it is desirable to derive formulae for the matrix elements which do not require the appropriate linear combinations of determinants to be known beforehand. Formulae such as these were derived by Gerratt (1971) using the fractional parentage decomposition of the total wavefunction [see also Gerratt and Lipscomb (1968)]. These formulae can also be derived from a fractional parentage decomposition of coordinate wavefunctions, of the kind described in Part 2 of Chapter VII. In the case of singly occupied orbitals considered here the derivation becomes more transparent if we start from the explicit form for the coordinate function.

We construct a coordinate wavefunction for a configuration of nonorthogonal orbitals as in (7.96) but with an additional normalization factor $N_t^{[\lambda]}$,‡

$$\Phi_{rt}^{[\lambda]}(K) = N_t^{[\lambda]} \omega_{rt}^{[\lambda]} \Phi_0(K), \qquad (8.241)$$

‡In the preceding sections of this chapter the Young tableau r is everywhere used to denote the symmetry of the function with respect to permutations of the orbitals. In the present and subsequent sections we revert to the notation introduced in Section 2.9: The tableau r characterizes the symmetry with respect to permutations of electron coordinates, and tableau t the symmetry with respect to permutations of the orbitals.

$N_t^{[\lambda]}$ being defined by the expression

$$N_t^{[\lambda]} = \langle \omega_{rt}^{[\lambda]} \Phi_0 | \omega_{rt}^{[\lambda]} \Phi_0 \rangle^{-1/2}. \tag{8.242}$$

Since the value of an integral over all space is unaffected by a relabeling of the variables,

$$\langle P\Phi_0 | \omega_{rt}^{[\lambda]} \Phi_0 \rangle \equiv P^{-1} \langle P\Phi_0 | \omega_{rt}^{[\lambda]} \Phi_0 \rangle = \langle \Phi_0 | P^{-1} \omega_{rt}^{[\lambda]} \Phi_0 \rangle. \tag{8.243}$$

Because of the orthogonality of the representation, $\Gamma_{rt}^{[\lambda]}(P^{-1}) = \Gamma_{tr}^{[\lambda]}(P)$, and using (2.32), we obtain

$$\langle \omega_{rt}^{[\lambda]} \Phi_0 | \omega_{rt}^{[\lambda]} \Phi_0 \rangle = \langle \Phi_0 | \omega_{tr}^{[\lambda]} \omega_{rt}^{[\lambda]} \Phi_0 \rangle = (N!/f_\lambda)^{1/2} \langle \Phi_0 | \omega_{tt}^{[\lambda]} \Phi_0 \rangle. \tag{8.244}$$

From this it follows that the normalization factor is equal to

$$N_t^{[\lambda]} = \{ \sum_P \Gamma_{tt}^{[\lambda]}(P) \langle \Phi_0 | P\Phi_0 \rangle \}^{-1/2}. \tag{8.245}$$

It is not difficult to write the integral on the right-hand side of this equation in the form of a product of overlap integrals if we use the fact that

$$P\Phi_0 = \bar{P}^{-1} \Phi_0, \tag{8.246}$$

where $\bar{P}^{-1}$ is the permutation inverse to P but acting upon the orbital labels. Thus

$$\langle \Phi_0 | P\Phi_0 \rangle = s_{1p_1} s_{2p_2} \cdots s_{Np_N}. \tag{8.247}$$

The set of integrals $s_{ap_a} = \langle \phi_a(i_a) | \phi_{p_a}(i_a) \rangle$ which occurs in this equation is defined by the ordered set of indices p_a which result from the permutation P^{-1} applied to the unpermuted set of orbital labels:

$$\{p_1, p_2, \ldots, p_N\} = P^{-1}\{1, 2, \ldots, N\}. \tag{8.248}$$

For antisymmetric representations

$$\Gamma^{[1^N]}(P) = \Gamma^{[1^N]}(P^{-1}) = (-1)^p, \tag{8.249}$$

where p is the parity of the permutations P and P^{-1}. Substituting (8.249) and (8.247) into (8.245), we obtain Löwdin's well-known result

$$N^{[1^N]} = \{ \sum_P (-1)^p s_{1p_1} s_{2p_2} \cdots s_{Np_N} \}^{-1/2} = \{ \det | s_{mn} | \}^{-1/2}. \tag{8.250}$$

By dividing the sum over P first into transpositions and then into cycles of increasing length, we can write the normalisation factor (8.245) in the form of a series of terms in powers of smallness (we assume that the set of non-orthogonal orbitals is normalized):

$$\begin{aligned} N_t^{[\lambda]} = [1 &+ \sum_{P\epsilon\{2\}} \Gamma_{tt}^{[\lambda]}(P_{ab}) s_{ab}^2 + \sum_{P\epsilon\{3\}} \Gamma_{tt}^{[\lambda]}(P_{abc}) s_{ac} s_{cb} s_{ba} \\ &+ \sum_{P\epsilon\{2^2\}} \Gamma_{tt}^{[\lambda]}(P_{ab}P_{cd}) s_{ab}^2 s_{cd}^2 + \sum_{P\epsilon\{4\}} \Gamma_{tt}^{[\lambda]}(P_{abcd}) s_{ad} s_{dc} s_{cb} s_{ba} + \cdots]^{-1/2}. \end{aligned} \tag{8.251}$$

In writing this equation, it should be remembered that the second set of indices for the overlap integrals is obtained from the first set by permutation operations which are the inverses of those which occur in the arguments of the matrix elements ($P_{abc}^{-1} = P_{acb}$, $P_{abcd}^{-1} = P_{adcb}$, etc.).

The overlap integral between coordinate functions defined by different configurations of singly occupied nonorthogonal orbitals is defined similarly to $N_t^{[\lambda]}$, and is given by

$$\begin{aligned}\langle \Phi_{rt}^{[\lambda]}(K) \,|\, \Phi_{\bar{r}\bar{t}}^{[\bar{\lambda}]}(\bar{K})\rangle &= N_t^{[\lambda]}(K) N_{\bar{t}}^{[\bar{\lambda}]}(\bar{K}) \langle \Phi_0(K) \,|\, \omega_{tr}^{[\lambda]} \omega_{\bar{r}\bar{t}}^{[\bar{\lambda}]} \Phi_0(\bar{K})\rangle \\ &= \delta_{\lambda\bar{\lambda}} \delta_{r\bar{r}} N_t^{[\lambda]}(K) N_{\bar{t}}^{[\lambda]}(\bar{K}) \sum_P \Gamma_{t\bar{t}}^{[\lambda]}(P) \langle \Phi_0(K) \,|\, P\Phi_0(\bar{K})\rangle,\end{aligned} \tag{8.252}$$

where

$$\langle \Phi_0(K) \,|\, P\Phi_0(\bar{K})\rangle = s_{1\bar{p}_1} s_{2\bar{p}_2} \cdots s_{N\bar{p}_N}. \tag{8.253}$$

The ordered set of second indices for the overlap integrals in this equation is determined by applying the permutation P^{-1} to the unpermuted orbital labels in the configuration $\bar{K}$:

$$\{\bar{p}_1, \bar{p}_2, \ldots, \bar{p}_N\} = P^{-1}\{\bar{1}, \bar{2}, \ldots, \bar{N}\}. \tag{8.254}$$

According to (8.53), the electronic part of the Hamiltonian for a molecule can be written as a sum of two operators $\hat{F}$ and $\hat{G}$ which are symmetric with respect to permutations of the electron coordinates. We now turn to the determination of the matrix elements of these operators, assuming them to be independent of spin.

a. *The operator* $\hat{F}$. By Eq. (6.57), we can write the total wavefunction $\Psi_t^{[\lambda]}$ in the form of a sum of products of coordinate wavefunctions (8.241) with spin functions. Since, by assumption, the operator F does not depend upon spin variables,

$$\begin{aligned}\langle Kt \,|\, \hat{F} \,|\, \bar{K}\bar{t}\rangle^{[\lambda]} &\equiv \langle \Psi_t^{[\lambda]}(K) \,|\, \hat{F} \,|\, \Psi_{\bar{t}}^{[\lambda]}(\bar{K})\rangle \\ &= (1/f_\lambda) \sum_r \langle \Phi_{rt}^{[\lambda]}(K) \,|\, \hat{F} \,|\, \Phi_{r\bar{t}}^{[\lambda]}(\bar{K})\rangle.\end{aligned} \tag{8.255}$$

We transpose the Young operator $\omega_{rt}^{[\lambda]}$ to the right-hand side of the matrix elements in the sum over r and use Eq. (2.32), noting that the resulting terms do not depend upon r. This gives a preliminary expression which we shall simplify presently:

$$\langle Kt \,|\, \hat{F} \,|\, \bar{K}\bar{t}\rangle^{[\lambda]} = N_t^{[\lambda]}(K) N_{\bar{t}}^{[\lambda]}(\bar{K}) \sum_P \Gamma_{t\bar{t}}^{[\lambda]}(P) \left\langle \Phi_0(K) \,\middle|\, \sum_{i=1}^{N} \hat{f}(i) \,\middle|\, P\Phi_0(\bar{K})\right\rangle. \tag{8.256}$$

Consider the term in this summation in which the operator acts upon the coordinates of electron i_a [the index a shows that electron i occurs in orbital ϕ_a in the function $\Phi_0(K)$]:

$$\sum_P \Gamma_{t\bar{t}}^{[\lambda]}(P) \langle \Phi_0(K) \,|\, \hat{f}(i_a) \,|\, P\Phi_0(\bar{K})\rangle. \tag{8.257}$$

We divide the $N!$ permutations P into N sets of permutations of the form $P'P_{\bar{a}}^{(i_a)}$ according to the position into which electron i_a in the function $\Phi_0(\bar{K})$

is sent. Each set consists of $(N-1)!$ permutations, the sets differing in the permutations P'. These last permutations do not affect electron i_a and belong to the group π_{N-1} which is obtained from π_N by removing electron i_a. The permutation $P_{\bar{a}}^{(i_a)}$ transfers electron i_a into orbital $\phi_{\bar{a}}$ while preserving the increasing order in which the other electrons are numbered:

$$P_{\bar{a}}^{(i_a)} = \begin{cases} P_{i_a i_a+1 \cdots i_{\bar{a}}}, & i_a < i_{\bar{a}}, \\ P_{i_a i_a-1 \cdots i_{\bar{a}}}, & i_a > i_{\bar{a}}. \end{cases} \tag{8.258}$$

The summation in (8.257) can be written as

$$\sum_{\bar{a}} \sum_{P'} \Gamma_{t\bar{t}}^{[\lambda]}(P'P_{\bar{a}}^{(i_a)}) \langle \Phi_0(K_0) | P'\Phi_0(\bar{K}_{\bar{a}}) \rangle f_{a\bar{a}}, \tag{8.259}$$

in which K_a denotes the configuration K with orbital ϕ_a missing, and $f_{a\bar{a}} = \langle \phi_a(i) | \hat{f}(i) | \phi_{\bar{a}}(i) \rangle$.

The matrix element (8.256) is obtained by summing expression (8.259) over all electrons i_a, or, equivalently, over all orbitals a of configuration K. As a result, we obtain the following expression for the matrix element of an operator F in terms of the one-electron integrals $f_{a\bar{a}}$:

$$\langle Kt | \hat{F} | \bar{K}\bar{t} \rangle^{[\lambda]} = N_t^{[\lambda]}(K) N_{\bar{t}}^{[\lambda]}(\bar{K}) \sum_{a,\bar{a}} N_{t\bar{t}}^{[\lambda]}(K_a, K_{\bar{a}}) f_{a\bar{a}}, \tag{8.260}$$

in which the summation over a and $\bar{a}$ is taken over all orbitals of the configuration K and $\bar{K}$, and the coefficient in front of $f_{a\bar{a}}$ is given by

$$N_{t\bar{t}}^{[\lambda]}(K_a, \bar{K}_{\bar{a}}) = \sum_{P'} \Gamma_{t\bar{t}}^{[\lambda]}(P'P_{\bar{a}}^{(i_a)}) \langle \Phi_0(K_a) | P'\Phi_0(\bar{K}_{\bar{a}}) \rangle. \tag{8.261}$$

The integral on the right-hand side of this equation is of the form of a product of overlap integrals and can be written as follows:

$$\langle \Phi_0(K_a) | P'\Phi_0(\bar{K}_{\bar{a}}) \rangle = \{ s_{1\bar{p}_1'} s_{2\bar{p}_2'} \cdots s_{N\bar{p}_N'} \}_{a\bar{a}}. \tag{8.262}$$

The first set of indices in this product of overlap integrals is generated by the ordered set $\{1, 2, \ldots, N\}_a$, which is obtained from the original set of orbitals of configuration K by removing orbital ϕ_a. The second set of indices is generated by the ordered set

$$\{\bar{p}_1', \bar{p}_2', \ldots, \bar{p}_N'\}_{\bar{a}} = P'^{-1}\{\bar{1}, \bar{2}, \ldots, \bar{N}\}_{\bar{a}}. \tag{8.263}$$

For antisymmetric coordinate states (states with maximum spin)

$$\Gamma^{[1^N]}(P'P_{\bar{a}}^{(i_a)}) = (-1)^{p'}(-1)^{i_a - i_{\bar{a}}}$$

and

$$N^{[1^N]}(K_a, \bar{K}_{\bar{a}}) = (-1)^{i_a - i_{\bar{a}}} \det | s_{m\bar{n}} |_{a\bar{a}},$$

where $\det | s_{m\bar{n}} |_{a\bar{a}}$ is the minor of $\det | s_{m\bar{n}} |$ obtained by deleting the ath row and $\bar{a}$th column. As a result, formula (8.260) reduces to

$$\langle K | \hat{F} | \bar{K} \rangle^{[1^N]} = \{ \sum_{a,\bar{a}} (-1)^{i_a - i_{\bar{a}}} \det | s_{m\bar{n}} |_{a\bar{a}} f_{a\bar{a}} \} (\det | s_{mn} | \det | s_{\bar{m}\bar{n}} |)^{-1/2}. \tag{8.264}$$

If we assume that m, $\bar{m}$, n, and $\bar{n}$ enumerate spin orbitals instead of orbitals, formula (8.264) describes a state with a single determinant as the total wavefunction.

If we impose orthogonality conditions upon the orbitals, then $N_{t\bar{t}}^{[\lambda]}(K_a, \bar{K}_{\bar{a}}) \neq 0$ only if $\Phi_0(K_a)$ and $\Phi_0(\bar{K}_{\bar{a}})$ differ solely in the ordering of the orbitals. Matrix elements (8.260) are therefore nonzero only for configurations which do not differ by more than one orbital. For example, let $\bar{K}_b{}^a$ be obtained from K by replacing the orbital ϕ_a by a new orbital ϕ_b and by a permutation $\bar{P}'$ of the other orbitals. Clearly the only nonzero term in (8.261) is the one where the permutation P' returns the electrons to their "own" orbitals,

$$P'\Phi_0(\bar{K}_a) = \Phi_0(K_a).$$

Noting further that the normalization factor $N_t^{[\lambda]} = 1$, we obtain the simple formula

$$\langle Kt \,|\, \hat{F} \,|\, \bar{K}_a{}^b \bar{t} \rangle^{[\lambda]} = \Gamma_{t\bar{t}}^{[\lambda]}(P' P_b^{(i_a)}) f_{ab}. \tag{8.265}$$

This expression is equivalent to expression (8.37c) in the footnote on p. 209 when $n_a = 1$ and $n_b = 0$. When comparing these two expressions, it should be borne in mind that $\bar{P}'$ in (8.37c) permutes the orbitals of the configuration K_a, whereas P' in (8.265) permutes the coordinates of the electrons in the function $\Phi_0(\bar{K}_a)$. If the orbitals are orthogonal, formula (8.260) for a matrix element diagonal in the configurations reduces to formula (8.35), as expected.

b. *The operator* $\hat{G}$. Since we assume that this operator as well as $\hat{F}$ does not depend upon spin variables, we can write an expression similar to (8.256) for its matrix elements,

$$\langle Kt \,|\, \hat{G} \,|\, \bar{K}\bar{t} \rangle^{[\lambda]} = N_t^{[\lambda]}(K) N_{\bar{t}}^{[\lambda]}(\bar{K}) \sum_P \Gamma_{t\bar{t}}^{[\lambda]}(P) \langle \Phi_0(K) \,|\, \sum_{i<j} \hat{g}(i,j) \,|\, P\Phi_0(\bar{K}) \rangle. \tag{8.266}$$

In what follows it will be convenient to number the electrons by the indices of the orbitals in which they occur in the function $\Phi_0(\bar{K})$. The indices i and j are therefore replaced by i_a and i_b.

Consider the term in (8.266) involving the operator $\hat{g}(i_a, i_b)$:

$$\sum_P \Gamma_{t\bar{t}}^{[\lambda]}(P) \langle \Phi_0(K) \,|\, \hat{g}(i_a, i_b) \,|\, P\Phi_0(\bar{K}) \rangle. \tag{8.267}$$

We divide the $N!$ permutations P into $N(N-1)$ sets according to the positions in the function $\Phi_0(\bar{K})$ into which electrons i_a and i_b are sent. Each consists of $(N-2)!$ permutations, and can be written in the form $P''P_{\bar{a}\bar{b}}^{(i_a i_b)}$ or $P''P_{\bar{a}\bar{b}}^{(i_b i_a)}$. The permutations P'' constitute the permutation group π_{N-2} obtained from π_N by removing electrons i_a and i_b. When the permutation $P_{\bar{a}\bar{b}}^{(i_a i_b)}$ is applied to $\Phi_0(\bar{K})$, electron i_a is carried into orbital $\phi_{\bar{a}}$ and electron i_b into $\phi_{\bar{b}}$, while preserving the increasing order in which the remaining electrons are numbered. The permutations $P_{\bar{a}\bar{b}}^{(i_b i_a)}$ possess a similar significance. The form of the permutations $P_{\bar{a}\bar{b}}^{(i_a i_b)}$ and $P_{\bar{a}\bar{b}}^{(i_b i_a)}$ is easily found in any particular case. It is obvious that

$$P_{\bar{a}\bar{b}}^{(i_b i_a)} = P_{ab} P_{\bar{a}\bar{b}}^{(i_a i_b)}, \qquad (8.268)$$

where P_{ab} is a transposition. The summation in (8.267) now assumes the following form:

$$\sum_{\bar{c}<d} \sum_{P'} \{\Gamma_{t\bar{t}}^{[\lambda]}(P'' P_{\bar{a}\bar{b}}^{(i_a i_b)}) \langle \Phi_0(K_{ab}) \phi_a(i_a) \phi_b(i_b) | \hat{g}(i_a, i_b) | P'' \Phi_0(\bar{K}_{\bar{a}\bar{b}}) \phi_{\bar{a}}(i_a) \phi_{\bar{b}}(i_b) \rangle$$
$$+ \Gamma_{t\bar{t}}^{[\lambda]}(P'' P_{\bar{a}\bar{b}}^{(i_b i_a)}) \langle \Phi_0(K_{ab}) \phi_a(i_a) \phi_b(i_b) | \hat{g}(i_a, i_b) | P'' \Phi_0(\bar{K}_{\bar{a}\bar{b}}) \phi_{\bar{b}}(i_a) \phi_{\bar{a}}(i_b) \rangle\}. \qquad (8.269)$$

We factorize out the two-electron matrix elements $g_{ab,\bar{a}\bar{b}}$ and $g_{ab,\bar{b}\bar{a}}$,

$$g_{ab,\bar{a}\bar{b}} = \langle \phi_a(i_a) \phi_b(i_b) | \hat{g}(i_a, i_b) | \phi_{\bar{a}}(i_a) \phi_{\bar{b}}(i_b) \rangle,$$

and define the quantities

$$N_{t\bar{t}}^{[\lambda]}(K_{ab}, \bar{K}_{\bar{a}\bar{b}}) = \sum_{P''} \Gamma_{t\bar{t}}^{[\lambda]}(P'' P_{\bar{a}\bar{b}}^{(i_a i_b)}) \langle \Phi_0(K_{ab}) | P'' \Phi_0(\bar{K}_{\bar{a}\bar{b}}) \rangle,$$
$$\tilde{N}_{t\bar{t}}^{[\lambda]}(K_{ab}, \bar{K}_{\bar{a}\bar{b}}) = \sum_{P''} \Gamma_{t\bar{t}}^{[\lambda]}(P'' P_{\bar{a}\bar{b}}^{(i_b i_a)}) \langle \Phi_0(K_{ab}) | P'' \Phi_0(\bar{K}_{\bar{a}\bar{b}}) \rangle. \qquad (8.270)$$

The overlap integrals in these equations can be written in terms of ordered products of overlap integrals between orbitals, similarly to (8.262),

$$\langle \Phi_0(K_{ab}) | P'' \Phi_0(\bar{K}_{\bar{a}\bar{b}}) \rangle = \{ s_{1\bar{p}_1''} s_{2\bar{p}_2''} \cdots s_{N\bar{p}_N''} \}_{ab,\bar{a}\bar{b}}, \qquad (8.271)$$

in which the second indices for the overlap integrals constitute the ordered set

$$\{\bar{p}_1'', \bar{p}_2'', \ldots, \bar{p}_N''\}_{\bar{a}\bar{b}} = P''^{-1}\{\bar{1}, \bar{2}, \ldots, \bar{N}\}_{\bar{a}\bar{b}}. \qquad (8.272)$$

The matrix element (8.266) is now obtained by summing expression (8.269) over all pairs of electrons i_a and i_b, or equivalently, over all pairs of orbitals a and b. Making use of the notation introduced in (8.270), we obtain the desired formula:

$$\langle Kt | \hat{G} | \bar{K}\bar{t} \rangle^{[\lambda]} = N_t^{[\lambda]}(K) N_{\bar{t}}^{[\lambda]}(\bar{K}) \sum_{a<b} \sum_{\bar{a}<\bar{b}} \{ N_{t\bar{t}}^{[\lambda]}(K_{ab}, \bar{K}_{\bar{a}\bar{b}}) g_{ab,\bar{a}\bar{b}}$$
$$+ \tilde{N}_{t\bar{t}}^{[\lambda]}(K_{ab}, \bar{K}_{\bar{a}\bar{b}}) g_{ab,\bar{b}\bar{a}} \}. \qquad (8.273)$$

If the coordinate wavefunctions are antisymmetric, the matrix elements of the permutations are determined by the parity of the permutations, so that

$$\Gamma^{[1^N]}(P_{\bar{a}}^{(i_a)}) = (-1)^{i_a - i_{\bar{a}}},$$
$$\Gamma^{[1^N]}(P_{\bar{a}\bar{b}}^{(i_a i_b)}) = (-1)^{i_a - i_{\bar{a}} + i_b - i_{\bar{b}}} = (-1)^{i_a + i_b + i_{\bar{a}} + i_{\bar{b}}},$$
$$\Gamma^{[1^N]}(P_{\bar{a}\bar{b}}^{(i_b i_a)}) = (-1)^{i_a + i_b + i_{\bar{a}} + i_{\bar{b}} + 1}.$$

As a result, formula (8.273) reduces to

$$\langle Kt | \hat{G} | \bar{K}\bar{t} \rangle^{[1^N]} = \{ \sum_{a<b} \sum_{\bar{a}<\bar{b}} (-1)^{i_a + i_b + i_{\bar{a}} + i_{\bar{b}}} \det | s_{m\bar{n}} |_{ab,\bar{a}\bar{b}} (g_{ab,\bar{a}\bar{b}} - g_{ab,\bar{b}\bar{a}}) \}$$
$$\times (\det | s_{mn} | \det | s_{\bar{m}\bar{n}} |)^{-1/2}, \qquad (8.274)$$

in which $\det | s_{m\bar{n}} |_{ab,\bar{a}\bar{b}}$ is the minor of $\det | s_{m\bar{n}} |$ obtained by striking out rows a and b and columns $\bar{a}$ and $\bar{b}$. If the indices in (8.274) enumerate spin orbitals instead of orbitals, the expression becomes identical to that derived by Löwdin (1955a) for states described by single-determinant wavefunctions.

If the orbitals form an orthonormal set, the coefficients (8.271) are nonzero only if the configurations K_{ab} and $\bar{K}_{\bar{a}\bar{b}}$ differ solely in the ordering of the orbitals. From this it follows that the matrix element (8.273) is nonvanishing only if the configurations K and $\bar{K}$ differ by not more than two orbitals; i.e., matrix elements which are nondiagonal in the configurations can be formed from $\bar{K} = \bar{K}_a{}^b = \bar{P}K_a{}^b$ or from $\bar{K} = \bar{K}_{ab}^{cd} = \bar{P}K_{ab}^{cd}$, where $\bar{P}$ is a permutation of the orbitals. As a result, we obtain for configurations of singly occupied orthonormal orbitals, instead of formulae (8.47)–(8.48) with coefficients (8.52), the following simpler expressions:

$$\langle Kt\,|\,\hat{G}\,|\,\bar{K}_a{}^b\bar{t}\rangle^{[\lambda]} = \sum_c \{\Gamma_{t\bar{t}}^{[\lambda]}(P''P_{bc}^{(i_a i_c)})g_{ac,\,bc} + \Gamma_{t\bar{t}}^{[\lambda]}(P''P_{bc}^{(i_c i_a)})g_{ac,\,cb}\}, \tag{8.275}$$

$$\langle Kt\,|\,\hat{G}\,|\,\bar{K}_{ab}^{cd}\bar{t}\rangle^{[\lambda]} = \Gamma_{t\bar{t}}^{[\lambda]}(P''P_{cd}^{(i_a i_b)})g_{ab,\,cd} + \Gamma_{t\bar{t}}^{[\lambda]}(P''P_{cd}^{(i_b i_a)})g_{ab,\,dc}, \tag{8.276}$$

in which the permutation P'' restores the ordering of the electrons,

$$P''\Phi_0(\bar{K}_{ab}) = \Phi_0(K_{ab}).$$

The reader is referred to the remarks on p. 212 concerning formula (8.47) as regards the ordering of the indices of the the two-electron integrals in formulae (8.275)–(8.276).

The formulae for matrix elements which are diagonal in the configurations become simplified further if the orbitals are orthonormal. In this case $a = \bar{a}$ and $b = \bar{b}$ and only Coulomb and exchange integrals $\alpha_{ab} = g_{ab,\,ab}$ and $\beta_{ab} = g_{ab,\,ba}$ survive in the summation in (8.273). The only nonzero terms in the sums over P'' in (8.270) are those for which $P'' = 1$. Furthermore, it is obvious that $P_{ab}^{(i_a i_b)} = 1$ and $P_{ab}^{(i_b i_a)} = P_{ab}$. As a result, we obtain the simple expression

$$\langle Kt\,|\,\hat{G}\,|\,K\bar{t}\rangle^{[\lambda]} = \sum_{a<b} (\alpha_{ab} + \Gamma_{t\bar{t}}^{[\lambda]}(P_{ab})\beta_{ab}). \tag{8.277}$$

The coefficients which precede the exchange integrals are just the matrix elements of the transpositions P_{ab} in the standard representation. Matrices of all the transpositions for the groups π_5 to π_6 are given in Appendix 5.

Expression (8.273) can be rewritten in a more compact form. For this purpose we note that from the form of the coefficients (8.270) it follows that

$$\tilde{N}_{t\bar{t}}(K_{ab}, \bar{K}_{\bar{a}\bar{b}}) = N_{t\bar{t}}^{[\lambda]}(K_{ab}, \bar{K}_{\bar{b}\bar{a}}). \tag{8.278}$$

Since $g_{ab,\,\bar{a}\bar{b}} = g_{ba,\,\bar{b}\bar{a}}$, expression (8.273) is equivalent to

$$\langle Kt\,|\,\hat{G}\,|\,\bar{K}\bar{t}\rangle^{[\lambda]} = \tfrac{1}{2}N_t^{[\lambda]}(K)N_{\bar{t}}^{[\lambda]}(\bar{K}) \sum_{a,b}\sum_{\bar{a},\bar{b}} \gamma_{ab}\gamma_{\bar{a}\bar{b}}N_{t\bar{t}}^{[\lambda]}(K_{ab}, K_{\bar{a}\bar{b}})g_{ab,\,\bar{a}\bar{b}}, \tag{8.279}$$

in which the occurrence of the symbols γ_{ab} and $\gamma_{\bar{a}\bar{b}}$,

$$\gamma_{ab} = 1 - \delta_{ab}, \tag{8.280}$$

reflects the fact that since no orbital occurs more than once in the two configurations K and $\bar{K}$, the indices a and b can never assume the same value, nor can $\bar{a}$ and $\bar{b}$.

It is useful to note that the coefficients $N_{t\bar{t}}^{[\lambda]}(K_a, \bar{K}_{\bar{a}})$ and $N_{t\bar{t}}^{[\lambda]}(K_{ab}, \bar{K}_{\bar{a}\bar{b}})$ are related by the equation (Gerratt, 1971)

$$N_{t\bar{t}}^{[\lambda]}(K_a, \bar{K}_{\bar{a}}) = \sum_{\bar{b}} \gamma_{ab}\gamma_{\bar{a}\bar{b}} N_{t\bar{t}}^{[\lambda]}(K_{ab}, \bar{K}_{\bar{a}\bar{b}}) \langle \phi_b | \phi_{\bar{b}} \rangle, \tag{8.281}$$

in which b may not equal a but is otherwise arbitrary. This equation can be proved by writing the permutations P' in (8.261) in the form $P''P_{\bar{b}}^{(i_b)}$ and replacing the sum over P' by the equivalent summation over P'' and $\bar{b}$. In a similar way, it is easily shown that

$$N_{t\bar{t}}^{[\lambda]}(K_{ab}, \bar{K}_{\bar{a}\bar{b}}) = \sum_{\bar{c}} \gamma_{abc}\gamma_{\bar{a}\bar{b}\bar{c}} N_{t\bar{t}}^{[\lambda]}(K_{abc}, \bar{K}_{\bar{a}\bar{b}\bar{c}}) \langle \phi_c | \phi_{\bar{c}} \rangle, \tag{8.282}$$

for any $c \neq a, b$. The symbol γ_{abc} is equal to zero if any two of its indices are equal and equal to unity in all other cases, i.e.,

$$\gamma_{abc} = (1 - \delta_{ab})(1 - \delta_{ac})(1 - \delta_{bc}). \tag{8.283}$$

The quantities $N_{t\bar{t}}^{[\lambda]}K_{abc}, \bar{K}_{\bar{a}\bar{b}\bar{c}})$ are defined similarly to $N_{t\bar{t}}^{[\lambda]}(K_a, \bar{K}_{\bar{a}})$ and $N_{t\bar{t}}^{[\lambda]}(K_{ab}, \bar{K}_{\bar{a}\bar{b}})$,

$$N_{t\bar{t}}^{[\lambda]}(K_{abc}, \bar{K}_{\bar{a}\bar{b}\bar{c}}) = \sum_{P'''} \Gamma_{t\bar{t}}^{[\lambda]}(P'''P_{\bar{a}\bar{b}\bar{c}}^{(i_a i_b i_c)}) \langle \Phi_0(K_{abc}) | P''' \Phi_0(\bar{K}_{\bar{a}\bar{b}\bar{c}}) \rangle. \tag{8.284}$$

One could further define quantities $N_{t\bar{t}}^{[\lambda]}(K_{abcd}, \bar{K}_{\bar{a}\bar{b}\bar{c}\bar{d}})$, corresponding to the removal of four electrons, etc. The chain of equations (8.281)–(8.284) can be used to calculate the $N_{t\bar{t}}^{[\lambda]}$ quantities by recurrence.

8.14. The Self-Consistent Field Equations for the Method of "Different Orbitals for Different Spins"

The SCF equations for a configuration of nonorthogonal orbitals in the DODS method were first derived by Goddard (1967b). Goddard chose as his variational function one which corresponds to a particular mode of coupling the electron spins to form the total spin S. There are always several independent wavefunctions corresponding to a state with total spin S for the configurations under consideration, except for the case of maximum spin. These independent functions are naturally all taken into account when constructing the function to be varied. The SCF equations in the DODS method for configurations of nonorthogonal orbitals which include all the independent wavefunctions for a given spin were derived by Gerratt (1968, 1971). In this section we present a systematic derivation of these equations and also consider the algebraic approximation, including the variation of the basis functions with respect to a parameter (Kaplan and Maksimov, 1973).

We seek a coordinate wavefunction for a state with permutational symmetry of a Young diagram $[\lambda]$, corresponding to a total spin S for the system, in the form of a superposition of coordinate functions $\Phi_{rt}^{[\lambda]}$ [see (8.241)],‡

$$\Phi_r^{[\lambda]}(K) = \sum_t A_t \Phi_{rt}^{[\lambda]}(K). \tag{8.285}$$

The coefficients A_t and orbitals ϕ_a must satisfy the equations which arise from the requirement that the expectation value of the Hamiltonian in a state with the coordinate wavefunction (8.285) be a minimum

$$\bar{E} = \sum_{t,\bar{t}} A_t{}^* A_{\bar{t}} \langle Kt \,|\hat{\mathcal{H}}|\, K\bar{t}\rangle^{[\lambda]}. \tag{8.286}$$

The expressions for the matrix elements of the Hamiltonian $\hat{\mathcal{H}} = \hat{F} + \hat{G}$ follow from the formulae derived in the previous section for the operators $\hat{F}$ [(8.260)] and $\hat{G}$ [(8.279)]. We write a matrix element in the following form:

$$\langle Kt \,|\, \hat{\mathcal{H}} \,|\, K\bar{t}\rangle^{[\lambda]} = N_t^{[\lambda]} N_{\bar{t}}^{[\lambda]} \{\sum_{a,b} N_{t\bar{t}}^{[\lambda]}(K_a, K_b) f_{ab}$$
$$+ \tfrac{1}{2} \sum_{a,b} \sum_{c,d} \gamma_{ac}\gamma_{bd} N_{t\bar{t}}^{[\lambda]}(K_{ac}, K_{bd}) g_{ac,\,bd}\} \tag{8.287}$$

Certain coupling conditions which arise from the requirement that the function (8.285) and the orbitals ϕ_a be normalized are imposed upon the coefficients A_t and the ϕ_a,

$$\sum_{t,\bar{t}} A_t A_{\bar{t}} \langle \Phi_{rt}^{[\lambda]}(K) \,|\, \Phi_{r\bar{t}}^{[\lambda]}(K)\rangle = 1 \tag{8.288}$$

$$\langle \phi_a | \phi_a \rangle = 1, \qquad a = 1, 2, \ldots, N. \tag{8.289}$$

The variational equations are simplified if the normalization factors $N_t^{[\lambda]}$ are included in the variational coefficients by introducing new coefficients B_t instead of the A_t,

$$B_t = A_t N_t^{[\lambda]}. \tag{8.290}$$

This means that we do not in fact normalize the functions $\Phi_{rt}^{[\lambda]}$, these being bounded by the requirement that the function (8.285) be normalized. We rewrite the expectation value of the energy (8.286) in terms of the coefficients B_t in the form

$$\bar{E} = \sum_{t,\bar{t}} B_t{}^* B_{\bar{t}} \mathcal{H}_{t\bar{t}}^{[\lambda]}, \tag{8.291}$$

in which a matrix element $\mathcal{H}_{t\bar{t}}^{[\lambda]}$ is given by (8.287) without the factors $N_t^{[\lambda]}$. The overlap integrals which occur in (8.288) are given by expression (8.252) and form a square matrix of dimension f_λ. Absorbing the factors $N_t^{[\lambda]}$ into the variational coefficients and denoting the remaining parts of the integrals by $S_{t\bar{t}}^{[\lambda]}$, we obtain instead of (8.288)

$$\sum_{t,\bar{t}} B_t{}^* B_{\bar{t}} S_{t\bar{t}}^{[\lambda]} = 1. \tag{8.292}$$

The B_t can be varied independently of the orbitals, although the sets of equations which are obtained must be solved simultaneously. We obtain the

‡The sum over the index t in (8.285) is equivalent to the sum over the index r in (8.200); see footnote on p. 272 with regard to this remark.

well-known Ritz equations for the coefficients B_t

$$\sum_{\bar{t}} (\mathcal{H}_{t\bar{t}}^{[\lambda]} - \Lambda S_{t\bar{t}}^{[\lambda]}) B_{\bar{t}} = 0, \tag{8.293}$$

in which the Lagrange multiplier Λ, which can be identified as the energy of the system, is determined by solving the secular equation

$$| \mathcal{H}_{t\bar{t}}^{[\lambda]} - \Lambda S_{t\bar{t}}^{[\lambda]} | = 0. \tag{8.294}$$

The equations which must be satisfied by the orbitals are found by minimizing expression (8.291) under the conditions (8.292) and (8.289). In order to simplify the form of the functional, we put

$$\begin{aligned} \mathfrak{M}_{ab}^{[\lambda]} &= \sum_{t,\bar{t}} B_t^* B_{\bar{t}} N_{t\bar{t}}^{[\lambda]}(K_a, K_b) \\ \mathfrak{M}_{ac,bd}^{[\lambda]} &= \gamma_{ac}\gamma_{bd} \sum_{t,\bar{t}} B_t^* B_{\bar{t}} N_{t\bar{t}}^{[\lambda]}(K_{ac}, K_{bd}). \end{aligned} \tag{8.295}$$

The condition (8.292) is now recast by writing the permutations P in the $S_{t\bar{t}}^{[\lambda]}$ in the form $P'P_b^{(i_a)}$ and replacing the summation over P by an equivalent summation over P' and b (see previous section). Noting the definition of the coefficients $N_{t\bar{t}}^{[\lambda]}(K_a, K_b)$, (8.261), we obtain

$$\sum_{t,\bar{t}} \sum_{b} B_t^* B_{\bar{t}} N_{t\bar{t}}^{[\lambda]}(K_a, K_b) s_{ab} = 1. \tag{8.296}$$

This expression holds for any value of a. We sum it over all a, divide by N, and use the notation introduced in (8.295). As a result, condition (8.292) becomes

$$(1/N) \sum_{a,b} \mathfrak{M}_{ab}^{[\lambda]} s_{ab} = 1. \tag{8.297}$$

The problem is now reduced to finding an unconditional minimim in the functional J which is obtained by subtracting from (8.291) the coupling conditions (8.297) and (8.289), these being respectively multiplied by the Lagrange multipliers Λ and ϵ_a:

$$J = \sum_{a,b} \{\mathfrak{M}_{ab}^{[\lambda]}(f_{ab} - N^{-1}\Lambda s_{ab}) + \tfrac{1}{2} \sum_{c,d} \mathfrak{M}_{ac,bd}^{[\lambda]} g_{ac,bd}\} - \sum_a \epsilon_a s_{aa}. \tag{8.298}$$

We wish to find the first variation of this functional with respect to the orbitals:

$$\begin{aligned} \delta J = \sum_{a,b} \{ & \delta\mathfrak{M}_{ab}^{[\lambda]}(f_{ab} - N^{-1}\Lambda s_{ab}) + \mathfrak{M}_{ab}^{[\lambda]}(\langle \delta\phi_a | \hat{f} - N^{-1}\Lambda | \phi_b \rangle \\ & + \langle \phi_a | \hat{f} - N^{-1}\Lambda | \delta\phi_b \rangle) + \tfrac{1}{2} \sum_{c,d} (\delta\mathfrak{M}_{ac,bd}^{[\lambda]} g_{ac,bd} \\ & + \mathfrak{M}_{ac,bd}^{[\lambda]} \delta g_{ac,bd})\} - \sum_a \epsilon_a (\langle \delta\phi_a | \phi_a \rangle + \langle \phi_a | \delta\phi_a \rangle). \end{aligned} \tag{8.299}$$

Before writing out an explicit form for the variation of the $\mathfrak{M}^{[\lambda]}$ we note that these quantities satisfy relations similar to (8.281) and (8.282). Thus

$$\mathfrak{M}_{ab}^{[\lambda]} = \sum_d \mathfrak{M}_{ac,bd}^{[\lambda]} \langle \phi_c | \phi_d \rangle \tag{8.300}$$

$$\mathfrak{M}_{ac,bd}^{[\lambda]} = \sum_f \mathfrak{M}_{ace,bdf} \langle \phi_e | \phi_f \rangle, \tag{8.301}$$

in which

$$\mathfrak{M}^{[\lambda]}_{ace,bdf} = \gamma_{ace}\gamma_{bdf} \sum_{t,\bar{t}} B_t{}^* B_{\bar{t}} N^{[\lambda]}_{t\bar{t}}(K_{ace}, K_{bdf}), \tag{8.302}$$

similarly to (8.295).

From Eqs. (8.300) and (8.301), it follows that

$$\delta\mathfrak{M}^{[\lambda]}_{ab} = \sum_{c,d} \gamma_{cd}\mathfrak{M}^{[\lambda]}_{ac,bd}(\langle\delta\phi_c|\phi_d\rangle + \langle\phi_c|\delta\phi_d\rangle), \tag{8.303}$$

$$\delta\mathfrak{M}^{[\lambda]}_{ac,bd} = \sum_{e,f} \gamma_{ef}\mathfrak{M}^{[\lambda]}_{ace,bdf}(\langle\delta\phi_e|\phi_f\rangle + \langle\delta\phi_e|\delta\phi_f\rangle). \tag{8.304}$$

The presence of γ_{cd} and γ_{ef} in (8.303) and (8.304) is due to the fact that before (8.301) and (8.302) are varied, we make use of the coupling condition (8.289) for the diagonal elements of the matrix of overlap integrals.

In order to vary a two-electron matrix element $g_{ac,bd}$, it is convenient to rewrite the element in terms of the one-electron operator $\hat{g}_{cd}$, defined as follows:

$$\begin{aligned} g_{ac,bd} &= \langle\phi_a(i)|\hat{g}_{cd}(i)|\phi_b(i)\rangle \\ &= \langle\phi_c(i)|\hat{g}_{ab}(i)|\phi_d(i)\rangle, \end{aligned} \tag{8.305}$$

from which it follows that

$$\hat{g}_{cd}(i) = \int \phi_c{}^*(j)\hat{g}(i,j)\phi_d(j)\, dV_j. \tag{8.306}$$

Using both forms (8.305) for $g_{ac,bd}$, we obtain for the variation

$$\begin{aligned} \delta g_{ac,bd} &= \langle\delta\phi_a|\hat{g}_{cd}|\phi_b\rangle + \langle\phi_a|\hat{g}_{cd}|\delta\phi_b\rangle \\ &\quad + \langle\delta\phi_c|\hat{g}_{ab}|\phi_d\rangle + \langle\phi_c|\hat{g}_{ab}|\delta\phi_d\rangle. \end{aligned} \tag{8.307}$$

We now substitute the explicit forms of the variations into (8.299). In order to separate the general coefficient of $\delta\phi_a$, it is necessary to make the replacements $a \leftrightarrow c$ and $b \leftrightarrow d$ when substituting expression (8.303) and the second half of (8.307), and the replacements $a \leftrightarrow e$ and $b \leftrightarrow f$ when substituting (8.304). Making use of the Hermitian character of the operators $\hat{f}$ and $\hat{g}_{cd}$ and of the equations

$$\mathfrak{M}^{[\lambda]}_{ac,bd} = \mathfrak{M}^{[\lambda]}_{ca,db}, \qquad \mathfrak{M}^{[\lambda]}_{ace,bdf} = \mathfrak{M}^{[\lambda]}_{eca,fdb} \qquad \sum_{c,d} \mathfrak{M}^{[\lambda]}_{ac,bd} S_{cd} = (N-1)\mathfrak{M}_{ab}, \tag{8.308}$$

which are consequences of the properties of the $\mathfrak{M}^{[\lambda]}$, δJ can be written in the following form:

$$\begin{aligned} \delta J &= \sum_a [\sum_b \langle\delta\phi_a|\hat{\mathfrak{F}}_{ab}|\phi_b\rangle - \langle\delta\phi_a|\epsilon_a\phi_a\rangle] \\ &\quad + \sum_a [\sum_b \langle\delta\phi_a|\hat{\mathfrak{F}}_{ab}|\phi_b\rangle - \langle\delta\phi_a|\epsilon_a\phi_a\rangle]^*, \end{aligned} \tag{8.309}$$

in which the operator $\hat{\mathfrak{F}}_{ab}$ is given by the expression

$$\begin{aligned} \hat{\mathfrak{F}}_{ab} &= \mathfrak{M}^{[\lambda]}_{ab}\hat{f} + \sum_{c,d} \mathfrak{M}^{[\lambda]}_{ac,bd}\, \hat{g}_{cd} + \gamma_{ab}[-\Lambda\mathfrak{M}^{[\lambda]}_{ab} \\ &\quad + \sum_{c,d} \mathfrak{M}^{[\lambda]}_{ac.bd} f_{cd} + \tfrac{1}{2} \sum_{c,d} \sum_{e,f} \mathfrak{M}^{[\lambda]}_{ace,bdf} g_{ce,df}]. \end{aligned} \tag{8.310}$$

Setting δJ equal to zero and making use of the linear independence of the variations $\delta\phi_a$, we obtain the desired system of N integro-differential equations

$$\sum_b \hat{\mathcal{H}}_{ab}\phi_b = \epsilon_a\phi_a. \tag{8.311}$$

These equations depend upon the coefficients B_t and must be solved simultaneously with Eqs. (8.293).

A practial method of solving Eqs. (8.311) is to represent the desired orbitals as expansions in a finite basis of ν basis functions,

$$\phi_a = \sum_p c_{pa}\chi_p, \qquad \phi_a^* = \sum_p c_{pa}^*\chi_p^*. \tag{8.312}$$

The range of the indices p in these equations is from one to ν, $\nu \geq N$.‡ This expansion is substantially improved if we increase its flexibility by introducing variational parameters ξ_p into the basis functions χ_p. A procedure of this kind for SCF equations in a central field has been described by Olive (1969). By varying the coefficients c_{pa} and the parameters ξ_p simultaneously, it is possible, with the same number of terms, to increase the accuracy of the expansion (8.312).

Since the orbitals are given an explicit form by the expansion (8.312), the integro-differential equations for the orbitals (8.311) now become algebraic equations for the coefficients c_{pa} and the parameters ξ_p. In order to derive these equations, we have to substitute in expression (8.309) for the variation δJ the representation (8.312) of the orbitals and also the explicit form of the variation $\delta\phi_a$. In order to determine $\delta\phi_a$, an orbital ϕ_a is regarded as an explicit function of the coefficients c_{pa} and an implicit function of the parameters ξ_p,§

$$\delta\phi_a = \sum_p \left(\frac{\partial\phi_a}{\partial c_{pa}}\delta c_{pa} + \frac{\partial\phi_a}{\partial\chi_p}\frac{\partial\chi_p}{\partial\xi_p}\delta\xi_p\right) = \sum_p\left(\chi_p\delta c_{pa} + c_{pa}\frac{\partial\chi_p}{\partial\xi_p}\delta\xi_p\right). \tag{8.313}$$

The variation δJ now assumes the following form:

$$\begin{aligned}\delta J = &\sum_{a,p}\delta c_{pa}[\sum_{b,q}\langle\chi_p|\hat{H}_{ab}|\chi_q\rangle c_{qb} - \sum_q\langle\chi_p|\epsilon_a\chi_q\rangle c_{qa}]\\ &+\sum_{a,p}\delta c_{pa}^*[\sum_{b,q}\langle\chi_p|\hat{H}_{ab}|\chi_q\rangle c_{qb} - \sum_q\langle\chi_p|\epsilon_a\chi_q\rangle c_{qa}]^*\\ &+\sum_p\delta\xi_p\sum_{a,q}\left\{\sum_b c_{pa}^*\left\langle\frac{\partial\chi_p}{\partial\xi_p}\middle|\hat{H}_{ab}\middle|\chi_q\right\rangle c_{qb} - c_{pa}\left\langle\frac{\partial\chi_p}{\partial\xi_p}\middle|\epsilon_a\chi_q\right\rangle c_{qa}\right.\\ &\left.+\left[\sum_b c_{pa}^*\left\langle\frac{\partial\chi_p}{\partial\xi_p}\middle|\hat{H}_{ab}\middle|\chi_q\right\rangle c_{qb} - c_{pa}^*\left\langle\frac{\partial\chi_p}{\partial\xi_p}\middle|\epsilon_a\chi_q\right\rangle c_{qa}\right]^*\right\},\end{aligned} \tag{8.314}$$

where $\hat{H}_{ab}$ denotes an operator $\hat{\mathcal{H}}_{ab}$ in which the orbitals are replaced by their expansions (8.312).

‡The first letters of the alphabet, $a, b, \ldots$, are used to denote orbitals, and the letters p, q to denote basis functions.

§The transition to the case when a basis function χ_p contains several parameters ξ_{pi}, enumerated by the index i, is accomplished by making the replacement $\delta\xi_p(\partial/\delta\xi_p) \longrightarrow \sum_i \delta\xi_{pi}(\partial/\partial\xi_{pi})$.

Setting the variation (8.314) equal to zero and using the independence of the variations δc_{pa} and $\delta \xi_p$, we obtain two systems of algebraic equations (Kaplan and Maksimov, 1970),

$$\begin{gathered} \sum_b \sum_q \langle \chi_p | \hat{H}_{ab} | \chi_q \rangle c_{qb} - \epsilon_a \sum_q \langle \chi_p | \chi_q \rangle c_{qa} = 0, \\ \operatorname{Re} \left\{ \sum_{a,b} \sum_q c_{pa}^* \left\langle \frac{\partial \chi_p}{\partial \xi_p} \middle| \hat{H}_{ab} \middle| \chi_q \right\rangle c_{qb} - \sum_a \sum_q \epsilon_a c_{pa}^* \left\langle \frac{\partial \chi_p}{\partial \xi_p} \middle| \chi_q \right\rangle c_{qa} \right\} = 0. \end{gathered} \tag{8.315}$$

The coefficients c_{pa} form a rectangular matrix **C** of order $\nu \times N$. By introducing the square matrices **T**, **f**, and $\mathbf{g}_{cd}$ of order ν, with matrix elements

$$t_{pq} = \langle \chi_p | \chi_q \rangle, \qquad f_{pq} = \langle \chi_p | \hat{f} | \chi_q \rangle, \tag{8.316}$$

$$(\hat{g}_{cd})_{pq} = \sum_{p',q'} c_{p'c}^* \langle \chi_p \chi_{p'} | \hat{g} | \chi_q \chi_{q'} \rangle c_{q'd},$$

we can define a matrix $\mathbf{H}_{ab}$, the matrix analog of the operator $\hat{\mathcal{H}}_{ab}$ [(8.310)]

$$\begin{aligned} (H_{ab})_{pq} &= \langle \chi_p | \hat{\mathcal{H}}_{ab} | \chi_a \rangle \\ &= M_{ab}^{[\lambda]} f_{pq} + \sum_{c,d} M_{ac,bd}^{[\lambda]} (\hat{g}_{cd})_{pq} \\ &\quad + \gamma_{ab} [-\Lambda M_{ab}^{[\lambda]} + \sum_{c,d} M_{ac,bd}^{[\lambda]} f_{cd} + \tfrac{1}{2} \sum_{c,d} \sum_{e,f} M_{ace,bdf}^{[\lambda]} g_{ce,df}] t_{pq}. \end{aligned} \tag{8.317}$$

All quantities which occur in this equation are defined in terms of orbitals which are expanded as in (8.312).‡ The overlap integrals between orbitals which occur in the $M^{[\lambda]}$ are expressed in terms of overlap integrals between basis functions by means of the familar relation

$$s_{ab} = \sum_{p,q} c_{pa}^* t_{pq} c_{qb}. \tag{8.318}$$

Matrices defined in the "asymmetric" representation in which χ_p on the left is replaced by $\partial \chi_p / \partial \xi_p$ are denoted by a prime. Thus

$$(H'_{ab})_{pq} = \langle (\partial \chi_p / \partial \xi_p) | \hat{\mathcal{H}}_{ab} | \chi_q \rangle. \tag{8.319}$$

With these definitions Eqs. (8.315) assume the forms

$$\sum_b \sum_q (H_{ab})_{pq} c_{qb} - \epsilon_a \sum_q t_{pq} c_{qa} = 0, \tag{8.320}$$

$$\operatorname{Re} \{ \sum_{a,b} \sum_q c_{pa}^* (H'_{ab})_{pq} c_{qb} - \sum_a \sum_q \epsilon_a c_{pa}^* t'_{pq} c_{qa} \} = 0. \tag{8.321}$$

The νN equations (8.320) and ν equations (8.321) involve νN coefficients c_{pa} which constitute the matrix **C**, ν parameters ξ_p, and N Lagrange multipliers ϵ_a. These equations must therefore be supplemented by N additional equations obtained from the coupling conditions (8.289),

$$\sum_{p,q} c_{pa}^* t_{pq} c_{qa} = 1. \tag{8.322}$$

Equations (8.320)–(8.322) can be solved by the quadratically convergent iteration method proposed by Hinze and Roothaan (1967) and Wessel (1967).

‡This is emphasized by substituting italic $M^{[\lambda]}$ in place of script $\mathfrak{M}^{[\lambda]}$.

APPENDIX 1

Character Tables for Point Groups

In the tables given in this appendix each class is characterized by one of its elements. A number standing in front of a symbol for an element of the group in the top rows of a table denotes the number of elements in the class. The notation for the operations of a point group is described in Section 3.8. The following notation for the irreducible representations has been adopted: One-dimensional irreducible representations are denoted by the letters A or B, two-dimensional representations by E, and three-dimensional representations by F. Functions of a basis for an A representation are symmetric with respect to rotations about the principal axis of symmetry of order n, while functions of a basis for a B representation are antisymmetric. The symbol A_1 always denotes the totally symmetric (unit) representation. Representations which differ in the behavior of their basis functions with respect to rotations about symmetry axes of order less than n, or with respect to the reflections σ_v and σ_d, are distinguished by a subscript added to the symbol for the representation. Representations whose basis functions are of different symmetry with respect to reflections in a plane σ_h are distinguished by primes. Finally, the subscripts g and u denote the symmetry of basis functions with respect to an inversion.

The coordinates x, y, and z beside a symbol denote an irreducible representation according to which the corresponding components of a polar vector transform. Similarly, R_x, R_y, and R_z denote an irreducible representation according to which the components of an axial vector transform.

It should be noted that when decomposing a representation into irreducible representations of $\mathbf{C}_{\infty v}$ and $\mathbf{D}_{\infty h}$, it is unnecessary to sum over the operations E and I since these are taken into account when one integrates over ϕ from zero to 2π. Characters of point groups not given are easily obtained from the relations

$$\mathbf{C}_{3h} = \mathbf{C}_3 \times \mathbf{C}_s, \quad \mathbf{D}_{2h} = \mathbf{D}_2 \times \mathbf{C}_i, \quad \mathbf{S}_6 = \mathbf{C}_3 \times \mathbf{C}_i,$$

$$\mathbf{C}_{4h} = \mathbf{C}_4 \times \mathbf{C}_i, \quad \mathbf{D}_{4h} = \mathbf{D}_4 \times \mathbf{C}_i, \quad \mathbf{T}_h = \mathbf{T} \times \mathbf{C}_i,$$

$$\mathbf{C}_{5h} = \mathbf{C}_5 \times \mathbf{C}_s, \quad \mathbf{D}_{5h} = \mathbf{D}_5 \times \mathbf{C}_s, \quad \mathbf{O}_h = \mathbf{O} \times \mathbf{C}_i,$$

$$\mathbf{C}_{6h} = \mathbf{C}_6 \times \mathbf{C}_i, \quad \mathbf{D}_{6h} = \mathbf{D}_6 \times \mathbf{C}_i.$$

$\mathbf{C}_i$	$\mathbf{C}_2$	$\mathbf{C}_s$	E	I
			E	C_2
			E	σ
$A_g\ \{R_x, R_y, R_z\}$	$A\ \{z, R_z\}$	$A' \begin{Bmatrix} x, y \\ R_z \end{Bmatrix}$	1	1
$A_u\ \{x, y, z\}$	$B \begin{Bmatrix} x, y \\ R_x, R_y \end{Bmatrix}$	$A'' \begin{Bmatrix} z \\ R_x, R_y \end{Bmatrix}$	1	−1

$\mathbf{C}_3$	E	C_3	C_3^2	
$A\ \{z, R_z\}$	1	1	1	
$E \begin{Bmatrix} x, y \\ R_x, R_y \end{Bmatrix}$	1	ε	ε^2	$(\varepsilon = e^{2\pi i/3})$
	1	ε^2	ε	

$\mathbf{C}_{2h}$	$\mathbf{C}_{2v}$	$\mathbf{D}_2$	E	C_2	σ_h	I
			E	C_2	σ_v	$\bar{\sigma}_v$
			E	C_2^z	C_2^y	C_2^x
$A_g\ \{R_z\}$	$A_1\ \{z\}$	A	1	1	1	1
$A_u\ \{z\}$	$A_2\ \{R_z\}$	$B_1\ \{z, R_z\}$	1	1	−1	−1
$B_g\ \{R_x, R_y\}$	$B_2\ \{y, R_x\}$	$B_3\ \{x, R_x\}$	1	−1	−1	1
$B_u\ \{x, y\}$	$B_1\ \{x, R_y\}$	$B_2\ \{y, R_y\}$	1	−1	1	−1

$\mathbf{C}_{3v}$	$\mathbf{D}_3$	E	$2C_3$	$3\sigma_v$
		E	$2C_3$	$3U_2$
$A_1\ \{z\}$	A_1	1	1	1
$A_2\ \{R_z\}$	$A_2\ \{z, R_z\}$	1	1	−1
$E \begin{Bmatrix} x, y \\ R_x, R_y \end{Bmatrix}$	$E \begin{Bmatrix} x, y \\ R_x, R_y \end{Bmatrix}$	2	−1	0

$\mathbf{C}_4$	$\mathbf{S}_4$	E	C_4	C_2	C_4^3
		E	S_4	C_2	S_4^3
$A\ \{z, R_z\}$	$A\ \{R_z\}$	1	1	1	1
B	$B\ \{z\}$	1	−1	1	−1
$E \begin{Bmatrix} x, y \\ R_x, R_y \end{Bmatrix}$	$E \begin{Bmatrix} x, y \\ R_x, R_y \end{Bmatrix}$	1	i	−1	$-i$
		1	$-i$	−1	i

$\mathbf{C}_5$	E	C_5	C_5^2	C_5^3	C_5^4	
A	1	1	1	1	1	
$E_1 \begin{Bmatrix} x, y \\ R_x, R_y \end{Bmatrix}$	1	ε	ε^2	ε^3	ε^4	
	1	ε^4	ε^3	ε^2	ε	$(\varepsilon = e^{2\pi i/5})$
E_2	1	ε^2	ε^4	ε	ε^3	
	1	ε^3	ε	ε^4	ε^2	

$\mathbf{C}_6$	E	C_6	C_3	C_2	C_3^2	C_6^5	
$A\ \{z, R_z\}$	1	1	1	1	1	1	
B	1	−1	1	−1	1	−1	
E_1	1	ε^2	$-\varepsilon$	1	ε^2	$-\varepsilon$	$(\varepsilon = e^{2\pi i/6})$
	1	$-\varepsilon$	ε^2	1	$-\varepsilon$	ε^2	
$E_2 \begin{Bmatrix} x, y \\ R_x, R_y \end{Bmatrix}$	1	ε	ε^2	−1	$-\varepsilon$	$-\varepsilon^2$	
	1	$-\varepsilon^2$	$-\varepsilon$	−1	ε^2	ε	

$\mathbf{C}_{4v}$	$\mathbf{D}_4$	$\mathbf{D}_{2d}$	E E E	C_2 C_2 C_2	$2C_4$ $2C_4$ $2S_4$	$2\sigma_v$ $2U_2$ $2U_2$	$2\bar{\sigma}_v$ $2\bar{U}_2$ $2\sigma_d$
$A_1\ \{z\}$	A_1	A_1	1	1	1	1	1
$A_2\ \{R_z\}$	$A_2\ \{z, R_z\}$	$A_2\ \{R_z\}$	1	1	1	-1	-1
B_1	B_2	B_1	1	1	-1	1	-1
B_2	B_1	$B_2\ \{z\}$	1	1	-1	-1	1
$E\begin{Bmatrix}x, y\\R_x, R_y\end{Bmatrix}$	$E\begin{Bmatrix}x, y\\R_x, R_y\end{Bmatrix}$	$E\begin{Bmatrix}x, y\\R_x, R_y\end{Bmatrix}$	2	-2	0	0	0

$\mathbf{C}_{5v}$	$\mathbf{D}_5$	E E	$2C_5$ $2C_5$	$2C_5^2$ $2C_5^2$	$5\sigma_v$ $5U_2$
$A_1\ \{z\}$	A_1	1	1	1	1
$A_2\ \{R_z\}$	$A_2\ \{z, R_z\}$	1	1	1	-1
$E_1\begin{Bmatrix}x, y\\R_x, R_y\end{Bmatrix}$	$E_1\begin{Bmatrix}x, y\\R_x, R_y\end{Bmatrix}$	2	$2\cos\alpha$	$2\cos 2\alpha$	$0\ \left(\alpha = \frac{2\pi}{5}\right)$
E_2	E_2	2	$2\cos 2\alpha$	$2\cos 4\alpha$	0

$\mathbf{D}_6$	$\mathbf{C}_{6v}$	$\mathbf{D}_{3h} = \mathbf{D}_3 \times \mathbf{C}_s$	$\mathbf{D}_{3d} = \mathbf{D}_3 \times \mathbf{C}_i$	E E E E	C_2 C_2 σ_h I	$2C_3$ $2C_3$ $2C_3$ $2C_3$	$2C_6$ $2C_6$ $2S_3$ $2S_6$	$3U_2$ $3\sigma_v$ $3U_2$ $3U_2$	$3\bar{U}_2$ $3\bar{\sigma}_v$ $3\sigma_v$ $3\sigma_d$
A_1	$A_1\ \{z\}$	A_1'	A_{1g}	1	1	1	1	1	1
$A_2\ \{z, R_z\}$	$A_2\ \{R_z\}$	$A_2'\ \{R_z\}$	$A_{2g}\ \{R_z\}$	1	1	1	1	-1	-1
B_1	B_2	A_1''	A_{1u}	1	-1	1	-1	1	-1
B_2	B_1	$A_2''\ \{z\}$	$A_{2u}\ \{z\}$	1	-1	1	-1	-1	1
$E_1\begin{Bmatrix}x, y\\R_x, R_y\end{Bmatrix}$	$E_1\begin{Bmatrix}x, y\\R_x, R_y\end{Bmatrix}$	$E''\ \{R_x, R_y\}$	$E_u\ \{x, y\}$	2	-2	-1	1	0	0
E_2	E_2	$E'\ \{x, y\}$	$E_g\ \{R_x, R_y\}$	2	2	-1	-1	0	0

T	E	$3C_2$	$4C_3$	$4C_3^2$	
A	1	1	1	1	
E	$\begin{cases}1\\1\end{cases}$	$\begin{matrix}1\\1\end{matrix}$	$\begin{matrix}\varepsilon\\\varepsilon^2\end{matrix}$	$\begin{matrix}\varepsilon^2\\\varepsilon\end{matrix}$	$(\varepsilon = e^{2\pi i/3})$
$F\begin{Bmatrix}x, y, z\\R_x, R_y, R_z\end{Bmatrix}$	3	−1	0	0	

O	T_d	E / E	$8C_3$ / $8C_3$	$3C_2$ / $3C_2$	$6C_2$ / $6\sigma_d$	$6C_4$ / $6S_4$
A_1	A_1	1	1	1	1	1
A_2	A_2	1	1	1	−1	−1
E	E	2	−1	2	0	0
$F_1\begin{Bmatrix}x, y, z\\R_x, R_y, R_z\end{Bmatrix}$	$F_1\{R_x, R_y, R_z\}$	3	0	−1	−1	1
F_2	$F_2\{x, y, z\}$	3	0	−1	1	−1

$C_{\infty v}$	E	$2C_\varphi$	σ_v
$A_1\{z\}$	1	1	1
$A_2\{R_z\}$	1	1	−1
$E_1\begin{Bmatrix}x, y\\R_x, R_y\end{Bmatrix}$	2	$2\cos\varphi$	0
E_2	2	$2\cos 2\varphi$	0
. . . .	. . .	. . .	. . .
E_m	2	$2\cos m\varphi$	0
. . . .	. . .	. . .	. . .

$D_{\infty h}$	E	$2C_\varphi$	U_2	I	$2IC_\varphi$	IU_2
A_{1g}	1	1	1	1	1	1
A_{1u}	1	1	1	−1	−1	−1
$A_{2g}\{R_z\}$	1	1	−1	1	1	−1
$A_{2u}\{z\}$	1	1	−1	−1	−1	1
$E_{1g}\{R_x, R_y\}$	2	$2\cos\varphi$	0	2	$2\cos\varphi$	0
$E_{1u}\{x, y\}$	2	$2\cos\varphi$	0	−2	$-2\cos\varphi$	0
E_{2g}	2	$2\cos 2\varphi$	0	2	$2\cos 2\varphi$	0
E_{2u}	2	$2\cos 2\varphi$	0	−2	$-2\cos 2\varphi$	0
. . . .	. . .	. . .	. . .	. . .	. . .	. . .

APPENDIX 2

Matrices of Orthogonal Irreducible Representations of the Point Groups

In order to determine the matrices for the irreducible representations of the point groups, it is sufficient to know the representation matrices for those groups whose operations consist of pure rotations. All other point groups are either isomorphic with groups of pure rotations or can be expressed as direct products of these groups with the groups $\mathbf{C}_s$ or $\mathbf{C}_i$. Matrices for two-dimensional and three-dimensional orthogonal irreducible representations are given in this appendix for the groups $\mathbf{D}_n$ ($n = 3$–6), $\mathbf{T}$, and $\mathbf{O}$. The symmetry operations for these groups can be found from Fig. A.1. The z axis for the $\mathbf{D}_n$ groups coincides with the principal axis of symmetry and is perpendicular to the plane of the paper. Rotations are everywhere taken to be clockwise. C_n^x is a rotation about an axis of order n, the axis coinciding with the X axis, $C_3^{(\alpha)}$ is a rotation about an axis of order three which passes through vertex α, and $C_2^{(k)}$ is a rotation about an axis of order two which passes through the points k, these lying in the middle of opposite edges of a cube. Blank spaces in the matrices denote zeros.

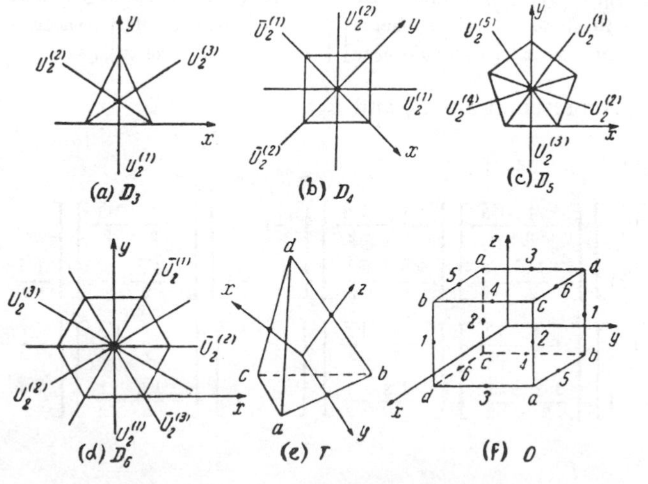

Figure A.1

Group $\mathbf{D}_3$, E representation:

$$\underset{E}{\begin{bmatrix} 1 & \\ & 1 \end{bmatrix}} \underset{C_3}{\begin{bmatrix} -\frac{1}{2} & -\frac{\sqrt{3}}{2} \\ \frac{\sqrt{3}}{2} & -\frac{1}{2} \end{bmatrix}} \underset{C_3^2}{\begin{bmatrix} -\frac{1}{2} & \frac{\sqrt{3}}{2} \\ -\frac{\sqrt{3}}{2} & -\frac{1}{2} \end{bmatrix}}$$

$$\underset{U_2^{(1)}}{\begin{bmatrix} 1 & \\ & -1 \end{bmatrix}} \underset{U_2^{(2)}}{\begin{bmatrix} -\frac{1}{2} & -\frac{\sqrt{3}}{2} \\ -\frac{\sqrt{3}}{2} & \frac{1}{2} \end{bmatrix}} \underset{U_2^{(3)}}{\begin{bmatrix} -\frac{1}{2} & \frac{\sqrt{3}}{2} \\ \frac{\sqrt{3}}{2} & \frac{1}{2} \end{bmatrix}}$$

Group $\mathbf{D}_4$, E representation:

$$\underset{E}{\begin{bmatrix} 1 & \\ & 1 \end{bmatrix}} \underset{C_4}{\begin{bmatrix} & 1 \\ -1 & \end{bmatrix}} \underset{C_2}{\begin{bmatrix} -1 & \\ & -1 \end{bmatrix}} \underset{C_4^3}{\begin{bmatrix} & -1 \\ 1 & \end{bmatrix}}$$

$$\underset{U_2^{(1)}}{\begin{bmatrix} & -1 \\ -1 & \end{bmatrix}} \underset{U_2^{(2)}}{\begin{bmatrix} & 1 \\ 1 & \end{bmatrix}} \underset{\bar{U}_2^{(1)}}{\begin{bmatrix} 1 & \\ & -1 \end{bmatrix}} \underset{\bar{U}_2^{(2)}}{\begin{bmatrix} -1 & \\ & 1 \end{bmatrix}}$$

Group $\mathbf{D}_5$, representations E_1 and E_2:

$$\begin{array}{cccccc} & E & C_5 & C_5^2 & C_5^3 & C_5^4 \\ E_1 & \begin{bmatrix} 1 & \\ & 1 \end{bmatrix} & \begin{bmatrix} \cos\alpha & \sin\alpha \\ -\sin\alpha & \cos\alpha \end{bmatrix} & \begin{bmatrix} \cos 2\alpha & \sin 2\alpha \\ -\sin 2\alpha & \cos 2\alpha \end{bmatrix} & \begin{bmatrix} \cos 2\alpha & -\sin 2\alpha \\ \sin 2\alpha & \cos 2\alpha \end{bmatrix} & \begin{bmatrix} \cos\alpha & -\sin\alpha \\ \sin\alpha & \cos\alpha \end{bmatrix} \\ E_2 & \begin{bmatrix} 1 & \\ & 1 \end{bmatrix} & \begin{bmatrix} \cos 2\alpha & \sin 2\alpha \\ -\sin 2\alpha & \cos 2\alpha \end{bmatrix} & \begin{bmatrix} \cos 4\alpha & \sin 4\alpha \\ -\sin 4\alpha & \cos 4\alpha \end{bmatrix} & \begin{bmatrix} \cos 4\alpha & -\sin 4\alpha \\ \sin 4\alpha & \cos 4\alpha \end{bmatrix} & \begin{bmatrix} \cos 2\alpha & -\sin 2\alpha \\ \sin 2\alpha & \cos 2\alpha \end{bmatrix} \end{array}$$

$$\begin{array}{cccccc} & U_2^{(1)} & U_2^{(2)} & U_2^{(3)} & U_2^{(4)} & U_2^{(5)} \\ E_1 & \begin{bmatrix} -\cos\alpha & -\sin\alpha \\ -\sin\alpha & \cos\alpha \end{bmatrix} & \begin{bmatrix} -\cos 2\alpha & \sin 2\alpha \\ \sin 2\alpha & \cos 2\alpha \end{bmatrix} & \begin{bmatrix} -1 & \\ & 1 \end{bmatrix} & \begin{bmatrix} -\cos 2\alpha & -\sin 2\alpha \\ -\sin 2\alpha & \cos 2\alpha \end{bmatrix} & \begin{bmatrix} -\cos\alpha & \sin\alpha \\ \sin\alpha & \cos\alpha \end{bmatrix} \\ E_2 & \begin{bmatrix} \cos 2\alpha & \sin 2\alpha \\ \sin 2\alpha & -\cos 2\alpha \end{bmatrix} & \begin{bmatrix} \cos 4\alpha & -\sin 4\alpha \\ -\sin 4\alpha & -\cos 4\alpha \end{bmatrix} & \begin{bmatrix} 1 & \\ & -1 \end{bmatrix} & \begin{bmatrix} \cos 4\alpha & \sin 4\alpha \\ \sin 4\alpha & -\cos 4\alpha \end{bmatrix} & \begin{bmatrix} \cos 2\alpha & -\sin 2\alpha \\ -\sin 2\alpha & -\cos 2\alpha \end{bmatrix} \end{array}$$

Group $\mathbf{D}_6$, representations E_1 and E_2:

$$\begin{array}{ccccccc} & E & C_3 & C_3^2 & C_2 & C_6 & C_6^5 \\ E_1 & \begin{bmatrix} 1 & \\ & 1 \end{bmatrix} & \begin{bmatrix} -\frac{1}{2} & \frac{\sqrt{3}}{2} \\ -\frac{\sqrt{3}}{2} & -\frac{1}{2} \end{bmatrix} & \begin{bmatrix} -\frac{1}{2} & -\frac{\sqrt{3}}{2} \\ \frac{\sqrt{3}}{2} & -\frac{1}{2} \end{bmatrix} & \begin{bmatrix} -1 & \\ & -1 \end{bmatrix} & \begin{bmatrix} \frac{1}{2} & \frac{\sqrt{3}}{2} \\ -\frac{\sqrt{3}}{2} & \frac{1}{2} \end{bmatrix} & \begin{bmatrix} \frac{1}{2} & -\frac{\sqrt{3}}{2} \\ \frac{\sqrt{3}}{2} & \frac{1}{2} \end{bmatrix} \\ E_2 & \begin{bmatrix} 1 & \\ & 1 \end{bmatrix} & \begin{bmatrix} -\frac{1}{2} & -\frac{\sqrt{3}}{2} \\ \frac{\sqrt{3}}{2} & -\frac{1}{2} \end{bmatrix} & \begin{bmatrix} -\frac{1}{2} & \frac{\sqrt{3}}{2} \\ -\frac{\sqrt{3}}{2} & -\frac{1}{2} \end{bmatrix} & \begin{bmatrix} 1 & \\ & 1 \end{bmatrix} & \begin{bmatrix} -\frac{1}{2} & \frac{\sqrt{3}}{2} \\ -\frac{\sqrt{3}}{2} & -\frac{1}{2} \end{bmatrix} & \begin{bmatrix} -\frac{1}{2} & -\frac{\sqrt{3}}{2} \\ \frac{\sqrt{3}}{2} & -\frac{1}{2} \end{bmatrix} \end{array}$$

	$U_2^{(1)}$	$U_2^{(2)}$	$U_2^{(3)}$	$\overline{U}_2^{(1)}$	$\overline{U}_2^{(2)}$	$\overline{U}_2^{(3)}$
E_1	$\begin{bmatrix} -1 & \\ & 1 \end{bmatrix}$	$\begin{bmatrix} \frac{1}{2} & -\frac{\sqrt{3}}{2} \\ -\frac{\sqrt{3}}{2} & -\frac{1}{2} \end{bmatrix}$	$\begin{bmatrix} \frac{1}{2} & \frac{\sqrt{3}}{2} \\ \frac{\sqrt{3}}{2} & -\frac{1}{2} \end{bmatrix}$	$\begin{bmatrix} -\frac{1}{2} & -\frac{\sqrt{3}}{2} \\ -\frac{\sqrt{3}}{2} & \frac{1}{2} \end{bmatrix}$	$\begin{bmatrix} 1 & \\ & -1 \end{bmatrix}$	$\begin{bmatrix} -\frac{1}{2} & \frac{\sqrt{3}}{2} \\ \frac{\sqrt{3}}{2} & \frac{1}{2} \end{bmatrix}$
E_2	$\begin{bmatrix} 1 & \\ & -1 \end{bmatrix}$	$\begin{bmatrix} -\frac{1}{2} & -\frac{\sqrt{3}}{2} \\ -\frac{\sqrt{3}}{2} & \frac{1}{2} \end{bmatrix}$	$\begin{bmatrix} -\frac{1}{2} & \frac{\sqrt{3}}{2} \\ \frac{\sqrt{3}}{2} & \frac{1}{2} \end{bmatrix}$	$\begin{bmatrix} -\frac{1}{2} & \frac{\sqrt{3}}{2} \\ \frac{\sqrt{3}}{2} & \frac{1}{2} \end{bmatrix}$	$\begin{bmatrix} 1 & \\ & -1 \end{bmatrix}$	$\begin{bmatrix} -\frac{1}{2} & -\frac{\sqrt{3}}{2} \\ -\frac{\sqrt{3}}{2} & \frac{1}{2} \end{bmatrix}$

Group **T**, representation F:

E	C_2^z	C_2^y	C_2^x
$\begin{bmatrix} 1 & & \\ & 1 & \\ & & 1 \end{bmatrix}$	$\begin{bmatrix} -1 & & \\ & -1 & \\ & & 1 \end{bmatrix}$	$\begin{bmatrix} -1 & & \\ & 1 & \\ & & -1 \end{bmatrix}$	$\begin{bmatrix} 1 & & \\ & -1 & \\ & & -1 \end{bmatrix}$
C_3^a	$(C_3^a)^2$	C_3^b	$(C_3^b)^2$
$\begin{bmatrix} & 1 & \\ & & -1 \\ -1 & & \end{bmatrix}$	$\begin{bmatrix} & & -1 \\ 1 & & \\ & -1 & \end{bmatrix}$	$\begin{bmatrix} & -1 & \\ & & 1 \\ -1 & & \end{bmatrix}$	$\begin{bmatrix} & & -1 \\ -1 & & \\ & 1 & \end{bmatrix}$
C_3^c	$(C_3^c)^2$	C_3^d	$(C_3^d)^2$
$\begin{bmatrix} & 1 & \\ & & 1 \\ 1 & & \end{bmatrix}$	$\begin{bmatrix} & & 1 \\ 1 & & \\ & 1 & \end{bmatrix}$	$\begin{bmatrix} & -1 & \\ & & -1 \\ 1 & & \end{bmatrix}$	$\begin{bmatrix} & & 1 \\ -1 & & \\ & -1 & \end{bmatrix}$

Group **O**, representations E, F_1, and F_2:

	E	C_2^x	C_2^y	C_2^z
E	$\begin{bmatrix} 1 & \\ & 1 \end{bmatrix}$	$\begin{bmatrix} 1 & \\ & 1 \end{bmatrix}$	$\begin{bmatrix} 1 & \\ & 1 \end{bmatrix}$	$\begin{bmatrix} 1 & \\ & 1 \end{bmatrix}$
F_1	$\begin{bmatrix} 1 & & \\ & 1 & \\ & & 1 \end{bmatrix}$	$\begin{bmatrix} 1 & & \\ & -1 & \\ & & -1 \end{bmatrix}$	$\begin{bmatrix} -1 & & \\ & 1 & \\ & & -1 \end{bmatrix}$	$\begin{bmatrix} -1 & & \\ & -1 & \\ & & 1 \end{bmatrix}$
F_2	$\begin{bmatrix} 1 & & \\ & 1 & \\ & & 1 \end{bmatrix}$	$\begin{bmatrix} 1 & & \\ & -1 & \\ & & -1 \end{bmatrix}$	$\begin{bmatrix} -1 & & \\ & 1 & \\ & & -1 \end{bmatrix}$	$\begin{bmatrix} -1 & & \\ & -1 & \\ & & 1 \end{bmatrix}$

	C_4^x	C_4^y	C_4^z	$(C_4^x)^3$
E	$\begin{bmatrix} -1 & \\ & 1 \end{bmatrix}$	$\begin{bmatrix} \frac{1}{2} & \frac{\sqrt{3}}{2} \\ \frac{\sqrt{3}}{2} & -\frac{1}{2} \end{bmatrix}$	$\begin{bmatrix} \frac{1}{2} & -\frac{\sqrt{3}}{2} \\ -\frac{\sqrt{3}}{2} & -\frac{1}{2} \end{bmatrix}$	$\begin{bmatrix} -1 & \\ & 1 \end{bmatrix}$
F_1	$\begin{bmatrix} 1 & & \\ & & -1 \\ & 1 & \end{bmatrix}$	$\begin{bmatrix} & & 1 \\ & 1 & \\ -1 & & \end{bmatrix}$	$\begin{bmatrix} & -1 & \\ 1 & & \\ & & 1 \end{bmatrix}$	$\begin{bmatrix} 1 & & \\ & & 1 \\ & -1 & \end{bmatrix}$
F_2	$\begin{bmatrix} -1 & & \\ & & 1 \\ & -1 & \end{bmatrix}$	$\begin{bmatrix} & & -1 \\ & -1 & \\ 1 & & \end{bmatrix}$	$\begin{bmatrix} & 1 & \\ -1 & & \\ & & -1 \end{bmatrix}$	$\begin{bmatrix} -1 & & \\ & & -1 \\ & 1 & \end{bmatrix}$

$$\begin{array}{ccccc}
 & (C_4^y)^3 & (C_4^z)^3 & C_2^{(1)} & C_2^{(2)} \\
E & \begin{bmatrix} \frac{1}{2} & \frac{\sqrt{3}}{2} \\ \frac{\sqrt{3}}{2} & -\frac{1}{2} \end{bmatrix} & \begin{bmatrix} \frac{1}{2} & -\frac{\sqrt{3}}{2} \\ -\frac{\sqrt{3}}{2} & -\frac{1}{2} \end{bmatrix} & \begin{bmatrix} \frac{1}{2} & -\frac{\sqrt{3}}{2} \\ -\frac{\sqrt{3}}{2} & -\frac{1}{2} \end{bmatrix} & \begin{bmatrix} \frac{1}{2} & -\frac{\sqrt{3}}{2} \\ -\frac{\sqrt{3}}{2} & -\frac{1}{2} \end{bmatrix} \\
F_1 & \begin{bmatrix} & & -1 \\ & 1 & \\ 1 & & \end{bmatrix} & \begin{bmatrix} & 1 & \\ -1 & & \\ & & 1 \end{bmatrix} & \begin{bmatrix} & -1 & \\ -1 & & \\ & & -1 \end{bmatrix} & \begin{bmatrix} & 1 & \\ 1 & & \\ & & -1 \end{bmatrix} \\
F_2 & \begin{bmatrix} & & 1 \\ & -1 & \\ -1 & & \end{bmatrix} & \begin{bmatrix} & -1 & \\ 1 & & \\ & & -1 \end{bmatrix} & \begin{bmatrix} & 1 & \\ 1 & & \\ & & 1 \end{bmatrix} & \begin{bmatrix} & -1 & \\ -1 & & \\ & & 1 \end{bmatrix}
\end{array}$$

$$\begin{array}{ccccc}
 & C_2^{(3)} & C_2^{(4)} & C_2^{(5)} & C_2^{(6)} \\
E & \begin{bmatrix} \frac{1}{2} & \frac{\sqrt{3}}{2} \\ \frac{\sqrt{3}}{2} & -\frac{1}{2} \end{bmatrix} & \begin{bmatrix} \frac{1}{2} & \frac{\sqrt{3}}{2} \\ \frac{\sqrt{3}}{2} & -\frac{1}{2} \end{bmatrix} & \begin{bmatrix} -1 & \\ & 1 \end{bmatrix} & \begin{bmatrix} -1 & \\ & 1 \end{bmatrix} \\
F_1 & \begin{bmatrix} & & -1 \\ & -1 & \\ -1 & & \end{bmatrix} & \begin{bmatrix} & & 1 \\ & -1 & \\ 1 & & \end{bmatrix} & \begin{bmatrix} -1 & & \\ & & -1 \\ & -1 & \end{bmatrix} & \begin{bmatrix} -1 & & \\ & & 1 \\ & 1 & \end{bmatrix} \\
F_2 & \begin{bmatrix} & & 1 \\ & 1 & \\ 1 & & \end{bmatrix} & \begin{bmatrix} & & -1 \\ & 1 & \\ -1 & & \end{bmatrix} & \begin{bmatrix} 1 & & \\ & & 1 \\ & 1 & \end{bmatrix} & \begin{bmatrix} 1 & & \\ & & -1 \\ & -1 & \end{bmatrix}
\end{array}$$

$$\begin{array}{ccccc}
 & C_3^{(a)} & C_3^{(b)} & C_3^{(c)} & C_3^{(d)} \\
E & \begin{bmatrix} -\frac{1}{2} & \frac{\sqrt{3}}{2} \\ -\frac{\sqrt{3}}{2} & -\frac{1}{2} \end{bmatrix} & \begin{bmatrix} -\frac{1}{2} & -\frac{\sqrt{3}}{2} \\ \frac{\sqrt{3}}{2} & -\frac{1}{2} \end{bmatrix} & \begin{bmatrix} -\frac{1}{2} & \frac{\sqrt{3}}{2} \\ -\frac{\sqrt{3}}{2} & -\frac{1}{2} \end{bmatrix} & \begin{bmatrix} -\frac{1}{2} & -\frac{\sqrt{3}}{2} \\ \frac{\sqrt{3}}{2} & -\frac{1}{2} \end{bmatrix} \\
F_1 & \begin{bmatrix} & & -1 \\ 1 & & \\ & -1 & \end{bmatrix} & \begin{bmatrix} & -1 & \\ & & -1 \\ 1 & & \end{bmatrix} & \begin{bmatrix} & & 1 \\ 1 & & \\ & 1 & \end{bmatrix} & \begin{bmatrix} & -1 & \\ & & 1 \\ -1 & & \end{bmatrix} \\
F_2 & \begin{bmatrix} & & -1 \\ 1 & & \\ & -1 & \end{bmatrix} & \begin{bmatrix} & -1 & \\ & & -1 \\ 1 & & \end{bmatrix} & \begin{bmatrix} & & 1 \\ 1 & & \\ & 1 & \end{bmatrix} & \begin{bmatrix} & -1 & \\ & & 1 \\ -1 & & \end{bmatrix}
\end{array}$$

$$\begin{array}{ccccc}
 & (C_3^{(a)})^2 & (C_3^{(b)})^2 & (C_3^{(c)})^2 & (C_3^{(d)})^2 \\
E & \begin{bmatrix} -\frac{1}{2} & -\frac{\sqrt{3}}{2} \\ \frac{\sqrt{3}}{2} & -\frac{1}{2} \end{bmatrix} & \begin{bmatrix} -\frac{1}{2} & \frac{\sqrt{3}}{2} \\ -\frac{\sqrt{3}}{2} & -\frac{1}{2} \end{bmatrix} & \begin{bmatrix} -\frac{1}{2} & -\frac{\sqrt{3}}{2} \\ \frac{\sqrt{3}}{2} & -\frac{1}{2} \end{bmatrix} & \begin{bmatrix} -\frac{1}{2} & \frac{\sqrt{3}}{2} \\ -\frac{\sqrt{3}}{2} & -\frac{1}{2} \end{bmatrix} \\
F_1 & \begin{bmatrix} & 1 & \\ & & -1 \\ -1 & & \end{bmatrix} & \begin{bmatrix} & & 1 \\ -1 & & \\ & -1 & \end{bmatrix} & \begin{bmatrix} & 1 & \\ & & 1 \\ 1 & & \end{bmatrix} & \begin{bmatrix} & & -1 \\ -1 & & \\ & 1 & \end{bmatrix} \\
F_2 & \begin{bmatrix} & 1 & \\ & & -1 \\ -1 & & \end{bmatrix} & \begin{bmatrix} & & 1 \\ -1 & & \\ & -1 & \end{bmatrix} & \begin{bmatrix} & 1 & \\ & & 1 \\ 1 & & \end{bmatrix} & \begin{bmatrix} & & -1 \\ -1 & & \\ & 1 & \end{bmatrix}
\end{array}$$

APPENDIX 3

Tables for the Reduction of the Representations $U_{2j+1}^{[\lambda]}$ to the Group R_3

The reduction of the representations $U_{2j+1}^{[\lambda]}$ for states of systems with one to four particles is given in this appendix. For larger numbers of particles the reduction can be carried out by the method described in Section 4.4.

Table 1

$j = 1/2$

N	$[\lambda]$	J	$\delta_\lambda(2)$
1	[1]	1/2	2
2	[2] $[1^2] \equiv [0]$	1 0	3 1
3	[3] $[21] \equiv [1]$	3/2 1/2	4 2
4	[4] $[31] \equiv [2]$ $[2^2] \equiv [0]$	2 1 0	5 3 1

Table 2

$j=1$

N	$[\lambda]$	J	δ_λ (3)
1	[1]	1	3
2	[2] $[1^2]$	0, 2 1	6 3
3	[3] [21] $[1^3]\equiv[0]$	1, 3 1, 2 0	10 8 1
4	[4] [31] $[2^2]\equiv[2]$ $[21^2]\equiv[1^2]$	0, 2, 4 1, 2, 3 0, 2 1	15 15 6 3

Table 3

$j=3/2$

N	$[\lambda]$	J	δ_λ (4)
1	[1]	3/2	4
2	[2] $[1^2]$	1, 3 0, 2	10 6
3	[3] [21] $[1^3]\equiv[1]$	3/2, 5/2, 9/2 1/2, 3/2, 5/2, 7/2 3/2	20 20 4
4	[4] [31] $[2^2]$ $[21^2]$ $[1^4]\equiv[0]$	0, 2, 3, 4, 6 $1^2, 2, 3^2, 4, 5$ $0, 2^2, 4$ 1, 2, 3 0	35 45 20 15 1

Table 4

$j=2$

N	$[\lambda]$	J	δ_λ (5)
1	[1]	2	5
2	[2] $[1^2]$	0, 2, 4 1, 3	15 10
3	[3] [21] $[1^3]\equiv[1^2]$	0, 2, 3, 4, 6 $1, 2^2, 3, 4, 5$ 1, 3	35 40 10
4	[4] [31] $[2^2]$ $[21^2]$ $[1^4]\equiv[1]$	$0, 2^2, 4^2, 5, 6, 8$ $1^2, 2^2, 3^3, 4^2, 5^2, 6, 7$ $0^2, 2^2, 3, 4^2, 6$ $1^2, 2, 3^2, 4, 5$ 2	70 105 50 45 5

Table 5

$j = 5/2$

N	$[\lambda]$	J	$\delta_\lambda(6)$
1	[1]	5/2	6
2	[2]	5, 3, 1	21
	$[1^2]$	4, 2, 0	15
3	[3]	3/2, 5/2, 7/2, 9/2, 11/2, 15/2	56
	[21]	1/2, 3/2, $(5/2)^2$, $(7/2)^2$, 9/2, 11/2, 13/2	70
	$[1^3]$	3/2, 5/2, 9/2	20
4	[4]	0, 2^2, 3, 4^2, 5, 6^2, 7, 8, 10	126
	[31]	1^3, 2^2, 3^4, 4^3, 5^4, 6^2, 7^2, 8, 9	210
	$[2^2]$	0^2, 2^3, 3, 4^3, 5, 6^2, 8	105
	$[21^2]$	1^2, 2^2, 3^3, 4^2, 5^2, 6, 7	105
	$[1^4]\equiv[1^2]$	4, 2, 0	15

Table 6

$j = 3$

N	$[\lambda]$	J	$\delta_\lambda(7)$
1	[1]	3	7
2	[2]	0, 2, 4, 6	28
	$[1^2]$	1, 3, 5	21
3	[3]	1, 3^2, 4, 5, 6, 7, 9	84
	[21]	1, 2^2, 3^2, 4^2, 5^2, 6, 7, 8	112
	$[1^3]$	0, 2, 3, 4, 6	35
4	[4]	0^2, 2^2, 3, 4^3, 5, 6^3, 7, 8^2, 9, 10, 12	210
	[31]	1^3, 2^3, 3^5, 4^4, 5^5, 6^4, 7^4, 8^2, 9^2, 10, 11	378
	$[2^2]$	0^2, 2^4, 3, 4^4, 5^2, 6^3, 7, 8^2, 10	196
	$[21^2]$	1^3, 2^2, 3^4, 4^3, 5^4, 6^2, 7^2, 8, 9	210
	$[1^4]\equiv[1^3]$	0, 2, 3, 4, 6	35

APPENDIX 4

Character Tables for the Permutation Groups π_2 to π_8

π_2	$\{1^2\}$	$\{2\}$
Order of class	1	1
$[2]$	1	1
$[1^2]$	1	−1

π_3	$\{1^3\}$	$\{12\}$	$\{3\}$
Order of class	1	3	2
$[3]$	1	1	1
$[21]$	2	0	−1
$[1^3]$	1	−1	1

π_4	$\{1^4\}$	$\{1^22\}$	$\{13\}$	$\{2^2\}$	$\{4\}$
Order of class	1	6	8	3	6
$[4]$	1	1	1	1	1
$[31]$	3	1	0	−1	−1
$[2^2]$	2	0	−1	2	0
$[21^2]$	3	−1	0	−1	1
$[1^4]$	1	−1	1	1	−1

π_5	$\{1^5\}$	$\{1^32\}$	$\{1^23\}$	$\{14\}$	$\{12^2\}$	$\{23\}$	$\{5\}$
Order of class	1	10	20	30	15	20	24
$[5]$	1	1	1	1	1	1	1
$[41]$	4	2	1	0	0	−1	−1
$[32]$	5	1	−1	−1	1	1	0
$[31^2]$	6	0	0	0	−2	0	1
$[2^21]$	5	−1	−1	1	1	−1	0
$[21^3]$	4	−2	1	0	0	1	−1
$[1^5]$	1	−1	1	−1	1	−1	1

π_6	$\{1^6\}$	$\{1^42\}$	$\{1^33\}$	$\{1^24\}$	$\{1^22^2\}$	$\{123\}$	$\{15\}$	$\{24\}$	$\{2^3\}$	$\{3^2\}$	$\{6\}$
Order of class	1	15	40	90	45	120	144	90	15	40	120
$[6]$	1	1	1	1	1	1	1	1	1	1	1
$[51]$	5	3	2	1	1	0	0	−1	−1	−1	−1
$[42]$	9	3	0	−1	1	0	−1	1	3	0	0
$[41^2]$	10	2	1	0	−2	−1	0	0	−2	1	1
$[3^2]$	5	1	−1	−1	1	1	0	−1	−3	2	0
$[321]$	16	0	−2	0	0	0	1	0	0	−2	0
$[31^3]$	10	−2	1	0	−2	1	0	0	2	1	−1
$[2^3]$	5	−1	−1	1	1	−1	0	−1	3	2	0
$[2^21^2]$	9	−3	0	1	1	0	−1	1	−3	0	0
$[21^4]$	5	−3	2	−1	1	0	0	−1	1	−1	1
$[1^6]$	1	−1	1	−1	1	−1	1	1	−1	1	−1

π_7	$\{1^7\}$	$\{1^5 2\}$	$\{1^4 3\}$	$\{1^3 4\}$	$\{1^3 2^2\}$	$\{1^2 23\}$	$\{1^2 5\}$	$\{16\}$	$\{124\}$	$\{12^3\}$	$\{13^2\}$	$\{25\}$	$\{2^2 3\}$	$\{34\}$	$\{7\}$
Order of class	1	21	70	210	105	420	504	840	630	105	280	504	210	420	720
$[7]$	1	1	1	1	1	1	1	1	1	1	1	1	1	1	1
$[61]$	6	4	3	2	2	1	1	0	0	0	0	−1	−1	−1	−1
$[52]$	14	6	2	0	2	0	−1	−1	0	2	−1	1	2	0	0
$[51^2]$	15	5	3	1	−1	−1	0	0	−1	−3	0	0	−1	1	1
$[43]$	14	4	−1	−2	2	1	−1	0	0	0	2	−1	−1	1	0
$[421]$	35	5	−1	−1	−1	−1	0	1	1	1	−1	0	−1	−1	0
$[3^2 1]$	21	1	−3	−1	1	1	1	0	−1	−3	0	1	1	−1	0
$[41^3]$	20	0	2	0	−4	0	0	0	0	0	2	0	2	0	−1
$[32^2]$	21	−1	−3	1	1	−1	1	0	−1	3	0	−1	1	1	0
$[321^2]$	35	−5	−1	1	−1	1	0	−1	1	−1	−1	0	−1	1	0
$[2^3 1]$	14	−4	−1	2	2	−1	−1	0	0	0	2	1	−1	−1	0
$[31^4]$	15	−5	3	−1	−1	1	0	0	−1	3	0	0	−1	−1	1
$[2^2 1^3]$	14	−6	2	0	2	0	−1	1	0	−2	−1	−1	2	0	0
$[21^5]$	6	−4	3	−2	2	−1	1	0	0	0	0	1	−1	1	−1
$[1^7]$	1	−1	1	−1	1	−1	1	−1	1	−1	1	−1	1	−1	1

π_8	$\{1^8\}$	$\{1^6 2\}$	$\{1^5 3\}$	$\{1^4 4\}$	$\{1^4 2^2\}$	$\{1^3 2 3\}$	$\{1^3 5\}$	$\{1^2 6\}$	$\{1^2 2 4\}$	$\{1^2 2^3\}$	$\{1^2 3^2\}$	$\{125\}$	$\{1 2^2 3\}$	$\{134\}$	$\{17\}$	$\{2^2 4\}$	$\{26\}$	$\{2 3^2\}$	$\{2^4\}$	$\{35\}$	$\{4^2\}$	$\{8\}$
Order of class	1	28	112	420	210	1120	1344	3360	2520	420	1120	4032	1680	3360	5760	1260	3360	1120	105	2688	1260	5040
[8]	1	1	1	1	1	1	1	1	1	1	1	1	1	1	1	1	1	1	1	1	1	1
[71]	7	5	4	3	3	2	2	1	1	1	1	0	0	0	0	−1	−1	−1	−1	−1	−1	−1
[62]	20	10	5	2	4	1	0	−1	0	2	−1	0	1	−1	−1	2	1	1	4	0	0	0
$[61^2]$	21	9	6	3	1	0	1	0	−1	−3	0	−1	−2	0	0	−1	0	0	−3	1	1	1
[53]	28	10	1	−2	4	1	−2	−1	0	2	1	0	1	1	0	−2	−1	1	−4	1	0	0
[521]	64	16	4	0	0	−2	−1	0	0	0	−2	1	0	0	1	0	0	−2	0	−1	0	0
$[51^3]$	35	5	5	1	−5	−1	0	0	−1	−3	2	0	1	1	0	1	0	2	3	0	−1	−1
$[4^2]$	14	4	−1	−2	2	1	−1	0	0	0	2	−1	−1	1	0	2	0	−2	6	−1	2	0
[431]	70	10	−5	−4	2	1	0	1	0	−2	1	0	−1	−1	0	0	1	1	−2	0	−2	0
$[42^2]$	56	4	−4	0	0	−2	1	1	0	4	−1	−1	0	0	0	0	−1	1	8	1	0	0
$[421^3]$	90	0	0	0	−6	0	0	0	2	0	0	0	0	0	−1	0	0	0	−6	0	2	0
$[3^2 2]$	42	0	−6	0	2	0	2	0	−2	0	0	0	2	0	0	0	0	0	−6	−1	2	0
$[3^2 1^2]$	56	−4	−4	0	0	2	1	−1	0	−4	−1	1	0	0	0	0	−1	−1	8	1	0	0
$[32^2 1]$	70	−10	−5	4	2	−1	0	−1	0	2	1	0	−1	1	0	0	1	−1	−2	0	−2	0
$[2^4]$	14	−4	−1	2	2	−1	−1	0	0	0	2	1	−1	−1	0	−2	0	2	6	−1	2	0
$[41^4]$	35	−5	5	−1	−5	1	0	0	−1	3	2	0	1	−1	0	−1	0	−2	3	0	−1	1
$[321^3]$	64	−16	4	0	0	2	−1	0	0	0	−2	−1	0	0	1	0	0	2	0	−1	0	0
$[2^3 1^2]$	28	−10	1	2	4	−1	−2	1	0	−2	1	0	1	−1	0	2	−1	−1	−4	1	0	0
$[31^5]$	21	−9	6	−3	1	0	1	0	−1	3	0	1	−2	0	0	1	0	0	−3	1	1	−1
$[2^2 1^4]$	20	−10	5	−2	4	−1	0	1	0	−2	−1	0	1	1	−1	−2	1	−1	4	0	0	0
$[21^6]$	7	−5	4	−3	3	−2	2	−1	1	−1	1	0	0	0	0	1	−1	1	−1	−1	−1	1
$[1^8]$	1	−1	1	−1	1	−1	1	−1	1	−1	1	−1	1	−1	1	−1	1	−1	1	1	1	−1

APPENDIX 5

Matrices of the Orthogonal Irreducible Representations for the Permutation Groups π_3 to π_6

In this appendix are given the representation matrices for the standard orthogonal Young–Yamanouchi representation. The rules by which they are determined have been described in Section 2.5. The rows and columns of the matrices are numbered by Young tableaux ordered according to the extent to which the numbers they contain diverge from the natural order (when read along the rows, moving from top to bottom).

Matrices for one-dimensional representations do not appear since these can be found from character tables (Appendix 4). All the matrices for the irreducible representations of π_3 and π_4 are given. In other cases only the matrices for the transpositions $\Gamma^{[\lambda]}(P_{i-1,i})$ are given. According to the theorem of Section 2.1, the matrices for the other permutations can always be represented as products of matrices of this type. Table 9 gives the matrices $\Gamma^{[2^3]}(P)$ for the subgroup of π_6 which is isomorphic with $\mathbf{D}_6$. Blank spaces in the matrices denote zeros.

Table 1

$[\lambda]=[21]$

$r^{(1)}$

1	2
3	

$r^{(2)}$

1	3
2	

I $\begin{bmatrix} 1 & \\ & 1 \end{bmatrix}$ P_{12} $\begin{bmatrix} 1 & \\ & -1 \end{bmatrix}$ P_{23} $\begin{bmatrix} -\frac{1}{2} & \frac{\sqrt{3}}{2} \\ \frac{\sqrt{3}}{2} & \frac{1}{2} \end{bmatrix}$ P_{13} $\begin{bmatrix} -\frac{1}{2} & -\frac{\sqrt{3}}{2} \\ -\frac{\sqrt{3}}{2} & \frac{1}{2} \end{bmatrix}$ P_{123} $\begin{bmatrix} -\frac{1}{2} & \frac{\sqrt{3}}{2} \\ -\frac{\sqrt{3}}{2} & -\frac{1}{2} \end{bmatrix}$ P_{132} $\begin{bmatrix} -\frac{1}{2} & -\frac{\sqrt{3}}{2} \\ \frac{\sqrt{3}}{2} & -\frac{1}{2} \end{bmatrix}$

Table 2

$[\lambda]=[2^2]$

$r^{(1)}$: $\begin{array}{|c|c|}\hline 1 & 2 \\ \hline 3 & 4 \\ \hline \end{array}$ $r^{(2)}$: $\begin{array}{|c|c|}\hline 1 & 3 \\ \hline 2 & 4 \\ \hline \end{array}$

$$I = \begin{bmatrix} 1 & \\ & 1 \end{bmatrix} \quad P_{12} = \begin{bmatrix} 1 & \\ & -1 \end{bmatrix} \quad P_{23} = \begin{bmatrix} -\frac{1}{2} & \frac{\sqrt{3}}{2} \\ \frac{\sqrt{3}}{2} & \frac{1}{2} \end{bmatrix} \quad P_{34} = \begin{bmatrix} 1 & \\ & -1 \end{bmatrix} \quad P_{13} = \begin{bmatrix} -\frac{1}{2} & -\frac{\sqrt{3}}{2} \\ -\frac{\sqrt{3}}{2} & \frac{1}{2} \end{bmatrix}$$

$$P_{24} = \begin{bmatrix} -\frac{1}{2} & -\frac{\sqrt{3}}{2} \\ -\frac{\sqrt{3}}{2} & \frac{1}{2} \end{bmatrix} \quad P_{14} = \begin{bmatrix} -\frac{1}{2} & \frac{\sqrt{3}}{2} \\ \frac{\sqrt{3}}{2} & \frac{1}{2} \end{bmatrix} \quad P_{123} = \begin{bmatrix} -\frac{1}{2} & \frac{\sqrt{3}}{2} \\ -\frac{\sqrt{3}}{2} & -\frac{1}{2} \end{bmatrix} \quad P_{132} = \begin{bmatrix} -\frac{1}{2} & -\frac{\sqrt{3}}{2} \\ \frac{\sqrt{3}}{2} & -\frac{1}{2} \end{bmatrix} \quad P_{234} = \begin{bmatrix} -\frac{1}{2} & -\frac{\sqrt{3}}{2} \\ \frac{\sqrt{3}}{2} & -\frac{1}{2} \end{bmatrix}$$

$$P_{243} = \begin{bmatrix} -\frac{1}{2} & \frac{\sqrt{3}}{2} \\ -\frac{\sqrt{3}}{2} & -\frac{1}{2} \end{bmatrix} \quad P_{134} = \begin{bmatrix} -\frac{1}{2} & \frac{\sqrt{3}}{2} \\ -\frac{\sqrt{3}}{2} & -\frac{1}{2} \end{bmatrix} \quad P_{143} = \begin{bmatrix} -\frac{1}{2} & -\frac{\sqrt{3}}{2} \\ \frac{\sqrt{3}}{2} & -\frac{1}{2} \end{bmatrix} \quad P_{124} = \begin{bmatrix} -\frac{1}{2} & -\frac{\sqrt{3}}{2} \\ \frac{\sqrt{3}}{2} & -\frac{1}{2} \end{bmatrix} \quad P_{142} = \begin{bmatrix} -\frac{1}{2} & \frac{\sqrt{3}}{2} \\ -\frac{\sqrt{3}}{2} & -\frac{1}{2} \end{bmatrix}$$

$$P_{1234} = \begin{bmatrix} -\frac{1}{2} & -\frac{\sqrt{3}}{2} \\ -\frac{\sqrt{3}}{2} & \frac{1}{2} \end{bmatrix} \quad P_{1243} = \begin{bmatrix} -\frac{1}{2} & \frac{\sqrt{3}}{2} \\ \frac{\sqrt{3}}{2} & \frac{1}{2} \end{bmatrix} \quad P_{1324} = \begin{bmatrix} 1 & \\ & -1 \end{bmatrix} \quad P_{1342} = \begin{bmatrix} -\frac{1}{2} & \frac{\sqrt{3}}{2} \\ \frac{\sqrt{3}}{2} & \frac{1}{2} \end{bmatrix} \quad P_{1432} = \begin{bmatrix} -\frac{1}{2} & -\frac{\sqrt{3}}{2} \\ -\frac{\sqrt{3}}{2} & \frac{1}{2} \end{bmatrix}$$

$$P_{1423} = \begin{bmatrix} 1 & \\ & -1 \end{bmatrix} \quad P_{12}P_{34} = \begin{bmatrix} 1 & \\ & 1 \end{bmatrix} \quad P_{13}P_{24} = \begin{bmatrix} 1 & \\ & 1 \end{bmatrix} \quad P_{14}P_{23} = \begin{bmatrix} 1 & \\ & 1 \end{bmatrix}$$

Table 3

$[\lambda]=[2\,1^2]$

$r^{(1)}$: $\begin{array}{|c|c|}\hline 1 & 2 \\ \hline 3 \\ \cline{1-1} 4 \\ \cline{1-1} \end{array}$ $r^{(2)}$: $\begin{array}{|c|c|}\hline 1 & 3 \\ \hline 2 \\ \cline{1-1} 4 \\ \cline{1-1} \end{array}$ $r^{(3)}$: $\begin{array}{|c|c|}\hline 1 & 4 \\ \hline 2 \\ \cline{1-1} 3 \\ \cline{1-1} \end{array}$

$$I = \begin{bmatrix} 1 & & \\ & 1 & \\ & & 1 \end{bmatrix} \quad P_{12} = \begin{bmatrix} 1 & & \\ & -1 & \\ & & -1 \end{bmatrix} \quad P_{23} = \begin{bmatrix} -\frac{1}{2} & \frac{\sqrt{3}}{2} & \\ \frac{\sqrt{3}}{2} & \frac{1}{2} & \\ & & -1 \end{bmatrix} \quad P_{34} = \begin{bmatrix} -1 & & \\ & -\frac{1}{3} & \frac{2\sqrt{2}}{3} \\ & \frac{2\sqrt{2}}{3} & \frac{1}{3} \end{bmatrix}$$

Table 3 (Cont.)

$$P_{13} = \begin{bmatrix} -\frac{1}{2} & -\frac{\sqrt{3}}{2} & \\ -\frac{\sqrt{3}}{2} & \frac{1}{2} & \\ & & -1 \end{bmatrix} \quad P_{24} = \begin{bmatrix} -\frac{1}{2} & \frac{\sqrt{3}}{6} & -\frac{\sqrt{6}}{3} \\ \frac{\sqrt{3}}{6} & -\frac{5}{6} & -\frac{\sqrt{2}}{3} \\ -\frac{\sqrt{6}}{3} & -\frac{\sqrt{2}}{3} & \frac{1}{3} \end{bmatrix} \quad P_{14} = \begin{bmatrix} -\frac{1}{2} & -\frac{\sqrt{3}}{6} & \frac{\sqrt{6}}{3} \\ -\frac{\sqrt{3}}{6} & -\frac{5}{6} & -\frac{\sqrt{2}}{3} \\ \frac{\sqrt{6}}{3} & -\frac{\sqrt{2}}{3} & \frac{1}{3} \end{bmatrix} \quad P_{123} = \begin{bmatrix} -\frac{1}{2} & \frac{\sqrt{3}}{2} & \\ -\frac{\sqrt{3}}{2} & -\frac{1}{2} & \\ & & 1 \end{bmatrix}$$

$$P_{132} = \begin{bmatrix} -\frac{1}{2} & -\frac{\sqrt{3}}{2} & \\ \frac{\sqrt{3}}{2} & -\frac{1}{2} & \\ & & -1 \end{bmatrix} \quad P_{234} = \begin{bmatrix} \frac{1}{2} & -\frac{\sqrt{3}}{6} & \frac{\sqrt{6}}{3} \\ -\frac{\sqrt{3}}{2} & -\frac{1}{6} & \frac{\sqrt{2}}{3} \\ & -\frac{2\sqrt{2}}{3} & -\frac{1}{3} \end{bmatrix} \quad P_{243} = \begin{bmatrix} \frac{1}{2} & -\frac{\sqrt{3}}{2} & \\ -\frac{\sqrt{3}}{6} & -\frac{1}{6} & -\frac{2\sqrt{2}}{3} \\ \frac{\sqrt{6}}{3} & \frac{\sqrt{2}}{3} & -\frac{1}{3} \end{bmatrix} \quad P_{134} = \begin{bmatrix} \frac{1}{2} & \frac{\sqrt{3}}{6} & -\frac{\sqrt{6}}{3} \\ \frac{\sqrt{3}}{2} & -\frac{1}{6} & \frac{\sqrt{2}}{3} \\ & -\frac{2\sqrt{2}}{3} & -\frac{1}{3} \end{bmatrix}$$

$$P_{143} = \begin{bmatrix} \frac{1}{2} & \frac{\sqrt{3}}{2} & \\ \frac{\sqrt{3}}{6} & -\frac{1}{6} & -\frac{2\sqrt{2}}{3} \\ -\frac{\sqrt{6}}{3} & \frac{\sqrt{2}}{3} & -\frac{1}{3} \end{bmatrix} \quad P_{124} = \begin{bmatrix} -\frac{1}{2} & \frac{\sqrt{3}}{6} & -\frac{\sqrt{6}}{3} \\ -\frac{\sqrt{3}}{6} & \frac{5}{6} & \frac{\sqrt{2}}{3} \\ \frac{\sqrt{6}}{3} & \frac{\sqrt{2}}{3} & -\frac{1}{3} \end{bmatrix} \quad P_{142} = \begin{bmatrix} -\frac{1}{2} & -\frac{\sqrt{3}}{6} & \frac{\sqrt{6}}{3} \\ \frac{\sqrt{3}}{3} & \frac{5}{6} & \frac{\sqrt{2}}{3} \\ -\frac{\sqrt{6}}{3} & \frac{\sqrt{2}}{3} & -\frac{1}{3} \end{bmatrix} \quad P_{1234} = \begin{bmatrix} \frac{1}{2} & -\frac{\sqrt{3}}{6} & \frac{\sqrt{6}}{3} \\ \frac{\sqrt{3}}{2} & \frac{1}{6} & -\frac{\sqrt{2}}{3} \\ & \frac{2\sqrt{2}}{3} & \frac{1}{3} \end{bmatrix}$$

$$P_{1243} = \begin{bmatrix} \frac{1}{2} & -\frac{\sqrt{3}}{2} & \\ \frac{\sqrt{3}}{6} & \frac{1}{6} & \frac{2\sqrt{2}}{3} \\ -\frac{\sqrt{6}}{3} & -\frac{\sqrt{2}}{3} & \frac{1}{3} \end{bmatrix} \quad P_{1324} = \begin{bmatrix} & \frac{\sqrt{3}}{3} & \frac{\sqrt{6}}{3} \\ -\frac{\sqrt{3}}{3} & \frac{2}{3} & -\frac{\sqrt{2}}{3} \\ -\frac{\sqrt{6}}{3} & -\frac{\sqrt{2}}{3} & \frac{1}{3} \end{bmatrix} \quad P_{1342} = \begin{bmatrix} \frac{1}{2} & \frac{\sqrt{3}}{6} & -\frac{\sqrt{6}}{3} \\ -\frac{\sqrt{3}}{2} & \frac{1}{6} & -\frac{\sqrt{2}}{3} \\ & \frac{2\sqrt{2}}{3} & \frac{1}{3} \end{bmatrix} \quad P_{1432} = \begin{bmatrix} \frac{1}{2} & \frac{\sqrt{3}}{2} & \\ -\frac{\sqrt{3}}{6} & \frac{1}{6} & -\frac{2\sqrt{2}}{3} \\ \frac{\sqrt{6}}{3} & -\frac{\sqrt{2}}{3} & -\frac{1}{3} \end{bmatrix}$$

$$P_{1423} = \begin{bmatrix} & -\frac{\sqrt{3}}{3} & -\frac{\sqrt{6}}{3} \\ \frac{\sqrt{3}}{3} & \frac{2}{3} & -\frac{\sqrt{2}}{3} \\ \frac{\sqrt{6}}{3} & -\frac{\sqrt{2}}{3} & \frac{1}{3} \end{bmatrix} \quad P_{12} \cdot P_{34} = \begin{bmatrix} -1 & & \\ & \frac{1}{3} & -\frac{2\sqrt{2}}{3} \\ & -\frac{2\sqrt{2}}{3} & -\frac{1}{3} \end{bmatrix}$$

$$P_{13} \cdot P_{24} = \begin{bmatrix} & \frac{\sqrt{3}}{3} & \frac{\sqrt{6}}{3} \\ \frac{\sqrt{3}}{3} & -\frac{2}{3} & \frac{\sqrt{2}}{3} \\ \frac{\sqrt{6}}{3} & \frac{\sqrt{2}}{3} & -\frac{1}{3} \end{bmatrix} \quad P_{14} \cdot P_{23} = \begin{bmatrix} & -\frac{\sqrt{3}}{3} & -\frac{\sqrt{6}}{3} \\ -\frac{\sqrt{3}}{3} & -\frac{2}{3} & \frac{\sqrt{2}}{3} \\ -\frac{\sqrt{6}}{3} & \frac{\sqrt{2}}{3} & -\frac{1}{3} \end{bmatrix}$$

Table 4

$[\lambda]=[2^2 1]$

$r^{(1)}$	$r^{(2)}$	$r^{(3)}$	$r^{(4)}$	$r^{(5)}$
$\begin{array}{\|c\|c\|}\hline 1 & 2 \\ \hline 3 & 4 \\ \hline 5 & \\ \cline{1-1}\end{array}$	$\begin{array}{\|c\|c\|}\hline 1 & 2 \\ \hline 3 & 5 \\ \hline 4 & \\ \cline{1-1}\end{array}$	$\begin{array}{\|c\|c\|}\hline 1 & 3 \\ \hline 2 & 4 \\ \hline 5 & \\ \cline{1-1}\end{array}$	$\begin{array}{\|c\|c\|}\hline 1 & 3 \\ \hline 2 & 5 \\ \hline 4 & \\ \cline{1-1}\end{array}$	$\begin{array}{\|c\|c\|}\hline 1 & 4 \\ \hline 2 & 5 \\ \hline 3 & \\ \cline{1-1}\end{array}$

$$P_{12} = \begin{bmatrix} 1 & & & & \\ & 1 & & & \\ & & -1 & & \\ & & & -1 & \\ & & & & -1 \end{bmatrix} \qquad P_{23} = \begin{bmatrix} -\frac{1}{2} & & \frac{\sqrt{3}}{2} & & \\ & -\frac{1}{2} & & \frac{\sqrt{3}}{2} & \\ \frac{\sqrt{3}}{2} & & \frac{1}{2} & & \\ & \frac{\sqrt{3}}{2} & & \frac{1}{2} & \\ & & & & -1 \end{bmatrix}$$

$$P_{34} = \begin{bmatrix} 1 & & & & \\ & -1 & & & \\ & & -1 & & \\ & & & -\frac{1}{3} & \frac{2\sqrt{2}}{3} \\ & & & \frac{2\sqrt{2}}{3} & \frac{1}{3} \end{bmatrix} \qquad P_{45} = \begin{bmatrix} -\frac{1}{2} & \frac{\sqrt{3}}{2} & & & \\ \frac{\sqrt{3}}{2} & \frac{1}{2} & & & \\ & & -\frac{1}{2} & \frac{\sqrt{3}}{2} & \\ & & \frac{\sqrt{3}}{2} & \frac{1}{2} & \\ & & & & -1 \end{bmatrix}$$

Table 5

$[\lambda]=[2\,1^3]$

$r^{(1)}$	$r^{(2)}$	$r^{(3)}$	$r^{(4)}$
$\begin{array}{\|c\|c\|}\hline 1 & 2 \\ \hline 3 & \\ \cline{1-1} 4 & \\ \cline{1-1} 5 & \\ \cline{1-1}\end{array}$	$\begin{array}{\|c\|c\|}\hline 1 & 3 \\ \hline 2 & \\ \cline{1-1} 4 & \\ \cline{1-1} 5 & \\ \cline{1-1}\end{array}$	$\begin{array}{\|c\|c\|}\hline 1 & 4 \\ \hline 2 & \\ \cline{1-1} 3 & \\ \cline{1-1} 5 & \\ \cline{1-1}\end{array}$	$\begin{array}{\|c\|c\|}\hline 1 & 5 \\ \hline 2 & \\ \cline{1-1} 3 & \\ \cline{1-1} 4 & \\ \cline{1-1}\end{array}$

$$P_{12} = \begin{bmatrix} 1 & & & \\ & -1 & & \\ & & -1 & \\ & & & -1 \end{bmatrix} \qquad P_{23} = \begin{bmatrix} -\frac{1}{2} & \frac{\sqrt{3}}{2} & & \\ \frac{\sqrt{3}}{2} & \frac{1}{2} & & \\ & & -1 & \\ & & & -1 \end{bmatrix}$$

Table 5 (Cont.)

$$P_{34} \qquad\qquad P_{45}$$

$$\begin{bmatrix} -1 & & & \\ & -\frac{1}{3} & \frac{2\sqrt{2}}{3} & \\ & \frac{2\sqrt{2}}{3} & \frac{1}{3} & \\ & & & -1 \end{bmatrix} \begin{bmatrix} -1 & & & \\ & -1 & & \\ & & -\frac{1}{4} & \frac{\sqrt{15}}{4} \\ & & \frac{\sqrt{15}}{4} & \frac{1}{4} \end{bmatrix}$$

$$P_{13} \qquad\qquad P_{14} \qquad\qquad P_{15}$$

$$\begin{bmatrix} -\frac{1}{2} & -\frac{\sqrt{3}}{2} & & \\ -\frac{\sqrt{3}}{2} & \frac{1}{2} & & \\ & & -1 & \\ & & & -1 \end{bmatrix} \begin{bmatrix} -\frac{1}{2} & -\frac{\sqrt{3}}{6} & \frac{\sqrt{6}}{3} & \\ -\frac{\sqrt{3}}{6} & -\frac{5}{6} & -\frac{\sqrt{2}}{6} & \\ \frac{\sqrt{6}}{3} & -\frac{\sqrt{2}}{3} & \frac{1}{3} & \\ & & & -1 \end{bmatrix} \begin{bmatrix} -\frac{1}{2} & -\frac{\sqrt{3}}{6} & \frac{\sqrt{6}}{12} & -\frac{\sqrt{10}}{4} \\ -\frac{\sqrt{3}}{6} & -\frac{5}{6} & -\frac{\sqrt{2}}{12} & \frac{\sqrt{30}}{12} \\ \frac{\sqrt{6}}{12} & -\frac{\sqrt{2}}{12} & -\frac{11}{12} & -\frac{\sqrt{15}}{12} \\ -\frac{\sqrt{10}}{4} & \frac{\sqrt{30}}{12} & -\frac{\sqrt{15}}{12} & \frac{1}{4} \end{bmatrix}$$

$$P_{24} \qquad\qquad P_{25} \qquad\qquad P_{35}$$

$$\begin{bmatrix} -\frac{1}{2} & \frac{\sqrt{3}}{6} & -\frac{\sqrt{6}}{3} & \\ \frac{\sqrt{3}}{6} & -\frac{5}{6} & -\frac{\sqrt{2}}{3} & \\ -\frac{\sqrt{6}}{3} & -\frac{\sqrt{2}}{3} & \frac{1}{3} & \\ & & & -1 \end{bmatrix} \begin{bmatrix} -\frac{1}{2} & \frac{\sqrt{3}}{6} & -\frac{\sqrt{6}}{12} & \frac{\sqrt{10}}{4} \\ \frac{\sqrt{3}}{6} & -\frac{5}{6} & -\frac{\sqrt{2}}{12} & \frac{\sqrt{30}}{12} \\ -\frac{\sqrt{6}}{12} & -\frac{\sqrt{2}}{12} & -\frac{11}{12} & -\frac{\sqrt{15}}{12} \\ \frac{\sqrt{10}}{4} & \frac{\sqrt{30}}{12} & -\frac{\sqrt{15}}{12} & \frac{1}{4} \end{bmatrix} \begin{bmatrix} -1 & & & \\ & -\frac{1}{3} & \frac{\sqrt{2}}{6} & -\frac{\sqrt{30}}{6} \\ & \frac{\sqrt{2}}{6} & -\frac{11}{12} & -\frac{\sqrt{15}}{12} \\ & -\frac{\sqrt{30}}{6} & -\frac{\sqrt{15}}{12} & \frac{1}{4} \end{bmatrix}$$

$$P_{13} \qquad\qquad P_{14}$$

$$\begin{bmatrix} -\frac{1}{2} & & -\frac{\sqrt{3}}{2} & & \\ & -\frac{1}{2} & & -\frac{\sqrt{3}}{2} & \\ -\frac{\sqrt{3}}{2} & & \frac{1}{2} & & \\ & -\frac{\sqrt{3}}{2} & & \frac{1}{2} & \\ & & & & -1 \end{bmatrix} \begin{bmatrix} -\frac{1}{2} & & \frac{\sqrt{3}}{2} & & \\ & -\frac{1}{2} & & -\frac{\sqrt{3}}{6} & \frac{\sqrt{6}}{3} \\ \frac{\sqrt{3}}{2} & & \frac{1}{2} & & \\ & -\frac{\sqrt{3}}{6} & & -\frac{5}{6} & -\frac{\sqrt{2}}{3} \\ & \frac{\sqrt{6}}{3} & & -\frac{\sqrt{2}}{3} & \frac{1}{3} \end{bmatrix}$$

$$P_{15} \qquad\qquad P_{24}$$

$$\begin{bmatrix} -\frac{1}{2} & & & -\frac{1}{2} & -\frac{\sqrt{2}}{2} \\ & -\frac{1}{2} & -\frac{1}{2} & \frac{\sqrt{3}}{3} & -\frac{\sqrt{6}}{6} \\ & -\frac{1}{2} & -\frac{1}{2} & -\frac{\sqrt{3}}{3} & \frac{\sqrt{6}}{6} \\ -\frac{1}{2} & \frac{\sqrt{3}}{3} & -\frac{\sqrt{3}}{3} & \frac{1}{6} & \frac{\sqrt{2}}{6} \\ -\frac{\sqrt{2}}{2} & -\frac{\sqrt{6}}{6} & \frac{\sqrt{6}}{6} & \frac{\sqrt{2}}{6} & \frac{1}{3} \end{bmatrix} \begin{bmatrix} -\frac{1}{2} & & -\frac{\sqrt{3}}{2} & & \\ & -\frac{1}{2} & & \frac{\sqrt{3}}{6} & -\frac{\sqrt{6}}{3} \\ -\frac{\sqrt{3}}{2} & & \frac{1}{2} & & \\ & \frac{\sqrt{3}}{6} & & -\frac{5}{6} & -\frac{\sqrt{2}}{3} \\ & -\frac{\sqrt{6}}{3} & & -\frac{\sqrt{2}}{3} & \frac{1}{3} \end{bmatrix}$$

Table 5 (Cont.)

$$P_{25} = \begin{bmatrix} -\frac{1}{2} & & & \frac{1}{2} & \frac{\sqrt{2}}{2} \\ & -\frac{1}{2} & \frac{1}{2} & -\frac{\sqrt{3}}{3} & \frac{\sqrt{6}}{6} \\ & \frac{1}{2} & -\frac{1}{2} & -\frac{\sqrt{3}}{3} & \frac{\sqrt{6}}{6} \\ \frac{1}{2} & -\frac{\sqrt{3}}{3} & -\frac{\sqrt{3}}{3} & \frac{1}{6} & \frac{\sqrt{2}}{6} \\ \frac{\sqrt{2}}{2} & \frac{\sqrt{6}}{6} & \frac{\sqrt{6}}{6} & \frac{\sqrt{2}}{6} & \frac{1}{3} \end{bmatrix} \qquad P_{35} = \begin{bmatrix} -\frac{1}{2} & -\frac{\sqrt{3}}{2} & & & \\ -\frac{\sqrt{3}}{2} & \frac{1}{2} & & & \\ & & -\frac{1}{2} & \frac{\sqrt{3}}{6} & -\frac{\sqrt{6}}{3} \\ & & \frac{\sqrt{3}}{6} & -\frac{5}{6} & -\frac{\sqrt{2}}{3} \\ & & -\frac{\sqrt{6}}{3} & -\frac{\sqrt{2}}{3} & \frac{1}{3} \end{bmatrix}$$

Table 6

$[\lambda] = [2^3]$

$$r^{(1)}: \begin{array}{|c|c|}\hline 1 & 2 \\ \hline 3 & 4 \\ \hline 5 & 6 \\ \hline \end{array} \quad r^{(2)}: \begin{array}{|c|c|}\hline 1 & 2 \\ \hline 3 & 5 \\ \hline 4 & 6 \\ \hline \end{array} \quad r^{(3)}: \begin{array}{|c|c|}\hline 1 & 3 \\ \hline 2 & 4 \\ \hline 5 & 6 \\ \hline \end{array} \quad r^{(4)}: \begin{array}{|c|c|}\hline 1 & 3 \\ \hline 2 & 5 \\ \hline 4 & 6 \\ \hline \end{array} \quad r^{(5)}: \begin{array}{|c|c|}\hline 1 & 4 \\ \hline 2 & 5 \\ \hline 3 & 6 \\ \hline \end{array}$$

$$P_{12} = \begin{bmatrix} 1 & & & & \\ & 1 & & & \\ & & -1 & & \\ & & & -1 & \\ & & & & -1 \end{bmatrix} \qquad P_{23} = \begin{bmatrix} -\frac{1}{2} & & \frac{\sqrt{3}}{2} & & \\ & -\frac{1}{2} & & \frac{\sqrt{3}}{2} & \\ \frac{\sqrt{3}}{2} & & \frac{1}{2} & & \\ & \frac{\sqrt{3}}{2} & & \frac{1}{2} & \\ & & & & -1 \end{bmatrix}$$

$$P_{34} = \begin{bmatrix} 1 & & & & \\ & -1 & & & \\ & & -1 & & \\ & & & -\frac{1}{3} & \frac{2\sqrt{2}}{3} \\ & & & \frac{2\sqrt{2}}{3} & \frac{1}{3} \end{bmatrix} \qquad P_{45} = \begin{bmatrix} -\frac{1}{2} & \frac{\sqrt{3}}{2} & & & \\ \frac{\sqrt{3}}{2} & \frac{1}{2} & & & \\ & & -\frac{1}{2} & \frac{\sqrt{3}}{2} & \\ & & \frac{\sqrt{3}}{2} & \frac{1}{2} & \\ & & & & -1 \end{bmatrix}$$

$$P_{13} = \begin{bmatrix} -\frac{1}{2} & & -\frac{\sqrt{3}}{2} & & \\ & -\frac{1}{2} & & -\frac{\sqrt{3}}{2} & \\ -\frac{\sqrt{3}}{2} & & \frac{1}{2} & & \\ & -\frac{\sqrt{3}}{2} & & \frac{1}{2} & \\ & & & & -1 \end{bmatrix} \qquad P_{14} = \begin{bmatrix} -\frac{1}{2} & & \frac{\sqrt{3}}{2} & & \\ & -\frac{1}{2} & & -\frac{\sqrt{3}}{6} & \frac{\sqrt{6}}{3} \\ \frac{\sqrt{3}}{2} & & \frac{1}{2} & & \\ & -\frac{\sqrt{3}}{6} & & -\frac{5}{6} & -\frac{\sqrt{2}}{3} \\ & \frac{\sqrt{6}}{3} & & -\frac{\sqrt{2}}{3} & \frac{1}{3} \end{bmatrix}$$

Table 6 (Cont.)

$$P_{15} \qquad\qquad\qquad\qquad\qquad\qquad P_{16}$$

$$\begin{bmatrix} -\frac{1}{2} & & & -\frac{1}{2} & -\frac{\sqrt{2}}{2} \\ & -\frac{1}{2} & -\frac{1}{2} & \frac{\sqrt{3}}{3} & -\frac{\sqrt{6}}{6} \\ & -\frac{1}{2} & -\frac{1}{2} & -\frac{\sqrt{3}}{3} & \frac{\sqrt{6}}{6} \\ -\frac{1}{2} & \frac{\sqrt{3}}{3} & -\frac{\sqrt{3}}{3} & \frac{1}{6} & \frac{\sqrt{2}}{6} \\ -\frac{\sqrt{2}}{2} & -\frac{\sqrt{6}}{6} & \frac{\sqrt{6}}{6} & \frac{\sqrt{2}}{6} & \frac{1}{3} \end{bmatrix} \begin{bmatrix} -\frac{1}{2} & & & \frac{1}{2} & \frac{\sqrt{2}}{2} \\ & -\frac{1}{2} & \frac{1}{2} & \frac{\sqrt{3}}{3} & -\frac{\sqrt{6}}{6} \\ & \frac{1}{2} & -\frac{1}{2} & \frac{\sqrt{3}}{3} & -\frac{\sqrt{6}}{6} \\ \frac{1}{2} & \frac{\sqrt{3}}{3} & \frac{\sqrt{3}}{3} & \frac{1}{6} & \frac{\sqrt{2}}{6} \\ \frac{\sqrt{2}}{2} & -\frac{\sqrt{6}}{6} & -\frac{\sqrt{6}}{6} & \frac{\sqrt{2}}{6} & \frac{1}{3} \end{bmatrix}$$

$$P_{24} \qquad\qquad\qquad\qquad\qquad\qquad P_{25}$$

$$\begin{bmatrix} -\frac{1}{2} & & -\frac{\sqrt{3}}{2} & & \\ & -\frac{1}{2} & & \frac{\sqrt{3}}{6} & -\frac{\sqrt{6}}{3} \\ -\frac{\sqrt{3}}{2} & & \frac{1}{2} & & \\ & \frac{\sqrt{3}}{6} & & -\frac{5}{6} & -\frac{\sqrt{2}}{3} \\ & -\frac{\sqrt{6}}{3} & & -\frac{\sqrt{2}}{3} & \frac{1}{3} \end{bmatrix} \begin{bmatrix} -\frac{1}{2} & & & \frac{1}{2} & \frac{\sqrt{2}}{2} \\ & -\frac{1}{2} & \frac{1}{2} & -\frac{\sqrt{3}}{3} & \frac{\sqrt{6}}{6} \\ & \frac{1}{2} & -\frac{1}{2} & -\frac{\sqrt{3}}{3} & \frac{\sqrt{6}}{6} \\ \frac{1}{2} & -\frac{\sqrt{3}}{3} & -\frac{\sqrt{3}}{3} & \frac{1}{6} & \frac{\sqrt{2}}{6} \\ \frac{\sqrt{2}}{2} & \frac{\sqrt{6}}{6} & \frac{\sqrt{6}}{6} & \frac{\sqrt{2}}{6} & \frac{1}{3} \end{bmatrix}$$

$$P_{26} \qquad\qquad\qquad\qquad\qquad\qquad P_{35}$$

$$\begin{bmatrix} -\frac{1}{2} & & & -\frac{1}{2} & -\frac{\sqrt{2}}{2} \\ & -\frac{1}{2} & -\frac{1}{2} & -\frac{\sqrt{3}}{3} & \frac{\sqrt{6}}{6} \\ & -\frac{1}{2} & -\frac{1}{2} & \frac{\sqrt{3}}{3} & -\frac{\sqrt{6}}{6} \\ -\frac{1}{2} & -\frac{\sqrt{3}}{3} & \frac{\sqrt{3}}{3} & \frac{1}{6} & \frac{\sqrt{2}}{6} \\ -\frac{\sqrt{2}}{2} & \frac{\sqrt{6}}{6} & -\frac{\sqrt{6}}{6} & \frac{\sqrt{2}}{6} & \frac{1}{3} \end{bmatrix} \begin{bmatrix} -\frac{1}{2} & -\frac{\sqrt{3}}{2} & & & \\ -\frac{\sqrt{3}}{2} & \frac{1}{2} & & & \\ & & -\frac{1}{2} & \frac{\sqrt{3}}{6} & -\frac{\sqrt{6}}{3} \\ & & \frac{\sqrt{3}}{6} & -\frac{5}{6} & -\frac{\sqrt{2}}{3} \\ & & -\frac{\sqrt{6}}{3} & -\frac{\sqrt{2}}{3} & \frac{1}{3} \end{bmatrix}$$

$$P_{36} \qquad\qquad\qquad\qquad\qquad\qquad P_{46}$$

$$\begin{bmatrix} -\frac{1}{2} & \frac{\sqrt{3}}{2} & & & \\ \frac{\sqrt{3}}{2} & \frac{1}{2} & & & \\ & & -\frac{1}{2} & -\frac{\sqrt{3}}{6} & \frac{\sqrt{6}}{3} \\ & & -\frac{\sqrt{3}}{6} & -\frac{5}{6} & -\frac{\sqrt{2}}{3} \\ & & \frac{\sqrt{6}}{3} & -\frac{\sqrt{2}}{3} & \frac{1}{3} \end{bmatrix} \begin{bmatrix} -\frac{1}{2} & -\frac{\sqrt{3}}{2} & & & \\ -\frac{\sqrt{3}}{2} & \frac{1}{2} & & & \\ & & -\frac{1}{2} & -\frac{\sqrt{3}}{2} & \\ & & -\frac{\sqrt{3}}{2} & \frac{1}{2} & \\ & & & & -1 \end{bmatrix}$$

$$P_{56}$$

$$\begin{bmatrix} 1 & & & & \\ & -1 & & & \\ & & 1 & & \\ & & & -1 & \\ & & & & -1 \end{bmatrix}$$

Table 7

$[\lambda] = [2^2\, 1^2]$

$r^{(1)}$: $\begin{array}{|c|c|}\hline 1 & 2 \\ \hline 3 & 4 \\ \hline 5 \\ \cline{1-1} 6 \\ \cline{1-1}\end{array}$
$r^{(2)}$: $\begin{array}{|c|c|}\hline 1 & 2 \\ \hline 3 & 5 \\ \hline 4 \\ \cline{1-1} 6 \\ \cline{1-1}\end{array}$
$r^{(3)}$: $\begin{array}{|c|c|}\hline 1 & 2 \\ \hline 3 & 6 \\ \hline 4 \\ \cline{1-1} 5 \\ \cline{1-1}\end{array}$
$r^{(4)}$: $\begin{array}{|c|c|}\hline 1 & 3 \\ \hline 2 & 4 \\ \hline 5 \\ \cline{1-1} 6 \\ \cline{1-1}\end{array}$
$r^{(5)}$: $\begin{array}{|c|c|}\hline 1 & 3 \\ \hline 2 & 5 \\ \hline 4 \\ \cline{1-1} 6 \\ \cline{1-1}\end{array}$
$r^{(6)}$: $\begin{array}{|c|c|}\hline 1 & 3 \\ \hline 2 & 6 \\ \hline 4 \\ \cline{1-1} 5 \\ \cline{1-1}\end{array}$
$r^{(7)}$: $\begin{array}{|c|c|}\hline 1 & 4 \\ \hline 2 & 5 \\ \hline 3 \\ \cline{1-1} 6 \\ \cline{1-1}\end{array}$
$r^{(8)}$: $\begin{array}{|c|c|}\hline 1 & 4 \\ \hline 2 & 6 \\ \hline 3 \\ \cline{1-1} 5 \\ \cline{1-1}\end{array}$
$r^{(9)}$: $\begin{array}{|c|c|}\hline 1 & 5 \\ \hline 2 & 6 \\ \hline 3 \\ \cline{1-1} 4 \\ \cline{1-1}\end{array}$

$$P_{12} = \begin{bmatrix} 1 &&&&&&&& \\ & 1 &&&&&&& \\ && 1 &&&&&& \\ &&& -1 &&&&& \\ &&&& -1 &&&& \\ &&&&& -1 &&& \\ &&&&&& -1 && \\ &&&&&&& -1 & \\ &&&&&&&& -1 \end{bmatrix}$$

$$P_{23} = \begin{bmatrix} -\frac{1}{2} &&& \frac{\sqrt{3}}{2} &&&&& \\ & -\frac{1}{2} &&& \frac{\sqrt{3}}{2} &&&& \\ && -\frac{1}{2} &&& \frac{\sqrt{3}}{2} &&& \\ \frac{\sqrt{3}}{2} &&& \frac{1}{2} &&&&& \\ & \frac{\sqrt{3}}{2} &&& \frac{1}{2} &&&& \\ && \frac{\sqrt{3}}{2} &&& \frac{1}{2} &&& \\ &&&&&& -1 && \\ &&&&&&& -1 & \\ &&&&&&&& -1 \end{bmatrix}$$

$$P_{34} = \begin{bmatrix} 1 &&&&&&&& \\ & -1 &&&&&&& \\ && -1 &&&&&& \\ &&& -1 &&&&& \\ &&&& -\frac{1}{3} && \frac{2\sqrt{2}}{3} && \\ &&&&& -\frac{1}{3} && \frac{2\sqrt{2}}{3} & \\ &&&& \frac{2\sqrt{2}}{3} && \frac{1}{3} && \\ &&&&& \frac{2\sqrt{2}}{3} && \frac{1}{3} & \\ &&&&&&&& -1 \end{bmatrix}$$

Table 7 (Cont.)

P_{45}

$$\begin{bmatrix}
-\frac{1}{2} & \frac{\sqrt{3}}{2} & & & & & & & \\
\frac{\sqrt{3}}{2} & \frac{1}{2} & & & & & & & \\
 & & -1 & & & & & & \\
 & & & -\frac{1}{2} & \frac{\sqrt{3}}{2} & & & & \\
 & & & \frac{\sqrt{3}}{2} & \frac{1}{2} & & & & \\
 & & & & & -1 & & & \\
 & & & & & & -1 & & \\
 & & & & & & & -\frac{1}{4} & \frac{\sqrt{15}}{4} \\
 & & & & & & & \frac{\sqrt{15}}{4} & \frac{1}{4}
\end{bmatrix}$$

P_{56}

$$\begin{bmatrix}
-1 & & & & & & & & \\
 & -\frac{1}{3} & \frac{2\sqrt{2}}{3} & & & & & & \\
 & \frac{2\sqrt{2}}{3} & \frac{1}{3} & & & & & & \\
 & & & -1 & & & & & \\
 & & & & -\frac{1}{3} & \frac{2\sqrt{2}}{3} & & & \\
 & & & & \frac{2\sqrt{2}}{3} & \frac{1}{3} & & & \\
 & & & & & & -\frac{1}{3} & \frac{2\sqrt{2}}{3} & \\
 & & & & & & \frac{2\sqrt{2}}{3} & \frac{1}{3} & \\
 & & & & & & & & -1
\end{bmatrix}$$

P_{13}

$$\begin{bmatrix}
-\frac{1}{2} & & & -\frac{\sqrt{3}}{2} & & & & & \\
 & -\frac{1}{2} & & & -\frac{\sqrt{3}}{2} & & & & \\
 & & -\frac{1}{2} & & & -\frac{\sqrt{3}}{2} & & & \\
-\frac{\sqrt{3}}{2} & & & \frac{1}{2} & & & & & \\
 & -\frac{\sqrt{3}}{2} & & & \frac{1}{2} & & & & \\
 & & -\frac{\sqrt{3}}{2} & & & \frac{1}{2} & & & \\
 & & & & & & -1 & & \\
 & & & & & & & -1 & \\
 & & & & & & & & -1
\end{bmatrix}$$

Table 7 (Cont.)

P_{14}

$$\begin{bmatrix}
-\frac{1}{2} & & & \frac{\sqrt{3}}{2} & & & & & \\
& -\frac{1}{2} & & & -\frac{\sqrt{3}}{6} & & \frac{\sqrt{6}}{3} & & \\
& & -\frac{1}{2} & & & -\frac{\sqrt{3}}{6} & & \frac{\sqrt{6}}{3} & \\
\frac{\sqrt{3}}{2} & & & \frac{1}{2} & & & & & \\
& -\frac{\sqrt{3}}{6} & & & -\frac{5}{6} & & -\frac{\sqrt{2}}{3} & & \\
& & -\frac{\sqrt{3}}{6} & & & -\frac{5}{6} & & -\frac{\sqrt{2}}{3} & \\
& \frac{\sqrt{6}}{3} & & & -\frac{\sqrt{2}}{3} & & \frac{1}{3} & & \\
& & \frac{\sqrt{6}}{3} & & & -\frac{\sqrt{2}}{3} & & \frac{1}{3} & \\
& & & & & & & & -1
\end{bmatrix}$$

P_{15}

$$\begin{bmatrix}
-\frac{1}{2} & & & & -\frac{1}{2} & & -\frac{\sqrt{2}}{2} & & \\
& -\frac{1}{2} & & -\frac{1}{2} & \frac{\sqrt{3}}{3} & & -\frac{\sqrt{6}}{6} & & \\
& & -\frac{1}{2} & & & -\frac{\sqrt{3}}{6} & & \frac{\sqrt{6}}{12} & -\frac{\sqrt{10}}{4} \\
& -\frac{1}{2} & & -\frac{1}{2} & -\frac{\sqrt{3}}{3} & & \frac{\sqrt{6}}{6} & & \\
-\frac{1}{2} & \frac{\sqrt{3}}{3} & & -\frac{\sqrt{3}}{3} & \frac{1}{6} & & \frac{\sqrt{2}}{6} & & \\
& & -\frac{\sqrt{3}}{6} & & & -\frac{5}{6} & & -\frac{\sqrt{2}}{12} & \frac{\sqrt{30}}{12} \\
-\frac{\sqrt{2}}{2} & -\frac{\sqrt{6}}{6} & & \frac{\sqrt{6}}{2} & \frac{\sqrt{2}}{6} & & \frac{1}{3} & & \\
& & \frac{\sqrt{6}}{12} & & & -\frac{\sqrt{2}}{12} & & -\frac{11}{12} & -\frac{\sqrt{15}}{12} \\
& & -\frac{\sqrt{10}}{4} & & & \frac{\sqrt{30}}{12} & & -\frac{\sqrt{15}}{12} & \frac{1}{4}
\end{bmatrix}$$

P_{16}

$$\begin{bmatrix}
-\frac{1}{2} & & & & -\frac{1}{6} & \frac{\sqrt{2}}{3} & -\frac{\sqrt{2}}{6} & \frac{2}{3} & \\
& -\frac{1}{2} & & -\frac{1}{6} & -\frac{\sqrt{3}}{9} & -\frac{\sqrt{6}}{9} & \frac{\sqrt{6}}{18} & \frac{\sqrt{3}}{9} & \frac{\sqrt{5}}{3} \\
& & -\frac{1}{2} & \frac{\sqrt{2}}{3} & -\frac{\sqrt{6}}{9} & \frac{5\sqrt{3}}{18} & \frac{\sqrt{3}}{9} & -\frac{5\sqrt{6}}{36} & \frac{\sqrt{10}}{12} \\
& -\frac{1}{6} & \frac{\sqrt{2}}{3} & -\frac{1}{2} & -\frac{\sqrt{3}}{9} & \frac{2\sqrt{6}}{9} & \frac{\sqrt{6}}{18} & -\frac{2\sqrt{3}}{9} & \\
-\frac{1}{6} & -\frac{\sqrt{3}}{9} & -\frac{\sqrt{6}}{9} & -\frac{\sqrt{3}}{9} & -\frac{13}{18} & -\frac{2\sqrt{2}}{9} & -\frac{\sqrt{2}}{18} & -\frac{1}{9} & -\frac{\sqrt{15}}{9} \\
\frac{\sqrt{2}}{3} & -\frac{\sqrt{6}}{9} & \frac{5\sqrt{3}}{18} & \frac{2\sqrt{6}}{9} & -\frac{2\sqrt{2}}{9} & \frac{1}{18} & -\frac{1}{9} & \frac{5\sqrt{2}}{36} & -\frac{\sqrt{30}}{36} \\
-\frac{\sqrt{2}}{6} & \frac{\sqrt{6}}{18} & \frac{\sqrt{3}}{9} & \frac{\sqrt{6}}{18} & -\frac{\sqrt{2}}{18} & -\frac{1}{9} & -\frac{7}{9} & -\frac{5\sqrt{2}}{18} & \frac{\sqrt{30}}{18} \\
\frac{2}{3} & \frac{\sqrt{3}}{9} & -\frac{5\sqrt{6}}{36} & -\frac{2\sqrt{3}}{9} & -\frac{1}{9} & \frac{5\sqrt{2}}{36} & -\frac{5\sqrt{2}}{18} & \frac{7}{36} & \frac{\sqrt{15}}{36} \\
& \frac{\sqrt{5}}{3} & \frac{\sqrt{10}}{12} & & -\frac{\sqrt{15}}{9} & -\frac{\sqrt{30}}{36} & \frac{\sqrt{30}}{18} & \frac{\sqrt{15}}{36} & \frac{1}{4}
\end{bmatrix}$$

Table 7 (Cont.)

P_{24}

$$
\begin{bmatrix}
-\frac{1}{2} & & & -\frac{\sqrt{3}}{2} & & & & & \\
 & -\frac{1}{2} & & & \frac{\sqrt{3}}{6} & & -\frac{\sqrt{6}}{3} & & \\
 & & -\frac{1}{2} & & & \frac{\sqrt{3}}{6} & & -\frac{\sqrt{6}}{3} & \\
-\frac{\sqrt{3}}{2} & & & \frac{1}{2} & & & & & \\
 & \frac{\sqrt{3}}{6} & & & -\frac{5}{6} & & -\frac{\sqrt{2}}{3} & & \\
 & & \frac{\sqrt{3}}{6} & & & -\frac{5}{6} & & -\frac{\sqrt{2}}{3} & \\
 & -\frac{\sqrt{6}}{3} & & & -\frac{\sqrt{2}}{3} & & \frac{1}{3} & & \\
 & & -\frac{\sqrt{6}}{3} & & & -\frac{\sqrt{2}}{3} & & \frac{1}{3} & \\
 & & & & & & & & -1
\end{bmatrix}
$$

P_{25}

$$
\begin{bmatrix}
-\frac{1}{2} & & & & \frac{1}{2} & & \frac{\sqrt{2}}{2} & & \\
 & -\frac{1}{2} & & \frac{1}{2} & -\frac{\sqrt{3}}{3} & & \frac{\sqrt{6}}{6} & & \\
 & & -\frac{1}{2} & & & \frac{\sqrt{3}}{6} & & -\frac{\sqrt{6}}{12} & \frac{\sqrt{10}}{4} \\
 & \frac{1}{2} & & -\frac{1}{2} & -\frac{\sqrt{3}}{3} & & \frac{\sqrt{6}}{6} & & \\
\frac{1}{2} & -\frac{\sqrt{3}}{3} & & -\frac{\sqrt{3}}{3} & \frac{1}{6} & & \frac{\sqrt{2}}{6} & & \\
 & & \frac{\sqrt{3}}{6} & & & -\frac{5}{6} & & -\frac{\sqrt{2}}{12} & \frac{\sqrt{30}}{12} \\
\frac{\sqrt{2}}{2} & \frac{\sqrt{6}}{6} & & \frac{\sqrt{6}}{6} & \frac{\sqrt{2}}{6} & & \frac{1}{3} & & \\
 & & -\frac{\sqrt{6}}{12} & & & -\frac{\sqrt{2}}{12} & & -\frac{11}{12} & -\frac{\sqrt{15}}{12} \\
 & & \frac{\sqrt{10}}{4} & & & \frac{\sqrt{30}}{12} & & -\frac{\sqrt{15}}{12} & \frac{1}{4}
\end{bmatrix}
$$

P_{26}

$$
\begin{bmatrix}
-\frac{1}{2} & & & & \frac{1}{6} & -\frac{\sqrt{2}}{3} & \frac{\sqrt{2}}{6} & -\frac{2}{3} & \\
 & -\frac{1}{2} & & \frac{1}{6} & \frac{\sqrt{3}}{9} & \frac{\sqrt{6}}{9} & -\frac{\sqrt{6}}{18} & -\frac{\sqrt{3}}{9} & -\frac{\sqrt{5}}{3} \\
 & & -\frac{1}{2} & -\frac{\sqrt{2}}{3} & \frac{\sqrt{6}}{9} & -\frac{5\sqrt{3}}{18} & -\frac{\sqrt{3}}{9} & \frac{5\sqrt{6}}{36} & -\frac{\sqrt{10}}{12} \\
 & \frac{1}{6} & -\frac{\sqrt{2}}{3} & -\frac{1}{2} & -\frac{\sqrt{3}}{9} & \frac{2\sqrt{6}}{9} & \frac{\sqrt{6}}{18} & -\frac{2\sqrt{3}}{9} & \\
\frac{1}{6} & \frac{\sqrt{3}}{9} & \frac{\sqrt{6}}{9} & -\frac{\sqrt{3}}{9} & -\frac{13}{18} & -\frac{2\sqrt{2}}{9} & -\frac{\sqrt{2}}{18} & -\frac{1}{9} & -\frac{\sqrt{15}}{9} \\
-\frac{\sqrt{2}}{3} & \frac{\sqrt{6}}{9} & -\frac{5\sqrt{3}}{18} & \frac{2\sqrt{6}}{9} & -\frac{2\sqrt{2}}{9} & \frac{1}{18} & -\frac{1}{9} & \frac{5\sqrt{2}}{36} & -\frac{\sqrt{30}}{36} \\
\frac{\sqrt{2}}{6} & -\frac{\sqrt{6}}{18} & -\frac{\sqrt{3}}{9} & \frac{\sqrt{6}}{18} & -\frac{\sqrt{2}}{18} & -\frac{1}{9} & -\frac{7}{9} & -\frac{5\sqrt{2}}{18} & \frac{\sqrt{30}}{18} \\
-\frac{2}{3} & -\frac{\sqrt{3}}{9} & \frac{5\sqrt{6}}{36} & -\frac{2\sqrt{3}}{9} & -\frac{1}{9} & \frac{5\sqrt{2}}{36} & -\frac{5\sqrt{2}}{18} & \frac{7}{36} & \frac{\sqrt{15}}{36} \\
 & -\frac{\sqrt{5}}{3} & -\frac{\sqrt{10}}{12} & & -\frac{\sqrt{15}}{9} & -\frac{\sqrt{30}}{36} & \frac{\sqrt{30}}{18} & \frac{\sqrt{15}}{36} & \frac{1}{4}
\end{bmatrix}
$$

Table 7 (Cont.)

P_{35}

$$\begin{bmatrix}
-\frac{1}{2} & -\frac{\sqrt{3}}{2} & & & & & & & \\
-\frac{\sqrt{3}}{2} & \frac{1}{2} & & & & & & & \\
 & & -1 & & & & & & \\
 & & & -\frac{1}{2} & \frac{\sqrt{3}}{6} & & -\frac{\sqrt{6}}{3} & & \\
 & & & \frac{\sqrt{3}}{6} & -\frac{5}{6} & & -\frac{\sqrt{2}}{3} & & \\
 & & & & & -\frac{1}{3} & & \frac{\sqrt{2}}{6} & -\frac{\sqrt{30}}{6} \\
 & & & -\frac{\sqrt{6}}{3} & -\frac{\sqrt{2}}{3} & & \frac{1}{3} & & \\
 & & & & & \frac{\sqrt{2}}{6} & & -\frac{11}{12} & -\frac{\sqrt{15}}{12} \\
 & & & & & -\frac{\sqrt{30}}{6} & & -\frac{\sqrt{15}}{12} & \frac{1}{4}
\end{bmatrix}$$

P_{36}

$$\begin{bmatrix}
-\frac{1}{2} & -\frac{\sqrt{3}}{6} & \frac{\sqrt{6}}{3} & & & & & & \\
-\frac{\sqrt{3}}{6} & -\frac{5}{6} & -\frac{\sqrt{2}}{3} & & & & & & \\
\frac{\sqrt{6}}{3} & -\frac{\sqrt{2}}{3} & \frac{1}{3} & & & & & & \\
 & & & -\frac{1}{2} & \frac{\sqrt{3}}{18} & -\frac{\sqrt{6}}{9} & -\frac{\sqrt{6}}{9} & \frac{4\sqrt{3}}{9} & \\
 & & & \frac{\sqrt{3}}{18} & -\frac{7}{18} & \frac{\sqrt{2}}{9} & \frac{\sqrt{2}}{9} & \frac{2}{9} & \frac{2\sqrt{15}}{9} \\
 & & & -\frac{\sqrt{6}}{9} & \frac{\sqrt{2}}{9} & -\frac{7}{9} & \frac{2}{9} & -\frac{5\sqrt{2}}{18} & \frac{\sqrt{30}}{18} \\
 & & & -\frac{\sqrt{6}}{9} & \frac{\sqrt{2}}{9} & \frac{2}{9} & -\frac{7}{9} & -\frac{5\sqrt{2}}{18} & \frac{\sqrt{30}}{18} \\
 & & & \frac{4\sqrt{3}}{9} & \frac{2}{9} & -\frac{5\sqrt{2}}{18} & -\frac{5\sqrt{2}}{18} & \frac{7}{36} & \frac{\sqrt{15}}{36} \\
 & & & & \frac{2\sqrt{15}}{9} & \frac{\sqrt{30}}{18} & \frac{\sqrt{30}}{18} & \frac{\sqrt{15}}{36} & \frac{1}{4}
\end{bmatrix}$$

P_{46}

$$\begin{bmatrix}
-\frac{1}{2} & \frac{\sqrt{3}}{6} & -\frac{\sqrt{6}}{3} & & & & & & \\
\frac{\sqrt{3}}{6} & -\frac{5}{6} & -\frac{\sqrt{2}}{3} & & & & & & \\
-\frac{\sqrt{6}}{3} & -\frac{\sqrt{2}}{3} & \frac{1}{3} & & & & & & \\
 & & & -\frac{1}{2} & \frac{\sqrt{3}}{6} & -\frac{\sqrt{6}}{3} & & & \\
 & & & \frac{\sqrt{3}}{6} & -\frac{5}{6} & -\frac{\sqrt{2}}{3} & & & \\
 & & & -\frac{\sqrt{6}}{3} & -\frac{\sqrt{2}}{3} & \frac{1}{3} & & & \\
 & & & & & & -\frac{1}{3} & \frac{\sqrt{2}}{6} & -\frac{\sqrt{30}}{6} \\
 & & & & & & \frac{\sqrt{2}}{6} & -\frac{11}{12} & -\frac{\sqrt{15}}{12} \\
 & & & & & & -\frac{\sqrt{30}}{6} & -\frac{\sqrt{15}}{12} & \frac{1}{4}
\end{bmatrix}$$

Table 8

$[\lambda] = [2\,1^4]$

$$r^{(1)}\ \begin{array}{|c|c|}\hline 1 & 2 \\ \hline 3 \\ \cline{1-1} 4 \\ \cline{1-1} 5 \\ \cline{1-1} 6 \\ \cline{1-1}\end{array} \qquad r^{(2)}\ \begin{array}{|c|c|}\hline 1 & 3 \\ \hline 2 \\ \cline{1-1} 4 \\ \cline{1-1} 5 \\ \cline{1-1} 6 \\ \cline{1-1}\end{array} \qquad r^{(3)}\ \begin{array}{|c|c|}\hline 1 & 4 \\ \hline 2 \\ \cline{1-1} 3 \\ \cline{1-1} 5 \\ \cline{1-1} 6 \\ \cline{1-1}\end{array} \qquad r^{(4)}\ \begin{array}{|c|c|}\hline 1 & 5 \\ \hline 2 \\ \cline{1-1} 3 \\ \cline{1-1} 4 \\ \cline{1-1} 6 \\ \cline{1-1}\end{array} \qquad r^{(5)}\ \begin{array}{|c|c|}\hline 1 & 6 \\ \hline 2 \\ \cline{1-1} 3 \\ \cline{1-1} 4 \\ \cline{1-1} 5 \\ \cline{1-1}\end{array}$$

$$P_{12} = \begin{bmatrix} 1 & & & & \\ & -1 & & & \\ & & -1 & & \\ & & & -1 & \\ & & & & -1 \end{bmatrix} \qquad P_{23} = \begin{bmatrix} -\frac{1}{2} & \frac{\sqrt{3}}{2} & & & \\ \frac{\sqrt{3}}{2} & \frac{1}{2} & & & \\ & & -1 & & \\ & & & -1 & \\ & & & & -1 \end{bmatrix}$$

$$P_{34} = \begin{bmatrix} -1 & & & & \\ & -\frac{1}{3} & \frac{2\sqrt{2}}{3} & & \\ & \frac{2\sqrt{2}}{3} & \frac{1}{3} & & \\ & & & -1 & \\ & & & & -1 \end{bmatrix} \qquad P_{45} = \begin{bmatrix} -1 & & & & \\ & -1 & & & \\ & & -\frac{1}{4} & \frac{\sqrt{15}}{4} & \\ & & \frac{\sqrt{15}}{4} & \frac{1}{4} & \\ & & & & -1 \end{bmatrix}$$

$$P_{56} = \begin{bmatrix} -1 & & & & \\ & -1 & & & \\ & & -1 & & \\ & & & -\frac{1}{5} & \frac{2\sqrt{6}}{5} \\ & & & \frac{2\sqrt{6}}{5} & \frac{1}{5} \end{bmatrix}$$

$$P_{13} = \begin{bmatrix} -\frac{1}{2} & -\frac{\sqrt{3}}{2} & & & \\ -\frac{\sqrt{3}}{2} & \frac{1}{2} & & & \\ & & -1 & & \\ & & & -1 & \\ & & & & -1 \end{bmatrix} \qquad P_{14} = \begin{bmatrix} -\frac{1}{2} & -\frac{\sqrt{3}}{6} & \frac{\sqrt{6}}{3} & & \\ -\frac{\sqrt{3}}{6} & -\frac{5}{6} & -\frac{\sqrt{2}}{3} & & \\ \frac{\sqrt{6}}{3} & -\frac{\sqrt{2}}{3} & \frac{1}{3} & & \\ & & & -1 & \\ & & & & -1 \end{bmatrix}$$

Table 8 (Cont.)

P_{15}

$$\begin{bmatrix} -\frac{1}{2} & -\frac{\sqrt{3}}{6} & \frac{\sqrt{6}}{12} & -\frac{\sqrt{10}}{4} & \\ -\frac{\sqrt{3}}{6} & -\frac{5}{6} & -\frac{\sqrt{2}}{12} & \frac{\sqrt{30}}{12} & \\ \frac{\sqrt{6}}{12} & -\frac{\sqrt{2}}{12} & -\frac{11}{12} & -\frac{\sqrt{15}}{12} & \\ -\frac{\sqrt{10}}{4} & \frac{\sqrt{30}}{12} & -\frac{\sqrt{15}}{12} & \frac{1}{4} & \\ & & & & -1 \end{bmatrix}$$

P_{16}

$$\begin{bmatrix} -\frac{1}{2} & -\frac{\sqrt{3}}{6} & \frac{\sqrt{6}}{12} & -\frac{\sqrt{10}}{20} & \frac{\sqrt{15}}{5} \\ -\frac{\sqrt{3}}{6} & -\frac{5}{6} & -\frac{\sqrt{2}}{12} & \frac{\sqrt{30}}{60} & -\frac{\sqrt{5}}{5} \\ \frac{\sqrt{6}}{12} & -\frac{\sqrt{2}}{12} & -\frac{11}{12} & -\frac{\sqrt{15}}{60} & \frac{\sqrt{10}}{10} \\ -\frac{\sqrt{10}}{20} & \frac{\sqrt{30}}{60} & -\frac{\sqrt{15}}{60} & -\frac{19}{20} & -\frac{\sqrt{6}}{10} \\ \frac{\sqrt{15}}{5} & -\frac{\sqrt{5}}{5} & \frac{\sqrt{10}}{10} & -\frac{\sqrt{6}}{10} & \frac{1}{5} \end{bmatrix}$$

P_{24}

$$\begin{bmatrix} -\frac{1}{2} & \frac{\sqrt{3}}{6} & -\frac{\sqrt{6}}{3} & & \\ \frac{\sqrt{3}}{6} & -\frac{5}{6} & -\frac{\sqrt{2}}{3} & & \\ -\frac{\sqrt{6}}{3} & -\frac{\sqrt{2}}{3} & \frac{1}{3} & & \\ & & & -1 & \\ & & & & -1 \end{bmatrix}$$

P_{25}

$$\begin{bmatrix} -\frac{1}{2} & \frac{\sqrt{3}}{6} & -\frac{\sqrt{6}}{12} & \frac{\sqrt{10}}{4} & \\ \frac{\sqrt{3}}{6} & -\frac{5}{6} & -\frac{\sqrt{2}}{12} & \frac{\sqrt{30}}{12} & \\ -\frac{\sqrt{6}}{12} & -\frac{\sqrt{2}}{12} & -\frac{11}{12} & -\frac{\sqrt{15}}{12} & \\ \frac{\sqrt{10}}{4} & \frac{\sqrt{30}}{12} & -\frac{\sqrt{15}}{12} & \frac{1}{4} & \\ & & & & -1 \end{bmatrix}$$

P_{26}

$$\begin{bmatrix} -\frac{1}{2} & \frac{\sqrt{3}}{6} & -\frac{\sqrt{6}}{12} & \frac{\sqrt{10}}{20} & -\frac{\sqrt{15}}{5} \\ \frac{\sqrt{3}}{6} & -\frac{5}{6} & -\frac{\sqrt{2}}{12} & \frac{\sqrt{30}}{60} & -\frac{\sqrt{5}}{5} \\ -\frac{\sqrt{6}}{12} & -\frac{\sqrt{2}}{12} & -\frac{11}{12} & -\frac{\sqrt{15}}{60} & \frac{\sqrt{10}}{10} \\ \frac{\sqrt{10}}{20} & \frac{\sqrt{30}}{60} & -\frac{\sqrt{15}}{60} & -\frac{19}{20} & -\frac{\sqrt{6}}{10} \\ -\frac{\sqrt{15}}{5} & -\frac{\sqrt{5}}{5} & \frac{\sqrt{10}}{10} & -\frac{\sqrt{6}}{10} & \frac{1}{5} \end{bmatrix}$$

P_{35}

$$\begin{bmatrix} -1 & & & & \\ & -\frac{1}{3} & \frac{\sqrt{2}}{6} & -\frac{\sqrt{30}}{6} & \\ & \frac{\sqrt{2}}{6} & -\frac{11}{12} & -\frac{\sqrt{15}}{12} & \\ & -\frac{\sqrt{30}}{6} & -\frac{\sqrt{15}}{12} & \frac{1}{4} & \\ & & & & -1 \end{bmatrix}$$

P_{36}

$$\begin{bmatrix} -1 & & & & \\ & -\frac{1}{3} & \frac{\sqrt{2}}{6} & -\frac{\sqrt{30}}{30} & \frac{2\sqrt{5}}{5} \\ & \frac{\sqrt{2}}{6} & -\frac{11}{12} & -\frac{\sqrt{15}}{60} & \frac{\sqrt{10}}{10} \\ & -\frac{\sqrt{30}}{30} & -\frac{\sqrt{15}}{60} & -\frac{19}{20} & -\frac{\sqrt{6}}{10} \\ & \frac{2\sqrt{5}}{5} & \frac{\sqrt{10}}{10} & -\frac{\sqrt{6}}{10} & \frac{1}{5} \end{bmatrix}$$

P_{46}

$$\begin{bmatrix} -1 & & & & \\ & -1 & & & \\ & & -\frac{1}{4} & \frac{\sqrt{15}}{20} & -\frac{3\sqrt{10}}{10} \\ & & \frac{\sqrt{15}}{20} & -\frac{19}{20} & -\frac{\sqrt{6}}{10} \\ & & -\frac{3\sqrt{10}}{10} & -\frac{\sqrt{6}}{10} & \frac{1}{5} \end{bmatrix}$$

Table 9

Matrices of the Representation $\Gamma^{[2^3]}$ for the Subgroup of π_6 Which Is Isomorphic with the Point Group $\mathbf{D}_6$

$I \equiv E$

$$\begin{bmatrix} 1 & & & & \\ & 1 & & & \\ & & 1 & & \\ & & & 1 & \\ & & & & 1 \end{bmatrix}$$

$P_{14} \cdot P_{25} \cdot P_{36} \equiv C_2$

$$\begin{bmatrix} \frac{1}{4} & \frac{\sqrt{3}}{4} & \frac{\sqrt{3}}{4} & \frac{3}{4} & \\ \frac{\sqrt{3}}{4} & \frac{3}{4} & -\frac{1}{4} & -\frac{\sqrt{3}}{4} & \\ \frac{\sqrt{3}}{4} & -\frac{1}{4} & \frac{3}{4} & -\frac{\sqrt{3}}{4} & \\ \frac{3}{4} & -\frac{\sqrt{3}}{4} & -\frac{\sqrt{3}}{4} & \frac{1}{4} & \\ & & & & 1 \end{bmatrix}$$

$P_{135} \cdot P_{246} \equiv C_3$

$$\begin{bmatrix} 1 & & & & \\ & & 1 & & \\ & & & \frac{\sqrt{3}}{3} & \frac{\sqrt{6}}{3} \\ & \frac{\sqrt{3}}{3} & & \frac{2}{3} & -\frac{\sqrt{2}}{3} \\ & \frac{\sqrt{6}}{3} & & -\frac{\sqrt{2}}{3} & \frac{1}{3} \end{bmatrix}$$

$P_{153} \cdot P_{264} \equiv C_3^2$

$$\begin{bmatrix} 1 & & & & \\ & & & \frac{\sqrt{3}}{3} & \frac{\sqrt{6}}{3} \\ & 1 & & & \\ & & \frac{\sqrt{3}}{3} & \frac{2}{3} & -\frac{\sqrt{2}}{3} \\ & & \frac{\sqrt{6}}{3} & -\frac{\sqrt{2}}{3} & \frac{1}{3} \end{bmatrix}$$

$P_{654321} \equiv C_6^5$

$$\begin{bmatrix} \frac{1}{4} & \frac{\sqrt{3}}{4} & \frac{\sqrt{3}}{4} & \frac{3}{4} & \\ \frac{\sqrt{3}}{4} & -\frac{1}{4} & \frac{3}{4} & -\frac{\sqrt{3}}{4} & \\ \frac{\sqrt{3}}{4} & -\frac{1}{4} & -\frac{1}{4} & \frac{\sqrt{3}}{12} & \frac{\sqrt{6}}{3} \\ \frac{3}{4} & \frac{\sqrt{3}}{12} & -\frac{\sqrt{3}}{4} & -\frac{1}{12} & -\frac{\sqrt{2}}{3} \\ & \frac{\sqrt{6}}{3} & & -\frac{\sqrt{2}}{3} & \frac{1}{3} \end{bmatrix}$$

$P_{123456} \equiv C_6$

$$\begin{bmatrix} \frac{1}{4} & \frac{\sqrt{3}}{4} & \frac{\sqrt{3}}{4} & \frac{3}{4} & \\ \frac{\sqrt{3}}{4} & -\frac{1}{4} & -\frac{1}{4} & \frac{\sqrt{3}}{12} & \frac{\sqrt{6}}{3} \\ \frac{\sqrt{3}}{4} & \frac{3}{4} & -\frac{1}{4} & -\frac{\sqrt{3}}{4} & \\ \frac{3}{4} & -\frac{\sqrt{3}}{4} & \frac{\sqrt{3}}{12} & -\frac{1}{12} & -\frac{\sqrt{2}}{3} \\ & & \frac{\sqrt{6}}{3} & -\frac{\sqrt{2}}{3} & \frac{1}{3} \end{bmatrix}$$

$P_{12} \cdot P_{36} \cdot P_{45} \equiv U_2^{(1)}$

$$\begin{bmatrix} 1 & & & & \\ & 1 & & & \\ & & & \frac{\sqrt{3}}{3} & \frac{\sqrt{6}}{3} \\ & & \frac{\sqrt{3}}{3} & \frac{2}{3} & -\frac{\sqrt{2}}{3} \\ & & \frac{\sqrt{6}}{3} & -\frac{\sqrt{2}}{3} & \frac{1}{3} \end{bmatrix}$$

$P_{23} \cdot P_{14} \cdot P_{56} \equiv U_2^{(2)}$

$$\begin{bmatrix} 1 & & & & \\ & & & \frac{\sqrt{3}}{3} & \frac{\sqrt{6}}{3} \\ & & 1 & & \\ & \frac{\sqrt{3}}{3} & & \frac{2}{3} & -\frac{\sqrt{2}}{3} \\ & \frac{\sqrt{6}}{3} & & -\frac{\sqrt{2}}{3} & \frac{1}{3} \end{bmatrix}$$

Table 9 (Cont.)

$P_{16}\cdot P_{25}\cdot P_{34}\equiv U_2^{(3)}$

$$\begin{bmatrix} 1 & & & & \\ & & 1 & & \\ & 1 & & & \\ & & & 1 & \\ & & & & 1 \end{bmatrix}$$

$P_{13}\cdot P_{46}\equiv \overline{U}_2^{(1)}$

$$\begin{bmatrix} \frac{1}{4} & \frac{\sqrt{3}}{4} & \frac{\sqrt{3}}{4} & \frac{3}{4} & \\ \frac{\sqrt{3}}{4} & -\frac{1}{4} & \frac{3}{4} & -\frac{\sqrt{3}}{4} & \\ \frac{\sqrt{3}}{4} & \frac{3}{4} & -\frac{1}{4} & -\frac{\sqrt{3}}{4} & \\ \frac{3}{4} & -\frac{\sqrt{3}}{4} & -\frac{\sqrt{3}}{4} & \frac{1}{4} & \\ & & & & 1 \end{bmatrix}$$

$P_{15}\cdot P_{24}\equiv \overline{U}_2^{(2)}$

$$\begin{bmatrix} \frac{1}{4} & \frac{\sqrt{3}}{4} & \frac{\sqrt{3}}{4} & \frac{3}{4} & \\ \frac{\sqrt{3}}{4} & \frac{3}{4} & -\frac{1}{4} & -\frac{\sqrt{3}}{4} & \\ \frac{\sqrt{3}}{4} & -\frac{1}{4} & -\frac{1}{4} & \frac{\sqrt{3}}{12} & \frac{\sqrt{6}}{3} \\ \frac{3}{4} & -\frac{\sqrt{3}}{4} & \frac{\sqrt{3}}{12} & -\frac{1}{12} & -\frac{\sqrt{2}}{3} \\ & & \frac{\sqrt{6}}{3} & -\frac{\sqrt{2}}{3} & \frac{1}{3} \end{bmatrix}$$

$P_{26}\cdot P_{35}\equiv \overline{U}_2^{(3)}$

$$\begin{bmatrix} \frac{1}{4} & \frac{\sqrt{3}}{4} & \frac{\sqrt{3}}{4} & \frac{3}{4} & \\ \frac{\sqrt{3}}{4} & -\frac{1}{4} & -\frac{1}{4} & \frac{\sqrt{3}}{12} & \frac{\sqrt{6}}{3} \\ \frac{\sqrt{3}}{4} & -\frac{1}{4} & \frac{3}{4} & -\frac{\sqrt{3}}{4} & \\ \frac{3}{4} & \frac{\sqrt{3}}{12} & -\frac{\sqrt{3}}{4} & -\frac{1}{12} & -\frac{\sqrt{2}}{3} \\ & \frac{\sqrt{6}}{3} & & -\frac{\sqrt{2}}{3} & \frac{1}{3} \end{bmatrix}$$

APPENDIX 6

Tables of the Matrices $\langle r'\lambda''|P_{ab}^{N-1\,N}|r\rangle^{[\lambda]}$ for Values of N from 3 to 6

The matrices needed in calculating the matrix elements of the Hamiltonian for a many-electron system (see Section 8.5) are given in this appendix. The rows of these matrices are defined by the set of basis functions which possess a definite symmetry under permutations of the $(N-1)$th and Nth electrons. Matrices of this kind are related to the matrices in the standard representation by the following formula:

$$\langle r'\lambda''|P_{ab}^{N-1\,N}|r\rangle^{[\lambda]} = \sum_{\bar{r}} \langle r'\lambda''|\bar{r}\rangle^{[\lambda]} \cdot \langle \bar{r}|P_{ab}^{N-1\,N}|r\rangle^{[\lambda]}. \tag{A.1}$$

The transformation matrix which occurs in this equation takes one from the standard representation to a representation which is reduced with respect to the subgroup $\pi_{N-2} \times \pi_2$ and is found by the method set out in Section 2.11 [see Eq. (2.65)]. The matrices in the standard representation which are required are calculated with aid of the transposition matrices given in Appendix 5.

Here $P_{ab}^{N-1\,N}$ denotes a permutation which sends electron $N-1$ into orbital ϕ_a and electron N into orbital ϕ_b. It is assumed that as a result of applying this permutation, the ordering of the electrons in the orbitals of the configuration K_{ab} is preserved, K_{ab} being obtained from the configuration K by removing orbitals ϕ_a and ϕ_b. The general form of such permutations is given by formulae (7.82). In the case of singly occupied orbitals, the requirement that one preserve the increasing order in which the electrons are numbered is unnecessary, and other choices for the set of permutations $P_{ab}^{N-1\,N}$ are possible.

The rows of the matrices are numbered by the Young tableaux r' for the $(N-2)$th electrons and by the Young diagrams $[\lambda'']$ for the $(N-1)$th and Nth electrons. The columns are enumerated by complete Young tableaux r. All the Young tableaux are ordered according to the extent to which the numbers they contain deviate from the natural order. Only the diagrams $[\lambda'']$

which enumerate the rows are given for each matrix. Blank spaces in the matrices denote zeros.

Table 1

$N=3$

$[\lambda]=[21], \quad S=1/2$

$$\begin{matrix}\lambda'' & & P^{23}_{12} & P^{23}_{13} & P^{23}_{23} \\ \begin{matrix}[2]\\[1^2]\end{matrix} & & \begin{bmatrix} -1 & \\ & -1 \end{bmatrix} & \begin{bmatrix} \frac{1}{2} & -\frac{\sqrt{3}}{2} \\ -\frac{\sqrt{3}}{2} & -\frac{1}{2} \end{bmatrix} & \begin{bmatrix} \frac{1}{2} & \frac{\sqrt{3}}{2} \\ -\frac{\sqrt{3}}{2} & \frac{1}{2} \end{bmatrix}\end{matrix}$$

Table 2

$N=4$

(a) $[\lambda]=[2^2], \quad S=0$

$$\begin{matrix}\lambda'' & & P^{34}_{12} & P^{34}_{13} & P^{34}_{14} \\ \begin{matrix}[2]\\[1^2]\end{matrix} & & \begin{bmatrix} 1 & \\ & 1 \end{bmatrix} & \begin{bmatrix} -\frac{1}{2} & \frac{\sqrt{3}}{2} \\ \frac{\sqrt{3}}{2} & \frac{1}{2} \end{bmatrix} & \begin{bmatrix} -\frac{1}{2} & -\frac{\sqrt{3}}{2} \\ \frac{\sqrt{3}}{2} & -\frac{1}{2} \end{bmatrix}\end{matrix}$$

$$\begin{matrix} P^{34}_{23} & P^{34}_{24} & P^{34}_{34} \\ \begin{bmatrix} -\frac{1}{2} & -\frac{\sqrt{3}}{2} \\ \frac{\sqrt{3}}{2} & -\frac{1}{2} \end{bmatrix} & \begin{bmatrix} -\frac{1}{2} & \frac{\sqrt{3}}{2} \\ \frac{\sqrt{3}}{2} & \frac{1}{2} \end{bmatrix} & \begin{bmatrix} 1 & \\ & 1 \end{bmatrix}\end{matrix}$$

(b) $[\lambda]=[21^2], \quad S=1$

$$\begin{matrix}\lambda'' & & P^{34}_{12} & P^{34}_{13} & P^{34}_{14} \\ \begin{matrix}[1^2]\\[2]\\[1^2]\end{matrix} & & \begin{bmatrix} & \frac{\sqrt{3}}{3} & \frac{\sqrt{6}}{3} \\ 1 & & \\ & \frac{\sqrt{6}}{3} & -\frac{\sqrt{3}}{3} \end{bmatrix} & \begin{bmatrix} \frac{1}{2} & \frac{\sqrt{3}}{6} & -\frac{\sqrt{6}}{3} \\ -\frac{1}{2} & \frac{\sqrt{3}}{2} & \\ \frac{\sqrt{2}}{2} & \frac{\sqrt{6}}{6} & \frac{\sqrt{3}}{3} \end{bmatrix} & \begin{bmatrix} -\frac{1}{2} & -\frac{\sqrt{3}}{2} & \\ \frac{1}{2} & -\frac{\sqrt{3}}{6} & \frac{\sqrt{6}}{3} \\ -\frac{\sqrt{2}}{2} & \frac{\sqrt{6}}{6} & \frac{\sqrt{3}}{3} \end{bmatrix}\end{matrix}$$

$$\begin{matrix} P^{34}_{23} & P^{34}_{24} & P^{34}_{34} \\ \begin{bmatrix} \frac{1}{2} & -\frac{\sqrt{3}}{6} & \frac{\sqrt{6}}{3} \\ -\frac{1}{2} & -\frac{\sqrt{3}}{2} & \\ \frac{\sqrt{2}}{2} & -\frac{\sqrt{6}}{6} & -\frac{\sqrt{3}}{3} \end{bmatrix} & \begin{bmatrix} -\frac{1}{2} & \frac{\sqrt{3}}{2} & \\ \frac{1}{2} & \frac{\sqrt{3}}{6} & -\frac{\sqrt{6}}{3} \\ -\frac{\sqrt{2}}{2} & -\frac{\sqrt{6}}{6} & -\frac{\sqrt{3}}{3} \end{bmatrix} & \begin{bmatrix} 1 & & \\ & \frac{\sqrt{3}}{3} & \frac{\sqrt{6}}{3} \\ & -\frac{\sqrt{6}}{3} & \frac{\sqrt{3}}{3} \end{bmatrix}\end{matrix}$$

Table 3

$N=5$

(a) $[\lambda]=[2^2 1]$, $S=1/2$

λ'': [2], [1²], [2], [1²], [1²]

$$P_{12}^{45} = \begin{bmatrix} -1 & & & & \\ & & & -\frac{\sqrt{3}}{3} & -\frac{\sqrt{6}}{3} \\ & -1 & & & \\ & & -\frac{\sqrt{3}}{3} & -\frac{2}{3} & \frac{\sqrt{2}}{3} \\ & & \frac{\sqrt{6}}{3} & -\frac{\sqrt{2}}{3} & \frac{1}{3} \end{bmatrix} \qquad P_{13}^{45} = \begin{bmatrix} \frac{1}{2} & & -\frac{\sqrt{3}}{2} & & \\ & -\frac{1}{2} & & -\frac{\sqrt{3}}{6} & \frac{\sqrt{6}}{3} \\ & \frac{1}{2} & & -\frac{\sqrt{3}}{2} & \\ -\frac{1}{2} & -\frac{\sqrt{3}}{3} & -\frac{\sqrt{3}}{6} & -\frac{1}{3} & -\frac{\sqrt{2}}{3} \\ \frac{\sqrt{2}}{2} & -\frac{\sqrt{6}}{6} & \frac{\sqrt{6}}{6} & -\frac{\sqrt{2}}{6} & -\frac{1}{3} \end{bmatrix}$$

$$P_{14}^{45} = \begin{bmatrix} \frac{1}{2} & & \frac{\sqrt{3}}{2} & & \\ & \frac{1}{2} & & \frac{\sqrt{3}}{2} & \\ & -\frac{1}{2} & & \frac{\sqrt{3}}{6} & -\frac{\sqrt{6}}{3} \\ -\frac{1}{2} & \frac{\sqrt{3}}{3} & \frac{\sqrt{3}}{6} & -\frac{1}{3} & -\frac{\sqrt{2}}{3} \\ \frac{\sqrt{2}}{2} & \frac{\sqrt{6}}{6} & -\frac{\sqrt{6}}{6} & -\frac{\sqrt{2}}{6} & -\frac{1}{3} \end{bmatrix} \qquad P_{15}^{45} = \begin{bmatrix} -\frac{1}{4} & \frac{\sqrt{3}}{4} & -\frac{\sqrt{3}}{4} & \frac{3}{4} & \\ \frac{\sqrt{3}}{4} & \frac{1}{4} & \frac{3}{4} & \frac{\sqrt{3}}{4} & \\ -\frac{\sqrt{3}}{4} & -\frac{1}{4} & \frac{1}{4} & \frac{\sqrt{3}}{12} & \frac{\sqrt{6}}{3} \\ \frac{3}{4} & -\frac{\sqrt{3}}{12} & -\frac{\sqrt{3}}{4} & \frac{1}{12} & \frac{\sqrt{2}}{3} \\ & \frac{\sqrt{6}}{3} & & -\frac{\sqrt{2}}{3} & \frac{1}{3} \end{bmatrix}$$

$$P_{23}^{45} = \begin{bmatrix} \frac{1}{2} & & \frac{\sqrt{3}}{2} & & \\ & -\frac{1}{2} & & \frac{\sqrt{3}}{6} & -\frac{\sqrt{6}}{3} \\ & \frac{1}{2} & & \frac{\sqrt{3}}{2} & \\ -\frac{1}{2} & -\frac{\sqrt{3}}{3} & \frac{\sqrt{3}}{6} & \frac{1}{3} & \frac{\sqrt{2}}{3} \\ \frac{\sqrt{2}}{2} & -\frac{\sqrt{6}}{6} & -\frac{\sqrt{6}}{6} & \frac{\sqrt{2}}{6} & \frac{1}{3} \end{bmatrix} \qquad P_{24}^{45} = \begin{bmatrix} \frac{1}{2} & & -\frac{\sqrt{3}}{2} & & \\ & \frac{1}{2} & & -\frac{\sqrt{3}}{2} & \\ & -\frac{1}{2} & & -\frac{\sqrt{3}}{6} & \frac{\sqrt{6}}{3} \\ -\frac{1}{2} & \frac{\sqrt{3}}{3} & -\frac{\sqrt{3}}{6} & \frac{1}{3} & \frac{\sqrt{2}}{3} \\ \frac{\sqrt{2}}{2} & \frac{\sqrt{6}}{6} & \frac{\sqrt{6}}{6} & \frac{\sqrt{2}}{6} & \frac{1}{3} \end{bmatrix}$$

Table 3 (Cont.)

$$P_{25}^{45} = \begin{bmatrix} -\frac{1}{4} & \frac{\sqrt{3}}{4} & \frac{\sqrt{3}}{4} & -\frac{3}{4} & \\ \frac{\sqrt{3}}{4} & \frac{1}{4} & -\frac{3}{4} & -\frac{\sqrt{3}}{4} & \\ -\frac{\sqrt{3}}{4} & -\frac{1}{4} & -\frac{1}{4} & -\frac{\sqrt{3}}{12} & -\frac{\sqrt{6}}{3} \\ \frac{3}{4} & -\frac{\sqrt{3}}{12} & \frac{\sqrt{3}}{4} & -\frac{1}{12} & -\frac{\sqrt{2}}{3} \\ & \frac{\sqrt{6}}{3} & & \frac{\sqrt{2}}{3} & -\frac{1}{3} \end{bmatrix} \qquad P_{34}^{45} = \begin{bmatrix} -1 & & & & \\ & -1 & & & \\ & & & -\frac{\sqrt{3}}{3} & -\frac{\sqrt{6}}{3} \\ & & -\frac{\sqrt{3}}{3} & \frac{2}{3} & -\frac{\sqrt{2}}{3} \\ & & \frac{\sqrt{6}}{3} & \frac{\sqrt{2}}{3} & -\frac{1}{3} \end{bmatrix}$$

$$P_{35}^{45} = \begin{bmatrix} \frac{1}{2} & -\frac{\sqrt{3}}{2} & & & \\ -\frac{\sqrt{3}}{2} & -\frac{1}{2} & & & \\ & & -\frac{1}{2} & -\frac{\sqrt{3}}{6} & \frac{\sqrt{6}}{3} \\ & & \frac{\sqrt{3}}{2} & -\frac{1}{6} & \frac{\sqrt{2}}{3} \\ & & & \frac{2\sqrt{2}}{3} & \frac{1}{3} \end{bmatrix} \qquad P_{45}^{45} = \begin{bmatrix} \frac{1}{2} & \frac{\sqrt{3}}{2} & & & \\ -\frac{\sqrt{3}}{2} & \frac{1}{2} & & & \\ & & \frac{1}{2} & \frac{\sqrt{3}}{2} & \\ & & -\frac{\sqrt{3}}{2} & \frac{1}{2} & \\ & & & & 1 \end{bmatrix}$$

(b) $[\lambda] = [21^3]$, $S = 3/2$

$$\begin{matrix} \lambda'' \\ [1^2] \\ [1^2] \\ [2] \\ [1^2] \end{matrix} \qquad P_{12}^{45} = \begin{bmatrix} & \frac{\sqrt{3}}{3} & \frac{\sqrt{6}}{3} & \\ & -\frac{1}{3} & \frac{\sqrt{2}}{6} & \frac{\sqrt{30}}{6} \\ -1 & & & \\ & -\frac{\sqrt{5}}{3} & \frac{\sqrt{10}}{6} & -\frac{\sqrt{6}}{6} \end{bmatrix} \qquad P_{13}^{45} = \begin{bmatrix} \frac{1}{2} & \frac{\sqrt{3}}{6} & -\frac{\sqrt{6}}{3} & \\ -\frac{\sqrt{3}}{6} & -\frac{1}{6} & -\frac{\sqrt{2}}{6} & -\frac{\sqrt{30}}{6} \\ \frac{1}{2} & -\frac{\sqrt{3}}{2} & & \\ -\frac{\sqrt{15}}{6} & -\frac{\sqrt{5}}{6} & -\frac{\sqrt{10}}{6} & \frac{\sqrt{6}}{6} \end{bmatrix}$$

$$P_{14}^{45} = \begin{bmatrix} -\frac{1}{2} & -\frac{\sqrt{3}}{2} & & \\ \frac{\sqrt{3}}{6} & -\frac{1}{6} & -\frac{\sqrt{2}}{6} & \frac{\sqrt{30}}{6} \\ -\frac{1}{2} & \frac{\sqrt{3}}{6} & -\frac{\sqrt{6}}{3} & \\ \frac{\sqrt{15}}{6} & -\frac{\sqrt{5}}{6} & -\frac{\sqrt{10}}{6} & -\frac{\sqrt{6}}{6} \end{bmatrix} \qquad P_{15}^{45} = \begin{bmatrix} \frac{1}{2} & \frac{\sqrt{3}}{2} & & \\ -\frac{\sqrt{3}}{6} & \frac{1}{6} & \frac{2\sqrt{2}}{3} & \\ \frac{1}{2} & -\frac{\sqrt{3}}{6} & \frac{\sqrt{6}}{12} & -\frac{\sqrt{10}}{4} \\ -\frac{\sqrt{15}}{6} & \frac{\sqrt{5}}{6} & -\frac{\sqrt{10}}{12} & -\frac{\sqrt{6}}{4} \end{bmatrix}$$

$$P_{23}^{45} = \begin{bmatrix} \frac{1}{2} & -\frac{\sqrt{3}}{6} & \frac{\sqrt{6}}{3} & \\ -\frac{\sqrt{3}}{6} & \frac{1}{6} & \frac{\sqrt{2}}{6} & \frac{\sqrt{30}}{6} \\ \frac{1}{2} & \frac{\sqrt{3}}{2} & & \\ -\frac{\sqrt{15}}{6} & \frac{\sqrt{5}}{6} & \frac{\sqrt{10}}{6} & -\frac{\sqrt{6}}{6} \end{bmatrix} \qquad P_{24}^{45} = \begin{bmatrix} -\frac{1}{2} & \frac{\sqrt{3}}{2} & & \\ \frac{\sqrt{3}}{6} & \frac{1}{6} & \frac{\sqrt{2}}{6} & -\frac{\sqrt{30}}{6} \\ -\frac{1}{2} & -\frac{\sqrt{3}}{6} & \frac{\sqrt{6}}{3} & \\ \frac{\sqrt{15}}{6} & \frac{\sqrt{5}}{6} & \frac{\sqrt{10}}{6} & \frac{\sqrt{6}}{6} \end{bmatrix}$$

$$P_{25}^{45} = \begin{bmatrix} \frac{1}{2} & -\frac{\sqrt{3}}{2} & & \\ -\frac{\sqrt{3}}{6} & -\frac{1}{6} & -\frac{2\sqrt{2}}{3} & \\ \frac{1}{2} & \frac{\sqrt{3}}{6} & -\frac{\sqrt{6}}{12} & \frac{\sqrt{10}}{4} \\ -\frac{\sqrt{15}}{6} & -\frac{\sqrt{5}}{6} & \frac{\sqrt{10}}{12} & \frac{\sqrt{6}}{4} \end{bmatrix} \qquad P_{34}^{45} = \begin{bmatrix} 1 & & & \\ & \frac{1}{3} & -\frac{\sqrt{2}}{6} & \frac{\sqrt{30}}{6} \\ & -\frac{\sqrt{3}}{3} & -\frac{\sqrt{6}}{3} & \\ & \frac{\sqrt{5}}{3} & -\frac{\sqrt{10}}{6} & -\frac{\sqrt{6}}{6} \end{bmatrix}$$

$$P_{35}^{45} = \begin{bmatrix} -1 & & & \\ & -\frac{1}{3} & \frac{2\sqrt{2}}{3} & \\ & \frac{\sqrt{3}}{3} & \frac{\sqrt{6}}{12} & -\frac{\sqrt{10}}{4} \\ & -\frac{\sqrt{5}}{3} & -\frac{\sqrt{10}}{12} & -\frac{\sqrt{6}}{4} \end{bmatrix} \qquad P_{45}^{45} = \begin{bmatrix} 1 & & & \\ & 1 & & \\ & & \frac{\sqrt{6}}{4} & \frac{\sqrt{10}}{4} \\ & & -\frac{\sqrt{10}}{4} & \frac{\sqrt{6}}{4} \end{bmatrix}$$

Table 4

$N = 6$

(a) $[\lambda] = [2^3]$, $S = 0$

$$\begin{matrix} \lambda'' \\ [2] \\ [1^2] \\ [2] \\ [1^2] \\ [1^2] \end{matrix} \quad P_{12}^{56} = \begin{bmatrix} 1 & & & & \\ & & & \frac{\sqrt{3}}{3} & \frac{\sqrt{6}}{3} \\ & 1 & & & \\ & & \frac{\sqrt{3}}{3} & \frac{2}{3} & -\frac{\sqrt{2}}{3} \\ & & \frac{\sqrt{6}}{3} & -\frac{\sqrt{2}}{3} & \frac{1}{3} \end{bmatrix} \qquad P_{13}^{56} = \begin{bmatrix} -\frac{1}{2} & & \frac{\sqrt{3}}{2} & & \\ & \frac{1}{2} & & \frac{\sqrt{3}}{6} & -\frac{\sqrt{6}}{3} \\ & -\frac{1}{2} & & \frac{\sqrt{3}}{2} & \\ \frac{1}{2} & \frac{\sqrt{3}}{3} & \frac{\sqrt{3}}{6} & \frac{1}{3} & \frac{\sqrt{2}}{3} \\ \frac{\sqrt{2}}{2} & -\frac{\sqrt{6}}{6} & \frac{\sqrt{6}}{6} & -\frac{\sqrt{2}}{6} & -\frac{1}{3} \end{bmatrix}$$

Table 4 (Cont.)

P^{56}_{14}

$$\begin{bmatrix} -\frac{1}{2} & & -\frac{\sqrt{3}}{2} & & \\ & -\frac{1}{2} & & -\frac{\sqrt{3}}{2} & \\ & \frac{1}{2} & & -\frac{\sqrt{3}}{6} & \frac{\sqrt{6}}{3} \\ \frac{1}{2} & -\frac{\sqrt{3}}{3} & -\frac{\sqrt{3}}{6} & \frac{1}{3} & \frac{\sqrt{2}}{3} \\ \frac{\sqrt{2}}{2} & \frac{\sqrt{6}}{6} & -\frac{\sqrt{6}}{6} & -\frac{\sqrt{2}}{6} & -\frac{1}{3} \end{bmatrix}$$

P^{56}_{15}

$$\begin{bmatrix} \frac{1}{4} & -\frac{\sqrt{3}}{4} & \frac{\sqrt{3}}{4} & -\frac{3}{4} & \\ -\frac{\sqrt{3}}{4} & -\frac{1}{4} & -\frac{3}{4} & -\frac{\sqrt{3}}{4} & \\ \frac{\sqrt{3}}{4} & \frac{1}{4} & -\frac{1}{4} & -\frac{\sqrt{3}}{12} & -\frac{\sqrt{6}}{3} \\ -\frac{3}{4} & \frac{\sqrt{3}}{12} & \frac{\sqrt{3}}{4} & -\frac{1}{12} & -\frac{\sqrt{2}}{3} \\ & \frac{\sqrt{6}}{3} & & -\frac{\sqrt{2}}{3} & \frac{1}{3} \end{bmatrix}$$

P^{56}_{23}

$$\begin{bmatrix} -\frac{1}{2} & & -\frac{\sqrt{3}}{2} & & \\ & \frac{1}{2} & & -\frac{\sqrt{3}}{6} & \frac{\sqrt{6}}{3} \\ & -\frac{1}{2} & & -\frac{\sqrt{3}}{2} & \\ \frac{1}{2} & \frac{\sqrt{3}}{3} & -\frac{\sqrt{3}}{6} & -\frac{1}{3} & -\frac{\sqrt{2}}{3} \\ \frac{\sqrt{2}}{2} & -\frac{\sqrt{6}}{6} & -\frac{\sqrt{6}}{6} & \frac{\sqrt{2}}{6} & \frac{1}{3} \end{bmatrix}$$

P^{56}_{24}

$$\begin{bmatrix} -\frac{1}{2} & & \frac{\sqrt{3}}{2} & & \\ & -\frac{1}{2} & & \frac{\sqrt{3}}{2} & \\ & \frac{1}{2} & & \frac{\sqrt{3}}{6} & -\frac{\sqrt{6}}{3} \\ \frac{1}{2} & -\frac{\sqrt{3}}{3} & \frac{\sqrt{3}}{6} & -\frac{1}{3} & -\frac{\sqrt{2}}{3} \\ \frac{\sqrt{2}}{2} & \frac{\sqrt{6}}{6} & \frac{\sqrt{6}}{6} & \frac{\sqrt{2}}{6} & \frac{1}{3} \end{bmatrix}$$

P^{56}_{25}

$$\begin{bmatrix} \frac{1}{4} & -\frac{\sqrt{3}}{4} & -\frac{\sqrt{3}}{4} & \frac{3}{4} & \\ -\frac{\sqrt{3}}{4} & -\frac{1}{4} & \frac{3}{4} & \frac{\sqrt{3}}{4} & \\ \frac{\sqrt{3}}{4} & \frac{1}{4} & \frac{1}{4} & \frac{\sqrt{3}}{12} & \frac{\sqrt{6}}{3} \\ -\frac{3}{4} & \frac{\sqrt{3}}{12} & -\frac{\sqrt{3}}{4} & \frac{1}{12} & \frac{\sqrt{2}}{3} \\ & \frac{\sqrt{6}}{3} & & \frac{\sqrt{2}}{3} & -\frac{1}{3} \end{bmatrix}$$

P^{56}_{26}

$$\begin{bmatrix} \frac{1}{4} & \frac{\sqrt{3}}{4} & -\frac{\sqrt{3}}{4} & -\frac{3}{4} & \\ -\frac{\sqrt{3}}{4} & \frac{1}{4} & \frac{3}{4} & -\frac{\sqrt{3}}{4} & \\ \frac{\sqrt{3}}{4} & -\frac{1}{4} & \frac{1}{4} & -\frac{\sqrt{3}}{12} & -\frac{\sqrt{6}}{3} \\ -\frac{3}{4} & -\frac{\sqrt{3}}{12} & -\frac{\sqrt{3}}{4} & -\frac{1}{12} & -\frac{\sqrt{2}}{3} \\ & -\frac{\sqrt{6}}{3} & & -\frac{\sqrt{2}}{3} & \frac{1}{3} \end{bmatrix}$$

P^{56}_{34}

$$\begin{bmatrix} 1 & & & & \\ & 1 & & & \\ & & & \frac{\sqrt{3}}{3} & \frac{\sqrt{6}}{3} \\ & & \frac{\sqrt{3}}{3} & -\frac{2}{3} & \frac{\sqrt{2}}{3} \\ & & \frac{\sqrt{6}}{3} & \frac{\sqrt{2}}{3} & -\frac{1}{3} \end{bmatrix}$$

P^{56}_{35}

$$\begin{bmatrix} -\frac{1}{2} & \frac{\sqrt{3}}{2} & & & \\ \frac{\sqrt{3}}{2} & \frac{1}{2} & & & \\ & & \frac{1}{2} & \frac{\sqrt{3}}{6} & -\frac{\sqrt{6}}{3} \\ & & -\frac{\sqrt{3}}{2} & \frac{1}{6} & -\frac{\sqrt{2}}{3} \\ & & & \frac{2\sqrt{2}}{3} & \frac{1}{3} \end{bmatrix}$$

Table 4 (Cont.)

$$P_{36}^{56} = \begin{bmatrix} -\frac{1}{2} & -\frac{\sqrt{3}}{2} & & & \\ \frac{\sqrt{3}}{2} & -\frac{1}{2} & & & \\ & & \frac{1}{2} & -\frac{\sqrt{3}}{6} & \frac{\sqrt{6}}{3} \\ & & -\frac{\sqrt{3}}{2} & -\frac{1}{6} & \frac{\sqrt{2}}{3} \\ & & & -\frac{2\sqrt{2}}{3} & -\frac{1}{3} \end{bmatrix} \qquad P_{45}^{56} = \begin{bmatrix} -\frac{1}{2} & -\frac{\sqrt{3}}{2} & & & \\ \frac{\sqrt{3}}{2} & -\frac{1}{2} & & & \\ & & -\frac{1}{2} & -\frac{\sqrt{3}}{2} & \\ & & \frac{\sqrt{3}}{2} & -\frac{1}{2} & \\ & & & & 1 \end{bmatrix}$$

$$P_{46}^{56} = \begin{bmatrix} -\frac{1}{2} & \frac{\sqrt{3}}{2} & & & \\ \frac{\sqrt{3}}{2} & \frac{1}{2} & & & \\ & & -\frac{1}{2} & \frac{\sqrt{3}}{2} & \\ & & \frac{\sqrt{3}}{2} & \frac{1}{2} & \\ & & & & -1 \end{bmatrix} \qquad P_{56}^{56} = \begin{bmatrix} 1 & & & & \\ & 1 & & & \\ & & 1 & & \\ & & & 1 & \\ & & & & 1 \end{bmatrix} \qquad P_{16}^{56} = \begin{bmatrix} \frac{1}{4} & \frac{\sqrt{3}}{4} & \frac{\sqrt{3}}{4} & \frac{3}{4} & \\ -\frac{\sqrt{3}}{4} & \frac{1}{4} & -\frac{3}{4} & \frac{\sqrt{3}}{4} & \\ \frac{\sqrt{3}}{4} & -\frac{1}{4} & -\frac{1}{4} & \frac{\sqrt{3}}{12} & \frac{\sqrt{6}}{3} \\ -\frac{3}{4} & -\frac{\sqrt{3}}{12} & \frac{\sqrt{3}}{4} & \frac{1}{12} & \frac{\sqrt{2}}{3} \\ & -\frac{\sqrt{6}}{3} & & \frac{\sqrt{2}}{3} & -\frac{1}{3} \end{bmatrix}$$

(b) $[\lambda] = [2^2 1^2]$, $S = 1$

$$P_{12}^{56}$$

$$\begin{array}{c} \lambda'' \\ {[1^2]} \\ {[2]} \\ {[1^2]} \\ {[1^2]} \\ {[2]} \\ {[1^2]} \\ {[2]} \\ {[1^2]} \\ {[1^2]} \end{array} \quad \begin{bmatrix} & & & & \frac{1}{3} & \frac{\sqrt{2}}{3} & \frac{\sqrt{2}}{3} & \frac{2}{3} & \\ 1 & & & & & & & & \\ & & & & \frac{\sqrt{2}}{3} & -\frac{1}{3} & \frac{2}{3} & -\frac{\sqrt{2}}{3} & \\ & & & \frac{1}{3} & \frac{2\sqrt{3}}{9} & -\frac{\sqrt{6}}{9} & -\frac{\sqrt{6}}{9} & \frac{\sqrt{3}}{9} & \frac{\sqrt{5}}{3} \\ & 1 & & & & & & & \\ & & & \frac{\sqrt{2}}{3} & \frac{2\sqrt{6}}{9} & \frac{\sqrt{3}}{9} & -\frac{2\sqrt{3}}{9} & -\frac{\sqrt{6}}{18} & -\frac{\sqrt{10}}{6} \\ & & 1 & & & & & & \\ & & & \frac{1}{3} & \frac{\sqrt{3}}{9} & \frac{5\sqrt{6}}{18} & -\frac{\sqrt{6}}{18} & -\frac{5\sqrt{3}}{18} & \frac{\sqrt{5}}{6} \\ & & & \frac{\sqrt{5}}{3} & -\frac{\sqrt{15}}{9} & \frac{\sqrt{30}}{18} & \frac{\sqrt{30}}{18} & -\frac{\sqrt{15}}{18} & \frac{1}{6} \end{bmatrix}$$

Table 4 (Cont.)

$$P_{13}^{56}$$

$$\begin{bmatrix}
 & \frac{\sqrt{3}}{6} & \frac{\sqrt{6}}{6} & & \frac{1}{6} & \frac{\sqrt{2}}{6} & -\frac{\sqrt{2}}{3} & -\frac{2}{3} & \\
-\frac{1}{2} & & & \frac{\sqrt{3}}{2} & & & & & \\
 & \frac{\sqrt{6}}{6} & -\frac{\sqrt{3}}{6} & & \frac{\sqrt{2}}{6} & -\frac{1}{6} & -\frac{2}{3} & \frac{\sqrt{2}}{3} & \\
\frac{\sqrt{3}}{6} & \frac{1}{3} & -\frac{\sqrt{2}}{6} & \frac{1}{6} & \frac{\sqrt{3}}{9} & -\frac{\sqrt{6}}{18} & \frac{\sqrt{6}}{9} & -\frac{\sqrt{3}}{9} & -\frac{\sqrt{5}}{3} \\
 & -\frac{1}{2} & & & \frac{\sqrt{3}}{2} & & & & \\
\frac{\sqrt{6}}{6} & \frac{\sqrt{2}}{3} & \frac{1}{6} & \frac{\sqrt{2}}{6} & \frac{\sqrt{6}}{9} & \frac{\sqrt{3}}{18} & \frac{2\sqrt{3}}{9} & \frac{\sqrt{6}}{18} & \frac{\sqrt{10}}{6} \\
 & & -\frac{1}{2} & & & \frac{\sqrt{3}}{2} & & & \\
-\frac{\sqrt{3}}{6} & \frac{1}{6} & \frac{5\sqrt{2}}{12} & -\frac{1}{6} & \frac{\sqrt{3}}{18} & \frac{5\sqrt{6}}{36} & \frac{\sqrt{6}}{18} & \frac{5\sqrt{3}}{18} & -\frac{\sqrt{5}}{6} \\
\frac{\sqrt{15}}{6} & -\frac{\sqrt{5}}{6} & \frac{\sqrt{10}}{12} & \frac{\sqrt{5}}{6} & -\frac{\sqrt{15}}{18} & \frac{\sqrt{30}}{36} & -\frac{\sqrt{30}}{18} & \frac{\sqrt{15}}{18} & -\frac{1}{6}
\end{bmatrix}$$

$$P_{14}^{56}$$

$$\begin{bmatrix}
 & -\frac{\sqrt{3}}{6} & -\frac{\sqrt{6}}{6} & & -\frac{1}{2} & -\frac{\sqrt{2}}{2} & & & \\
-\frac{1}{2} & & & -\frac{\sqrt{3}}{2} & & & & & \\
 & -\frac{\sqrt{6}}{6} & \frac{\sqrt{3}}{6} & & -\frac{\sqrt{2}}{2} & \frac{1}{2} & & & \\
\frac{\sqrt{3}}{6} & -\frac{1}{3} & \frac{\sqrt{2}}{6} & -\frac{1}{6} & \frac{\sqrt{3}}{9} & -\frac{\sqrt{6}}{18} & \frac{\sqrt{6}}{9} & -\frac{\sqrt{3}}{9} & \frac{\sqrt{5}}{3} \\
 & \frac{1}{2} & & & -\frac{\sqrt{3}}{6} & & \frac{\sqrt{6}}{3} & & \\
\frac{\sqrt{6}}{6} & -\frac{\sqrt{2}}{3} & -\frac{1}{6} & -\frac{\sqrt{2}}{6} & \frac{\sqrt{6}}{9} & \frac{\sqrt{3}}{18} & \frac{2\sqrt{3}}{9} & \frac{\sqrt{6}}{18} & -\frac{\sqrt{10}}{6} \\
 & & \frac{1}{2} & & & -\frac{\sqrt{3}}{6} & & \frac{\sqrt{6}}{3} & \\
-\frac{\sqrt{3}}{6} & -\frac{1}{6} & -\frac{5\sqrt{2}}{12} & \frac{1}{6} & \frac{\sqrt{3}}{18} & \frac{5\sqrt{6}}{36} & \frac{\sqrt{6}}{18} & \frac{5\sqrt{3}}{18} & \frac{\sqrt{5}}{6} \\
\frac{\sqrt{15}}{6} & \frac{\sqrt{5}}{6} & -\frac{\sqrt{10}}{12} & -\frac{\sqrt{5}}{6} & -\frac{\sqrt{15}}{18} & \frac{\sqrt{30}}{36} & -\frac{\sqrt{30}}{18} & \frac{\sqrt{15}}{18} & \frac{1}{6}
\end{bmatrix}$$

Table 4 (Cont.)

$$P_{15}^{56}$$

$$\begin{bmatrix}
-\frac{1}{4} & -\frac{\sqrt{3}}{12} & \frac{\sqrt{6}}{6} & -\frac{\sqrt{3}}{4} & -\frac{1}{4} & \frac{\sqrt{2}}{2} & & & \\
\frac{1}{4} & -\frac{\sqrt{3}}{4} & & \frac{\sqrt{3}}{4} & -\frac{3}{4} & & & & \\
-\frac{\sqrt{2}}{4} & -\frac{\sqrt{6}}{12} & -\frac{\sqrt{3}}{6} & -\frac{\sqrt{6}}{4} & -\frac{\sqrt{2}}{4} & -\frac{1}{2} & & & \\
-\frac{\sqrt{3}}{4} & \frac{1}{12} & -\frac{\sqrt{2}}{6} & \frac{1}{4} & -\frac{\sqrt{3}}{36} & \frac{\sqrt{6}}{18} & -\frac{\sqrt{6}}{9} & \frac{4\sqrt{3}}{9} & \\
\frac{\sqrt{3}}{4} & \frac{1}{4} & & -\frac{1}{4} & -\frac{\sqrt{3}}{12} & & -\frac{\sqrt{6}}{3} & & \\
-\frac{\sqrt{6}}{4} & \frac{\sqrt{2}}{12} & \frac{1}{6} & \frac{\sqrt{2}}{4} & -\frac{\sqrt{6}}{36} & -\frac{\sqrt{3}}{18} & -\frac{2\sqrt{3}}{9} & -\frac{2\sqrt{6}}{9} & \\
 & & -\frac{1}{2} & & & \frac{\sqrt{3}}{6} & & -\frac{\sqrt{6}}{12} & \frac{\sqrt{10}}{4} \\
 & -\frac{1}{3} & \frac{5\sqrt{2}}{12} & & \frac{\sqrt{3}}{9} & -\frac{5\sqrt{6}}{36} & -\frac{\sqrt{6}}{18} & \frac{5\sqrt{3}}{36} & \frac{\sqrt{5}}{4} \\
 & \frac{\sqrt{5}}{3} & \frac{\sqrt{10}}{12} & & -\frac{\sqrt{15}}{9} & -\frac{\sqrt{30}}{36} & \frac{\sqrt{30}}{18} & \frac{\sqrt{15}}{36} & \frac{1}{4}
\end{bmatrix}$$

$$P_{16}^{56}$$

$$\begin{bmatrix}
\frac{1}{4} & \frac{\sqrt{3}}{4} & & \frac{\sqrt{3}}{4} & \frac{3}{4} & & & & \\
-\frac{1}{4} & \frac{\sqrt{3}}{12} & -\frac{\sqrt{6}}{6} & -\frac{\sqrt{3}}{4} & \frac{1}{4} & -\frac{\sqrt{2}}{2} & & & \\
\frac{\sqrt{2}}{4} & -\frac{\sqrt{6}}{12} & -\frac{\sqrt{3}}{6} & \frac{\sqrt{6}}{4} & -\frac{\sqrt{2}}{4} & -\frac{1}{2} & & & \\
\frac{\sqrt{3}}{4} & -\frac{1}{4} & & -\frac{1}{4} & \frac{\sqrt{3}}{12} & & \frac{\sqrt{6}}{3} & & \\
-\frac{\sqrt{3}}{4} & -\frac{1}{12} & \frac{\sqrt{2}}{6} & \frac{1}{4} & \frac{\sqrt{3}}{36} & -\frac{\sqrt{6}}{18} & \frac{\sqrt{6}}{9} & -\frac{4\sqrt{3}}{9} & \\
\frac{\sqrt{6}}{4} & \frac{\sqrt{2}}{12} & \frac{1}{6} & -\frac{\sqrt{2}}{4} & -\frac{\sqrt{6}}{36} & -\frac{\sqrt{3}}{18} & -\frac{2\sqrt{3}}{9} & -\frac{2\sqrt{6}}{9} & \\
 & -\frac{\sqrt{2}}{3} & -\frac{1}{6} & & \frac{\sqrt{6}}{9} & \frac{\sqrt{3}}{18} & -\frac{\sqrt{3}}{9} & -\frac{\sqrt{6}}{36} & -\frac{\sqrt{10}}{4} \\
 & \frac{2}{3} & -\frac{\sqrt{2}}{12} & & -\frac{2\sqrt{3}}{9} & \frac{\sqrt{6}}{36} & \frac{\sqrt{6}}{9} & -\frac{\sqrt{3}}{33} & -\frac{\sqrt{5}}{4} \\
 & & \frac{\sqrt{10}}{4} & & & -\frac{\sqrt{30}}{12} & & \frac{\sqrt{15}}{12} & -\frac{1}{4}
\end{bmatrix}$$

Table 4 (Cont.)

$$P^{56}_{23}$$

$$\begin{bmatrix}
 & \frac{\sqrt{3}}{6} & \frac{\sqrt{6}}{6} & & -\frac{1}{6} & -\frac{\sqrt{2}}{6} & \frac{\sqrt{2}}{3} & \frac{2}{3} & \\
-\frac{1}{2} & & & -\frac{\sqrt{3}}{2} & & & & & \\
 & \frac{\sqrt{6}}{6} & -\frac{\sqrt{3}}{6} & & -\frac{\sqrt{2}}{6} & \frac{1}{6} & \frac{2}{3} & -\frac{\sqrt{2}}{3} & \\
\frac{\sqrt{3}}{6} & \frac{1}{3} & -\frac{\sqrt{2}}{6} & -\frac{1}{6} & -\frac{\sqrt{3}}{9} & \frac{\sqrt{6}}{18} & -\frac{\sqrt{6}}{9} & \frac{\sqrt{3}}{9} & \frac{\sqrt{5}}{3} \\
 & -\frac{1}{2} & & & -\frac{\sqrt{3}}{2} & & & & \\
\frac{\sqrt{6}}{6} & \frac{\sqrt{2}}{3} & \frac{1}{6} & -\frac{\sqrt{2}}{6} & -\frac{\sqrt{6}}{9} & -\frac{\sqrt{3}}{18} & -\frac{2\sqrt{3}}{9} & -\frac{\sqrt{6}}{18} & -\frac{\sqrt{10}}{6} \\
 & & -\frac{1}{2} & & & -\frac{\sqrt{3}}{2} & & & \\
-\frac{\sqrt{3}}{6} & \frac{1}{6} & \frac{5\sqrt{2}}{12} & \frac{1}{6} & -\frac{\sqrt{3}}{18} & -\frac{5\sqrt{6}}{36} & -\frac{\sqrt{6}}{18} & -\frac{5\sqrt{3}}{18} & \frac{\sqrt{5}}{6} \\
\frac{\sqrt{15}}{6} & -\frac{\sqrt{5}}{6} & \frac{\sqrt{10}}{12} & -\frac{\sqrt{5}}{6} & \frac{\sqrt{15}}{18} & -\frac{\sqrt{30}}{36} & \frac{\sqrt{30}}{18} & -\frac{\sqrt{15}}{18} & \frac{1}{6}
\end{bmatrix}$$

$$P^{56}_{24}$$

$$\begin{bmatrix}
 & -\frac{\sqrt{3}}{6} & -\frac{\sqrt{6}}{6} & & \frac{1}{2} & \frac{\sqrt{2}}{2} & & & \\
-\frac{1}{2} & & & \frac{\sqrt{3}}{2} & & & & & \\
 & -\frac{\sqrt{6}}{6} & \frac{\sqrt{3}}{6} & & \frac{\sqrt{2}}{2} & -\frac{1}{2} & & & \\
\frac{\sqrt{3}}{6} & -\frac{1}{3} & \frac{\sqrt{2}}{6} & \frac{1}{6} & -\frac{\sqrt{3}}{9} & \frac{\sqrt{6}}{18} & -\frac{\sqrt{6}}{9} & \frac{\sqrt{3}}{9} & -\frac{\sqrt{5}}{3} \\
 & \frac{1}{2} & & & \frac{\sqrt{3}}{6} & & -\frac{\sqrt{6}}{3} & & \\
\frac{\sqrt{6}}{6} & -\frac{\sqrt{2}}{3} & -\frac{1}{6} & \frac{\sqrt{2}}{6} & -\frac{\sqrt{6}}{9} & -\frac{\sqrt{3}}{18} & -\frac{2\sqrt{3}}{9} & -\frac{\sqrt{6}}{18} & \frac{\sqrt{10}}{6} \\
 & & \frac{1}{2} & & & \frac{\sqrt{3}}{6} & & -\frac{\sqrt{6}}{3} & \\
-\frac{\sqrt{3}}{6} & -\frac{1}{6} & -\frac{5\sqrt{2}}{12} & -\frac{1}{6} & -\frac{\sqrt{3}}{18} & -\frac{5\sqrt{6}}{36} & -\frac{\sqrt{6}}{18} & -\frac{5\sqrt{3}}{18} & -\frac{\sqrt{5}}{6} \\
\frac{\sqrt{15}}{6} & \frac{\sqrt{5}}{6} & -\frac{\sqrt{10}}{12} & \frac{\sqrt{5}}{6} & \frac{\sqrt{15}}{18} & -\frac{\sqrt{30}}{36} & \frac{\sqrt{30}}{18} & -\frac{\sqrt{15}}{18} & -\frac{1}{6}
\end{bmatrix}$$

Table 4 (Cont.)

P_{25}^{56}

$-\frac{1}{4}$	$-\frac{\sqrt{3}}{12}$	$\frac{\sqrt{6}}{6}$	$\frac{\sqrt{3}}{4}$	$\frac{1}{4}$	$-\frac{\sqrt{2}}{2}$			
$\frac{1}{4}$	$-\frac{\sqrt{3}}{4}$		$-\frac{\sqrt{3}}{4}$	$\frac{3}{4}$				
$-\frac{\sqrt{2}}{4}$	$-\frac{\sqrt{6}}{12}$	$-\frac{\sqrt{3}}{6}$	$\frac{\sqrt{6}}{4}$	$\frac{\sqrt{2}}{4}$	$\frac{1}{2}$			
$-\frac{\sqrt{3}}{4}$	$\frac{1}{12}$	$-\frac{\sqrt{2}}{6}$	$-\frac{1}{4}$	$\frac{\sqrt{3}}{36}$	$-\frac{\sqrt{6}}{18}$	$\frac{\sqrt{6}}{9}$	$-\frac{4\sqrt{3}}{9}$	
$\frac{\sqrt{3}}{4}$	$\frac{1}{4}$		$\frac{1}{4}$	$\frac{\sqrt{3}}{12}$		$\frac{\sqrt{6}}{3}$		
$-\frac{\sqrt{6}}{4}$	$\frac{\sqrt{2}}{12}$	$\frac{1}{6}$	$-\frac{\sqrt{2}}{4}$	$\frac{\sqrt{6}}{36}$	$\frac{\sqrt{3}}{18}$	$\frac{2\sqrt{3}}{9}$	$\frac{2\sqrt{6}}{9}$	
		$-\frac{1}{2}$			$-\frac{\sqrt{3}}{6}$		$\frac{\sqrt{6}}{12}$	$-\frac{\sqrt{10}}{4}$
	$-\frac{1}{3}$	$\frac{5\sqrt{2}}{12}$		$-\frac{\sqrt{3}}{9}$	$\frac{5\sqrt{6}}{36}$	$\frac{\sqrt{6}}{18}$	$-\frac{5\sqrt{3}}{36}$	$-\frac{\sqrt{5}}{4}$
	$\frac{\sqrt{5}}{3}$	$\frac{\sqrt{10}}{12}$		$\frac{\sqrt{15}}{9}$	$\frac{\sqrt{30}}{36}$	$-\frac{\sqrt{30}}{18}$	$-\frac{\sqrt{15}}{36}$	$-\frac{1}{4}$

P_{26}^{56}

$\frac{1}{4}$	$\frac{\sqrt{3}}{4}$		$-\frac{\sqrt{3}}{4}$	$-\frac{3}{4}$				
$-\frac{1}{4}$	$\frac{\sqrt{3}}{12}$	$-\frac{\sqrt{6}}{6}$	$\frac{\sqrt{3}}{4}$	$-\frac{1}{4}$	$\frac{\sqrt{2}}{2}$			
$\frac{\sqrt{2}}{4}$	$-\frac{\sqrt{6}}{12}$	$-\frac{\sqrt{3}}{6}$	$-\frac{\sqrt{6}}{4}$	$\frac{\sqrt{2}}{4}$	$\frac{1}{2}$			
$\frac{\sqrt{3}}{4}$	$-\frac{1}{4}$		$\frac{1}{4}$	$-\frac{\sqrt{3}}{12}$		$-\frac{\sqrt{6}}{3}$		
$-\frac{\sqrt{3}}{4}$	$-\frac{1}{12}$	$\frac{\sqrt{2}}{6}$	$-\frac{1}{4}$	$-\frac{\sqrt{3}}{36}$	$\frac{\sqrt{6}}{18}$	$-\frac{\sqrt{6}}{9}$	$\frac{4\sqrt{3}}{9}$	
$\frac{\sqrt{6}}{4}$	$\frac{\sqrt{2}}{12}$	$\frac{1}{6}$	$\frac{\sqrt{2}}{4}$	$\frac{\sqrt{6}}{36}$	$\frac{\sqrt{3}}{18}$	$\frac{2\sqrt{3}}{9}$	$\frac{2\sqrt{6}}{9}$	
	$-\frac{\sqrt{2}}{3}$	$-\frac{1}{6}$		$-\frac{\sqrt{6}}{9}$	$-\frac{\sqrt{3}}{18}$	$\frac{\sqrt{3}}{9}$	$\frac{\sqrt{6}}{36}$	$\frac{\sqrt{10}}{4}$
	$\frac{2}{3}$	$-\frac{\sqrt{2}}{12}$		$\frac{2\sqrt{3}}{9}$	$-\frac{\sqrt{6}}{36}$	$-\frac{\sqrt{6}}{9}$	$\frac{\sqrt{3}}{36}$	$\frac{\sqrt{5}}{4}$
		$\frac{\sqrt{10}}{4}$			$\frac{\sqrt{30}}{12}$		$-\frac{\sqrt{15}}{12}$	$\frac{1}{4}$

Table 4 (Cont.)

P_{34}^{56}

$$\begin{bmatrix}
 & \frac{\sqrt{3}}{3} & \frac{\sqrt{6}}{3} & & & & & & \\
1 & & & & & & & & \\
 & \frac{\sqrt{6}}{3} & -\frac{\sqrt{3}}{3} & & & & & & \\
 & & & \frac{1}{3} & -\frac{2\sqrt{3}}{9} & \frac{\sqrt{6}}{9} & \frac{\sqrt{6}}{9} & -\frac{\sqrt{3}}{9} & \frac{\sqrt{5}}{3} \\
 & & & & \frac{\sqrt{3}}{3} & & \frac{\sqrt{6}}{3} & & \\
 & & & \frac{\sqrt{2}}{3} & -\frac{2\sqrt{6}}{9} & -\frac{\sqrt{3}}{9} & \frac{2\sqrt{3}}{9} & \frac{\sqrt{6}}{18} & -\frac{\sqrt{10}}{6} \\
 & & & & & \frac{\sqrt{3}}{3} & & \frac{\sqrt{6}}{3} & \\
 & & & -\frac{1}{3} & -\frac{\sqrt{3}}{9} & -\frac{5\sqrt{6}}{18} & \frac{\sqrt{6}}{18} & \frac{5\sqrt{3}}{18} & \frac{\sqrt{5}}{6} \\
 & & & \frac{\sqrt{5}}{3} & \frac{\sqrt{15}}{9} & -\frac{\sqrt{30}}{18} & -\frac{\sqrt{30}}{18} & \frac{\sqrt{15}}{18} & \frac{1}{6}
\end{bmatrix}$$

P_{35}^{56}

$$\begin{bmatrix}
\frac{1}{2} & \frac{\sqrt{3}}{6} & -\frac{\sqrt{6}}{3} & & & & & & \\
-\frac{1}{2} & \frac{\sqrt{3}}{2} & & & & & & & \\
\frac{\sqrt{2}}{2} & \frac{\sqrt{6}}{6} & \frac{\sqrt{3}}{3} & & & & & & \\
 & & & -\frac{1}{2} & \frac{\sqrt{3}}{18} & -\frac{\sqrt{6}}{9} & -\frac{\sqrt{6}}{9} & \frac{4\sqrt{3}}{9} & \\
 & & & \frac{1}{2} & \frac{\sqrt{3}}{6} & & -\frac{\sqrt{6}}{3} & & \\
 & & & -\frac{\sqrt{2}}{2} & \frac{\sqrt{6}}{18} & \frac{\sqrt{3}}{9} & -\frac{2\sqrt{3}}{9} & -\frac{2\sqrt{6}}{9} & \\
 & & & & & -\frac{\sqrt{3}}{3} & & -\frac{\sqrt{6}}{12} & \frac{\sqrt{10}}{4} \\
 & & & & -\frac{2\sqrt{3}}{9} & \frac{5\sqrt{6}}{18} & -\frac{\sqrt{6}}{18} & \frac{5\sqrt{3}}{36} & \frac{\sqrt{5}}{4} \\
 & & & & \frac{2\sqrt{15}}{9} & \frac{\sqrt{30}}{18} & \frac{\sqrt{30}}{18} & \frac{\sqrt{15}}{36} & \frac{1}{4}
\end{bmatrix}$$

Table 4 (Cont.)

$$P^{56}_{36}$$

$$\begin{bmatrix}
-\frac{1}{2} & -\frac{\sqrt{3}}{2} & & & & & & & \\
\frac{1}{2} & -\frac{\sqrt{3}}{6} & \frac{\sqrt{6}}{3} & & & & & & \\
-\frac{\sqrt{2}}{2} & \frac{\sqrt{6}}{6} & \frac{\sqrt{3}}{3} & & & & & & \\
 & & & \frac{1}{2} & -\frac{\sqrt{3}}{6} & & \frac{\sqrt{6}}{3} & & \\
 & & & -\frac{1}{2} & -\frac{\sqrt{3}}{18} & \frac{\sqrt{6}}{9} & \frac{\sqrt{6}}{9} & -\frac{4\sqrt{3}}{9} & \\
 & & & \frac{\sqrt{2}}{2} & \frac{\sqrt{6}}{18} & \frac{\sqrt{3}}{9} & -\frac{2\sqrt{3}}{9} & -\frac{2\sqrt{6}}{9} & \\
 & & & & -\frac{2\sqrt{6}}{9} & -\frac{\sqrt{3}}{9} & -\frac{\sqrt{3}}{9} & -\frac{\sqrt{6}}{36} & -\frac{\sqrt{10}}{4} \\
 & & & & \frac{4\sqrt{3}}{9} & -\frac{\sqrt{6}}{18} & \frac{\sqrt{6}}{9} & -\frac{\sqrt{3}}{36} & -\frac{\sqrt{5}}{4} \\
 & & & & & \frac{\sqrt{30}}{6} & & \frac{\sqrt{15}}{12} & -\frac{1}{4}
\end{bmatrix}$$

$$P^{56}_{45}$$

$$\begin{bmatrix}
\frac{1}{2} & -\frac{\sqrt{3}}{6} & \frac{\sqrt{6}}{3} & & & & & & \\
-\frac{1}{2} & -\frac{\sqrt{3}}{2} & & & & & & & \\
\frac{\sqrt{2}}{2} & -\frac{\sqrt{6}}{6} & -\frac{\sqrt{3}}{3} & & & & & & \\
 & & & \frac{1}{2} & -\frac{\sqrt{3}}{6} & \frac{\sqrt{6}}{3} & & & \\
 & & & -\frac{1}{2} & -\frac{\sqrt{3}}{2} & & & & \\
 & & & \frac{\sqrt{2}}{2} & -\frac{\sqrt{6}}{6} & -\frac{\sqrt{3}}{3} & & & \\
 & & & & & & & -\frac{\sqrt{6}}{4} & -\frac{\sqrt{10}}{4} \\
 & & & & & & -\frac{\sqrt{6}}{6} & \frac{5\sqrt{3}}{12} & -\frac{\sqrt{5}}{4} \\
 & & & & & & \frac{\sqrt{30}}{6} & \frac{\sqrt{15}}{12} & -\frac{1}{4}
\end{bmatrix}$$

Table 4 (Cont.)

P_{46}^{56}

$$\begin{bmatrix}
-\frac{1}{2} & \frac{\sqrt{3}}{2} & & & & & & & \\
\frac{1}{2} & \frac{\sqrt{3}}{6} & -\frac{\sqrt{6}}{3} & & & & & & \\
-\frac{\sqrt{2}}{2} & -\frac{\sqrt{6}}{6} & -\frac{\sqrt{3}}{3} & & & & & & \\
 & & & -\frac{1}{2} & \frac{\sqrt{3}}{2} & & & & \\
 & & & \frac{1}{2} & \frac{\sqrt{3}}{6} & -\frac{\sqrt{6}}{3} & & & \\
 & & & -\frac{\sqrt{2}}{2} & -\frac{\sqrt{6}}{6} & -\frac{\sqrt{3}}{3} & & & \\
 & & & & & & -\frac{\sqrt{3}}{3} & -\frac{\sqrt{6}}{12} & \frac{\sqrt{10}}{4} \\
 & & & & & & \frac{\sqrt{6}}{3} & -\frac{\sqrt{3}}{12} & \frac{\sqrt{5}}{4} \\
 & & & & & & & \frac{\sqrt{15}}{4} & \frac{1}{4}
\end{bmatrix}$$

P_{56}^{56}

$$\begin{bmatrix}
1 & & & & & & & & \\
 & \frac{\sqrt{3}}{3} & \frac{\sqrt{6}}{3} & & & & & & \\
 & -\frac{\sqrt{6}}{3} & \frac{\sqrt{3}}{3} & & & & & & \\
 & & & 1 & & & & & \\
 & & & & \frac{\sqrt{3}}{3} & \frac{\sqrt{6}}{3} & & & \\
 & & & & -\frac{\sqrt{6}}{3} & \frac{\sqrt{3}}{3} & & & \\
 & & & & & & \frac{\sqrt{3}}{3} & \frac{\sqrt{6}}{3} & \\
 & & & & & & -\frac{\sqrt{6}}{3} & \frac{\sqrt{3}}{3} & \\
 & & & & & & & & 1
\end{bmatrix}$$

Table 4 (Cont.)

(c) $[\lambda] = [21^4]$, $S = 2$

$$P_{12}^{56} \qquad \begin{array}{c} \lambda'' \\ [1^2] \\ [1^2] \\ [1^2] \\ [2] \\ [1^3] \end{array} \begin{bmatrix} 0 & \frac{\sqrt{3}}{3} & \frac{\sqrt{6}}{3} & 0 & 0 \\ 0 & -\frac{1}{3} & \frac{\sqrt{2}}{6} & \frac{\sqrt{30}}{6} & 0 \\ 0 & \frac{\sqrt{2}}{6} & -\frac{1}{6} & \frac{\sqrt{15}}{30} & \frac{3\sqrt{10}}{10} \\ 1 & 0 & 0 & 0 & 0 \\ 0 & \frac{\sqrt{2}}{2} & -\frac{1}{2} & \frac{\sqrt{15}}{10} & -\frac{\sqrt{10}}{10} \end{bmatrix}$$

$$P_{13}^{56} \qquad \begin{bmatrix} \frac{1}{2} & \frac{\sqrt{3}}{6} & -\frac{\sqrt{6}}{3} & 0 & 0 \\ -\frac{\sqrt{3}}{6} & -\frac{1}{6} & -\frac{\sqrt{2}}{6} & -\frac{\sqrt{30}}{6} & 0 \\ \frac{\sqrt{6}}{12} & \frac{\sqrt{2}}{12} & \frac{1}{6} & -\frac{\sqrt{15}}{30} & -\frac{3\sqrt{10}}{10} \\ -\frac{1}{2} & \frac{\sqrt{3}}{2} & 0 & 0 & 0 \\ \frac{\sqrt{6}}{4} & \frac{\sqrt{2}}{4} & \frac{1}{2} & -\frac{\sqrt{15}}{10} & \frac{\sqrt{10}}{10} \end{bmatrix}$$

$$P_{14}^{56} \qquad \begin{bmatrix} -\frac{1}{2} & -\frac{\sqrt{3}}{2} & 0 & 0 & 0 \\ \frac{\sqrt{3}}{6} & -\frac{1}{6} & -\frac{\sqrt{2}}{6} & \frac{\sqrt{30}}{6} & 0 \\ -\frac{\sqrt{6}}{12} & \frac{\sqrt{2}}{12} & \frac{1}{6} & \frac{\sqrt{15}}{30} & \frac{3\sqrt{10}}{10} \\ \frac{1}{2} & -\frac{\sqrt{3}}{6} & \frac{\sqrt{6}}{3} & 0 & 0 \\ -\frac{\sqrt{6}}{4} & \frac{\sqrt{2}}{4} & \frac{1}{2} & \frac{\sqrt{15}}{10} & -\frac{\sqrt{10}}{10} \end{bmatrix}$$

$$P_{15}^{56} \qquad \begin{bmatrix} \frac{1}{2} & \frac{\sqrt{3}}{2} & 0 & 0 & 0 \\ -\frac{\sqrt{3}}{6} & \frac{1}{6} & \frac{2\sqrt{2}}{3} & 0 & 0 \\ \frac{\sqrt{6}}{12} & -\frac{\sqrt{2}}{12} & \frac{1}{12} & \frac{\sqrt{15}}{20} & -\frac{3\sqrt{10}}{10} \\ -\frac{1}{2} & \frac{\sqrt{3}}{6} & -\frac{\sqrt{6}}{12} & \frac{\sqrt{10}}{4} & 0 \\ \frac{\sqrt{6}}{4} & -\frac{\sqrt{2}}{4} & \frac{1}{4} & \frac{3\sqrt{15}}{20} & \frac{\sqrt{10}}{10} \end{bmatrix}$$

$$P_{16}^{56} \qquad \begin{bmatrix} -\frac{1}{2} & -\frac{\sqrt{3}}{2} & 0 & 0 & 0 \\ \frac{\sqrt{3}}{6} & -\frac{1}{6} & -\frac{2\sqrt{2}}{3} & 0 & 0 \\ -\frac{\sqrt{6}}{12} & \frac{\sqrt{2}}{12} & -\frac{1}{12} & -\frac{\sqrt{15}}{4} & 0 \\ \frac{1}{2} & -\frac{\sqrt{3}}{6} & \frac{\sqrt{6}}{12} & -\frac{\sqrt{10}}{20} & \frac{\sqrt{15}}{5} \\ -\frac{\sqrt{6}}{4} & \frac{\sqrt{2}}{4} & -\frac{1}{4} & \frac{\sqrt{15}}{20} & \frac{\sqrt{10}}{5} \end{bmatrix}$$

$$P_{23}^{56} \qquad \begin{bmatrix} \frac{1}{2} & -\frac{\sqrt{3}}{6} & \frac{\sqrt{6}}{3} & 0 & 0 \\ -\frac{\sqrt{3}}{6} & \frac{1}{6} & \frac{\sqrt{2}}{6} & \frac{\sqrt{30}}{6} & 0 \\ \frac{\sqrt{6}}{12} & -\frac{\sqrt{2}}{12} & -\frac{1}{6} & \frac{\sqrt{15}}{30} & \frac{3\sqrt{10}}{10} \\ -\frac{1}{2} & -\frac{\sqrt{3}}{2} & 0 & 0 & 0 \\ \frac{\sqrt{6}}{4} & -\frac{\sqrt{2}}{4} & -\frac{1}{2} & \frac{\sqrt{15}}{10} & -\frac{\sqrt{10}}{10} \end{bmatrix}$$

$$P_{24}^{56} \qquad \begin{bmatrix} -\frac{1}{2} & \frac{\sqrt{3}}{2} & 0 & 0 & 0 \\ \frac{\sqrt{3}}{6} & \frac{1}{6} & \frac{\sqrt{2}}{6} & -\frac{\sqrt{30}}{6} & 0 \\ -\frac{\sqrt{6}}{12} & -\frac{\sqrt{2}}{12} & -\frac{1}{6} & -\frac{\sqrt{15}}{30} & -\frac{3\sqrt{10}}{10} \\ \frac{1}{2} & \frac{\sqrt{3}}{6} & -\frac{\sqrt{6}}{3} & 0 & 0 \\ -\frac{\sqrt{6}}{4} & -\frac{\sqrt{2}}{4} & -\frac{1}{2} & -\frac{\sqrt{15}}{10} & \frac{\sqrt{10}}{10} \end{bmatrix}$$

$$P_{25}^{56} \qquad \begin{bmatrix} \frac{1}{2} & -\frac{\sqrt{3}}{2} & 0 & 0 & 0 \\ -\frac{\sqrt{3}}{6} & -\frac{1}{6} & -\frac{2\sqrt{2}}{3} & 0 & 0 \\ \frac{\sqrt{6}}{12} & \frac{\sqrt{2}}{12} & -\frac{1}{12} & -\frac{\sqrt{15}}{20} & \frac{3\sqrt{10}}{10} \\ -\frac{1}{2} & -\frac{\sqrt{3}}{6} & \frac{\sqrt{6}}{12} & -\frac{\sqrt{10}}{4} & 0 \\ \frac{\sqrt{6}}{4} & \frac{\sqrt{2}}{4} & -\frac{1}{4} & -\frac{3\sqrt{15}}{20} & -\frac{\sqrt{10}}{10} \end{bmatrix}$$

Table 4 (Cont.)

$$P^{56}_{26} = \begin{bmatrix} -\frac{1}{2} & \frac{\sqrt{3}}{2} & & & \\ \frac{\sqrt{3}}{6} & \frac{1}{6} & \frac{2\sqrt{2}}{3} & & \\ -\frac{\sqrt{6}}{12} & -\frac{\sqrt{2}}{12} & \frac{1}{12} & \frac{\sqrt{15}}{4} & \\ \frac{1}{2} & \frac{\sqrt{3}}{6} & -\frac{\sqrt{6}}{12} & \frac{\sqrt{10}}{20} & -\frac{\sqrt{15}}{5} \\ -\frac{\sqrt{6}}{4} & -\frac{\sqrt{2}}{4} & \frac{1}{4} & -\frac{\sqrt{15}}{20} & -\frac{\sqrt{10}}{5} \end{bmatrix} \qquad P^{56}_{34} = \begin{bmatrix} 1 & & & & \\ & \frac{1}{3} & -\frac{\sqrt{2}}{6} & \frac{\sqrt{30}}{6} & \\ & -\frac{\sqrt{2}}{6} & \frac{1}{6} & \frac{\sqrt{15}}{30} & \frac{3\sqrt{10}}{10} \\ & \frac{\sqrt{3}}{3} & \frac{\sqrt{6}}{3} & & \\ & -\frac{\sqrt{2}}{2} & \frac{1}{2} & \frac{\sqrt{15}}{10} & -\frac{\sqrt{10}}{10} \end{bmatrix}$$

$$P^{56}_{35} = \begin{bmatrix} -1 & & & & \\ & -\frac{1}{3} & \frac{2\sqrt{2}}{3} & & \\ & \frac{\sqrt{2}}{6} & \frac{1}{12} & \frac{\sqrt{15}}{20} & -\frac{3\sqrt{10}}{10} \\ & -\frac{\sqrt{3}}{3} & -\frac{\sqrt{6}}{12} & \frac{\sqrt{10}}{4} & \\ & \frac{\sqrt{2}}{2} & \frac{1}{4} & \frac{3\sqrt{15}}{20} & \frac{\sqrt{10}}{10} \end{bmatrix} \qquad P^{56}_{36} = \begin{bmatrix} 1 & & & & \\ & \frac{1}{3} & -\frac{2\sqrt{2}}{3} & & \\ & -\frac{\sqrt{2}}{6} & -\frac{1}{12} & -\frac{\sqrt{15}}{4} & \\ & \frac{\sqrt{3}}{3} & \frac{\sqrt{6}}{12} & -\frac{\sqrt{10}}{20} & \frac{\sqrt{15}}{5} \\ & -\frac{\sqrt{2}}{2} & -\frac{1}{4} & \frac{\sqrt{15}}{20} & \frac{\sqrt{10}}{5} \end{bmatrix}$$

$$P^{56}_{45} = \begin{bmatrix} 1 & & & & \\ & 1 & & & \\ & & \frac{1}{4} & -\frac{\sqrt{15}}{20} & \frac{3\sqrt{10}}{10} \\ & & -\frac{\sqrt{6}}{4} & -\frac{\sqrt{10}}{4} & \\ & & \frac{3}{4} & -\frac{3\sqrt{15}}{20} & -\frac{\sqrt{10}}{10} \end{bmatrix} \qquad P^{56}_{46} = \begin{bmatrix} -1 & & & & \\ & -1 & & & \\ & & -\frac{1}{4} & \frac{\sqrt{15}}{4} & \\ & & \frac{\sqrt{6}}{4} & \frac{\sqrt{10}}{20} & -\frac{\sqrt{15}}{5} \\ & & -\frac{3}{4} & -\frac{\sqrt{15}}{20} & -\frac{\sqrt{10}}{5} \end{bmatrix} \qquad P^{56}_{56} = \begin{bmatrix} 1 & & & & \\ & 1 & & & \\ & & 1 & & \\ & & & \frac{\sqrt{10}}{5} & \frac{\sqrt{15}}{5} \\ & & & -\frac{\sqrt{15}}{5} & \frac{\sqrt{10}}{5} \end{bmatrix}$$

APPENDIX 7

Tables of the Matrices $\langle (r_1'r_2')\lambda'\lambda'' | P_{ac}^{N-1\,N} | r_1 r_2 \rangle^{[\lambda]}$ for Values of *N* from 3 to 6

The matrices needed to calculate the interaction energy between two subsystems by the formula of Section 8.6 are given in this appendix. These matrices are calculated with the aid of the transformation matrices which take one from the standard representation to the nonstandard representation which is induced by the sets of basis functions on the left- and right-hand sides of the matrix element. According to the general formula (2.59),

$$\langle (r_1'r_2')\lambda'\lambda'' | P_{ac}^{N-1\,N} | r_1 r_2 \rangle^{[\lambda]} = \sum_{r,\bar{r}} \langle (r_1'r_2')\lambda'\lambda'' | r \rangle^{[\lambda]} \langle r | P_{ac}^{N-1\,N} | \bar{r} \rangle^{[\lambda]} \langle \bar{r} | r_1 r_2 \rangle^{[\lambda]}. \qquad \text{(A.2)}$$

The transformation matrices are calculated by the method described in Section 2.11, where examples of the determination of such matrices are given [see Eq. (2.64)]. The permutations $P_{ac}^{N-1\,N}$ send electron $N-1$ into orbital ϕ_a of the first subsystem and electron N into orbital ϕ_c of the second subsystem while preserving the increasing order in which the first $N-2$ electrons are numbered.

The Young tableaux which enumerate the rows and columns of the matrices are everywhere ordered according to the extent to which the numbers they contain deviate from the natural order (see Appendix 5). Blank spaces in the matrices denote zeros.

Table 1

$$N=3;\quad N_1=2,\quad N_2=1$$

$$[\lambda]=[21]$$

$$r_1:\quad [2] \qquad [1^2]$$

$$\lambda''$$

$$\langle P^{23}_{13}\rangle = \begin{matrix}[2]\\ [1^2]\end{matrix}\begin{bmatrix} \frac{1}{2} & -\frac{\sqrt{3}}{2} \\ -\frac{\sqrt{3}}{2} & -\frac{1}{2} \end{bmatrix} \qquad \langle P^{23}_{23}\rangle = \begin{bmatrix} \frac{1}{2} & \frac{\sqrt{3}}{2} \\ -\frac{\sqrt{3}}{2} & \frac{1}{2} \end{bmatrix}$$

Table 2

$$N=4;\quad N_1=2,\quad N_2=2$$

(a) $[\lambda]=[2^2]$

$$r_1:\quad [2] \qquad [1^2]$$

$$r_2:\quad [2] \qquad [1^2]$$

$$\lambda''$$

$$\langle P^{34}_{13}\rangle = \begin{matrix}[2]\\ [1^2]\end{matrix}\begin{bmatrix} -\frac{1}{2} & \frac{\sqrt{3}}{2} \\ \frac{\sqrt{3}}{2} & \frac{1}{2} \end{bmatrix} \qquad \langle P^{34}_{14}\rangle = \begin{bmatrix} -\frac{1}{2} & -\frac{\sqrt{3}}{2} \\ \frac{\sqrt{3}}{2} & -\frac{1}{2} \end{bmatrix}$$

$$\langle P^{34}_{23}\rangle = \begin{bmatrix} -\frac{1}{2} & -\frac{\sqrt{3}}{2} \\ \frac{\sqrt{3}}{2} & -\frac{1}{2} \end{bmatrix} \qquad \langle P^{34}_{24}\rangle = \begin{bmatrix} -\frac{1}{2} & \frac{\sqrt{3}}{2} \\ \frac{\sqrt{3}}{2} & \frac{1}{2} \end{bmatrix}$$

(b) $[\lambda]=[21^2]$

$$r_1:\quad [2] \quad [1^2] \qquad [1^2]$$

$$r_2:\quad [1^2] \quad [2] \qquad [1^2]$$

$$\lambda''$$

$$\langle P^{34}_{13}\rangle = \begin{matrix}[1^2]\\ [2]\\ [1^2]\end{matrix}\begin{bmatrix} \frac{1}{2} & -\frac{1}{2} & -\frac{\sqrt{2}}{2} \\ -\frac{1}{2} & \frac{1}{2} & -\frac{\sqrt{2}}{2} \\ \frac{\sqrt{2}}{2} & \frac{\sqrt{2}}{2} & \end{bmatrix} \qquad \langle P^{34}_{14}\rangle = \begin{bmatrix} -\frac{1}{2} & -\frac{1}{2} & \frac{\sqrt{2}}{2} \\ \frac{1}{2} & \frac{1}{2} & \frac{\sqrt{2}}{2} \\ -\frac{\sqrt{2}}{2} & \frac{\sqrt{2}}{2} & \end{bmatrix}$$

$$\langle P^{34}_{23}\rangle = \begin{bmatrix} \frac{1}{2} & \frac{1}{2} & \frac{\sqrt{2}}{2} \\ -\frac{1}{2} & -\frac{1}{2} & \frac{\sqrt{2}}{2} \\ \frac{\sqrt{2}}{2} & -\frac{\sqrt{2}}{2} & \end{bmatrix} \qquad \langle P^{34}_{24}\rangle = \begin{bmatrix} -\frac{1}{2} & \frac{1}{2} & -\frac{\sqrt{2}}{2} \\ \frac{1}{2} & -\frac{1}{2} & -\frac{\sqrt{2}}{2} \\ -\frac{\sqrt{2}}{2} & -\frac{\sqrt{2}}{2} & \end{bmatrix}$$

Table 3

$N = 5;\ N_1 = 3,\ N_2 = 2$

(a) $[\lambda] = [2^2 1]$

$$
\begin{array}{llllll}
r_1: & [21]_1 & [21]_1 & [21]_2 & [21]_2 & [1^3] \\
r_2: & [2] & [1^2] & [2] & [1^2] & [1^2] \\
\lambda'' & & & & &
\end{array}
$$

$$
\langle P_{14}^{45} \rangle =
\begin{array}{c}
[2] \\ [1^2] \\ [2] \\ [1^2] \\ [1^2]
\end{array}
\begin{bmatrix}
\frac{1}{4} & -\frac{\sqrt{3}}{4} & \frac{\sqrt{3}}{4} & -\frac{3}{4} & \\
\frac{\sqrt{3}}{4} & \frac{1}{4} & \frac{3}{4} & \frac{\sqrt{3}}{4} & \\
-\frac{\sqrt{3}}{4} & -\frac{1}{4} & \frac{1}{4} & \frac{\sqrt{3}}{12} & -\frac{\sqrt{6}}{3} \\
\frac{1}{4} & \frac{5\sqrt{3}}{12} & -\frac{\sqrt{3}}{12} & -\frac{5}{12} & -\frac{\sqrt{2}}{3} \\
\frac{\sqrt{2}}{2} & -\frac{\sqrt{6}}{6} & -\frac{\sqrt{6}}{6} & \frac{\sqrt{2}}{6} & -\frac{1}{3}
\end{bmatrix}
$$

$$
\langle P_{15}^{45} \rangle =
\begin{bmatrix}
\frac{1}{4} & \frac{\sqrt{3}}{4} & \frac{\sqrt{3}}{4} & \frac{3}{4} & \\
\frac{\sqrt{3}}{4} & -\frac{1}{4} & \frac{3}{4} & -\frac{\sqrt{3}}{4} & \\
-\frac{\sqrt{3}}{4} & \frac{1}{4} & \frac{1}{4} & -\frac{\sqrt{3}}{12} & \frac{\sqrt{6}}{3} \\
\frac{1}{4} & -\frac{5\sqrt{3}}{12} & -\frac{\sqrt{3}}{12} & \frac{5}{12} & \frac{\sqrt{2}}{3} \\
\frac{\sqrt{2}}{2} & \frac{\sqrt{6}}{6} & -\frac{\sqrt{6}}{6} & -\frac{\sqrt{2}}{6} & \frac{1}{3}
\end{bmatrix}
$$

$$
\langle P_{24}^{45} \rangle =
\begin{bmatrix}
\frac{1}{4} & -\frac{\sqrt{3}}{4} & -\frac{\sqrt{3}}{4} & \frac{3}{4} & \\
\frac{\sqrt{3}}{4} & \frac{1}{4} & -\frac{3}{4} & -\frac{\sqrt{3}}{4} & \\
-\frac{\sqrt{3}}{4} & -\frac{1}{4} & -\frac{1}{4} & -\frac{\sqrt{3}}{12} & \frac{\sqrt{6}}{3} \\
\frac{1}{4} & \frac{5\sqrt{3}}{12} & \frac{\sqrt{3}}{12} & \frac{5}{12} & \frac{\sqrt{2}}{3} \\
\frac{\sqrt{2}}{2} & -\frac{\sqrt{6}}{6} & \frac{\sqrt{6}}{6} & -\frac{\sqrt{2}}{6} & \frac{1}{3}
\end{bmatrix}
$$

$$
\langle P_{25}^{45} \rangle =
\begin{bmatrix}
\frac{1}{4} & \frac{\sqrt{3}}{4} & -\frac{\sqrt{3}}{4} & -\frac{3}{4} & \\
\frac{\sqrt{3}}{4} & -\frac{1}{4} & -\frac{3}{4} & \frac{\sqrt{3}}{4} & \\
-\frac{\sqrt{3}}{4} & \frac{1}{4} & -\frac{1}{4} & \frac{\sqrt{3}}{12} & -\frac{\sqrt{6}}{3} \\
\frac{1}{4} & -\frac{5\sqrt{3}}{12} & \frac{\sqrt{3}}{12} & -\frac{5}{12} & -\frac{\sqrt{2}}{3} \\
\frac{\sqrt{2}}{2} & \frac{\sqrt{6}}{6} & \frac{\sqrt{6}}{6} & \frac{\sqrt{2}}{6} & -\frac{1}{3}
\end{bmatrix}
$$

Table 3 (Cont.)

$$\langle P_{34}^{45}\rangle = \begin{bmatrix} -\frac{1}{2} & \frac{\sqrt{3}}{2} & & & \\ -\frac{\sqrt{3}}{2} & -\frac{1}{2} & & & \\ & & -\frac{1}{2} & -\frac{\sqrt{3}}{6} & -\frac{\sqrt{6}}{3} \\ & & \frac{\sqrt{3}}{6} & \frac{5}{6} & -\frac{\sqrt{2}}{3} \\ & & \frac{\sqrt{6}}{3} & -\frac{\sqrt{2}}{3} & -\frac{1}{3} \end{bmatrix} \quad \langle P_{35}^{45}\rangle = \begin{bmatrix} -\frac{1}{2} & -\frac{\sqrt{3}}{2} & & & \\ -\frac{\sqrt{3}}{2} & \frac{1}{2} & & & \\ & & -\frac{1}{2} & \frac{\sqrt{3}}{6} & \frac{\sqrt{6}}{3} \\ & & \frac{\sqrt{3}}{6} & -\frac{5}{6} & \frac{\sqrt{2}}{3} \\ & & \frac{\sqrt{6}}{3} & \frac{\sqrt{2}}{3} & \frac{1}{3} \end{bmatrix}$$

(b) $[\lambda] = [21^3]$

r_1:	$[21]_1$	$[21]_2$	$[1^3]$	$[1^3]$
r_2:	$[1^2]$	$[1^2]$	$[2]$	$[1^2]$

λ''

$$\begin{matrix}[1^2]\\ [1^2]\\ [2]\\ [1^2]\end{matrix} \quad \langle P_{14}^{45}\rangle = \begin{bmatrix} -\frac{1}{2} & -\frac{\sqrt{3}}{2} & & \\ \frac{\sqrt{3}}{6} & -\frac{1}{6} & \frac{\sqrt{3}}{3} & \frac{\sqrt{5}}{3} \\ -\frac{1}{2} & \frac{\sqrt{3}}{6} & -\frac{1}{2} & \frac{\sqrt{15}}{6} \\ \frac{\sqrt{15}}{6} & -\frac{\sqrt{5}}{6} & -\frac{\sqrt{15}}{6} & \frac{1}{6} \end{bmatrix} \quad \langle P_{15}^{45}\rangle = \begin{bmatrix} \frac{1}{2} & \frac{\sqrt{3}}{2} & & \\ -\frac{\sqrt{3}}{6} & \frac{1}{6} & \frac{\sqrt{3}}{3} & -\frac{\sqrt{5}}{3} \\ \frac{1}{2} & -\frac{\sqrt{3}}{6} & -\frac{1}{2} & -\frac{\sqrt{15}}{6} \\ -\frac{\sqrt{15}}{6} & \frac{\sqrt{5}}{6} & -\frac{\sqrt{15}}{6} & -\frac{1}{6} \end{bmatrix}$$

$$\langle P_{24}^{45}\rangle = \begin{bmatrix} -\frac{1}{2} & \frac{\sqrt{3}}{2} & & \\ \frac{\sqrt{3}}{6} & \frac{1}{6} & -\frac{\sqrt{3}}{3} & -\frac{\sqrt{5}}{3} \\ -\frac{1}{2} & -\frac{\sqrt{3}}{6} & \frac{1}{2} & -\frac{\sqrt{15}}{6} \\ \frac{\sqrt{15}}{6} & \frac{\sqrt{5}}{6} & \frac{\sqrt{15}}{6} & -\frac{1}{6} \end{bmatrix} \quad \langle P_{25}^{45}\rangle = \begin{bmatrix} \frac{1}{2} & -\frac{\sqrt{3}}{2} & & \\ -\frac{\sqrt{3}}{6} & -\frac{1}{6} & -\frac{\sqrt{3}}{3} & \frac{\sqrt{5}}{3} \\ \frac{1}{2} & \frac{\sqrt{3}}{6} & \frac{1}{2} & \frac{\sqrt{15}}{6} \\ -\frac{\sqrt{15}}{6} & -\frac{\sqrt{5}}{6} & \frac{\sqrt{15}}{6} & \frac{1}{6} \end{bmatrix}$$

$$\langle P_{34}^{45}\rangle = \begin{bmatrix} 1 & & & \\ & \frac{1}{3} & \frac{\sqrt{3}}{3} & \frac{\sqrt{5}}{3} \\ & -\frac{\sqrt{3}}{3} & -\frac{1}{2} & \frac{\sqrt{15}}{6} \\ & \frac{\sqrt{5}}{3} & -\frac{\sqrt{15}}{6} & \frac{1}{6} \end{bmatrix} \quad \langle P_{35}^{45}\rangle = \begin{bmatrix} -1 & & & \\ & -\frac{1}{3} & \frac{\sqrt{3}}{3} & -\frac{\sqrt{5}}{3} \\ & \frac{\sqrt{3}}{3} & -\frac{1}{2} & -\frac{\sqrt{15}}{6} \\ & -\frac{\sqrt{5}}{6} & -\frac{\sqrt{15}}{6} & -\frac{1}{6} \end{bmatrix}$$

Table 4

$N=6;\quad N_1=4,\quad N_2=2$

(a) $[\lambda]=[2^3]$

r_1:	$[2^2]_1$	$[21^2]_1$	$[2^2]_2$	$[21^2]_2$	$[21^2]_3$
r_2:	$[2]$	$[1^2]$	$[2]$	$[1^2]$	$[1^2]$
λ''					

$$\langle P_{15}^{56}\rangle = \begin{matrix} [2] \\ [1^2] \\ [2] \\ [1^2] \\ [1^2] \end{matrix} \begin{bmatrix} \frac{1}{4} & -\frac{\sqrt{3}}{4} & \frac{\sqrt{3}}{4} & -\frac{3}{4} & \\ -\frac{\sqrt{3}}{4} & -\frac{1}{4} & -\frac{3}{4} & -\frac{\sqrt{3}}{4} & \\ \frac{\sqrt{3}}{4} & \frac{1}{4} & -\frac{1}{4} & -\frac{\sqrt{3}}{12} & -\frac{\sqrt{6}}{3} \\ -\frac{3}{4} & \frac{\sqrt{3}}{12} & \frac{\sqrt{3}}{4} & -\frac{1}{12} & -\frac{\sqrt{2}}{3} \\ & \frac{\sqrt{6}}{3} & & -\frac{\sqrt{2}}{3} & \frac{1}{3} \end{bmatrix}$$

$$\langle P_{16}^{56}\rangle = \begin{bmatrix} \frac{1}{4} & \frac{\sqrt{3}}{4} & \frac{\sqrt{3}}{4} & \frac{3}{4} & \\ -\frac{\sqrt{3}}{4} & \frac{1}{4} & -\frac{3}{4} & \frac{\sqrt{3}}{4} & \\ \frac{\sqrt{3}}{4} & -\frac{1}{4} & -\frac{1}{4} & \frac{\sqrt{3}}{12} & \frac{\sqrt{6}}{3} \\ -\frac{3}{4} & -\frac{\sqrt{3}}{12} & \frac{\sqrt{3}}{4} & \frac{1}{12} & \frac{\sqrt{2}}{3} \\ & -\frac{\sqrt{6}}{3} & & \frac{\sqrt{2}}{3} & -\frac{1}{3} \end{bmatrix}$$

$$\langle P_{25}^{56}\rangle = \begin{bmatrix} \frac{1}{4} & -\frac{\sqrt{3}}{4} & -\frac{\sqrt{3}}{4} & \frac{3}{4} & \\ -\frac{\sqrt{3}}{4} & -\frac{1}{4} & \frac{3}{4} & \frac{\sqrt{3}}{4} & \\ \frac{\sqrt{3}}{4} & \frac{1}{4} & \frac{1}{4} & \frac{\sqrt{3}}{12} & \frac{\sqrt{6}}{3} \\ -\frac{3}{4} & \frac{\sqrt{3}}{12} & -\frac{\sqrt{3}}{4} & \frac{1}{12} & \frac{\sqrt{2}}{3} \\ & \frac{\sqrt{6}}{3} & & \frac{\sqrt{2}}{3} & -\frac{1}{3} \end{bmatrix}$$

$$\langle P_{26}^{56}\rangle = \begin{bmatrix} \frac{1}{4} & \frac{\sqrt{3}}{4} & -\frac{\sqrt{3}}{4} & -\frac{3}{4} & \\ -\frac{\sqrt{3}}{4} & \frac{1}{4} & \frac{3}{4} & -\frac{\sqrt{3}}{4} & \\ \frac{\sqrt{3}}{4} & -\frac{1}{4} & \frac{1}{4} & -\frac{\sqrt{3}}{12} & -\frac{\sqrt{6}}{3} \\ -\frac{3}{4} & -\frac{\sqrt{3}}{12} & -\frac{\sqrt{3}}{4} & -\frac{1}{12} & -\frac{\sqrt{2}}{3} \\ & -\frac{\sqrt{6}}{3} & & -\frac{\sqrt{2}}{3} & \frac{1}{3} \end{bmatrix}$$

Table 4 (Cont.)

$$\langle P_{35}^{56}\rangle=\begin{bmatrix} -\frac{1}{2} & \frac{\sqrt{3}}{2} & & & \\ \frac{\sqrt{3}}{2} & \frac{1}{2} & & & \\ & & \frac{1}{2} & \frac{\sqrt{3}}{6} & -\frac{\sqrt{6}}{3} \\ & & -\frac{\sqrt{3}}{2} & \frac{1}{6} & -\frac{\sqrt{2}}{3} \\ & & & \frac{2\sqrt{2}}{3} & \frac{1}{3} \end{bmatrix}$$

$$\langle P_{36}^{56}\rangle=\begin{bmatrix} -\frac{1}{2} & -\frac{\sqrt{3}}{2} & & & \\ \frac{\sqrt{3}}{2} & -\frac{1}{2} & & & \\ & & \frac{1}{2} & -\frac{\sqrt{3}}{6} & \frac{\sqrt{6}}{3} \\ & & -\frac{\sqrt{3}}{2} & -\frac{1}{6} & \frac{\sqrt{2}}{3} \\ & & & -\frac{2\sqrt{2}}{3} & -\frac{1}{3} \end{bmatrix}$$

$$\langle P_{45}^{56}\rangle=\begin{bmatrix} -\frac{1}{2} & -\frac{\sqrt{3}}{2} & & & \\ \frac{\sqrt{3}}{2} & -\frac{1}{2} & & & \\ & & -\frac{1}{2} & -\frac{\sqrt{3}}{2} & \\ & & \frac{\sqrt{3}}{2} & -\frac{1}{2} & \\ & & & & 1 \end{bmatrix}$$

$$\langle P_{46}^{56}\rangle=\begin{bmatrix} -\frac{1}{2} & \frac{\sqrt{3}}{2} & & & \\ \frac{\sqrt{3}}{2} & \frac{1}{2} & & & \\ & & -\frac{1}{2} & \frac{\sqrt{3}}{2} & \\ & & \frac{\sqrt{3}}{2} & \frac{1}{2} & \\ & & & & -1 \end{bmatrix}$$

Table 4 (Cont.)

(b) $[\lambda] = [21^4]$

r_1:	$[21^2]_1$	$[21^2]_2$	$[21^2]_3$	$[1^4]$	$[1^4]$
r_2:	$[1^2]$	$[1^2]$	$[1^2]$	$[2]$	$[1^2]$
λ''					

$$\langle P_{15}^{56} \rangle = \begin{matrix} [1^2] \\ [1^2] \\ [1^2] \\ [2] \\ [1^2] \end{matrix} \begin{bmatrix} \frac{1}{2} & \frac{\sqrt{3}}{2} & & & \\ -\frac{\sqrt{3}}{6} & \frac{1}{6} & \frac{2\sqrt{2}}{3} & & \\ \frac{\sqrt{6}}{12} & -\frac{\sqrt{2}}{12} & \frac{1}{12} & -\frac{\sqrt{6}}{4} & -\frac{3}{4} \\ -\frac{1}{2} & \frac{\sqrt{3}}{6} & -\frac{\sqrt{6}}{12} & \frac{1}{2} & -\frac{\sqrt{6}}{4} \\ \frac{\sqrt{6}}{4} & -\frac{\sqrt{2}}{4} & \frac{1}{4} & \frac{\sqrt{6}}{4} & -\frac{1}{4} \end{bmatrix}$$

$$\langle P_{16}^{56} \rangle = \begin{bmatrix} -\frac{1}{2} & -\frac{\sqrt{3}}{2} & & & \\ \frac{\sqrt{3}}{6} & -\frac{1}{6} & -\frac{2\sqrt{2}}{3} & & \\ -\frac{\sqrt{6}}{12} & \frac{\sqrt{2}}{12} & -\frac{1}{12} & -\frac{\sqrt{6}}{4} & \frac{3}{4} \\ \frac{1}{2} & -\frac{\sqrt{3}}{6} & \frac{\sqrt{6}}{12} & \frac{1}{2} & -\frac{\sqrt{6}}{4} \\ -\frac{\sqrt{6}}{4} & \frac{\sqrt{2}}{4} & -\frac{1}{4} & \frac{\sqrt{6}}{4} & \frac{1}{4} \end{bmatrix}$$

$$\langle P_{25}^{56} \rangle = \begin{bmatrix} \frac{1}{2} & -\frac{\sqrt{3}}{2} & & & \\ -\frac{\sqrt{3}}{6} & -\frac{1}{6} & -\frac{2\sqrt{2}}{3} & & \\ \frac{\sqrt{6}}{12} & \frac{\sqrt{2}}{12} & -\frac{1}{12} & \frac{\sqrt{6}}{4} & \frac{3}{4} \\ -\frac{1}{2} & -\frac{\sqrt{3}}{6} & \frac{\sqrt{6}}{12} & -\frac{1}{2} & \frac{\sqrt{6}}{4} \\ \frac{\sqrt{6}}{4} & \frac{\sqrt{2}}{4} & -\frac{1}{4} & -\frac{\sqrt{6}}{4} & \frac{1}{4} \end{bmatrix}$$

$$\langle P_{26}^{56} \rangle = \begin{bmatrix} -\frac{1}{2} & \frac{\sqrt{3}}{2} & & & \\ \frac{\sqrt{3}}{6} & \frac{1}{6} & \frac{2\sqrt{2}}{3} & & \\ -\frac{\sqrt{6}}{12} & -\frac{\sqrt{2}}{12} & \frac{1}{12} & \frac{\sqrt{6}}{4} & -\frac{3}{4} \\ \frac{1}{2} & \frac{\sqrt{3}}{6} & -\frac{\sqrt{6}}{12} & -\frac{1}{2} & -\frac{\sqrt{6}}{4} \\ -\frac{\sqrt{6}}{4} & -\frac{\sqrt{2}}{4} & \frac{1}{4} & -\frac{\sqrt{6}}{4} & -\frac{1}{4} \end{bmatrix}$$

Table 4 (Cont.)

$$\langle P^{56}_{35}\rangle = \begin{bmatrix} -1 & & & & \\ & -\frac{1}{3} & \frac{2\sqrt{2}}{3} & & \\ & \frac{\sqrt{2}}{6} & \frac{1}{12} & -\frac{\sqrt{6}}{4} & -\frac{3}{4} \\ & -\frac{\sqrt{3}}{3} & -\frac{\sqrt{6}}{12} & \frac{1}{2} & -\frac{\sqrt{6}}{4} \\ & \frac{\sqrt{2}}{2} & \frac{1}{4} & \frac{\sqrt{6}}{4} & -\frac{1}{4} \end{bmatrix}$$

$$\langle P^{56}_{36}\rangle = \begin{bmatrix} 1 & & & & \\ & \frac{1}{3} & -\frac{2\sqrt{2}}{3} & & \\ & -\frac{\sqrt{2}}{6} & -\frac{1}{12} & -\frac{\sqrt{6}}{4} & \frac{3}{4} \\ & \frac{\sqrt{3}}{3} & \frac{\sqrt{6}}{12} & \frac{1}{2} & \frac{\sqrt{6}}{4} \\ & -\frac{\sqrt{2}}{2} & -\frac{1}{4} & \frac{\sqrt{6}}{4} & \frac{1}{4} \end{bmatrix}$$

$$\langle P^{56}_{45}\rangle = \begin{bmatrix} 1 & & & & \\ & 1 & & & \\ & & \frac{1}{4} & \frac{\sqrt{6}}{4} & \frac{3}{4} \\ & & -\frac{\sqrt{6}}{4} & -\frac{1}{2} & \frac{\sqrt{6}}{4} \\ & & \frac{3}{4} & -\frac{\sqrt{6}}{4} & \frac{1}{4} \end{bmatrix}$$

$$\langle P^{56}_{46}\rangle = \begin{bmatrix} -1 & & & & \\ & -1 & & & \\ & & -\frac{1}{4} & \frac{\sqrt{6}}{4} & -\frac{3}{4} \\ & & \frac{\sqrt{6}}{4} & -\frac{1}{2} & -\frac{\sqrt{6}}{4} \\ & & -\frac{3}{4} & -\frac{\sqrt{6}}{4} & -\frac{1}{4} \end{bmatrix}$$

Table 4 (Cont.)

(c) $[\lambda] = [2^2 1^2]$

$\langle P_{15}^{56} \rangle =$

r_1:	$[2^2]_1$	$[21^2]_1$	$[21^2]_1$	$[2^2]_2$	$[21^2]_2$	$[21^2]_2$	$[21^2]_3$	$[21^2]_3$	$[1^4]$
r_2:	$[1^2]$	$[2]$	$[1^2]$	$[1^2]$	$[2]$	$[1^2]$	$[2]$	$[1^2]$	$[1^2]$
λ''									
$[1^2]$	$-\frac{1}{4}$	$\frac{1}{4}$	$\frac{\sqrt{2}}{4}$	$-\frac{\sqrt{3}}{4}$	$\frac{\sqrt{3}}{4}$	$\frac{\sqrt{6}}{4}$			
$[2]$	$\frac{1}{4}$	$-\frac{1}{4}$	$\frac{\sqrt{2}}{4}$	$\frac{\sqrt{3}}{4}$	$-\frac{\sqrt{3}}{4}$	$\frac{\sqrt{6}}{4}$			
$[1^2]$	$-\frac{\sqrt{2}}{4}$	$-\frac{\sqrt{2}}{4}$		$-\frac{\sqrt{6}}{4}$	$-\frac{\sqrt{6}}{4}$				
$[1^2]$	$-\frac{\sqrt{3}}{4}$	$-\frac{\sqrt{3}}{12}$	$-\frac{\sqrt{6}}{12}$	$\frac{1}{4}$	$\frac{1}{12}$	$\frac{\sqrt{2}}{12}$	$\frac{\sqrt{2}}{3}$	$\frac{2}{3}$	
$[2]$	$\frac{\sqrt{3}}{4}$	$\frac{\sqrt{3}}{12}$	$-\frac{\sqrt{6}}{12}$	$-\frac{1}{4}$	$-\frac{1}{12}$	$\frac{\sqrt{2}}{12}$	$-\frac{\sqrt{2}}{3}$	$\frac{2}{3}$	
$[1^2]$	$-\frac{\sqrt{6}}{4}$	$\frac{\sqrt{6}}{12}$		$\frac{\sqrt{2}}{4}$	$-\frac{\sqrt{2}}{12}$		$-\frac{2}{3}$		
$[2]$		$-\frac{\sqrt{6}}{6}$	$-\frac{\sqrt{3}}{6}$		$\frac{\sqrt{2}}{6}$	$\frac{1}{6}$	$-\frac{1}{6}$	$-\frac{\sqrt{2}}{12}$	$\frac{\sqrt{10}}{4}$
$[1^2]$		$\frac{\sqrt{3}}{6}$	$\frac{\sqrt{6}}{4}$		$-\frac{1}{6}$	$-\frac{\sqrt{2}}{4}$	$\frac{\sqrt{2}}{12}$	$\frac{1}{4}$	$\frac{\sqrt{5}}{4}$
$[1^2]$		$\frac{\sqrt{15}}{6}$	$-\frac{\sqrt{30}}{12}$		$-\frac{\sqrt{5}}{6}$	$\frac{\sqrt{10}}{12}$	$\frac{\sqrt{10}}{12}$	$-\frac{\sqrt{5}}{12}$	$\frac{1}{4}$

$\langle P_{16}^{56} \rangle =$

$\frac{1}{4}$	$\frac{1}{4}$	$-\frac{\sqrt{2}}{4}$	$\frac{\sqrt{3}}{4}$	$\frac{\sqrt{3}}{4}$	$-\frac{\sqrt{6}}{4}$			
$-\frac{1}{4}$	$-\frac{1}{4}$	$-\frac{\sqrt{2}}{4}$	$-\frac{\sqrt{3}}{4}$	$-\frac{\sqrt{3}}{4}$	$-\frac{\sqrt{6}}{4}$			
$\frac{\sqrt{2}}{4}$	$-\frac{\sqrt{2}}{4}$		$\frac{\sqrt{6}}{4}$	$-\frac{\sqrt{6}}{4}$				
$\frac{\sqrt{3}}{4}$	$-\frac{\sqrt{3}}{12}$	$\frac{\sqrt{6}}{12}$	$-\frac{1}{4}$	$\frac{1}{12}$	$-\frac{\sqrt{2}}{12}$	$\frac{\sqrt{2}}{3}$	$-\frac{2}{3}$	
$-\frac{\sqrt{3}}{4}$	$\frac{\sqrt{3}}{12}$	$\frac{\sqrt{6}}{12}$	$\frac{1}{4}$	$-\frac{1}{12}$	$-\frac{\sqrt{2}}{12}$	$-\frac{\sqrt{2}}{3}$	$-\frac{2}{3}$	
$\frac{\sqrt{6}}{4}$	$\frac{\sqrt{6}}{12}$		$-\frac{\sqrt{2}}{4}$	$-\frac{\sqrt{2}}{12}$		$-\frac{2}{3}$		
	$-\frac{\sqrt{6}}{6}$	$\frac{\sqrt{3}}{6}$		$\frac{\sqrt{2}}{6}$	$-\frac{1}{6}$	$-\frac{1}{6}$	$\frac{\sqrt{2}}{12}$	$-\frac{\sqrt{10}}{4}$
	$\frac{\sqrt{3}}{6}$	$-\frac{\sqrt{6}}{4}$		$-\frac{1}{6}$	$\frac{\sqrt{2}}{4}$	$\frac{\sqrt{2}}{12}$	$-\frac{1}{4}$	$-\frac{\sqrt{5}}{4}$
	$\frac{\sqrt{15}}{6}$	$\frac{\sqrt{30}}{12}$		$-\frac{\sqrt{5}}{6}$	$-\frac{\sqrt{10}}{12}$	$\frac{\sqrt{10}}{12}$	$\frac{\sqrt{5}}{12}$	$-\frac{1}{4}$

Table 4 (Cont.)

$$\left\langle P_{25}^{56}\right\rangle = \begin{bmatrix}
-\frac{1}{4} & \frac{1}{4} & \frac{\sqrt{2}}{4} & \frac{\sqrt{3}}{4} & -\frac{\sqrt{3}}{4} & -\frac{\sqrt{6}}{4} & & & \\
\frac{1}{4} & -\frac{1}{4} & \frac{\sqrt{2}}{4} & -\frac{\sqrt{3}}{4} & \frac{\sqrt{3}}{4} & -\frac{\sqrt{6}}{4} & & & \\
-\frac{\sqrt{2}}{4} & -\frac{\sqrt{2}}{4} & & \frac{\sqrt{6}}{4} & \frac{\sqrt{6}}{4} & & & & \\
-\frac{\sqrt{3}}{4} & -\frac{\sqrt{3}}{12} & -\frac{\sqrt{6}}{12} & -\frac{1}{4} & -\frac{1}{12} & -\frac{\sqrt{2}}{12} & -\frac{\sqrt{2}}{3} & -\frac{2}{3} & \\
\frac{\sqrt{3}}{4} & -\frac{\sqrt{3}}{12} & -\frac{\sqrt{6}}{12} & \frac{1}{4} & \frac{1}{12} & -\frac{\sqrt{2}}{12} & \frac{\sqrt{2}}{3} & -\frac{2}{3} & \\
-\frac{\sqrt{6}}{4} & \frac{\sqrt{6}}{12} & & -\frac{\sqrt{2}}{4} & \frac{\sqrt{2}}{12} & & \frac{2}{3} & & \\
 & -\frac{\sqrt{6}}{6} & -\frac{\sqrt{3}}{6} & & -\frac{\sqrt{2}}{6} & -\frac{1}{6} & \frac{1}{6} & \frac{\sqrt{2}}{12} & -\frac{\sqrt{10}}{4} \\
 & \frac{\sqrt{3}}{6} & \frac{\sqrt{6}}{4} & & \frac{1}{6} & \frac{\sqrt{2}}{4} & -\frac{\sqrt{2}}{12} & -\frac{1}{4} & -\frac{\sqrt{5}}{4} \\
 & \frac{\sqrt{15}}{6} & -\frac{\sqrt{30}}{12} & & \frac{\sqrt{5}}{6} & -\frac{\sqrt{10}}{12} & -\frac{\sqrt{10}}{12} & \frac{\sqrt{5}}{12} & -\frac{1}{4}
\end{bmatrix}$$

$$\left\langle P_{26}^{56}\right\rangle = \begin{bmatrix}
\frac{1}{4} & \frac{1}{4} & -\frac{\sqrt{2}}{4} & -\frac{\sqrt{3}}{4} & -\frac{\sqrt{3}}{4} & \frac{\sqrt{6}}{4} & & & \\
-\frac{1}{4} & -\frac{1}{4} & -\frac{\sqrt{2}}{4} & \frac{\sqrt{3}}{4} & \frac{\sqrt{3}}{4} & \frac{\sqrt{6}}{4} & & & \\
\frac{\sqrt{2}}{4} & -\frac{\sqrt{2}}{4} & & -\frac{\sqrt{6}}{4} & \frac{\sqrt{6}}{4} & & & & \\
\frac{\sqrt{3}}{4} & -\frac{\sqrt{3}}{12} & \frac{\sqrt{6}}{12} & \frac{1}{4} & -\frac{1}{12} & \frac{\sqrt{2}}{12} & -\frac{\sqrt{2}}{3} & \frac{2}{3} & \\
-\frac{\sqrt{3}}{4} & \frac{\sqrt{3}}{12} & \frac{\sqrt{6}}{12} & -\frac{1}{4} & \frac{1}{12} & \frac{\sqrt{2}}{12} & \frac{\sqrt{2}}{3} & \frac{2}{3} & \\
\frac{\sqrt{6}}{4} & \frac{\sqrt{6}}{12} & & \frac{\sqrt{2}}{4} & \frac{\sqrt{2}}{12} & & \frac{2}{3} & & \\
 & -\frac{\sqrt{6}}{6} & \frac{\sqrt{3}}{6} & & -\frac{\sqrt{2}}{6} & \frac{1}{6} & \frac{1}{6} & -\frac{\sqrt{2}}{12} & \frac{\sqrt{10}}{4} \\
 & \frac{\sqrt{3}}{6} & -\frac{\sqrt{6}}{4} & & \frac{1}{6} & -\frac{\sqrt{2}}{4} & -\frac{\sqrt{2}}{12} & \frac{1}{4} & \frac{\sqrt{5}}{4} \\
 & \frac{\sqrt{15}}{6} & \frac{\sqrt{30}}{12} & & \frac{\sqrt{5}}{6} & \frac{\sqrt{10}}{12} & -\frac{\sqrt{10}}{12} & -\frac{\sqrt{5}}{12} & \frac{1}{4}
\end{bmatrix}$$

Table 4 (Cont.)

$$\langle P_{35}^{56} \rangle = \begin{bmatrix} \frac{1}{2} & -\frac{1}{2} & -\frac{\sqrt{2}}{2} & & & & & & \\ -\frac{1}{2} & \frac{1}{2} & -\frac{\sqrt{2}}{2} & & & & & & \\ \frac{\sqrt{2}}{2} & \frac{\sqrt{2}}{2} & & & & & & & \\ & & & -\frac{1}{2} & -\frac{1}{6} & -\frac{\sqrt{2}}{6} & \frac{\sqrt{2}}{3} & \frac{2}{3} & \\ & & & \frac{1}{2} & \frac{1}{6} & -\frac{\sqrt{2}}{6} & -\frac{\sqrt{2}}{3} & \frac{2}{3} & \\ & & & -\frac{\sqrt{2}}{2} & \frac{\sqrt{2}}{6} & & -\frac{2}{3} & & \\ & & & & -\frac{\sqrt{2}}{3} & -\frac{1}{3} & -\frac{1}{6} & -\frac{\sqrt{2}}{12} & \frac{\sqrt{10}}{4} \\ & & & & \frac{1}{3} & \frac{\sqrt{2}}{2} & \frac{\sqrt{2}}{12} & \frac{1}{4} & \frac{\sqrt{5}}{4} \\ & & & & \frac{\sqrt{5}}{3} & -\frac{\sqrt{10}}{6} & \frac{\sqrt{10}}{12} & -\frac{\sqrt{5}}{12} & \frac{1}{4} \end{bmatrix}$$

$$\langle P_{36}^{56} \rangle = \begin{bmatrix} -\frac{1}{2} & -\frac{1}{2} & \frac{\sqrt{2}}{2} & & & & & & \\ \frac{1}{2} & \frac{1}{2} & \frac{\sqrt{2}}{2} & & & & & & \\ -\frac{\sqrt{2}}{2} & \frac{\sqrt{2}}{2} & & & & & & & \\ & & & \frac{1}{2} & -\frac{1}{6} & \frac{\sqrt{2}}{6} & \frac{\sqrt{2}}{3} & -\frac{2}{3} & \\ & & & -\frac{1}{2} & \frac{1}{6} & \frac{\sqrt{2}}{6} & -\frac{\sqrt{2}}{3} & -\frac{2}{3} & \\ & & & \frac{\sqrt{2}}{2} & \frac{\sqrt{2}}{6} & & -\frac{2}{3} & & \\ & & & & -\frac{\sqrt{2}}{3} & \frac{1}{3} & -\frac{1}{6} & \frac{\sqrt{2}}{12} & -\frac{\sqrt{10}}{4} \\ & & & & \frac{1}{3} & -\frac{\sqrt{2}}{2} & \frac{\sqrt{2}}{12} & -\frac{1}{4} & -\frac{\sqrt{5}}{4} \\ & & & & \frac{\sqrt{5}}{3} & \frac{\sqrt{10}}{6} & \frac{\sqrt{10}}{12} & \frac{\sqrt{5}}{12} & -\frac{1}{4} \end{bmatrix}$$

Table 4 (Cont.)

$$
\langle P_{45}^{56} \rangle = \begin{bmatrix}
\frac{1}{2} & \frac{1}{2} & \frac{\sqrt{2}}{2} & & & & & & \\
-\frac{1}{2} & -\frac{1}{2} & \frac{\sqrt{2}}{2} & & & & & & \\
\frac{\sqrt{2}}{2} & -\frac{\sqrt{2}}{2} & & & & & & & \\
& & & \frac{1}{2} & \frac{1}{2} & \frac{\sqrt{2}}{2} & & & \\
& & & -\frac{1}{2} & -\frac{1}{2} & \frac{\sqrt{2}}{2} & & & \\
& & & \frac{\sqrt{2}}{2} & -\frac{\sqrt{2}}{2} & & & & \\
& & & & & & -\frac{1}{2} & -\frac{\sqrt{2}}{4} & -\frac{\sqrt{10}}{4} \\
& & & & & & \frac{\sqrt{2}}{4} & \frac{3}{4} & -\frac{\sqrt{5}}{4} \\
& & & & & & \frac{\sqrt{10}}{4} & -\frac{\sqrt{5}}{4} & -\frac{1}{4}
\end{bmatrix}
$$

$$
\langle P_{46}^{56} \rangle = \begin{bmatrix}
-\frac{1}{2} & \frac{1}{2} & -\frac{\sqrt{2}}{2} & & & & & & \\
\frac{1}{2} & -\frac{1}{2} & -\frac{\sqrt{2}}{2} & & & & & & \\
-\frac{\sqrt{2}}{2} & -\frac{\sqrt{2}}{2} & & & & & & & \\
& & & -\frac{1}{2} & \frac{1}{2} & -\frac{\sqrt{2}}{2} & & & \\
& & & \frac{1}{2} & -\frac{1}{2} & -\frac{\sqrt{2}}{2} & & & \\
& & & -\frac{\sqrt{2}}{2} & -\frac{\sqrt{2}}{2} & & & & \\
& & & & & & -\frac{1}{2} & \frac{\sqrt{2}}{4} & \frac{\sqrt{10}}{4} \\
& & & & & & \frac{\sqrt{2}}{4} & -\frac{3}{4} & \frac{\sqrt{5}}{4} \\
& & & & & & \frac{\sqrt{10}}{4} & \frac{\sqrt{5}}{4} & \frac{1}{4}
\end{bmatrix}
$$

Table 5

$N=6;\quad N_1=3,\quad N_2=3$

(a) $[\lambda]=[2^3]$

$$\begin{array}{llllll} r_1: & [21]_1 & [21]_1 & [21]_2 & [21]_2 & [1^3] \\ r_2: & [21]_1 & [21]_2 & [21]_1 & [21]_2 & [1^3] \\ \lambda'' & & & & & \end{array}$$

$$\langle P_{14}^{56} \rangle = \begin{array}{c} [2] \\ [1^2] \\ [2] \\ [1^2] \\ [1^2] \end{array} \begin{bmatrix} -\frac{1}{4} & -\frac{\sqrt{3}}{4} & -\frac{\sqrt{3}}{4} & -\frac{3}{4} & \\ -\frac{\sqrt{3}}{4} & \frac{1}{4} & -\frac{3}{4} & \frac{\sqrt{3}}{4} & \\ \frac{\sqrt{3}}{4} & -\frac{1}{4} & -\frac{1}{4} & \frac{\sqrt{3}}{12} & \frac{\sqrt{6}}{3} \\ \frac{\sqrt{3}}{4} & \frac{3}{4} & -\frac{1}{4} & -\frac{\sqrt{3}}{4} & \\ \frac{\sqrt{6}}{4} & -\frac{\sqrt{2}}{4} & -\frac{\sqrt{2}}{4} & \frac{\sqrt{6}}{12} & -\frac{\sqrt{3}}{3} \end{bmatrix}$$

$$\langle P_{15}^{56} \rangle = \begin{bmatrix} -\frac{1}{4} & \frac{\sqrt{3}}{4} & -\frac{\sqrt{3}}{4} & \frac{3}{4} & \\ -\frac{\sqrt{3}}{4} & -\frac{1}{4} & -\frac{3}{4} & -\frac{\sqrt{3}}{4} & \\ \frac{\sqrt{3}}{4} & \frac{1}{4} & -\frac{1}{4} & -\frac{\sqrt{3}}{12} & -\frac{\sqrt{6}}{3} \\ \frac{\sqrt{3}}{4} & -\frac{3}{4} & -\frac{1}{4} & \frac{\sqrt{3}}{4} & \\ \frac{\sqrt{6}}{4} & \frac{\sqrt{2}}{4} & -\frac{\sqrt{2}}{4} & -\frac{\sqrt{6}}{12} & \frac{\sqrt{3}}{3} \end{bmatrix}$$

$$\langle P_{16}^{56} \rangle = \begin{bmatrix} \frac{1}{2} & & \frac{\sqrt{3}}{2} & & \\ & -\frac{1}{2} & & -\frac{\sqrt{3}}{2} & \\ & \frac{1}{2} & & -\frac{\sqrt{3}}{6} & \frac{\sqrt{6}}{3} \\ -\frac{\sqrt{3}}{2} & & \frac{1}{2} & & \\ & \frac{\sqrt{2}}{2} & & -\frac{\sqrt{6}}{6} & -\frac{\sqrt{3}}{3} \end{bmatrix}$$

$$\langle P_{24}^{56} \rangle = \begin{bmatrix} -\frac{1}{4} & -\frac{\sqrt{3}}{4} & \frac{\sqrt{3}}{4} & \frac{3}{4} & \\ -\frac{\sqrt{3}}{4} & \frac{1}{4} & \frac{3}{4} & -\frac{\sqrt{3}}{4} & \\ \frac{\sqrt{3}}{4} & -\frac{1}{4} & \frac{1}{4} & -\frac{\sqrt{3}}{12} & -\frac{\sqrt{6}}{3} \\ \frac{\sqrt{3}}{4} & \frac{3}{4} & \frac{1}{4} & \frac{\sqrt{3}}{4} & \\ \frac{\sqrt{6}}{4} & -\frac{\sqrt{2}}{4} & \frac{\sqrt{2}}{4} & -\frac{\sqrt{6}}{12} & \frac{\sqrt{3}}{3} \end{bmatrix}$$

Table 5 (Cont.)

$$\langle P_{25}^{56}\rangle = \begin{bmatrix} -\frac{1}{4} & \frac{\sqrt{3}}{4} & \frac{\sqrt{3}}{4} & -\frac{3}{4} & \\ -\frac{\sqrt{3}}{4} & -\frac{1}{4} & \frac{3}{4} & \frac{\sqrt{3}}{4} & \\ \frac{\sqrt{3}}{4} & \frac{1}{4} & \frac{1}{4} & \frac{\sqrt{3}}{12} & \frac{\sqrt{6}}{3} \\ \frac{\sqrt{3}}{4} & -\frac{3}{4} & \frac{1}{4} & -\frac{\sqrt{3}}{4} & \\ \frac{\sqrt{6}}{4} & \frac{\sqrt{2}}{4} & \frac{\sqrt{2}}{4} & \frac{\sqrt{6}}{12} & -\frac{\sqrt{3}}{3} \end{bmatrix}$$

$$\langle P_{26}^{56}\rangle = \begin{bmatrix} \frac{1}{2} & & -\frac{\sqrt{3}}{2} & & \\ & -\frac{1}{2} & & \frac{\sqrt{3}}{2} & \\ & \frac{1}{2} & & \frac{\sqrt{3}}{6} & -\frac{\sqrt{6}}{3} \\ -\frac{\sqrt{3}}{2} & & -\frac{1}{2} & & \\ & \frac{\sqrt{2}}{2} & & \frac{\sqrt{6}}{6} & \frac{\sqrt{3}}{3} \end{bmatrix}$$

$$\langle P_{34}^{56}\rangle = \begin{bmatrix} \frac{1}{2} & \frac{\sqrt{3}}{2} & & & \\ \frac{\sqrt{3}}{2} & -\frac{1}{2} & & & \\ & & \frac{1}{2} & -\frac{\sqrt{3}}{6} & \frac{\sqrt{6}}{3} \\ & & \frac{1}{2} & \frac{\sqrt{3}}{2} & \\ & & \frac{\sqrt{2}}{2} & -\frac{\sqrt{6}}{6} & -\frac{\sqrt{3}}{3} \end{bmatrix} \qquad \langle P_{35}^{56}\rangle = \begin{bmatrix} \frac{1}{2} & -\frac{\sqrt{3}}{2} & & & \\ \frac{\sqrt{3}}{2} & \frac{1}{2} & & & \\ & & \frac{1}{2} & \frac{\sqrt{3}}{6} & -\frac{\sqrt{6}}{3} \\ & & \frac{1}{2} & -\frac{\sqrt{3}}{2} & \\ & & \frac{\sqrt{2}}{2} & \frac{\sqrt{6}}{6} & \frac{\sqrt{3}}{3} \end{bmatrix}$$

$$\langle P_{36}^{56}\rangle = \begin{bmatrix} -1 & & & & \\ & 1 & & & \\ & & & \frac{\sqrt{3}}{3} & \frac{\sqrt{6}}{3} \\ & & -1 & & \\ & & & \frac{\sqrt{6}}{3} & -\frac{\sqrt{3}}{3} \end{bmatrix}$$

Table 5 (Cont.)

(b) $[\lambda] = [21^4]$

r_1:	$[21]_1$	$[21]_2$	$[1^3]$	$[1^3]$	$[1^3]$
r_2:	$[1^3]$	$[1^3]$	$[21]_1$	$[21]_2$	$[1^3]$

λ''

$$\langle P_{14}^{56} \rangle = \begin{matrix} [1^2] \\ [1^2] \\ [1^2] \\ [2] \\ [1^2] \end{matrix} \begin{bmatrix} -\frac{1}{2} & -\frac{\sqrt{3}}{2} & & & \\ & & \frac{1}{2} & \frac{\sqrt{3}}{2} & \\ -\frac{\sqrt{3}}{4} & \frac{\sqrt{6}}{12} & -\frac{\sqrt{2}}{4} & \frac{\sqrt{6}}{12} & \frac{\sqrt{6}}{3} \\ \frac{1}{2} & -\frac{\sqrt{3}}{6} & \frac{1}{2} & -\frac{\sqrt{3}}{6} & \frac{\sqrt{3}}{3} \\ -\frac{\sqrt{6}}{4} & \frac{\sqrt{2}}{4} & \frac{\sqrt{6}}{4} & -\frac{\sqrt{2}}{4} & \end{bmatrix}$$

$$\langle P_{15}^{56} \rangle = \begin{bmatrix} \frac{1}{2} & \frac{\sqrt{3}}{2} & & & \\ & & \frac{1}{2} & -\frac{\sqrt{3}}{2} & \\ \frac{\sqrt{2}}{4} & -\frac{\sqrt{6}}{12} & -\frac{\sqrt{2}}{4} & -\frac{\sqrt{6}}{12} & -\frac{\sqrt{6}}{3} \\ -\frac{1}{2} & \frac{\sqrt{3}}{6} & \frac{1}{2} & \frac{\sqrt{3}}{6} & -\frac{\sqrt{3}}{3} \\ \frac{\sqrt{6}}{4} & -\frac{\sqrt{2}}{4} & \frac{\sqrt{6}}{4} & \frac{\sqrt{2}}{4} & \end{bmatrix}$$

$$\langle P_{16}^{56} \rangle = \begin{bmatrix} -\frac{1}{2} & -\frac{\sqrt{3}}{2} & & & \\ & & -1 & & \\ -\frac{\sqrt{2}}{4} & \frac{\sqrt{6}}{12} & & -\frac{\sqrt{6}}{6} & \frac{\sqrt{6}}{3} \\ \frac{1}{2} & -\frac{\sqrt{3}}{6} & & \frac{\sqrt{3}}{3} & \frac{\sqrt{3}}{3} \\ -\frac{\sqrt{6}}{4} & \frac{\sqrt{2}}{4} & & \frac{\sqrt{2}}{2} & \end{bmatrix}$$

$$\langle P_{24}^{56} \rangle = \begin{bmatrix} -\frac{1}{2} & \frac{\sqrt{3}}{2} & & & \\ & & -\frac{1}{2} & -\frac{\sqrt{3}}{2} & \\ -\frac{\sqrt{2}}{4} & -\frac{\sqrt{6}}{12} & \frac{\sqrt{2}}{4} & -\frac{\sqrt{6}}{12} & -\frac{\sqrt{6}}{3} \\ \frac{1}{2} & \frac{\sqrt{3}}{6} & -\frac{1}{2} & \frac{\sqrt{3}}{6} & -\frac{\sqrt{3}}{3} \\ -\frac{\sqrt{6}}{4} & -\frac{\sqrt{2}}{4} & -\frac{\sqrt{6}}{4} & \frac{\sqrt{2}}{4} & \end{bmatrix}$$

Table 5 (Cont.)

$$\langle P^{56}_{25}\rangle = \begin{bmatrix} \frac{1}{2} & -\frac{\sqrt{3}}{2} & & & \\ & & -\frac{1}{2} & \frac{\sqrt{3}}{2} & \\ \frac{\sqrt{2}}{4} & \frac{\sqrt{6}}{12} & \frac{\sqrt{2}}{4} & \frac{\sqrt{6}}{12} & \frac{\sqrt{6}}{3} \\ -\frac{1}{2} & -\frac{\sqrt{3}}{6} & -\frac{1}{2} & -\frac{\sqrt{3}}{6} & \frac{\sqrt{3}}{3} \\ \frac{\sqrt{6}}{4} & \frac{\sqrt{2}}{4} & -\frac{\sqrt{6}}{4} & -\frac{\sqrt{2}}{4} & \end{bmatrix}$$

$$\langle P^{56}_{26}\rangle = \begin{bmatrix} -\frac{1}{2} & \frac{\sqrt{3}}{2} & & & \\ & & 1 & & \\ -\frac{\sqrt{2}}{4} & -\frac{\sqrt{6}}{12} & & \frac{\sqrt{6}}{6} & -\frac{\sqrt{6}}{3} \\ \frac{1}{2} & \frac{\sqrt{3}}{6} & & -\frac{\sqrt{3}}{3} & -\frac{\sqrt{3}}{3} \\ -\frac{\sqrt{6}}{4} & -\frac{\sqrt{2}}{4} & & -\frac{\sqrt{2}}{2} & \end{bmatrix}$$

$$\langle P^{56}_{34}\rangle = \begin{bmatrix} 1 & & & & \\ & & \frac{1}{2} & \frac{\sqrt{3}}{2} & \\ & -\frac{\sqrt{6}}{6} & -\frac{\sqrt{2}}{4} & \frac{\sqrt{6}}{12} & \frac{\sqrt{6}}{3} \\ & \frac{\sqrt{3}}{3} & \frac{1}{2} & -\frac{\sqrt{3}}{6} & \frac{\sqrt{3}}{3} \\ & -\frac{\sqrt{2}}{2} & \frac{\sqrt{6}}{4} & -\frac{\sqrt{2}}{4} & \end{bmatrix} \qquad \langle P^{56}_{35}\rangle = \begin{bmatrix} -1 & & & & \\ & & \frac{1}{2} & -\frac{\sqrt{3}}{2} & \\ & \frac{\sqrt{6}}{6} & -\frac{\sqrt{2}}{4} & -\frac{\sqrt{6}}{12} & -\frac{\sqrt{6}}{3} \\ & -\frac{\sqrt{3}}{3} & \frac{1}{2} & \frac{\sqrt{3}}{6} & -\frac{\sqrt{3}}{3} \\ & \frac{\sqrt{2}}{2} & \frac{\sqrt{6}}{4} & \frac{\sqrt{2}}{4} & \end{bmatrix}$$

$$\langle P^{56}_{36}\rangle = \begin{bmatrix} 1 & & & & \\ & & 1 & & \\ & -\frac{\sqrt{6}}{6} & & -\frac{\sqrt{6}}{6} & \frac{\sqrt{6}}{3} \\ & \frac{\sqrt{3}}{3} & & \frac{\sqrt{3}}{3} & \frac{\sqrt{3}}{3} \\ & -\frac{\sqrt{2}}{2} & & \frac{\sqrt{2}}{2} & \end{bmatrix}$$

Table 5 (Cont.)

(c) $[\lambda]=[2^21^2]$

λ'' \ r_1:	$[21]_1$	$[21]_1$	$[21]_1$	$[21]_2$	$[21]_2$	$[21]_2$	$[1^3]$	$[1^3]$	$[1^3]$
r_2:	$[21]_1$	$[21]_2$	$[1^3]$	$[21]_1$	$[21]_2$	$[1^3]$	$[21]_1$	$[21]_2$	$[1^3]$

$$\langle P_{14}^{56}\rangle = \begin{matrix}[1^2]\\[2]\\[1^2]\\[1^2]\\[2]\\[1^2]\\[2]\\[1^2]\\[1^2]\end{matrix}\begin{bmatrix}
-\frac{1}{4} & -\frac{\sqrt{3}}{4} & & -\frac{\sqrt{3}}{4} & -\frac{3}{4} & & & & \\
-\frac{1}{4} & \frac{\sqrt{3}}{12} & -\frac{\sqrt{6}}{6} & -\frac{\sqrt{3}}{4} & \frac{1}{4} & -\frac{\sqrt{2}}{2} & & & \\
-\frac{\sqrt{2}}{4} & \frac{\sqrt{6}}{12} & \frac{\sqrt{3}}{6} & -\frac{\sqrt{6}}{4} & \frac{\sqrt{2}}{4} & \frac{1}{2} & & & \\
-\frac{\sqrt{3}}{12} & \frac{1}{12} & \frac{\sqrt{2}}{3} & \frac{1}{12} & -\frac{\sqrt{3}}{36} & -\frac{\sqrt{6}}{9} & \frac{\sqrt{2}}{3} & -\frac{\sqrt{6}}{9} & \frac{\sqrt{30}}{9} \\
\frac{1}{4} & \frac{\sqrt{3}}{4} & & -\frac{\sqrt{3}}{12} & -\frac{1}{4} & & \frac{\sqrt{6}}{6} & \frac{\sqrt{2}}{2} & \\
-\frac{\sqrt{2}}{4} & -\frac{\sqrt{6}}{4} & & \frac{\sqrt{6}}{12} & \frac{\sqrt{2}}{4} & & \frac{\sqrt{3}}{6} & \frac{1}{2} & \\
-\frac{\sqrt{2}}{4} & \frac{\sqrt{6}}{12} & \frac{\sqrt{3}}{6} & \frac{\sqrt{6}}{12} & -\frac{\sqrt{2}}{12} & -\frac{1}{6} & \frac{\sqrt{3}}{6} & -\frac{1}{6} & -\frac{\sqrt{5}}{3} \\
 & & -\frac{\sqrt{6}}{4} & & & \frac{\sqrt{2}}{4} & \frac{\sqrt{6}}{4} & -\frac{\sqrt{2}}{4} & \\
\frac{\sqrt{15}}{6} & -\frac{\sqrt{5}}{6} & \frac{\sqrt{10}}{12} & -\frac{\sqrt{5}}{6} & \frac{\sqrt{15}}{18} & -\frac{\sqrt{30}}{36} & \frac{\sqrt{10}}{12} & -\frac{\sqrt{30}}{36} & -\frac{\sqrt{6}}{9}
\end{bmatrix}$$

$$\langle P_{15}^{56}\rangle = \begin{bmatrix}
-\frac{1}{4} & \frac{\sqrt{3}}{4} & & -\frac{\sqrt{3}}{4} & \frac{3}{4} & & & & \\
-\frac{1}{4} & -\frac{\sqrt{3}}{12} & \frac{\sqrt{6}}{6} & -\frac{\sqrt{3}}{4} & -\frac{1}{4} & \frac{\sqrt{2}}{2} & & & \\
-\frac{\sqrt{2}}{4} & -\frac{\sqrt{6}}{12} & -\frac{\sqrt{3}}{6} & -\frac{\sqrt{6}}{4} & -\frac{\sqrt{2}}{4} & -\frac{1}{2} & & & \\
-\frac{\sqrt{3}}{12} & -\frac{1}{12} & -\frac{\sqrt{2}}{3} & \frac{1}{12} & \frac{\sqrt{3}}{36} & \frac{\sqrt{6}}{9} & \frac{\sqrt{2}}{3} & \frac{\sqrt{6}}{9} & -\frac{\sqrt{30}}{9} \\
\frac{1}{4} & -\frac{\sqrt{3}}{4} & & -\frac{\sqrt{3}}{12} & \frac{1}{4} & & \frac{\sqrt{6}}{6} & -\frac{\sqrt{2}}{2} & \\
-\frac{\sqrt{2}}{4} & \frac{\sqrt{6}}{4} & & \frac{\sqrt{6}}{12} & -\frac{\sqrt{2}}{4} & & \frac{\sqrt{3}}{6} & -\frac{1}{2} & \\
-\frac{\sqrt{2}}{4} & -\frac{\sqrt{6}}{12} & -\frac{\sqrt{3}}{6} & \frac{\sqrt{6}}{12} & \frac{\sqrt{2}}{12} & \frac{1}{6} & \frac{\sqrt{3}}{6} & \frac{1}{6} & \frac{\sqrt{5}}{3} \\
 & & \frac{\sqrt{6}}{4} & & & -\frac{\sqrt{2}}{4} & \frac{\sqrt{6}}{4} & \frac{\sqrt{2}}{4} & \\
\frac{\sqrt{15}}{6} & \frac{\sqrt{5}}{6} & -\frac{\sqrt{10}}{12} & -\frac{\sqrt{5}}{6} & -\frac{\sqrt{15}}{18} & \frac{\sqrt{30}}{36} & \frac{\sqrt{10}}{12} & \frac{\sqrt{30}}{36} & \frac{\sqrt{6}}{9}
\end{bmatrix}$$

Table 5 (Cont.)

$$\langle P^{56}_{16}\rangle = \begin{bmatrix} \frac{1}{2} & & & \frac{\sqrt{3}}{2} & & & & & \\ & -\frac{\sqrt{3}}{6} & -\frac{\sqrt{6}}{6} & & -\frac{1}{2} & -\frac{\sqrt{2}}{2} & & & \\ & -\frac{\sqrt{6}}{6} & \frac{\sqrt{3}}{6} & & -\frac{\sqrt{2}}{2} & \frac{1}{2} & & & \\ & -\frac{1}{6} & \frac{\sqrt{2}}{3} & & \frac{\sqrt{3}}{18} & -\frac{\sqrt{6}}{9} & & \frac{2\sqrt{6}}{9} & \frac{\sqrt{30}}{9} \\ -\frac{1}{2} & & & \frac{\sqrt{3}}{6} & & & -\frac{\sqrt{6}}{3} & & \\ \frac{\sqrt{2}}{2} & & & -\frac{\sqrt{6}}{6} & & & -\frac{\sqrt{3}}{3} & & \\ & -\frac{\sqrt{6}}{6} & \frac{\sqrt{3}}{6} & & \frac{\sqrt{2}}{6} & -\frac{1}{6} & & \frac{1}{3} & -\frac{\sqrt{5}}{3} \\ & & -\frac{\sqrt{6}}{4} & & & \frac{\sqrt{2}}{4} & & \frac{\sqrt{2}}{2} & \\ & \frac{\sqrt{5}}{3} & \frac{\sqrt{10}}{12} & & -\frac{\sqrt{15}}{9} & -\frac{\sqrt{30}}{36} & & \frac{\sqrt{30}}{18} & -\frac{\sqrt{6}}{9} \end{bmatrix}$$

$$\langle P^{56}_{24}\rangle = \begin{bmatrix} -\frac{1}{4} & -\frac{\sqrt{3}}{4} & & \frac{\sqrt{3}}{4} & \frac{3}{4} & & & & \\ -\frac{1}{4} & \frac{\sqrt{3}}{12} & -\frac{\sqrt{6}}{6} & \frac{\sqrt{3}}{4} & -\frac{1}{4} & \frac{\sqrt{2}}{2} & & & \\ -\frac{\sqrt{2}}{4} & \frac{\sqrt{6}}{12} & \frac{\sqrt{3}}{6} & \frac{\sqrt{6}}{4} & -\frac{\sqrt{2}}{4} & -\frac{1}{2} & & & \\ -\frac{\sqrt{3}}{12} & \frac{1}{12} & \frac{\sqrt{2}}{3} & -\frac{1}{12} & \frac{\sqrt{3}}{36} & \frac{\sqrt{6}}{9} & -\frac{\sqrt{2}}{3} & \frac{\sqrt{6}}{9} & -\frac{\sqrt{30}}{9} \\ \frac{1}{4} & \frac{\sqrt{3}}{4} & & \frac{\sqrt{3}}{12} & \frac{1}{4} & & -\frac{\sqrt{6}}{6} & -\frac{\sqrt{2}}{2} & \\ -\frac{\sqrt{2}}{4} & -\frac{\sqrt{6}}{4} & & -\frac{\sqrt{6}}{12} & -\frac{\sqrt{2}}{4} & & -\frac{\sqrt{3}}{6} & -\frac{1}{2} & \\ -\frac{\sqrt{2}}{4} & \frac{\sqrt{6}}{12} & \frac{\sqrt{3}}{6} & -\frac{\sqrt{6}}{12} & \frac{\sqrt{2}}{12} & \frac{1}{6} & -\frac{\sqrt{3}}{6} & \frac{1}{6} & \frac{\sqrt{5}}{3} \\ & & -\frac{\sqrt{6}}{4} & & & -\frac{\sqrt{2}}{4} & -\frac{\sqrt{6}}{4} & \frac{\sqrt{2}}{4} & \\ \frac{\sqrt{15}}{6} & -\frac{\sqrt{5}}{6} & \frac{\sqrt{10}}{12} & \frac{\sqrt{5}}{6} & -\frac{\sqrt{15}}{18} & \frac{\sqrt{30}}{36} & -\frac{\sqrt{10}}{12} & \frac{\sqrt{30}}{36} & \frac{\sqrt{6}}{9} \end{bmatrix}$$

Table 5 (Cont.)

$$\langle P^{56}_{25}\rangle = \begin{bmatrix} -\frac{1}{4} & \frac{\sqrt{3}}{4} & & \frac{\sqrt{3}}{4} & -\frac{3}{4} & & & & \\ -\frac{1}{4} & -\frac{\sqrt{3}}{12} & \frac{\sqrt{6}}{6} & \frac{\sqrt{3}}{4} & \frac{1}{4} & -\frac{\sqrt{2}}{2} & & & \\ -\frac{\sqrt{2}}{4} & -\frac{\sqrt{6}}{12} & -\frac{\sqrt{3}}{6} & \frac{\sqrt{6}}{4} & \frac{\sqrt{2}}{4} & \frac{1}{2} & & & \\ -\frac{\sqrt{3}}{12} & -\frac{1}{12} & -\frac{\sqrt{2}}{3} & -\frac{1}{12} & -\frac{\sqrt{3}}{36} & -\frac{\sqrt{6}}{9} & -\frac{\sqrt{2}}{3} & -\frac{\sqrt{6}}{9} & \frac{\sqrt{30}}{9} \\ \frac{1}{4} & -\frac{\sqrt{3}}{4} & & \frac{\sqrt{3}}{12} & -\frac{1}{4} & & -\frac{\sqrt{6}}{6} & \frac{\sqrt{2}}{2} & \\ -\frac{\sqrt{2}}{4} & \frac{\sqrt{6}}{4} & & -\frac{\sqrt{6}}{12} & \frac{\sqrt{2}}{4} & & -\frac{\sqrt{3}}{6} & \frac{1}{2} & \\ -\frac{\sqrt{2}}{4} & -\frac{\sqrt{6}}{12} & -\frac{\sqrt{3}}{6} & -\frac{\sqrt{6}}{12} & -\frac{\sqrt{2}}{12} & -\frac{1}{6} & -\frac{\sqrt{3}}{6} & -\frac{1}{6} & -\frac{\sqrt{5}}{3} \\ & & \frac{\sqrt{6}}{4} & & & \frac{\sqrt{2}}{4} & -\frac{\sqrt{6}}{4} & -\frac{\sqrt{2}}{4} & \\ \frac{\sqrt{15}}{6} & \frac{\sqrt{5}}{6} & -\frac{\sqrt{10}}{12} & \frac{\sqrt{5}}{6} & \frac{\sqrt{15}}{18} & -\frac{\sqrt{30}}{36} & -\frac{\sqrt{10}}{12} & -\frac{\sqrt{30}}{36} & -\frac{\sqrt{6}}{9} \end{bmatrix}$$

$$\langle P^{56}_{26}\rangle = \begin{bmatrix} \frac{1}{2} & & & -\frac{\sqrt{3}}{2} & & & & & \\ & -\frac{\sqrt{3}}{6} & -\frac{\sqrt{6}}{6} & & \frac{1}{2} & \frac{\sqrt{2}}{2} & & & \\ & -\frac{\sqrt{6}}{6} & \frac{\sqrt{3}}{6} & & \frac{\sqrt{2}}{2} & -\frac{1}{2} & & & \\ & -\frac{1}{6} & \frac{\sqrt{2}}{3} & & -\frac{\sqrt{3}}{18} & \frac{\sqrt{6}}{9} & & -\frac{2\sqrt{6}}{9} & -\frac{\sqrt{30}}{9} \\ -\frac{1}{2} & & & -\frac{\sqrt{3}}{6} & & & \frac{\sqrt{6}}{3} & & \\ \frac{\sqrt{2}}{2} & & & \frac{\sqrt{6}}{6} & & & \frac{\sqrt{3}}{3} & & \\ & -\frac{\sqrt{6}}{6} & \frac{\sqrt{3}}{3} & & -\frac{\sqrt{2}}{6} & \frac{1}{6} & & -\frac{1}{3} & \frac{\sqrt{5}}{3} \\ & & -\frac{\sqrt{6}}{4} & & & -\frac{\sqrt{2}}{4} & & -\frac{\sqrt{2}}{2} & \\ & \frac{\sqrt{5}}{3} & \frac{\sqrt{10}}{12} & & \frac{\sqrt{15}}{9} & \frac{\sqrt{30}}{36} & & -\frac{\sqrt{30}}{18} & \frac{\sqrt{6}}{9} \end{bmatrix}$$

Table 5 (Cont.)

$$
\left\langle P^{56}_{34} \right\rangle = \begin{bmatrix}
\frac{1}{2} & \frac{\sqrt{3}}{2} & & & & & & & \\
\frac{1}{2} & -\frac{\sqrt{3}}{6} & \frac{\sqrt{6}}{3} & & & & & & \\
\frac{\sqrt{2}}{2} & -\frac{\sqrt{6}}{6} & -\frac{\sqrt{3}}{3} & & & & & & \\
 & & & -\frac{1}{6} & \frac{\sqrt{3}}{18} & \frac{2\sqrt{6}}{9} & \frac{\sqrt{2}}{3} & -\frac{\sqrt{6}}{9} & \frac{\sqrt{30}}{9} \\
 & & & \frac{\sqrt{3}}{6} & \frac{1}{2} & & \frac{\sqrt{6}}{6} & \frac{\sqrt{2}}{2} & \\
 & & & -\frac{\sqrt{6}}{6} & -\frac{\sqrt{2}}{2} & & \frac{\sqrt{3}}{6} & \frac{1}{2} & \\
 & & & -\frac{\sqrt{6}}{6} & \frac{\sqrt{2}}{6} & \frac{1}{3} & \frac{\sqrt{3}}{6} & -\frac{1}{6} & -\frac{\sqrt{5}}{3} \\
 & & & & & -\frac{\sqrt{2}}{2} & \frac{\sqrt{6}}{4} & -\frac{\sqrt{2}}{4} & \\
 & & & \frac{\sqrt{5}}{3} & -\frac{\sqrt{15}}{9} & \frac{\sqrt{30}}{18} & \frac{\sqrt{10}}{12} & -\frac{\sqrt{30}}{36} & -\frac{\sqrt{6}}{9}
\end{bmatrix}
$$

$$
\left\langle P^{56}_{35} \right\rangle = \begin{bmatrix}
\frac{1}{2} & -\frac{\sqrt{3}}{2} & & & & & & & \\
\frac{1}{2} & \frac{\sqrt{3}}{6} & -\frac{\sqrt{6}}{3} & & & & & & \\
\frac{\sqrt{2}}{2} & \frac{\sqrt{6}}{6} & \frac{\sqrt{3}}{3} & & & & & & \\
 & & & -\frac{1}{6} & -\frac{\sqrt{3}}{18} & -\frac{2\sqrt{6}}{9} & \frac{\sqrt{2}}{3} & \frac{\sqrt{6}}{9} & -\frac{\sqrt{30}}{9} \\
 & & & \frac{\sqrt{3}}{6} & -\frac{1}{2} & & \frac{\sqrt{6}}{6} & -\frac{\sqrt{2}}{2} & \\
 & & & -\frac{\sqrt{6}}{6} & \frac{\sqrt{2}}{2} & & \frac{\sqrt{3}}{6} & -\frac{1}{2} & \\
 & & & -\frac{\sqrt{6}}{6} & -\frac{\sqrt{2}}{6} & -\frac{1}{3} & \frac{\sqrt{3}}{6} & \frac{1}{6} & \frac{\sqrt{5}}{3} \\
 & & & & & \frac{\sqrt{2}}{2} & \frac{\sqrt{6}}{4} & \frac{\sqrt{2}}{4} & \\
 & & & \frac{\sqrt{5}}{3} & \frac{\sqrt{15}}{9} & -\frac{\sqrt{30}}{18} & \frac{\sqrt{10}}{12} & \frac{\sqrt{30}}{36} & \frac{\sqrt{6}}{9}
\end{bmatrix}
$$

Table 5 (Cont.)

$$\langle P_{36}^{56}\rangle = \begin{bmatrix} -1 & & & & & & & & \\ & \frac{\sqrt{3}}{3} & \frac{\sqrt{6}}{3} & & & & & & \\ & \frac{\sqrt{6}}{3} & -\frac{\sqrt{3}}{3} & & & & & & \\ & & & & -\frac{\sqrt{3}}{9} & \frac{2\sqrt{6}}{9} & & \frac{2\sqrt{6}}{9} & \frac{\sqrt{30}}{9} \\ & & & -\frac{\sqrt{3}}{3} & & & -\frac{\sqrt{6}}{3} & & \\ & & & \frac{\sqrt{6}}{3} & & & -\frac{\sqrt{3}}{3} & & \\ & & & & -\frac{\sqrt{2}}{3} & \frac{1}{3} & & \frac{1}{3} & -\frac{\sqrt{5}}{3} \\ & & & & & -\frac{\sqrt{2}}{2} & & \frac{\sqrt{2}}{2} & \\ & & & & \frac{2\sqrt{15}}{9} & \frac{\sqrt{30}}{18} & & \frac{\sqrt{30}}{18} & -\frac{\sqrt{6}}{9} \end{bmatrix}$$

References

Adams, W. H. (1962). *Phys. Rev.* **127**,1650.
Adams, W. H. (1967). *Phys. Rev.* **156**, 109.
Allen, L., Clementi, E., and Gladney, H. (1963). *Rev. Mod. Phys.* **35**, 465.
Altmann, S. L. (1963a). *Phil. Trans. Roy. Soc. London Ser. A* **255**, 216.
Altmann, S. L. (1963b). *Rev. Mod. Phys.* **35**, 641.
Altmann, S. L. (1967). *Proc. Roy. Soc. Ser. A* **298**, 184.
Bacher, R. F. and Goudsmit, S., (1934). *Phys. Rev.* **46**, 948.
Bauer, E. (1933). "Introduction a la Théorie des Groupes et ses Applications á la Physique Quantique." Presses Univ. de France, Paris.
Berthier, G. (1954). *J. Chem. Phys.* **51**, 363.
Bessis, S., Espagnet, P., and Bratož, S. (1969). *Int. J. Quantum Chem.* **3**, 205.
Bethe, H. A. (1939). *Ann. Phys. (Leipzig)* **3**, 133.
Bethe, H. A. (1956). *Phys. Rev.* **103**, 1353.
Bethe, H. A., and Goldstone, J. (1957). *Proc. Roy. Soc. Ser. A.* **238**, 551.
Birss, F. W., and Fraga, S. (1963). *J. Chem. Phys.* **38**, 2552.
Bolotin, A. B., and Levinson, I. B. (1960). *Trudy Akad. Nauk Litovskoii SSR. Ser. B.* **3**, 21.
Born, M., and Oppenheimer, J. R. (1927). *Ann. Phys. (New York)* **84**, 457.
Brillouin, L. (1934). Les champs 'self-consistents' de Hartree et de Fock. *Actual. Sci. Ind.* **No. 159.**
Bratoz, S., and Durand, P. (1965). *J. Chim Phys. Physicochim. Biol.* **43**, 2670.
Brueckner, K. A. (1954). *Phys. Rev.* **96**, 508.
Brueckner, K. A. (1955). *Phys. Rev.* **97**, 1353.
Cade, P. E., Sales, K. D., and Wahl, A. C. (1966). *J. Chem. Phys.* **44**, 1973.
Chisholm, C. D. H., Dalgarno, A., and Innes, F. R. (1969). *Advan. At. Mol. Phys.* **5**, 297.
Chong, D. P. (1965). *J. Chem. Phys.* **43**, S 73.
Čižek, J., and Paldus, J. (1967). *J. Chem. Phys.* **47**, 3976.
Clementi, E. (1963a). *J. Chem. Phys.* **39**, 175.
Clementi, E. (1963b). *J. Chem. Phys.* **39**, 487.
Clementi, E. (1965). "Tables of Atomic Wavefunctions." Suppl. to paper "Ab Initio Computations in Atoms and Molecules." *IBM J. Res. Develop.* **9**, 2.
Clementi, E. (1967). *J. Chem. Phys.* **46**, 3842.
Condon, E., and Shortley, G. (1964). "The Theory of Atomic Spectra." Cambridge Univ. Press, London and New York.
Conroy, H. (1964). *J. Chem. Phys.* **41**, 1341.
Conroy, H. (1967). *J. Chem. Phys.* **47**, 912, 921, 930.
Coulson, C. A. (1937). *Trans. Faraday Soc.* **33**, 1479.
Coulson, C. A., and Fischer, I. (1949). *Phil. Mag.* **40**, 386.

Coulson, C. A., and Longuet-Higgins, H. C. (1948). *Proc. Roy. Soc. Ser. A* **192**, 16.
Cure, R., Kasper, J., Pitzer, K., and Satbianandan, K. (1966). *J. Chem. Phys.* **44**, 4636.
Das, G. (1967). *J. Chem. Phys.* **46**, 1568.
Das, G., Wahl, A. C. (1966). *J. Chem. Phys.* **44**, 87.
Daudel, R., and Durand, P. (1965). *In* "Modern Quantum Chemistry" (O. Sinanoğlu, ed.). Academic Press, New York.
Davydov, A. S. (1965). "Quantum Mechanics." Pergamon, Oxford. (Translated by D. ter Haar from the Russian edition "Kvantovaya Mekhanika." Fizmatgiz, Moscow, 1963.)
de-Shalit, A., and Talmi, I. (1963). "Nuclear Shell Theory." Academic Press, New York.
Dyadusha, G. G., and Kuprievich, V. A. (1965a). *Dokl. Akad. Nauk Ukr. SSR* p. 1161.
Dyadusha, G. G., and Kuprievich, V. A. (1965b). *Teor. Eksp. Khim.* **1**, 406.
Eckart, C. (1930a). *Rev. Mod. Phys.* **2**, 305.
Eckart, C. (1930b). *Phys. Rev.* **36**, 878.
Edmonds, A. R. (1957). "Angular Momentum in Quantum Mechanics." Princeton Univ. Press, Princeton, New Jersey.
Edmonds, A. R., and Flowers, B. H. (1952). *Proc. Roy. Soc. Ser. A* **214**, 515.
Ehrenfest, P., and Trkal, V. (1921). *Proc. Sec. Sci. Amsterdam* **23**, 162.
Elliott, J. P., Hope, J., and Jahn, H. A. (1953). *Phil. Trans. Roy. Soc. London Ser. A* **246**, 241.
Empedocles, P. B., and Linnett, J. W. (1964). *Proc. Roy. Soc. Ser. A* **282**, 166.
Eyring, H., Walter, J., and Kimball, G. E. (1944). "Quantum Chemistry." Wiley, New York.
Fano, U., and Racah, G. (1959). "Irreducible Tensorial Sets." Academic Press, New York.
Flodmark, S., and Blokker, E. (1967). *Int. J. Quantum Chem.* **1S**, 703.
Fock, V. A. (1930a). *Z. Phys.* **61**, 126.
Fock, V. A. (1930b). *Z. Phys.* **62**, 795.
Fock, V. A. (1935). *Izv. Akad. Nauk SSSR Phys.* **2**, 169.
Fock, V. A. (1935). *Z. Phys.* **98**, 145.
Fock, V. A., Veselov, M. G., and Petrashen, M. I. (1940). *Zh. Eksp. Teor. Fiz.* **10**, 723.
Fraga, S., and Mulliken, R. (1960). *Rev. Mod. Phys.* **32**, 254.
Fraga, S., and Ransil, B. J. (1961). *J. Chem. Phys.* **35**, 1967.
Freed, K. (1968). *Phys. Rev.* **173**, 1, 24.
Frenkel, J. (1934). "Wave Mechanics." Oxford Univ. Press (Clarendon), London and New York.
Gallup, G. A. (1968). *J. Chem. Phys.* **48**, 1752.
Gallup, G. A. (1969). *J. Chem. Phys.* **50**, 1206.
Geller, M., Taylor, H., and Levine, H. (1965). *J. Chem. Phys.* **43**, 1727.
Gerratt, J. (1968). *Annu. Rep. Chem. Soc. (London)* **3**.
Gerratt, J. (1971). *Advan. At. Mol. Phys.* **7**, 141.
Gerratt, J., and Lipscomb, W. N. (1968). *Proc. Nat. Acad. Sci. U. S.* **59**, 332.
Gerratt, J., and Mills, I. M. (1968). *J. Chem. Phys.* **49**, 1719.
Goddard, Jr., W. A. (1967a). *Phys. Rev.* **157**, 73.
Goddard, Jr., W. A. (1967b). *Phys. Rev.* **157**, 81.
Goddard, Jr., W. A. (1968). *J. Chem. Phys.* **48**, 450.
Godnev, I. N. (1945). *Zh. Fiz. Khim.* **19**, 637.
Gombas, P. (1950). "Theorie und Lösungsmethoden des Mehrteilchenproblems der Quantenmechanik." Birkhaeuser, Basel.
Goodisman, J., and Klemperer, W. (1963). *J. Chem. Phys.* **38**, 721.
Goscinski, O., and Öhrn, Y. (1968). *Int. J. Quantum Chem.* **2**, 845.
Griffith, J. S. (1961). "The Irreducible Tensor Method for Molecular Symmetry Groups." Prentice-Hall, Englewood Cliffs, New Jersey.

Hameed, S., Hui, S. S., Musher, J. I., Schulman, J. M. (1969). *J. Chem. Phys.* **51**, 502.
Hamermesh, M. (1962). "Group Theory." Addison-Wesley, Reading, Massachusetts.
Hartree, D. R. (1928). *Proc. Cambridge Phil. Soc.* **24**, 89, 111.
Hartree, D. R. (1957). "The Calculation of Atomic Structures." Wiley, New York.
Hartree, D. R., Hartree, W., and Swirles, B. (1939). *Phil. Trans. Roy. Soc. Ser. A* **238**, 223.
Hassitt, A. (1955). *Proc. Roy. Soc. Ser. A* **229**, 110.
Heitler, W., and London, F. (1927). *Z. Phys.* **44**, 455.
Helfand, I. M., Minlos, R. A., and Shapiro, Z. Ya. (1963). "Representations of the Rotation and Lorentz Groups." Macmillan, New York. (Translated by G. Cummins and T. Boddington from the Russian edition "Predstavleniya Gruppy Vrashchenii i Gruppy Lorentsa." Fizmatgiz, Moscow, 1958.)
Hellman, H. (1937). "Einführung in die Quantenchemie." Deuticke, Leipzig. (H. Hellmann, "Kvantovaya Khimiya." ONTI, Moscow, 1937.)
Herzberg, G. (1947). "Molecular Spectra and Molecular Structure," Vol. 2, "Infrared and Raman Spectra of Polyatomic Molecules." Van Nostrand-Reinhold, Princeton, New Jersey (2nd ed., 1959).
Herzberg, G. (1950). "Spectra of Diatomic Molecules." Van Nostrand-Reinhold, Princeton, New Jersey.
Hinze, J., and Roothaan, C. C. J. (1967). *Progr. Theor. Phys. Suppl.* **No. 40, 37.**
Hirst, D. M., and Linnett, J. W. (1961). *Proc. Chem. Soc. London p.* 427.
Hollister, C., and Sinanoğlu, O. (1965). In "Modern Quantum Chemistry" (O. Sinanoğlu, ed.). Academic Press, New York.
Horie, H. (1964). *J. Phys. Soc. Jap.* **19**, 1783.
Hund, F. (1927). *Z. Phys.* **40**, 742; **42**, 93.
Huzinaga, S. (1960). *Phys. Rev.* **120**, 866.
Huzinaga, S. (1961). *Phys. Rev.* **122**, 131.
Huzinaga, S. (1969). *Progr. Theor. Phys.* **41**, 307.
Hylleraas, E. (1928). *Z. Phys.* **48**, 469.
Hylleraas, E. (1929). *Z. Phys.* **54**, 347.
Ishidzu, T., Horie, H., Obi, S., Sata, M., Tanabe, Y., and Yanagawa, S. (1960). "Tables of Racah Coefficients." Pan-Pacific Press, Tokyo. (A. F. Nikiforov, V. B. Uvarov, and Yu. L. Levitan, "Tablitsy Koeffitsientov Raka." VTs. Akad. Nauk SSSR, Moscow, 1962.)
Ishiguro, E., Kayama, K., Kotani, M., and Mizuno, Y. (1957). *Proc. Phys. Soc. Jap.* **12**, 1355.
Itoh, T., and Yoshizumi, H. (1955). *J. Phys. Soc. Jap.* **10**, 201.
Jahn, H. A. (1950). *Proc. Roy. Soc. Ser. A* **201**, 516.
Jahn, H. A. (1951). *Proc. Roy. Soc. Ser. A* **205**, 192.
Jahn, H. A. (1954). *Phys. Rev.* **96**, 989.
Jahn, H. A., and van Wieringen, H. (1951). *Proc. Roy. Soc. Ser. A* **209**, 502.
James, H. M., and Coolidge, A. S. (1933). *J. Chem. Phys.* **1**, 825.
Judd, B. R. (1963). "Operator Techniques in Atomic Spectroscopy." McGraw-Hill, New York.
Judd, B. R. (1968). *Phys. Rev.* **173**, 40.
Kahalas, S. L., and Nesbet, R. K. (1963). *J. Chem. Phys.* **39**, 529.
Kaplan, I. G. (1959). *Zh. Eksp. Teor. Fiz.* **37**, 1050 [*Sov. Phys. JETP* **37**, 747 (1960)].
Kaplan, I. G. (1961a). *Zh. Eksp. Teor. Fiz.* **41**, 560 [*Sov. Phys. JETP* **14**, 401 (1962)].
Kaplan, I. G. (1961b). *Zh. Eksp. Teor. Fiz.* **41**, 790 [*Sov. Phys. JETP* **14**, 568 (1962)].
Kaplan, I. G. (1962a). Thesis, Inst. Khim. Fiz. Akad. Nauk SSSR, Moscow.
Kaplan, I. G. (1962b). "Tables of the Transformation Matrices for the Permutation Group which Occur in the Coordinate Coefficients of Fractional Parentage." Rotaprint.

Kaplan, I. G. (1963). *Litovskii Fiz. Sb.* **3**, 227.
Kaplan, I. G. (1965a). *Teor. Eksp. Khim.* **1**, 608.
Kaplan, I. G. (1965b). *Teor. Eksp. Khim.* **1**, 619.
Kaplan, I. G. (1966). *Zh. Eksp. Teor. Fiz.* **51**, 169 [*Sov. Phys. JETP* 24, 114 (1967)].
Kaplan, I. G. (1967). *Teor. Eksp. Khim.* **3**, 150.
Kaplan, I. G., and Maximov, A. F. (1969). *Int. Symp. Theory Electron. Shells of At. Mol., Vilna, USSR, June* 16–20, 1969, Abstracts of Papers. *Opt. Spektrosk.* **28**, 662 (1970). [*Opt. Spectrosc.* **28**, 358 (1970)].
Kaplan, I. G., and Maximov, A. F. (1973). *Teor. Eksp. Khim.* **9**, 147.
Kaplan, I. G., and Rodimova, O. B. (1968). *Zh. Eksp. Teor. Fiz.* **55**, 1881 [*Sov. Phys. JETP* **28**, 995 (1969)].
Kaplan, I. G., and Rodimova, O. B. (1970). *Teor. Eksp. Khim.* **6**, 435, 442.
Kaplan, I. G., and Rodimova, O. B. (1971). *Int. J. Quantum Chem.* **5**, 669.
Karo, A. M. (1959). *J. Chem. Phys.* **30**, 1241.
Kauzmann, W. (1957) "Quantum Chemistry," Academic Press, New York.
Kelly, H. P. (1963). *Phys. Rev.* **131**, 684.
Kelly, H. P. (1964). *Phys. Rev.* **136**, 896.
Kelly, H. P. (1966). *Phys. Rev.* **144**, 39.
Kibertas, V. V., and Yutsis, A. P. (1953). *Zh. Eksp. Teor. Fiz.* **25**, 264.
Kołos, W. (1968). *Int. J. Quantum Chem.* **2**, 471.
Kołos, W., and Roothaan, C. C. J. (1960). *Rev. Mod. Phys.* **32**, 219.
Kompaneyets, A. S. (1940). *Zh. Eksp. Teor. Fiz.* **10**, 1175.
Koster, G. F. (1958). *Phys. Rev.* **109**, 227.
Koster, G. F., Dimmock, J. O., Wheeler, R. G., and Statz, H. (1963). "Properties of the Thirty-Two Point Groups." MIT Press, Cambridge, Massachusetts.
Kotani, M. (1937). *Proc. Phys. Math. Soc. Jap.* **19**, 460.
Kotani, M. (1964). *J. Phys. Soc. Jap.* **19**, 2150.
Kotani, M., and Siga, M. (1937). *Proc. Phys. Math. Soc. Jap.* **19**, 471.
Kotani, M., Mizuno, Y., Kayama, K., and Ishiguro, E. (1957). *Proc. Phys. Soc. Jap.* **12**, 707.
Kramer, P. (1967). *Z. Phys.* **205**, 181.
Kramer, P. (1968). *Z. Phys.* **216**, 68.
Kukulin, V. I., Smirnov, Yu. F., and Majling, L. (1967). *Nucl. Phys. A* **103**, 681.
Kurdyumov, I. V., Smirnov, Ya F., Shitikova, K. V., and El-Samarai, S. Kh. (1969). *Izv. Akad. Nauk SSSR. Ser. Fiz.* **33**, 150.
Kutzelnigg, W. (1964). J. Chem. Phys. **40**, 3640.
Labzovskii, L. N. (1963). *Vyestnik Leningradskogo Universiteta*, No. 16, **12**, 127.
Ladner, R. C., and Goddard, Jr., W. A. (1969). *J. Chem. Phys.* **51**, 1073.
Landau, L. D., and Lifshitz, E. M. (1965). "Quantum Mechanics." Pergamon, Oxford. (Translated by J. B. Sykes and E. S. Bell from the Russian edition "Kvantovaya Mekhanika." Fizmatgiz, Moscow, 1963.)
Larsson, S. (1968). *Phys. Rev.* **169**, 49.
Lefebvre, R. (1957). *J. Chem. Phys.* **54**, 168.
Levinson, I. B. (1957). *Trudy Akad. Nauk Litovskoi SSR Ser. B* **4**, 17.
Levy, B., and Berthier, G. (1968). *Int. J. Quantum Chem.* **2**, 307.
Linnett, J. W. (1964). "The Electronic Structure of Molecules (A New Approach)." Methuen, London.
Littlewood, D. E. (1940). "The Theory of Group Characters and Matrix Representations of Groups." Oxford Univ. Press (Clarendon), London and New York.
Longuet-Higgins, H. C. (1963). *Mol. Phys.* **6**, 445.

Löwdin, P. O. (1950). *J. Chem. Phys.* **18**, 365.
Löwdin, P. O. (1955a). *Phys. Rev.* **97**, 1474.
Löwdin, P. O. (1955b). *Phys. Rev.* **97**, 1509.
Löwdin, P. O. (1959). *Advan. Chem. Phys.* **2**, 207
Löwdin, P. O. (1963). *Rev. Mod. Phys.* **35**, 496.
Lunnell, S. (1968). *Phys. Rev.* **173**, 85.
Lunnell, S., and Lindner, P. (1968). *J. Chem. Phys.* **48**, 2752. Lyubarskii, G. Ya. (1960). "The Application of Group Theory in Physics." Pergamon, Oxford. (Translated by S. Dedijer from the Russian edition "Teoriya Grupp i ee Primenenie v Fizike." Fizmatgiz, Moscow, 1958.)
McLean, A. D., Weiss, A., and Yoshimine, M. (1960). *Rev. Mod. Phys.* **32**, 211.
McWeeny, R. (1955). *Proc. Roy. Soc. Ser. A.* **232**, 114.
Matsen, F. A. (1964a). *Advan. Quantum Chem.* **1**, 60.
Matsen, F. A. (1964b). *J. Phys. Chem.* **68**, 3282.
Mattheiss, L. F. (1961). *Phys. Rev.* **123**, 1209.
Mestechkin, M. M. (1967). *Int. J. Quantum Chem.* **1**, 675.
Moskowitz, J. (1963). *J. Chem. Phys.* **38**, 677.
Mulder, G. J. (1966). *Mol. Phys.* **10**, 479.
Murnaghan, F. D. (1938). "The Theory of Group Representations." Johns Hopkins Press, Baltimore, Maryland.
Nesbet, R. K. (1961). *Phys. Rev.* **122**, 1497.
Nesbet, R. K. (1965). *Advan. Chem. Phys.* **9**, 321.
Nesbet, R. K. (1967). *Advan. Quantum Chem.* **3**, 1.
Neudachin, V. G., and Smirnov, Yu. F. (1959). *Zh. Eksp. Teor. Fiz.* **36**, 186 [*Sov. Phys. JETP* **9**, 127 (1959)].
Olive, J. P. (1969). *J. Chem. Phys.* **51**, 4340.
Overhauser, A. W. (1960). *Phys. Rev. Lett.* **4**, 415, 462.
Overhauser, A. W. (1962). *Phys. Rev.* **128**, 1437.
Parr, R. G. (1963). "Quantum Theory of Molecular Electronic Structure." Benjamin, New York.
Parr, R. G., Craig, D. P., and Ross, J. R. (1950). *J. Chem. Phys.* **18**, 1561.
Pauling, L. (1960). "The Nature of the Chemical Bond." Cornell Univ. Press, Ithaca, New York.
Pauncz, R. (1965). *J. Chem. Phys.* **43**, S69.
Pauncz, R. (1967). "Alternant Molecular Orbital Method." Saunders, Philadelphia, Pennsylvania.
Petrashen, M. I., and Trifonov, E. D. (1969). "Applications of Group Theory to Quantum Mechanics." Iliffe, London. (Translated by S. Chomet from the Russian edition "Primeneniya Teorii Grupp v Kvantovoi Mekhanike." Nauka, Moscow, 1967.)
Pontriagin, L. S. (1946). "Continuous Groups." Princeton Univ. Press, Princeton, New Jersey. (Translated by E. Lehmer from the first Russian edition "Nepreryvnye Gruppy." Gostekhizdat, Moscow, 1954.)
Pople, J. A., and Nesbet, R. K. (1954). *J. Chem. Phys.* **22**, 571.
Poshusta, R. D., and Kramling, R. W. (1968). *Phys. Rev.* **167**, 139.
Racah, G. (1942). *Phys. Rev.* **62**, 438.
Racah, G. (1943). *Phys. Rev.* **63**, 367.
Racah, G. (1949). *Phys. Rev.* **75**, 1352.
Racah, G. (1965). Group theory and spectroscopy. *Ergeb. Exakt. Naturwiss.* **37**, 28.
Redmond, P. J. (1954). *Proc. Roy. Soc. Ser. A* **222**, 84.
Roothaan, C. C. J. (1951). *Rev. Mod. Phys.* **23**, 69.

Roothaan, C. C. J. (1960). *Rev. Mod. Phys.* **32**, 179.
Rose, M. E. (1957). "Elementary Theory of Angular Momentum." Wiley, New York.
Rotenberg, M., Metropolis, N., Bivins, R., and Wooten, Jr., J. K. (1959). "The 3-j and 6-j Symbols." MIT Press, Cambridge, Massachusetts.
Rutherford, D. E. (1947). "Substitutional Analysis." Edinburgh Univ. Press, Edinburgh.
Schiff, L. I. (1955). "Quantum Mechanics." McGraw-Hill, New York.
Schwartz, C., and de-Shalit, A, (1954). *Phys. Rev.* **94**, 1257.
Serre, J. (1968). *Int. J. Quantum Chem.* **2S**, 107.
Silverstone, H. J., and Sinanoğlu, O. (1965). *In* "Modern Quantum Chemistry" (O. Sinanoğlu, ed.). Academic Press, New York.
Silverstone, H. J., and Sinanoğlu, O. (1966). *J. Chem. Phys.* **44**, 1899, 3608.
Sinanoğlu, O. (1964). *Advan. Chem. Phys.* **6**, 315.
Sinanoğlu, O. (1969). *Advan. Chem. Phys.* **14**, 237.
Slater, J. C. (1923). *Phys. Rev.* **34**, 1293.
Slater, J. C. (1951). *J. Chem. Phys.* **19**, 220.
Slater, J. C. (1955). *Phys. Rev.* **98**, 1039.
Slater, J. C. (1960). "Quantum Theory of Atomic Structure," Vol. 1. McGraw-Hill, New York.
Slater, J. C. (1963). "Quantum Theory of Molecules and Solids," Vol. 1, "Electronic Structure of Molecules." McGraw-Hill, New York.
Slater, J. C. (1967). *Int. J. Quantum Chem.* **1S**, 783.
Smirnov, V. I. (1964). "Course of Higher Mathematics," Vol. III, Pt. 1. Pergamon, Oxford. (Translated by D. E. Brown from the Russian edition "Kurs Vysshei Matematiki." Gostekhizdat, Moscow, 1957.)
Smirnov, Yu. F., Shitikova, K. V., and El-Samorai, S. Kh. (1968). *Yad. Fiz.* **8**, 470 [*Sov. J. Nucl. Phys.* **8**, 275 (1969)].
Sobel'man, I. I. (1972). "An Introduction to the Theory of Atomic Spectra." Pergamon, Oxford. (Translated by T. F. J. Le Vierge from the Russian edition "Vvedenie v Teoriu Atomnykh Spektrov." Fizmatgiz, Moscow, 1963).
Sokolov, A. V., and Shirokovsky, V. P. (1956). *Usp. Fiz. Nauk* **60**, 617.
Sugiura, Y. (1927). *Z. Phys.* **45**, 484.
Szasz, L. (1959). *Z. Naturforsch. A* **14**, 1014.
Tamir, I., and Pauncz, R. (1968). *Int. J. Quantum Chem.* **2**, 433.
Thouless, D. J. (1961). "The Quantum Mechanics of Many-Body Systems." Academic Press, New York.
Trifonov, E. D. (1959). *Dokl. Akad. Nauk SSSR* **129**, 74.
Veillard, A., and Clementi, E. (1967). *Theor. Chim. Acta* **7**, 133.
Veselov, M. G., and Labzovskii, L. N. (1968). *Teor. Eksp. Khim.* **4**, 107.
Wahl, A. C. (1964). *J. Chem. Phys.* **41**, 2600.
Wannier, G. (1937). *Phys. Rev.* **52**, 191.
Watson, R. E. (1960a). *Phys. Rev.* **118**, 1036.
Watson, R. E. (1960b). *Phys. Rev.* **119**, 1934.
Weinbaum, S. (1933). *J. Chem. Phys.* **1**, 593.
Wessel, W. R. (1967). *J. Chem. Phys.* **47**, 3253.
Weyl, H. (1946). "The Classical Groups." Princeton Univ. Press, Princeton, New Jersey.
Wigner, E. P. (1927). *Z. Phys.* **43**, 624.
Wigner, E. P. (1938). *Trans. Faraday Soc.* **34**, 678.
Wigner, E. P. (1959). "Group Theory." Academic Press, New York.

Wigner, E. P. (1965). On the matrices which reduce the Kronecker products of simply reducible groups. *In* "Quantum Theory of Angular Momentum" (L. C. Biedenharn and H. van Dam, eds.). Academic Press, New York.

Wilson, E. B. (1935). *J. Chem. Phys.* **3**, 276.

Yamanouchi, T. (1937). *Proc. Phys. Math. Soc. Jap.* **19**, 436.

Yutsis, A. P. (1952). *Zh. Eksp. Teor. Fiz.* **23**, 129.

Yutsis, A. P., and Bandzaitis, A. A. (1965). "Teoriya Momenta Kolichestva Dvizheniya." Vilna.

Yutsis, A. P., Levinson, I. B., and Vanagas, V. V. (1962a). "The Theory of Angular Momentum." Published for the NSF and NAS by the Israel Program for Scientific Translations, Jerusalem. (Translated by A. Sen and A. R. Sen from the Russian edition "Matematicheskii Apparat Teorii Momenta Kolichestva Dvizheniya." Gosudarstvennoe Izdatel'stvo Politicheskoi i Nauchnoi Literatury Litovskoi SSSR Vilna.

Yutsis, A. P., Vizbaraite, Ya. I., Strotskite, T. D., and Bandzaitis, A. A. (1962b). *Opt. Spektrosk.* **12**, 157.

Author Index

Numbers in italics refer to the pages on which the complete references are listed.

G

H

I

J

K

L

Subject Index

Physical Chemistry

A Series of Monographs

Ernest M. Loebl, Editor

Department of Chemistry

Polytechnic Institute of New York

Brooklyn, New York

1 W. Jost: Diffusion in Solids, Liquids, Gases, 1952
2 S. Mizushima: Structure of Molecules and Internal Rotation, 1954
3 H. H. G. Jellinek: Degradation of Vinyl Polymers, 1955
4 M. E. L. McBain and E. Hutchinson: Solubilization and Related Phenomena, 1955
5 C. H. Bamford, A. Elliott, and W. E. Hanby: Synthetic Polypeptides, 1956
6 George J. Janz: Thermodynamic Properties of Organic Compounds — Estimation Methods, Principles and Practice, Revised Edition, 1967
7 G. K. T. Conn and D. G. Avery: Infrared Methods, 1960
8 C. B. Monk: Electrolytic Dissociation, 1961
9 P. Leighton: Photochemistry of Air Pollution, 1961
10 P. J. Holmes: Electrochemistry of Semiconductors, 1962
11 H. Fujita: The Mathematical Theory of Sedimentation Analysis, 1962
12 K. Shinoda, T. Nakagawa, B. Tamamushi, and T. Isemura: Colloidal Surfactants, 1963
13 J. E. Wollrab: Rotational Spectra and Molecular Structure, 1967
14 A. Nelson Wright and C. A. Winkler: Active Nitrogen, 1968
15 R. B. Anderson: Experimental Methods in Catalytic Research, 1968
16 Milton Kerker: The Scattering of Light and Other Electromagnetic Radiation, 1969
17 Oleg V. Krylov: Catalysis by Nonmetals — Rules for Catalyst Selection, 1970
18 Alfred Clark: The Theory of Adsorption and Catalysis, 1970
19 Arnold Reisman: Phase Equilibria: Basic Principles, Applications, Experimental Techniques, 1970
20 J. J. Bikerman: Physical Surfaces, 1970
21 R. T. Sanderson: Chemical Bonds and Bond Energy, 1970
22 S. Petrucci, ed.: Ionic Interactions: From Dilute Solutions to Fused Salts (In Two Volumes), 1971
23 A. B. F. Duncan: Rydberg Series in Atoms and Molecules, 1971

24 J. R. Anderson: Chemisorption and Reactions on Metallic Films, 1971

25 E. A. Moelwyn-Hughes: Chemical Statics and Kinetics of Solution, 1971

26 Ivan Draganic and Zorica Draganic: The Radiation Chemistry of Water, 1971

27 M. B. Huglin: Light Scattering from Polymer Solutions, 1972

28 M. J. Blandamer: Introduction to Chemical Ultrasonics, 1973

29 A. I. Kitaigorodsky: Molecular Crystals and Molecules, 1973

30 Wendell Forst: Theory of Unimolecular Reactions, 1973

31 Jerry Goodisman: Diatomic Interaction Potential Theory. Volume 1, Fundamentals, 1973; Volume 2, Applications, 1973

32 Alfred Clark: The Chemisorptive Bond: Basic Concepts, 1974

33 Saul T. Epstein: The Variation Method in Quantum Chemistry, 1974

34 I. G. Kaplan: Symmetry of Many-Electron Systems, 1975

A
B 5
C 6
D 7
E 8
F 9
G 0
H 1
I 2
J 3